NATO ASI Series

Advanced Science Institutes Series

A series presenting the results of activities sponsored by the NATO Science Committee, which aims at the dissemination of advanced scientific and technological knowledge, with a view to strengthening links between scientific communities.

The Series is published by an international board of publishers in conjunction with the NATO Scientific Affairs Division

A Life Sciences B Physics	Plenum Publishing Corporation London and New York
C Mathematical and Physical Sciences D Behavioural and Social Sciences E Applied Sciences	Kluwer Academic Publishers Dordrecht, Boston and London
F Computer and Systems Sciences G Ecological Sciences H Cell Biology I Global Environmental Change	Springer-Verlag Berlin Heidelberg New York London Paris Tokyo Hong Kong Barcelona Budapest

NATO-PCO DATABASE

The electronic index to the NATO ASI Series provides full bibliographical references (with keywords and/or abstracts) to more than 30000 contributions from international scientists published in all sections of the NATO ASI Series. Access to the NATO-PCO DATABASE compiled by the NATO Publication Coordination Office is possible in two ways:

- via online FILE 128 (NATO-PCO DATABASE) hosted by ESRIN, Via Galileo Galilei, I-00044 Frascati, Italy.
- via CD-ROM „NATO Science & Technology Disk“ with user-friendly retrieval software in English, French and German (© WTV GmbH and DATAWARE Technologies Inc. 1992).

The CD-ROM can be ordered through any member of the Board of Publishers or through NATO-PCO, Overijse, Belgium.

Series I: Global Environmental Change, Vol. 9

The ASI Series Books Published as a Result of Activities of the Special Programme on Global Environmental Change

This book contains the proceedings of a NATO Advanced Research Workshop held within the activities of the NATO Special Programme on Global Environmental Change, which started in 1991 under the auspices of the NATO Science Committee.

The volumes published as a result of the activities of the Special Programme are:

Vol. 1: **Global Environmental Change.** Edited by R. W. Corell and P. A. Anderson. 1991.
Vol. 2: **The Last Deglaciation: Absolute and Radiocarbon Chronologies.** Edited by E. Bard and W. S. Broecker. 1992.
Vol. 3: **Start of a Glacial.** Edited by G. J. Kukla and E. Went. 1992.
Vol. 4: **Interactions of C, N, P and S Biogeochemical Cycles and Global Change.** Edited by R. Wollast, F. T. Mackenzie and L. Chou. 1993.
Vol. 5: **Energy and Water Cycles in the Climate System.** Edited by E. Raschke and D. Jacob. 1993.
Vol. 6: **Prediction of Interannual Climate Variations.** Edited by J. Shukla. 1993.
Vol. 7: **The Tropospheric Chemistry of Ozone in the Polar Regions.** Edited by H. Niki and K. H. Becker. 1993.
Vol. 8: **The Role of the Stratosphere in Global Change.** Edited by M.-L. Chanin. 1993.
Vol. 9: **High Spectral Resolution Infrared Remote Sensing for Earth's Weather and Climate Studies.** Edited by A. Chedin, M.T. Chahine and N.A. Scott. 1993.
Vol.10: **Towards a Model of Ocean Biogeochemical Processes.** Edited by G. T. Evans and M.J.R. Fasham. 1993.
Vol.11: **Modelling Oceanic Climate Interactions.** Edited by J. Willebrand and D.L.T. Anderson. 1993.

High Spectral Resolution Infrared Remote Sensing for Earth's Weather and Climate Studies

Edited by

Alain Chedin

Laboratoire de Météorologie Dynamique du CNRS
Ecole Polytechnique
F-91128 Palaiseau Cedex
France

Moustafa T. Chahine

Jet Propulsion Laboratory
4800 Oak Grove Drive
Pasadena, CA 91109
USA

Noëlle A. Scott

Laboratoire de Météorologie Dynamique
Ecole Polytechnique
F-91128 Palaiseau Cedex
France

Springer-Verlag
Berlin Heidelberg New York London Paris Tokyo
Hong Kong Barcelona Budapest
Published in cooperation with NATO Scientific Affairs Division

Proceedings of the NATO Advanced Research Workshop on High Spectral Resolution Infrared Remote Sensing for Earth's Weather and Climate Studies, held at Paris, France, March 23-26, 1992

ISBN-13: 978-3-642-84601-4 e-ISBN-13: 978-3-642-84599-4
DOI: 10.1007/978-3-642-84599-4

Library of Congress Cataloging-in-Publication Data
High spectral resolution infrated remote sensing for Earth's weather and climate studies / edited by Alain Chedin, Moustafa T. Chahine, Noelle Anne Scott.
p. cm. – (NATO ASI series. Series I, Global environmental change; vol. 9)
"Published in cooperation with NATO Scientific Affairs Division."
"Proceedings of the NATO Advanced Research Workshop on High Spectral Resolution Infrared Remote Sensing for Earth's Weather and Climate Studies, held at Paris, France, March 23-26, 1992" – T.p. verso.
ISBN-13: 978-3-642-84601-4
1. Climatology–Remote sensing–Congresses. 2. Infrared detectors–Congresses. I. Chedin, Alain. II. Chahine, Moustafa T. III. Scott, Moelle Anne, 1941- . IV. North Atlantic Treaty Organization. Scientific Affairs Division. V. NATO Advanced Research Workshop on High Spectral Resolution Infrared Remote Sensing for Earth's Weather and Climate Studies (1992: Paris, France) VI. Series.
QC980.H5918 1993 551.6'35–dc20 93-4816

Softcover reprint of the hardcover 1st edition 1993

Typesetting: Camera ready by authors
31/3145 - 5 4 3 2 1 0 - Printed on acid-free paper

Preface

This volume contains the Proceedings of the NATO Advanced Research Workshop on "High Spectral Resolution Infrared Remote Sensing for Earth's Weather and Climate Studies", held at Paris, France, March 23-26, 1992.

One of the major challenges facing Earth science in the next decade and beyond is the development of an accurate long term observational data set to study global change. To accomplish this, a wide range of observations will be required to provide both new measurements not previously achievable and measurements with a greater degree of accuracy and resolution than are now available.

Among the parameters that are currently retrieved from satellite vertical sounding observations, temperature and moisture profiles are the most important for the description of the thermodynamic state of the medium. Other parameters, like those describing the cloud cover (top pressure and temperature, amount, radiative properties), the surface state (temperature, presence of snow or ice, emissivity, ...) or the conditions close to the surface (temperature, humidity) are also key parameters for meteorology and climatology.

Presently limited by the type of the vertical sounding instruments on board the operational meteorological satellites, these retrievals do not reach the required level of accuracy. An analysis of these limitations brings into evidence the too low *spectral* resolution of the infrared sounders ($\nu/\Delta\nu \approx 50$ to 100) which main consequence is a degradation of the *vertical* resolution and of the accuracy of these retrieved quantities.

It is now well recognized that an improvement by one order of magnitude of the spectral resolution should allow to meet the specifications issued by national or international agencies : temperature with an average rms error of 1 K ; humidity with an average rms error of 10 % ; vertical resolution of 1 km, at least in the lower troposphere.

Next generation infrared vertical sounders like the Improved Atmospheric Sounder in the Infrared (IASI) or the Advanced Infrared Radiation Sounder (AIRS) are designed to provide both new and more accurate data about the atmosphere, land and oceans for applications to climate studies. Among the important observations that AIRS or IASI will contribute to the climate data set are day and night global measurements of :

- atmospheric temperature profiles with an average accuracy of 1 K and 1 km vertical resolution ;
- relative humidity profiles with an average accuracy of 10 % and a vertical resolution of about 2 km, and total precipitable water vapor ;
- fractional cloud cover, cloud infrared spectral emissivity, and cloud-top pressure and temperature ;
- total ozone burden of the atmosphere and possibly two to three layer total burden (troposphere and lower stratosphere) ;
- mapping of the distribution of minor atmospheric gases such as methane, carbon monoxide and nitrous oxide : total burden and possibly the resolution of about 3 layers ;
- sea surface temperature with an average accuracy better than 1 K ;
- land surface temperature and spectral infrared emissivity ;
- surface albedo ;
- snow and ice cover ;
- outgoing long wave radiation ;
- precipitation index (computed from a knowledge of the cloud height, surface temperature, and atmospheric thermal structure).

Such information is essential to the study of numerous inter-disciplinary questions to be addressed in Earth science :

- cloud-radiation interaction,
- global energy and water cycles,
- measure of the increase of the greenhouse gases,
- atmosphere-surface interaction,
- land properties,

and, last but not least :

- improvement of numerical weather forecasting.

The consideration of a much larger set of parameters describing the complexity of the Atmosphere-Ocean-Land system obviously requires a multispectral approach based on a simultaneous analysis of a large set of measurement channels. An increase by more than one order of magnitude of the number of channels presently implemented on the operational weather satellites (NOAA series) is expected. Specific signatures of such or such phenomenon radiatively active may then be easily isolated, interpreted and quantified.

High spectral resolution is also the key for analysing atmospheric chemicals which relevance to climate evolution is now obvious, in particular through their interaction with the distribution of ozone. Ground based, airborne and spaceborne instruments have already demonstrated the capability of high spectral resolution or limb nadir sounding (in the microwaves, in the far infrared or in the infrared) to monitor numerous trace gases for a better description of their respective cycles owing to coupled dynamical and photo-chemical models. Major space programs for the end of this century do include such instruments.

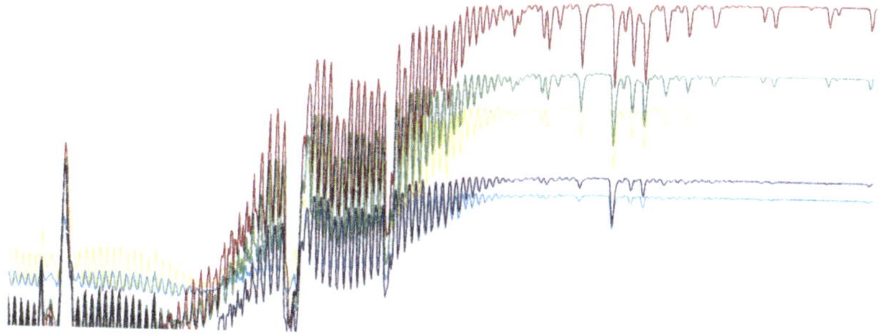

High spectral resolution infrared spectra of the Earth's atmosphere for 5 standard atmospheres, from polar to tropical

The NATO Advanced Research Workshop, from which this book results, had, as main objective, to review the benefits one may expect from the next generation sounders for a better understanding of the Earth's weather and climate. It has been organized in seven sessions covering status, theory and applications of high spectral IR remote sensing for Earth's weather and climate studies.

Session 1 : Review of future programs from 1994 (ATOVS) to EOS era. Papers have been presented on the programs of the major space agencies.

Session 2 : Next generation IR sounders : temperature and moisture analysis. Papers have been presented on new techniques for the retrieval of thermodynamic variables from high spectral resolution observations in the infrared.

Session 3 : Next generation IR sounders and cloud field analysis. The sounders from the next generation should greatly improve our present analysis of the cloudiness which importance for weather forecasting and climate modelling is well recognized.

Session 4 : Surface characteristics analysis. Surface skin temperature is an important parameter, governing in particular the exchanges between the oceans and the atmosphere.

Session 5 : Trace gas analysis. High spectral resolution is the key for analysing chemicals which relevance to climate evolution is now obvious (O_3, CH_4, CO, nitrogen oxides, ...).

Session 6 : Impact on weather and climate analysis. Assimilation of thousands of radiance observations in numerical weather models is a problem which has been reviewed during this session.

Session 7 : Forward modelling. Accurate physical simulation of high spectral resolution data is a difficult problem. The present state of the art and future plans have been presented.

Each chapter of the present book corresponds to one of these seven sessions.

Acknowledgements

Help from the Scientific and Environmental Affairs Division of NATO is warmly acknowledged. We are happy to thank the authors for their willing cooperation. All the papers presented here have been reviewed by anonymous referees.

Paris, March 1993

Alain Chedin

Moustapha T. Chahine

Noëlle A. Scott

Table of Contents

Chapter 1

Review of future programs from 1994 (ATOVS) to the EOS era

"Historical background for IASI"
B. BIZZARI - Italian Meteorological Office, Italy

"IASI, infrared interferometer for operations and research"
F. CAYLA - CNES, France

"Requirements for an Operational Interferometer Thermal Sounder"
M. PERRONE - EUMETSAT, Europe

"Development of a Fourier Transform Infrared sounder. IMG"
H. SHIMODA, T. OGAWA - TRIC, Japan

HISTORICAL BACKGROUND FOR IASI

Bizzarro BIZZARRI
Italian Space Agency
Via di Villa Patrizi, 13
00161 Rome, Italy

1. Introduction

IASI (Infrared Atmospheric Sounding Interferometer) is the European project for and improved infrared atmospheric sounder, a parallel project to the US AIRS (Atmospheric Infra Red Sounder). Both projects derive from the original concepts put forward in the US in the early 80's and supported by the development of airborne instruments such as AMTS (Advanced Moisture and Temperature Sounder) and HIS (High-resolution Interferometer Sounder). In Europe, the wait-and-see attitude was prevailing till a few years ago; but, when ESA and, afterwards, EUMETSAT, started to plan for a European contribution to the overall polar-orbiting satellite system, the scientific and operational communities undertook a transition to an active participation in this development, which is felt as the most important improvement in the near future for operational satellite meteorology.

As the historical background to AIRS is well documented, it may be of interest to keep trace of the origin and the evolution of the IASI project in Europe. This is the purpose of this paper.

2. Evolution of the requirement

Probably, the first step in the European consideration for an improved sounder occurred within the Atmospheric Instrumentation Definition Activities (AIDA) Working Group, active within ESA in 1982-83 to collect ideas about the next programme for meteorology and atmospheric physics to follow Meteosat. The famous "1 K / 1 km" requirement about sounding accuracy and vertical resolution respectively can be found in their final report, dated May 1983 (members: Bengtsson as chairman, Berroir, Bizzarri, Bolle, Fischer and Miller) (see ESA 1983).

The heritage of AIDA was then taken in ESA by the Meteorological Advisory Committee (METAC), active in 1985-1988 (members: Bolle as chairman, Ackermann, Bizzarri, Collier, Mohr, Morgan, Pastre and Rodriguez-Franco). It was within this group that a study on the possibilities of high spectral resolution to improve vertical resolution was commissioned.

NATO ASI Series, Vol. I 9
High Spectral Resolution Infrared Remote Sensing
for Earth's Weather and Climate Studies
Edited by A. Chedin, M.T. Chahine and N.A. Scott

The study (Stewart, Carli, Chedin and Fischer), dated September 1987, analyzed the findings available from AMTS and HIS, and gave preliminary design specifications for an interferometer for Meteosat Second Generation (MSG) (see ESA 1987).

In the meantime (1985-1989), the Technical Working Group (TWG) of IPOMS (International Polar Orbiting Meteorological Satellites, a group descending from the Summit of the Group of Seven) was progressing with the planning of the meteorological mission for the Polar Platform era. During a series of annual meetings the TWG (chaired by Bizzarri, variable membership), analysed the proposal from NOAA for the successor of HIRS-2 and AVHRR, a combined imager-sounder called AMRIR (Advanced Medium Resolution Infrared Radiometer), claimed to provide, in association with AMSU, soundings not worse than those to be provided by NOAA K,L,M by means of HIRS-2 and AMSU. This perspective was finally declared unsatisfactory by the TWG, which strongly indicated a requirement which could only be satisfied by high spectral resolution. After this, the AIRS proposal was put forward by NASA and fully endorsed by the TWG, which also invited Europe to participate in the development of this advanced technology (see TWG 1989).

3. The early proposals

As a follow-on of the METAC activity, the first proposal for an improved infrared sounder was put forward in the context of MSG. It was the Italian delegation to the EUMETSAT Scientific and Technical Group (STG) to propose in 1988 a high-spectral resolution sounder for MSG, supporting the proposal with the study of an instrument called IRIAS (Infra Red Interferometer for Atmospheric Sounding) (see EUM 1989a). A follow-on of this study was performed by ESA but, after the final decision to adopt a spinning concept for MSG, the possibility of flying a sounder on MSG (either spectrometer or radiometer) has been finally ruled out.

In the same 1988, the Announcement of Opportunities (A.O.) for the provision of instruments for the ESA and the NASA Polar Platforms were issued. The Italian Space Agency (ASI) put forward the proposal for ISAS (Infrared Spectrometer for Atmospheric Sounding), an interferometer which was intended to be a simplified version of AIRS, limited to the 3.4-5.0 micron band for improved sounding in the low troposphere. The proposal was not accepted by ESA on the basis of its immature technical status, and of the competition with the NASA AIRS development. Also, the limitation of the spectral range to the shorter wavelengths was not seen with favour by the experts.

At this point, therefore, the European attempts to participate in the development of improved infrared sounders were virtually dead. However, the NASA decision to select a grating concept for AIRS re-opened the matter. The Italian team (and industrial support) which had originally proposed ISAS and also compared the grating versus the interferometer approach, considered that:

* the development of the grating spectrometer in the expected time-frame (1998) was critical, as it requested important technological developments such as improved detector efficiency and extention of CCD capability to some 16 microns, and long life-time active cooling;

* the preliminary studies of the interferometer had shown that all enabling technologies were available, passive cooling could have been sufficient and the performances would have been the same of AIRS in the long wavelengths and better in the short.

Consequently, ASI decided to run an industrial Pre-Phase-A study of an extended version of ISAS (3.4-15.5 microns). This study was run across 1989 and formed the basis for the future proposal.

At the same time in France, under the pressure from a number of scientific teams attempting to define the mission of a national programme for operational climatology (Globsat), CNES had undertaken a number of in-house and in-industry studies relevant for the definition of an improved sounder called ASSA (Atmospheric Sounding by Spectral Analysis). As a difference from the Italian approach, which was driven by operational meteorology, the French interest was equally shared between meteorology and tropospheric chemistry. This, though ASSA was not defined whether to be a grating or an interferometer, was in fact favouring the interferometer concept, thanks to its potentially better performances in the shorter wavelengths which include more signals from trace gases.

4. The decisional process

In the course of 1989 the voice of ECMWF was becoming more and more convinced in showing the declining impact of TOVS sounding on Numerical Weather Prediction. The International TOVS Working Group, whilst evaluating the expected improvement from ATOVS on NOAA K,L,M, was clearly identifying the limits of the expected improvements. At the same time, EUMETSAT was facing the problem of ensuring continuity to the polar system to be first developed by ESA, and consideration started to be given to a EUMETSAT Polar System (EPS). It was therefore the appropriate moment to concretize the proposal for a European improved sounder.

At the EUMETSAT Council in December 1989 the Italian delegation presented the results of the ISAS Pre-Phase-A study (see EUM 1989b), re-iterated all the motivations for improved sounding and invited interested delegations to a joint development. An immediate response came from France, and a first brain-storming meeting between the Italian and French scientific and operational communities, with EUMETSAT participation, took place in Florence on 20 December 1989.

The decision to proceed with a joint study was taken in Rome, on 16 March 1990. The Statement of Work for the Phase-A study was agreed, and the name IASI (Infrared Atmospheric Sounding Interferometer) introduced. EUMETSAT was involved in the work with the attendance of a staff member to key meetings and with regular up-datings on the status of the project at STG meetings.

On 1st June 1990, in the context of setting-up the conditions for EUMETSAT to join the ESA POEM-1 programme for the flight of the first EPS mission, the EUMETSAT Council adopted a resolution (see EUM 1990) stressing, i.a., the urgent need of the meteorological community for an improved infrared sounder, to be embarked on POEM-1 if at all possible.

In response to this, the ESA Programme Board for Earth Observation, on 26 September 1990, mandated the Executive to aggressively pursue the matter of investigating possible accommodations of IASI on POEM-1 (see ESA 1990).

5. Status of the project

Most of the results of the Phase-A study, including possible accommodation on POEM-1, were available at the end of March 1991. They basically confirmed the feasibility of an instrument fulfilling the mission specifications within the prescribed boundary conditions of size, mass, power and cooling requirements, data rate and readiness to launch (1998). A preliminary design was available, and accommodation possibilities on POEM-1 have been envisaged.

The accommodation aspect of IASI has been included by ESA in the activities to be carried out within the POEM-1 Preparatory Programme, due to be accomplished in mid 1992.

The official and detailed presentation of the Phase-A results have occurred through a number of specialized meetings in May and June 1991, followed by a series of activities to account for remarks, anticipate Phase-B items and prepare for the management structure of the Phase-B study. One major activity was the organization of a User Conference, in Paris, 9-10 January 1992. The Conference confirmed the requirements for improved infrared sounding, was informed of the positive feasibility results for IASI to meet the mission specifications, agreed that the interface between the IASI system and the users would be "the unapodized spectral data at full resolution" and, in conclusion, requested ESA and EUMETSAT to cooperate for providing IASI with an early flight opportunity and long-term continuity.

6. Conclusion

The Phase-B study has officially started on 7 February 1992. It will still be a joint ASI/CNES activity, but an international Mission Team will include other than French/Italian experts, a major role being foreseen for EUMETSAT. Also, EUMETSAT and other international involvement is envisaged in a Ground Segment support team (the Ground Segment will in fact be in Phase-A for a number of months). ESA will be involved for the accommodation aspect and EUMETSAT for the interface with the data transmission system.

The final confirmation of the flight opportunity on POEM-1 has still to take place, possibly at the end of 1992. After that, the only problem will be ... to build the instrument in time!

REFERENCES

ESA (1983) - Proposal for a European Atmospheric Space Programme - Report of the AIDA Group, May 1983.

ESA (1987) - Atmospheric Sounding and Trace Gas Measurements with Interferometers from Geostationary Altitude - Report to the METAC Group, September 1987.

ESA (1990) - Declaration on the Preparatory Programme of the First Polar Orbit Earth Observation Mission using the Polar Platform (POEM-1) - ESA/PB-EO/XX/Dec.1 (final) drawn up on 26 September 1990.

EUM (1989a) - Pre-Phase-A study of IRIAS (Infra Red Interferometer for Atmospheric Sounding) - EUM/STG(89)29.

EUM (1989b) - Proposal for an Infrared Spectrometer for Atmospheric Sounding (ISAS) - EUM/C(89)42 + EUM/STG(89)53.

EUM (1990) - Resolution concerning the use of the ESA Polar Platform - EUM/C/Res.XXIV adopted on 31 May - 1 June 1990.

TWG (1989) - Report of the Technical Working Group of the International Polar Orbiting Meteorological Satellite Group - Ravenna, 15-19 May 1989.

IASI INFRARED INTERFEROMETER FOR OPERATIONS AND RESEARCH

F-R. CAYLA
seconded from METEO-FRANCE / CNRM /GMGEC to
Centre National d'Etudes Spatiales /TE/AE/INS/IO
18 Avenue E. Belin
31055 TOULOUSE Cedex
FRANCE

KEYWORD/ABSTRACTS : meteorology / atmospheric science / space instruments / sounders / Fourier transform radiometry

The paper presents the high-resolution infrared interferometer being developed, in co-operation between France and Italy, to meet the needs of meteorology and atmospheric sciences in temperature, humidity profiles, and to provide limited information on trace gas abundance's. The instrument is planned for launch as part of the meteorological payload onboard the European Polar Platform POEM-1. Presently in phase B, this Fourier transform spectrometer will provide spectra of high radiometric quality at 0.25 cm^{-1} resolution from 15.5 to 3.4 micrometers, with global coverage twice per day at 25 km horizontal resolution.

Foreword: This paper describes IASI as it stands at the end of its Phase A study. In order to give a consistent view some modifications already decided at time of final revision have not been taken into account. Although functional specifications and major trade-off should be considered as firm, actual figures, in particular as concern performances, will be refined in the subsequent development phases.

1. MISSION AND MEASUREMENT OBJECTIVES

1.1 . Mission Objectives

As explained in more detail in the same Advanced Research Workshop by B. Bizzarri, the decision to undertake, in co-operation between France and Italy, the development of an advanced infra-red sounder, answered perceived needs both from operational meteorology and atmospheric science community.
Operational meteorology is aware that present quality of global measurements is insufficient and that the ultimate quality, expected from the TOVS and even from ATOVS, will hinder progress in numerical weather forecasting. Both French and Italian meteorological services insisted upon the necessity of a dramatic improvement of the quality of temperature and humidity sounding available on a global scale.
Atmospheric research community, faced with the challenge to assess climate changes, define their causes, and forecast their evolution, uses the same tools than operational meteorology and will benefit from the improved knowledge of atmospheric dynamics resulting from better temperature and humidity sounding. In the exploration of mechanisms contributing to climate evolution, it needs a better knowledge of the atmospheric components contributing to the greenhouse effect or more generally to the radiative balance of earth and atmosphere. This

NATO ASI Series, Vol. I 9
High Spectral Resolution Infrared Remote Sensing
for Earth's Weather and Climate Studies
Edited by A. Chedin, M.T. Chahine and N.A. Scott

leads to a mission, in the framework of the World Climate Research Project, where measurements of trace gases and of properties of clouds and surfaces are to be obtained at a scale consistent with dynamic processes.

1.2 . Measurements Objectives

Translated into measurement specifications, the two above missions result in the following three groups of requirements, giving first expected accuracy, then when applicable vertical resolution, and finally horizontal resolution of products :

a - Temperature and humidity profiles

Temperature acc.	1 K	vertical res.	1 km	horizontal res.	50 km
Humidity (rel.)	10%	"	1 km	"	50 km

b - Trace gas column amount

Integrated column content of :

N_2O, CO, CH_4	Acc.	10%	horizontal res. 100 km	
O3 (potentially up to 3 layers)		5%	"	50 km

The accuracy figure quoted for O_3 is what is expected for the integrated column content. For determinations of concentration at several levels no accuracy figure can be given with confidence.

c - Other potential uses

The following parameters of surface and clouds could probably be determined with a high spectral resolution infrared sounder : Sea Surface Temperature, Land surface temperature and emissivity, cloud coverage and spectral properties

For this last group of possible products, definition of necessary algorithms is insufficient to allow specification of reasonable figures for accuracy. Spatial resolution is expected to be the highest possible with the instrument i.e. of the order of 25 km. At least for surface properties, significance of measurements at this scale, has to be demonstrated.

1.3 . Instrument Specifications

Translation of geophysical measurement specifications to radiometric measurement specifications is a difficult problem. In the case of IASI this translation was done under responsibility of CNES and ASI and relied heavily on the work done in the framework of the AIRS science team and on the demonstrated capability of HIS to provide temperature and humidity sounding of high quality.

1.3.1 . Spectral Range And Resolution

The spectral range is dictated in first approximation by the needs of temperature sounding for which both of the CO_2 absorption bands at 15 and 4.3 μm are useful. The domain in between this bands is dominated by the 6.3 μm water vapour absorption band. The domain is slightly expanded on the short wavelength side to include the 3.7 μm window and some absorption bands of trace gases.

Range :	645	to	2940	cm^{-1}
	15.5	to	3.4	μm

For technological reasons this spectral range is broken in 3 bands and for these bands the necessary apodised resolution is indicated :

Band 1	645	to	1100 cm^{-1}	at	0.5	cm^{-1}	resolution
Band 2	1100	to	2000 cm^{-1}	at	1.0	cm^{-1}	"
Band 3	2000	to	2940 cm^{-1}	at	1.5	cm^{-1}	"

1.3.2 . Radiometric noise

The radiometric noise level is specified separately for each band in terms of NEΔT at a reference temperature equal to 280K

The noise considered in the specification includes all contributions. These contributions are either conventional, i.e. associated with the detectors, amplifiers and A/D converters, or generated by pseudo-noises sources, i.e. non systematic differences between the measurements provided by the actual instrument and the measurements expected from a nominal instrument.

It specifies the performance of a channel or group of channels having a width equal to the spectral resolution of the band specified above.

Band 1	< 0.2 K	at	280 K
Band 2	< 0.2 K	at	280 K
Band 3	< 0.2 K	at	280 K

1.3.3 . Calibration

a - Spectral

The position of any channel should be known with an accuracy better than 0.01 cm^{-1}.

b - Radiometric

Equivalent temperature determination should have an accuracy of 0.5 K for a black target at 280 K.

1.3.4 . Sampling, Spatial Resolution

IASI is an instrument that will be flown as part of the meteorological payload on POEM-1; its sampling shall be compatible with AMSU-A's one. The AMSU-A FOV will be subdivided in 2x2 IASI FOV's. Each of IASI FOV's will be less than 25 km in diameter (baseline design is 18 km, a smaller size is under consideration).

1.3.5 . Integrated Imaging System

One of the major problems in the processing of sounding data is the correct handling of partly cloudy FOV's. This leads to inclusion of an integrated imaging system the purpose of which is to provide a good description of cloud properties inside the sounder FOV. Such a description should consist ideally of partial cloud coverage, top temperature and transmittance of any of the cloud layers present ; its derivation, as currently performed from AVHRR data, relies on multispectral treatment at pixel level and impose a high radiometric quality to the imager.

An obvious solution is simultaneous use of AVHRR data ; it has however been found by most of the groups that did attempt to use simultaneously HIRS and AVHRR that coregistration of both instrument was very difficult. The reason why is that relative positions of the FOV's belonging to both instrument must be determined from the data with an accuracy which is a small fraction of the sampling interval of HIRS.

The minimal requirement on the integrated imager was then to allow day and night accurate registration with AVHRR or its followers. Obvious cost considerations imposed to rely on of the shelf detectors. The design was then made around an existing HgCdTe detectors' matrix with CCD readout of 128x128 elements working at 150 K for wavelengths shorter than 5μm. Associated with IASI total FOV of 45x45 km at SSP this led to a pixel size of 400 m. Limitations on allowable data rate force to reduce resolution by averaging of adjacent pixels to a new pixel size of 1.6 km ; it was possible to reduce the somewhat wasteful loss of information on small scale structures by generating at this scale a description of variability (baseline is rms. of the 4x4 elementary pixels used in a large one) which could be used in Coakley Bretherton type of algorithms.

As a result of above considerations the specification of the integrated imager is the following:

One channel 3.7 to 4.0 μm.
One pseudo-channel : variability at 400 m scale
Pixel size 1.6 km. averaged from 4x4 400 m pixel
Coregistration with sounder FOV's better than 1 km.

1.3.6 . On Board Processing

The amount of on board processing should be the result of trade-off between transmission load and processing complexity. As the constraints imposed upon the maximum data-rate did not allow much flexibility the only solution which did not result in information loss and is still associated to a reasonably simple software was the generation of unapodised full resolution calibrated spectra. Following independent considerations that are exposed in section 4 this is the chosen user interface. On-board processing to this level will prevent that users are faced with different type of data on the direct read-out facility and on the links to the ground segment.

2. INSTRUMENT DESCRIPTION

The instrument description will be rather schematic. It will mostly rely on diagrams and just highlight in the text aspects of interest.
The figure 1 illustrates the breakdown of the instrument in its different functional units. They are ordered according to information path, ancillary functions associated to given units are shown off the main line. The selection of passive cooling was a design driver. The table at the bottom shows the budget in term of mass, power and bit rate.

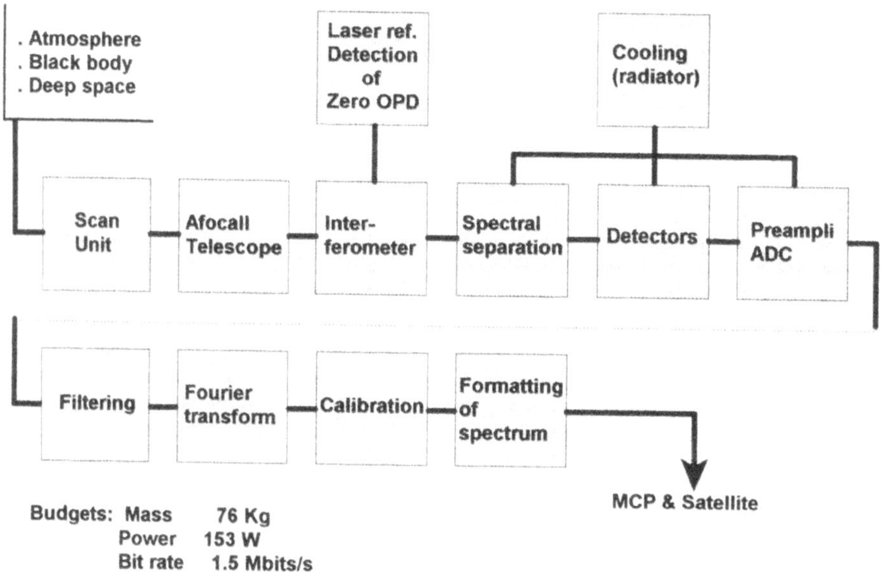

Figure 1: Functionnal breakdown of IASI

The following figures will illustrate design of sub-systems or group of sub systems.
The figure 2 shows the solution envisaged for the scanning system. This system combines two functions, one is a classical scan in step and dwell mode, following AMSU-A scan pattern, the other one specific to an interferometer compensates the FOV motion during the time of interferogram measurement. The chosen concept based on magnetic bearing will use the degrees of freedom on axis position to compensate field motion. As an aside, this compensation allows a reasonable image quality on the imager with a 150 ms integration time.

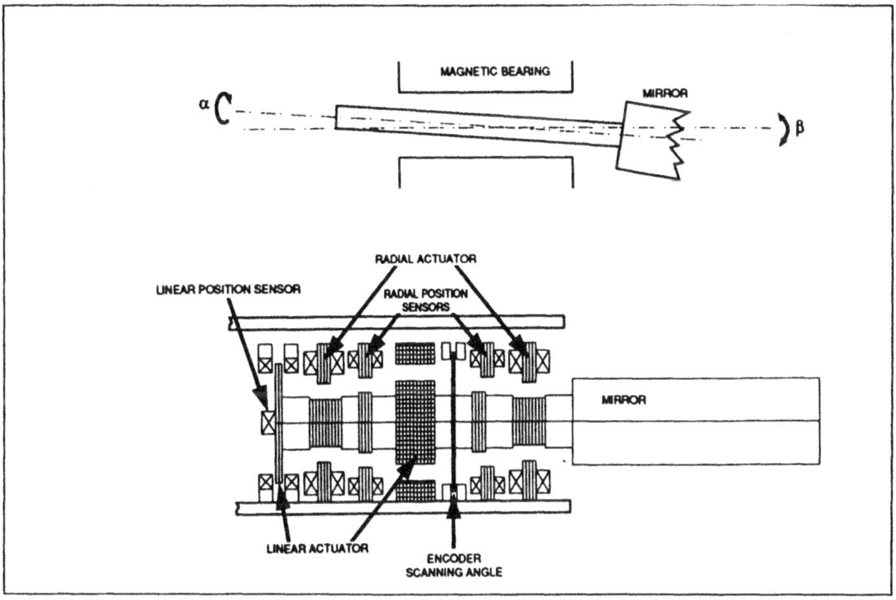

Figure 2: Description of the scanning mechanism

The figure 3 shows the optical layout of the instrument. The main purpose of the front-end telescope is reduction of angular spread of beam in the interferometer in order to maintain spectral resolution at high wave numbers. In order to prevent that inhomogeneity over the different detectors associated to the different bands, generates pseudo noise over inhomogeneous scenes, detectors are conjugated with the aperture stop.
The interferometer itself is a classical Michelson design with corner cubes replacing mirrors. Mechanism concept is based on elastic suspension of the corner cubes and the motor and allows a very good compensation of dynamical perturbations.
Instrument layout allows to associate a mechanism to each reflector giving thus a cold redundancy. Interferogram sampling is driven by the fringes of a reference laser source. The laser chosen is a laser diode with active frequency control based on absorption line in a gas cell. An additional system allows zero path difference determination.
Calibration is based on space and blackbody views scanned at each line. Occasionally a second space view with a very different incidence angle is used to monitor the behaviour of the scanning mirror.

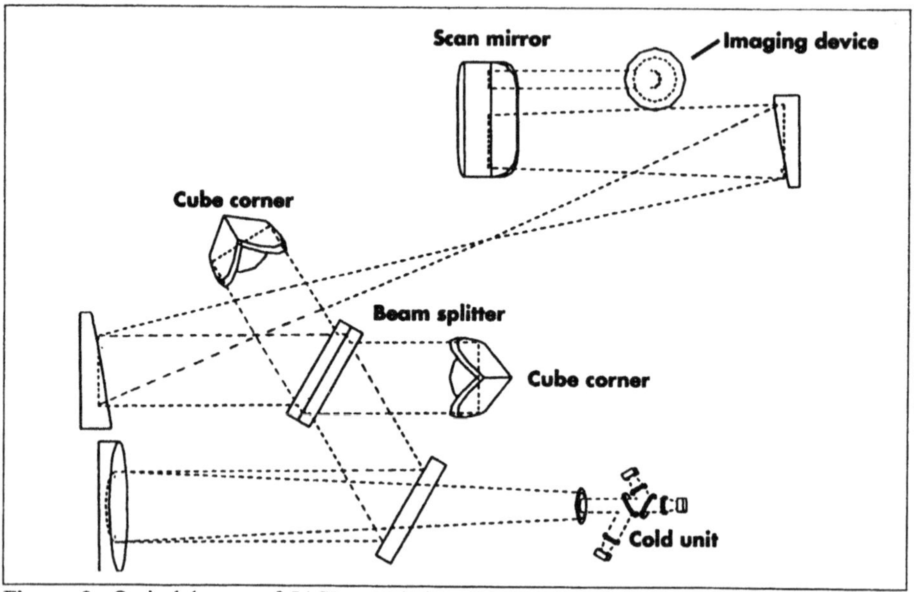

Figure 3: Optical layout of IASI - Includes after the scan mirror two confocal off-axis parabolic mirrors, the interferometer, a folding mirror, a third parabola and an assembly of llenses and dichroics forming the cold unit.

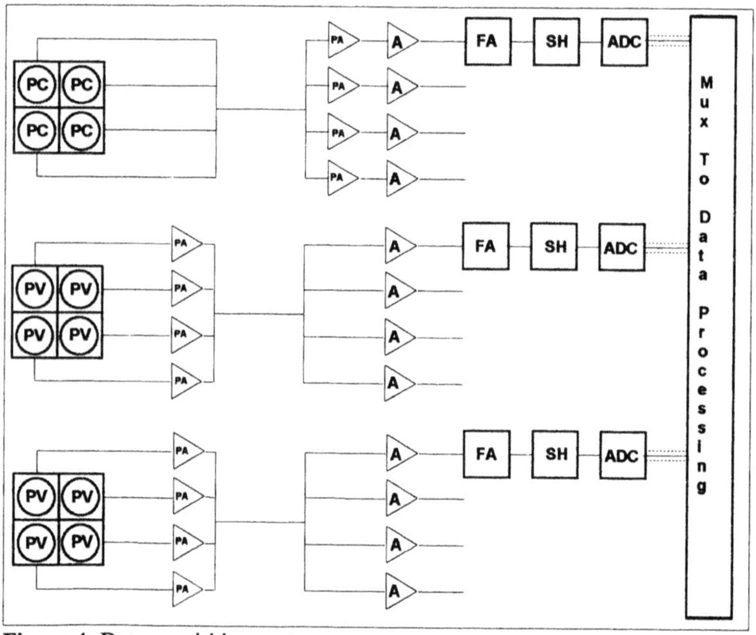

Figure 4: Data acquisition system

The figure 4 shows the data acquisition characteristics.

3. PERFORMANCES

3.1 . Spectral

Spectral performances are twofold. Design of the interferometer with its maximum optical path difference of 2 cm and the reduction of beam divergence with a front end afocal telescope allows to meet the required apodised spectral resolution of 0.5 cm^{-1} . The frequency calibration is dependant on two parameters first the stability of the reference laser second the geometrical stability of the position of IASI FOV's ; for both these parameters the constraints resulting from the specification are within the state of the art.

3.2 . Radiometric

3.2.1 . Radiometric noise

Following the definition of radiometric noise in the specifications, which covers both conventional and pseudo-noise the two type of contributions have been studied separately.
The following basic considerations explain the behaviour of conventional radiometric performances expressed in NEΔT as shown in figure 5 ,6 and 7 for the three bands :
a. noise in the interferogram measurements are in first approximation equally distributed on the spectrum after Fourier transformation .
b. assuming a constant quantic efficiency in the band, the spectral noise is constant in term of photons/s /cm^{-1}.
c. translated in radiance this noise will increase with wave number in the same manner than the photon energy.
d. NEΔT will increase even more due to the behaviour of the Planck function.
As a consequence, the sizing of the instrument, necessary to meet requirements at the high wave number boundaries of the bands, will, close to the low wave number boundary, result in conventional noise performances significantly better than the specifications. This property will be particularly useful for band 1.

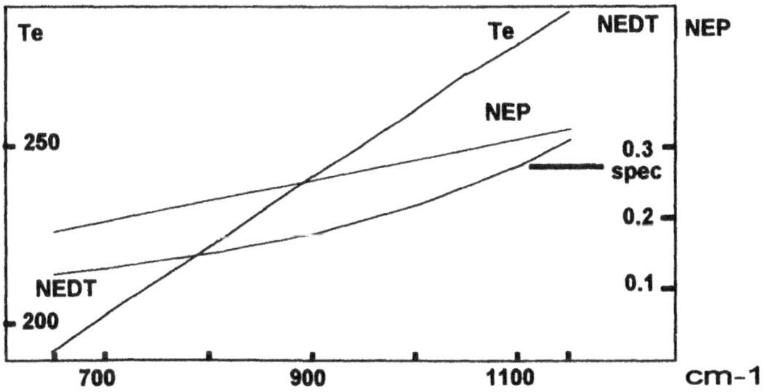

Figure 5: Radiometric performances of first band -
plots versus wavenumber of NEDT at 280 K, NEP and noted Te the temperature at which NEDT is equal to the specification (line noted **spec**)

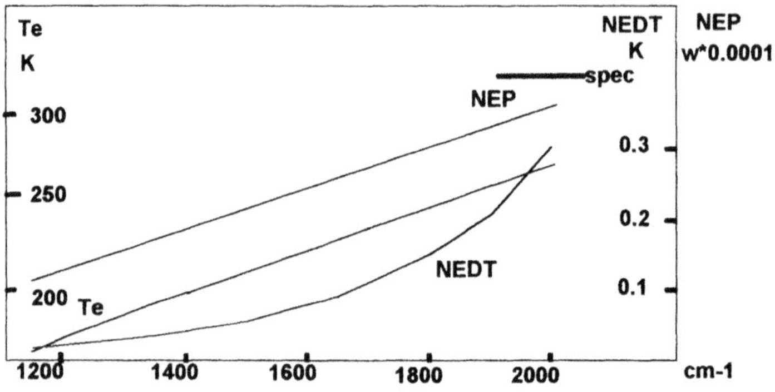

Figure 6: Radiometric performances of second band -
plots versus wavenumber of NEDT at 280 K, NEP and noted Te the temperature at which NEDT is equal to the specification (line noted **spec)**

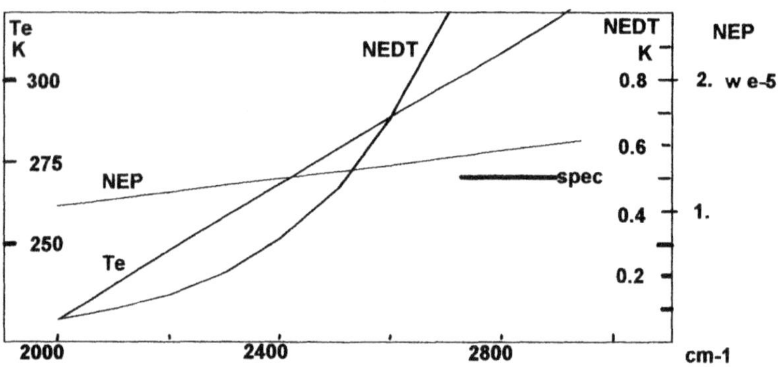

Figure 7: Radiometric performances of third band -
plots versus wavenumber of NEDT at 280 K, NEP and noted Te the temperature at which NEDT is equal to the specification (line noted **spec)**

The evaluation of pseudo-noises was performed in the following way : in a first step instrument parameters contributing to pseudo-noises where identified ; then sensitivity to these parameters was evaluated using appropriate first order models of the effect generating pseudo-noise ; next step was to specify according to sensitivity reasonable constraints for the instruments parameters and finally the pseudo-noise was re-evaluated with parameter values insides the limits defined in previous step.

Two sub-classes of pseudo noises were considered. The first class is common to all radiometers ; it groups the three following pseudo noises sources : the effects of incidence angle on the scanning mirror, the effect of the form and location of the instrument's response function and the effect of misregistration of the three bands. The second one is specific to interferometers and groups the effect of FOV displacement during the measurement of the interferogram, the noise generated by sampling errors and finally the noise occurring inside the digital processing of the interferogram.

The total performances obtained by combining all evaluations is given in figure 8. As compared with figure 6 the large difference in the low wave number side of band 1 is due to the effect of the uncertainties on instrument response function shape and position.

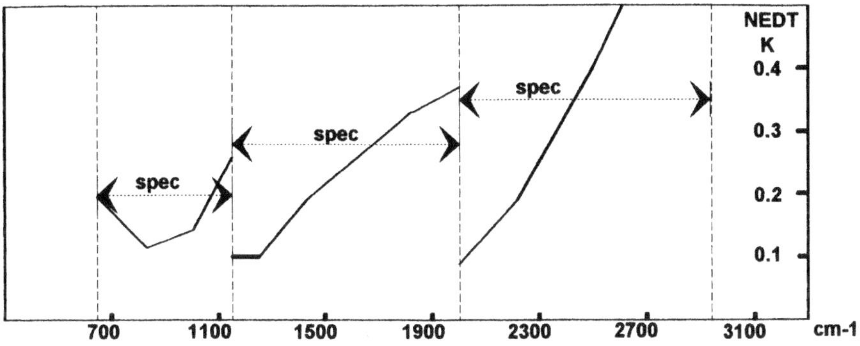

Figure 8: Radiometric performances all bands for channels of width 0.5 cm^{-1} -

3.2 Radiometric calibration

Radiometric calibration errors budget includes a large number of terms where black body temperature measurement (accuracy and representativeness) and emissivity are major terms. Additional contributions of front mirror incidence effects and various departures from linearity are expected to be rather well under control and the figure of 0.5 K seems to be reasonable.

4. USER INTERFACE

In order to define the desirable level of user interface, a first description of the chain of treatments was established showing the different levels at which the information appears. Each level being the result of transformation of the preceding level by an operator which is also indicated. The decomposition foreseen in IASI is the following :

Analogue Interferogram
    Analogue Filters
Filtered Analogue Interferogram
    Sampling A To D Conversion
Numerical Interferogram
    Digital Filters
Filtered Digital Interferogram
    Resampling / Decimation
Minimum Digital Interferogram
    Discrete Fourier Transform
Rotated Spectrum In Complex Plane
    Phase Error Correction
Corrected Spectrum In Complex Plane (Raw Spectrum)
    Calibration

Full Resolution Calibrated Spectrum
Apodisation - Convolution
Nominal Spectrum
Users' Processes
Products

This decomposition is not unique and differs from a more classical one in two points which call for explanations.
The first difference is the presence of digital filtering and resampling. These steps are imposed by the selection of a relatively high level of over sampling in the measurement of the interferograms. This option was selected on the following grounds:

- Over sampling of the interferogram eases the problem of analogue filtering.
- It allows a relatively simple detection and correction of spikes.
- Accuracy of A/D converter could be relaxed.
- Sensitivity to sampling jitters is reduced.

The second difference is the fact that Fourier transform is performed before applying an apodisation function. The reason for this choice is analysed below :

- The resultant apodisation function is the FT of desired spectral response
- The apodisation function applied to the interferogram is such that its product by the auto-apodisation function of the instrument is equal to the resultant apodisation function.
- The part of the auto-apodisation function which is caused by finite size of the FOV is dependent on wave number.
- As a consequence of the above propositions there is no unique apodisation function to apply to a band, or conversely applying a unique apodisation function will result in a variable spectral response function inside the bands.
- It was also felt that it was easier to take care of variable convolution in the spectrum domain.

Once defined the possible levels it was necessary to define criteria for choice. The following ones were used :
A .IASI data are new. Numerous potential usage's will develop during the instrument lifetime. This implies that it is today difficult to decide which part of the information is not useful.
As a consequence:
USER INTERFACE SHOULD PRESERVE ALL POTENTIAL INFORMATION

B .Conversion of raw interferograms to spectra requires a detailed knowledge of instrument physics, technology and status. USERS have more important problems to study.
As a consequence:
USER INTERFACE SHOULD NOT IMPOSE USERS INVOLVEMENT IN INTERFEROMETER PHYSICS AND TECHNOLOGY.

Each of this two criteria tends to displace the interface in reverse direction in the above table. Criterion A brings us upstream, B downstream ; a reasonable compromise seems to be reached at the level :

Full resolution calibrated spectrum
i.e. unapodised spectrum sampled at 0.25 cm^{-1} intervals

4.4 . Additional Information's

4.4.1 . Services

As the initial users requirement where for apodised spectra at 0.5 cm^{-1} resolution, Project is responsible to provide either access to nominal (apodised) spectra or corresponding software packages.

4.4.2 . Verification

The decision to perform calibration on board, generates legitimate concern inside the users community. In order to alleviate this concern, great care will be taken to provide full visibility on the performance of on board processing with special emphasis on the calibration steps. All data sets necessary to check on a sampled basis the performances of the processing will be down-linked, archived and available.

4.4.3 . Coding

In the above discussions calibrated data are meant as provided in engineering units. The transmission of the spectra with a reasonable efficiency will imply transformation of corresponding data according to a coding scheme still to be defined . It is inside project responsibility to provide to all users, software packages to translate coded data back to engineering units.

REQUIREMENTS FOR AN OPERATIONAL INTERFEROMETER THERMAL SOUNDER

Marilena Perrone
EUMETSAT
Am Elfengrund 45
6100 Darmstadt
Germany

1. INTRODUCTION

This paper summarizes the work of a group of consultant experts in the area of sounding retrieval and interferometer retrieval methodology convened to provide EUMETSAT with advice on the interferometer sounder issues. **The EUMETSAT Sounder Science Working Group (ESSWG), whose members are listed at the end of this paper**, has reviewed the meteorological requirements for an interferometer for sounding the atmosphere in the infrared region, to be developed for operational implementation on the EUMETSAT Polar System (EPS) intended as the a.m. satellite of the world's two polar orbiting meteorological satellite system. The Working Group has produced a report by the title "**Requirements for an Operational Interferometer Thermal Sounder**", November 1991, in which specifications are provided to satisfy the operational high vertical resolution temperature and moisture sounding mission. The assumption for the Working Group was that the operational interferometer sounder should be designed to incorporate low risk and proven technology with the minimum complexity needed to accomplish the mission throughout the operational lifetime of the spacecraft. For the first EPS application the instrument's spectral resolution and spectral coverage are limited to that needed to provide high vertical resolution temperature and moisture sounding. However, the interferometer sounder specifications also allow for the measurement of other parameters important for climate, such as surface properties, cloud distribution and properties, water vapour distribution and greenhouse gases concentrations. In parallel to this activity EUMETSAT has sponsored an industrial study for an Interferometer Thermal Sounder (ITS) performed by the Santa Barbara Research Center and the

NATO ASI Series, Vol. I 9
High Spectral Resolution Infrared Remote Sensing
for Earth's Weather and Climate Studies
Edited by A. Chedin, M.T. Chahine and N.A. Scott

University of Wisconsin, in order to investigate the feasibility of the approach and gain independent engineering assessment. The report by the ESSWG is intended to serve as a basis for the refinement of the operational interferometer sounder specifications through studies sponsored by EUMETSAT. It is organised so that the requirements for an operational interferometer sounder are specified in meteorological terms and then translated into radiometric, spectroscopic, and interferometric terms. It is preliminary in the sense that it states requirements in quantitative terms where possible, but also indicates the nature of the further studies and calculations which are needed. The structure of the following summary **mirrors the full ESSWG report to which the reader is referred to find the supporting details**. The report EUM/ESSWG/REP/91/1, November 1991, is available from EUMETSAT upon request.

2. METEOROLOGICAL REQUIREMENTS

2.1 Weather

The ESSWG report reviews the goals for observation of the atmospheric temperature and humidity fields which have found consensus among Numerical Weather Prediction (NWP) centers and have been adopted in the form of "WMO requirements". The origin of these goals derives partly from what NWP models could use and partly from what an idealised radiosonde network might provide. However the translation of these goals into the specification for an observing system, particularly for a remote sensing system, is not straightforward. The report emphasizes the need to carefully investigate the relationship between the goals and any real observing system. Observing systems which do not meet (or even approach) the goal in one or more respects may still be useful if they provide information of an accuracy higher than or comparable to our prior knowledge of the atmospheric state. Information on temperature and humidity profiles would be the primary contributions made by an advanced sounder towards meeting the observational requirements of NWP. However, the interferometer could also provide information to meet other requirements such as those for ozone, surface temperature (sea and land) and cloud properties, which will become more useful with data assimilation systems under development.

2.2 Climate

Although the operational interferometer sounder is primarily designed for the significant improvement of the data base for weather prediction, useful information concerning climate monitoring can, as from current meteorological satellites, be extracted as well. The ESSWG report quotes requirements compiled for climate monitoring and how the interferometer sounder can contribute to this issue emphasizing on those ones which can specifically be addressed by interferometric measurements. However, it should be noted that climate studies need long-term missions and special radiance patterns with spatial and temporal coverage.

The present lack of precision in the estimate of the concentrations of some of the minor atmospheric constituents and trace gases with impact on the climate (greenhouse gases) arises from the difficulty of inferring the planetary scale distributions of sources and sinks by successive integrations, along the relevant space and time scales, of a network of current observations which are certainly precise but limited to a few square metres of the terrestrial surface. Here, satellite observations have a fundamental role to play. In particular, a down looking spectrometer is the only way to quantify the global budgets of ozone and source constituents like methane, carbon monoxide, aerosols, sulphur compounds and nitrogen oxides. There are, however at present, only a few studies investigating changes in spectral outgoing radiances due to changes in trace gas abundances. As a first guess, a precision of the order of 1 $mWm^{-2}sr^{-1}$ $(cm^{-1})^{-1}$ should be sufficient to detect the CFC inputs, estimated with a current rate of annual atmospheric accumulation of 4%. For CO_2 (0.5% increase per year), CH_4 (0.9% increase per year) and N_2O (0.25% increase per year) long-term monitoring of at least 10 years is required to evaluate the signal with the anticipated measurement precision of 1 $mWm^{-2}sr^{-1}$ $(cm^{-1})^{-1}$.

3. OBSERVATIONAL EXPERIENCE

Experience with satellite-borne interferometers began aboard Nimbus-3 in 1969 with the Infrared Radiation Interferometer Spectrometer (IRIS). In spite of the limited spectral resolution of this instrument the advantage of measuring the continuous spectra of the outgoing radiance allowed studies concerning for example the cleanest parts of the atmospheric window, the land surface emissivities, characteristics of cirrus clouds, the interference of boundary layer structures over the ocean, the derivation of gross geographic features of methane and nitrous oxide. Current interferometers have taken full advantage of recent developments in detector, laser, optics and computer technology, leading to substantial improvements in spectral resolution, radiometric precision and spatial coverage rate. With these improvements, interferometers are a serious candidate for future operational meteorological satellite missions. Depending upon the objective, several approaches in the interferometer development have been followed. Concerning nadir-like emission observations, the High resolution Interferometer Sounder (HIS) concept has been developed at the University of Wisconsin. A HIS aircraft version flown aboard the NASA U2/ER2 aircraft on more than 40 aircraft missions demonstrated that highly accurate high-resolution radiance measurements (0.35 cm^{-1} resolution in the 600-1100 cm^{-1} region) can be made, resulting in an improved vertical resolution capability and accuracy, the possibility of tracing fine vertical scale water vapour features, more accurate measurements of cloud altitudes and the possibility of deriving spectral emissivities of the underlying surfaces.

Interferometers have also been designed at several institutions for solar spectra measurements in the near and middle IR from ground, aircraft and balloon, to derive atmospheric trace gas abundances, in particular those of substantial interest for the stratospheric ozone photochemistry. Comparable efforts have been made to measure the thermal emission of the Earth's limb from balloon, aircraft and space platforms.

4 INFORMATION CONTENT

The meteorological requirements for NWP are specified in terms most appropriate to conventional measurements such as radiosondes, and it is not easy to assess whether a satellite sounding system of a given instrumental specification meets these requirements. Satellite sounders do not measure temperature and humidity profiles; they measure outgoing radiance intensities. The process of using retrieval procedures may lead to temperature and humidity errors which are significantly correlated in the vertical and quite different in structure from radiosonde errors. This is so for radiometers and spectrometers, but the problem is more acute in the case of an interferometer since the actual measurement (the interferogram) is further removed from the meteorological variables of interest.

For a Fourier Transform Spectrometer (FTS) the relation between the varying component of the intensity at the detector and the power spectrum of the radiation entering the instrument is a Fourier Transform of the spectrum. In practice various factors intervene to make the recovered spectrum an imperfect representation of the true spectrum. The most important ones are: aperture effect, tilt and aberrations, channeling, phase and compensation, truncation and the effect of noise in its various forms including the scene variation noise. Acknowledging the presence of these error sources is extremely important when examining trade off options. The other important issue is the fact that in the interferogram only a finite number of samples corresponding to a given maximum path difference are measured and, as a consequence, the information beyond the truncation point is lost. The way to deal with truncation effects depends on the final use of the spectral data. As the truncation error and noise play a conflicting role, a compromise between resolution and variance of the final spectral estimates is needed and the process of optimization of the information content is usually highly empirical and tuned to a particular end use of the data.

5 INSTRUMENT REQUIREMENTS

5.1 Introduction

In this section of the report, the spectral requirements for an ITS are reviewed on the basis of spectroscopic considerations, and verified through retrieval simulations. The radiometric accuracy and spatial resolution and sampling requirements are established on the basis of retrieval simulation and prior experience with Polar and Geostationary infrared sounding instruments. The feasibility of meeting the requirements has already been established through the instrument feasibility Phase A study conducted by industry for EUMETSAT.

The experience with aircraft and ground-based measurements coupled with theoretical studies and engineering feasibility analyses have produced the following specifications for an Interferometer Thermal Sounder:

Spectral Bands and Radiometric Performance

BAND Nr	RANGE (μm)	RANGE (cm^{-1})	RESOLUTION (cm^{-1})	OPTICAL PATH DIFF. (cm)	NEdN (mW/m^2-cm^{-1}-sr)
1	8.70 - 16.1	620 - 1150	0.625	± 0.8	0.25
2	5.75 - 8.26	1210 - 1740	1.250	± 0.4	0.10
3	3.68 - 4.65	2150 - 2710	2.500	± 0.2	0.01
4	0.67 - 0.71	14000 - 15000	1000	filter	NEdA = 0.1 %

TABLE 1: An Example ITS Instrument Measurement Specification

The spatial resolution is 10 Km diameter circular and the spacing between field of view centres is 16 Km. The sample spacing is designed to match the MTS/MHS. A 3 x 3 detector array is to be used for each spectral band and the scan time is to be synchronized with the MTS/MHS.

5.2 Spectral Characteristics

For achieving high vertical resolution temperature and moisture profiles, the spectral resolution of band 1 of Table 1 above, is most critical. This is because the high vertical resolution of the atmospheric temperature profile results from the ability of the instrument to sense in between the CO_2 absorption lines within the 15μm CO_2 absorption band. As shown in Figure 1, the weighting function sharpness for a spectral region in between two relatively strong lines is much greater than the sharpness of the weighting function resulting from the same absorption produced at the centre of a relatively weak line. The narrow half width (i.e., 0.05cm^{-1}) of the lines and relatively large and uniform spacing of the CO_2 absorption lines (i.e., approximately 1.5cm^{-1}) within this band makes it possible to resolve the regions in between all the intense CO_2 lines with an instrument possessing a fixed spectral resolution across the absorbing region (i.e., a resolution of about 0.7cm^{-1}). On the other hand, water vapour absorption lines are more randomly spaced. However, there are thousands of lines in the spectral domain of interest so that a wide variety of relatively sharp weighting functions can be achieved with an instrument with moderate spectral resolution across this region. Because water vapour weighting functions are a factor of two or more sharper than CO_2 weighting functions, and because the water vapour radiance depends upon both the water vapour and temperature profile structure, the accuracy of the retrieved water vapour profile is limited by the vertical resolution of the CO_2 radiance weighting functions rather than the water vapour radiance weighting functions.

Figure 2, shows, as an example, interferograms and corresponding spectra (unapodized) for measurements with the High resolution Interferometer Sounder (HIS) aboard the NASA U2 aircraft over Tucson, Arizona. The unapodized full spectral resolution of the HIS is 0.27 cm^{-1}. Shown in Figure 2a are the 1.6 cm and 0.8 cm Optical Path Difference

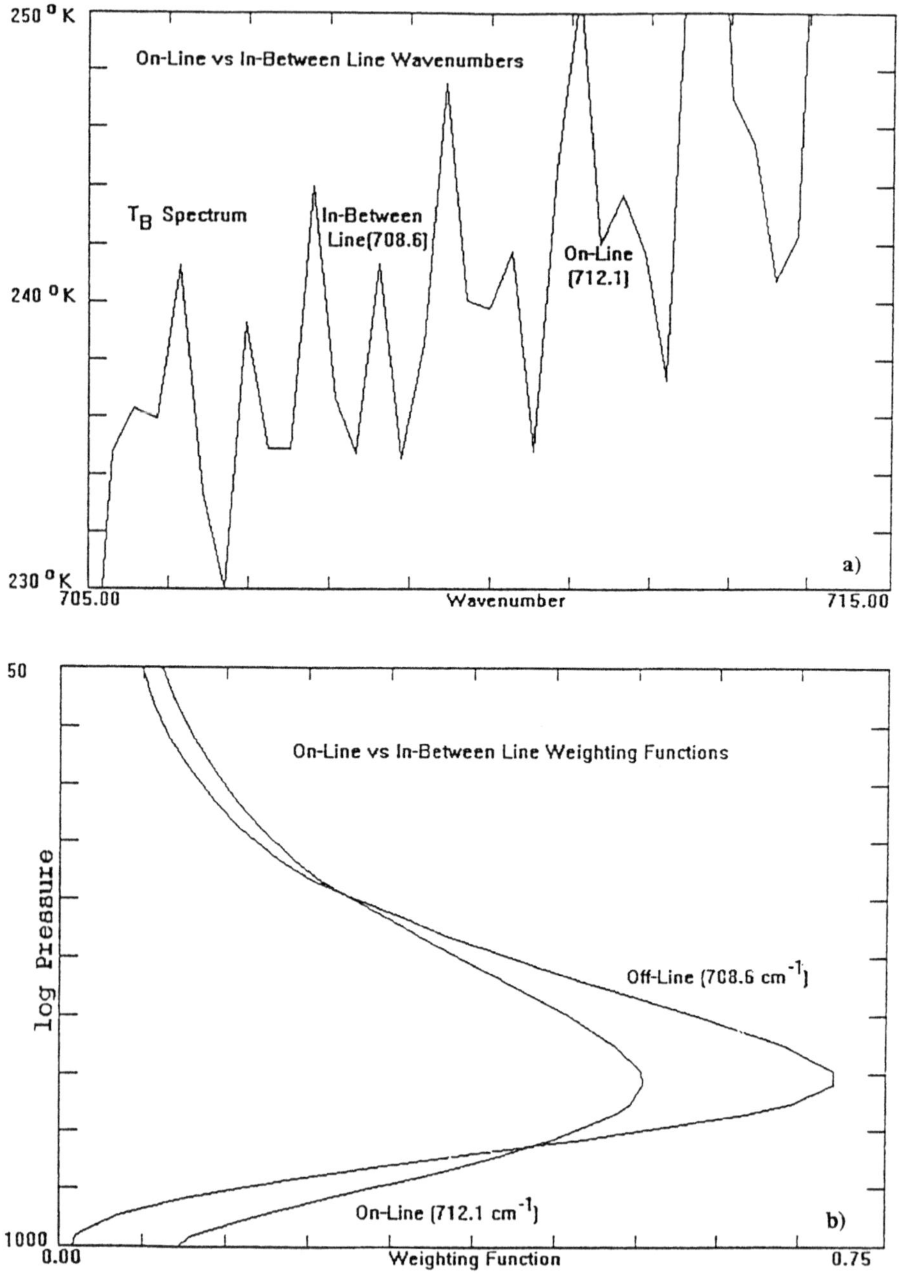

Figure 1: (a) Top, spectrum of brightness temperature within a small spectral region of the 15 μm CO_2 band. Shown are two wavenumbers, one in-between two absorption lines and one on an absorption line for whose temperature profile weighting functions are shown in panel (b), bottom.

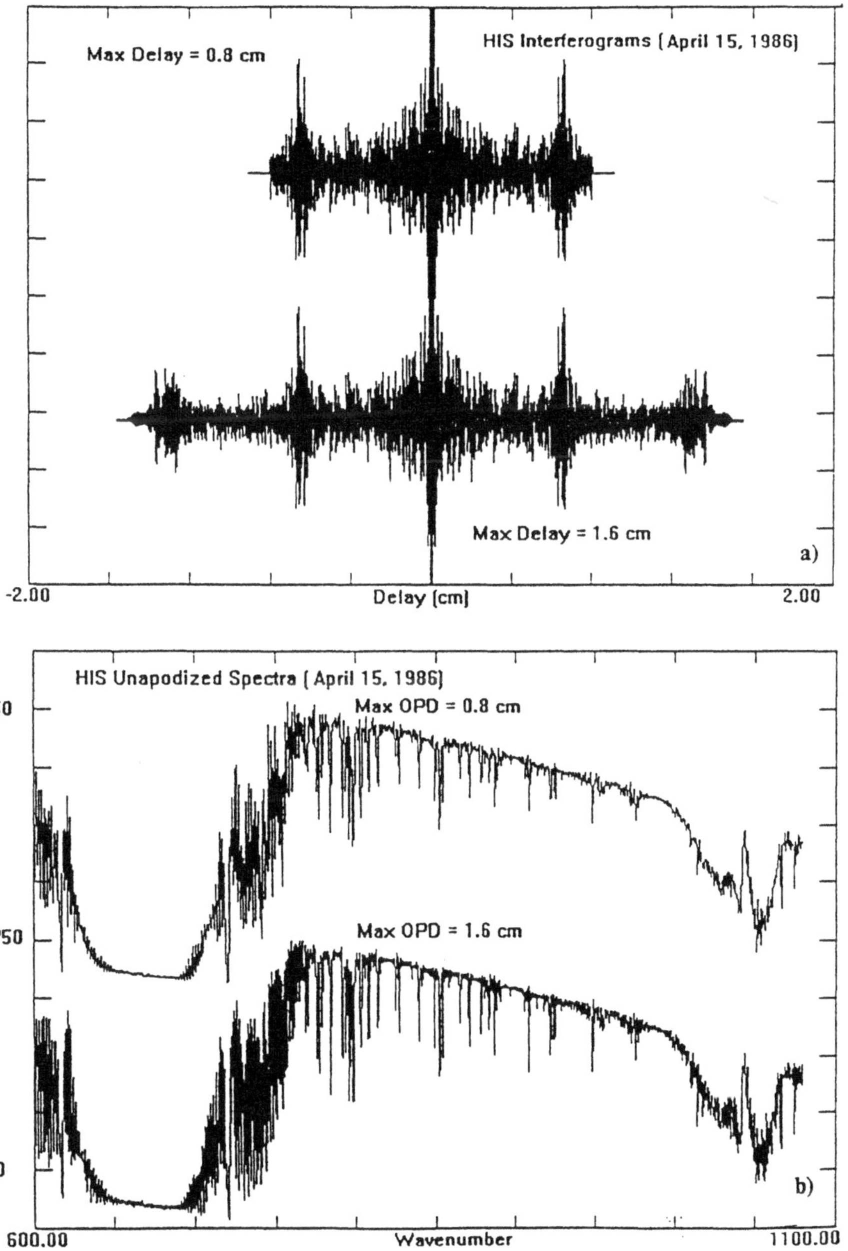

Figure 2: Interferograms (a) and unapodized spectra (b) corresponding to two delay limits (maximum optical path differences): 0.8 cm and 1.6 cm. The radiance scale is in $mw/m^2 sr\ cm^{-1}$ (y-axes).

(OPD). Shown in Figure 2b are the corresponding spectra for unapodized spectral resolution of 0.625 cm^{-1} and 0.312 cm^{-1}. In viewing the interferograms, two significant resonance regions are apparent, away from Zero optical Path Difference (ZPD), one centered at 0.65 cm and another at 1.3 cm delay. These are produced by the uniform 0.65 cm (i.e., 1/0.65 = 1.5 cm^{-1}) spacing of the most intense CO_2 lines within this spectral region. The amplitude of the spectra is virtually the same between line centres but significantly changed at line centre spectral positions. This is expected because 0.8 cm OPD is sufficient to resolve the gaps between the absorption lines. However, the lower resolution weakens the ability to resolve the shape of the line which is much narrower than that resolved by even the original 1.8 cm OPD HIS measurements. It must be recalled that the most important information on the temperature profile is expected from spectral frequencies in-between absorption lines, rather than on absorption lines, because of the much higher vertical resolution achieved in the far wings of the lines as opposed to the line centre.

In defining the spectral resolution, or corresponding maximum OPD requirements of a temperature profiling interferometer, the key question is the information content added by measurements within secondary resonance regions away from ZPD since they contain the information on the CO_2 emission line contributions. To see whether the instrument needs to sample out to 1.6 cm OPD or is 0.8 cm OPD sufficient one has to consider the effects of noise and truncation(and hence resolution). Through profile retrieval simulations it is shown that the optimum spectral resolution for Band 1 is the specified 0.8 cm OPD. There is little improvement with a doubling of the resolution which would require a more complex interferometer. Degrading the resolution to the point of not being able to distinguish the contributions from the wings of the absorption lines from those from the line centres produces a significant degradation in sounding accuracy.

Regarding Bands 2 and 3, the spectral resolution is much less critical. There is a weak dependence of the water vapour retrieval performance on the spectral resolution and associated noise performance in Band 2. However, the spectral resolution provided by the maximum OPD specification of 0.4 cm appears to optimize the resolution of the information content available within this spectral region.

The vertical temperature profile information provided by Band 3 results primarily from the higher order Planck radiance dependence on temperature at the short wavelengths rather than from any improvement in the sharpness of the weighting function. Moreover, the CO_2 absorption lines in this spectral region do not exhibit a uniform line spacing except in the R-branch of the 4.3μm CO_2 band.

5.3 Spatial Characteristics

Studies with NOAA satellite HIRS/AVHRR and GOES satellite VAS sounding radiance data indicate that the ability to sound successfully (in terms of yield and accuracy) down to the Earth's surface under partly cloudy sky conditions degrades rapidly with decreasing spatial resolution of the infrared sounding instrument. Furthermore, the importance of the use of simultaneous microwave sounding radiance data in the sounding retrieval process is extensively documented. Thus, the interferometer sounder should have the highest spatial resolution achievable with the NEdN requirement being satisfied (in the case of the ITS the spatial resolution is 10 Km diameter) and at the same time be synchronized as much as possible with the companion microwave sounding radiometer (i.e., the MTS/MHS in the case of the EPS).

The visible channel fields of view will match those of the interferometer to enable daytime assessments of cloud contamination on an individual field of view basis. The use of the visible channel for this purpose is crucial for the case of daytime low level stratus or fog since there is little thermal contrast between the cloud and the surface background to enable the detection from IR data alone. The synchronization of the IR instrument with the microwave sounder and the use of the visible channel for cloud detection have their heritage with the TOVS sounding system which has been operational since 1979.

5.4 Radiometric Accuracy

The radiometric accuracy of an operational interferometer sounder must meet or exceed the requirements stated in Table 1. Figure 3 is presented to translate the NEdN

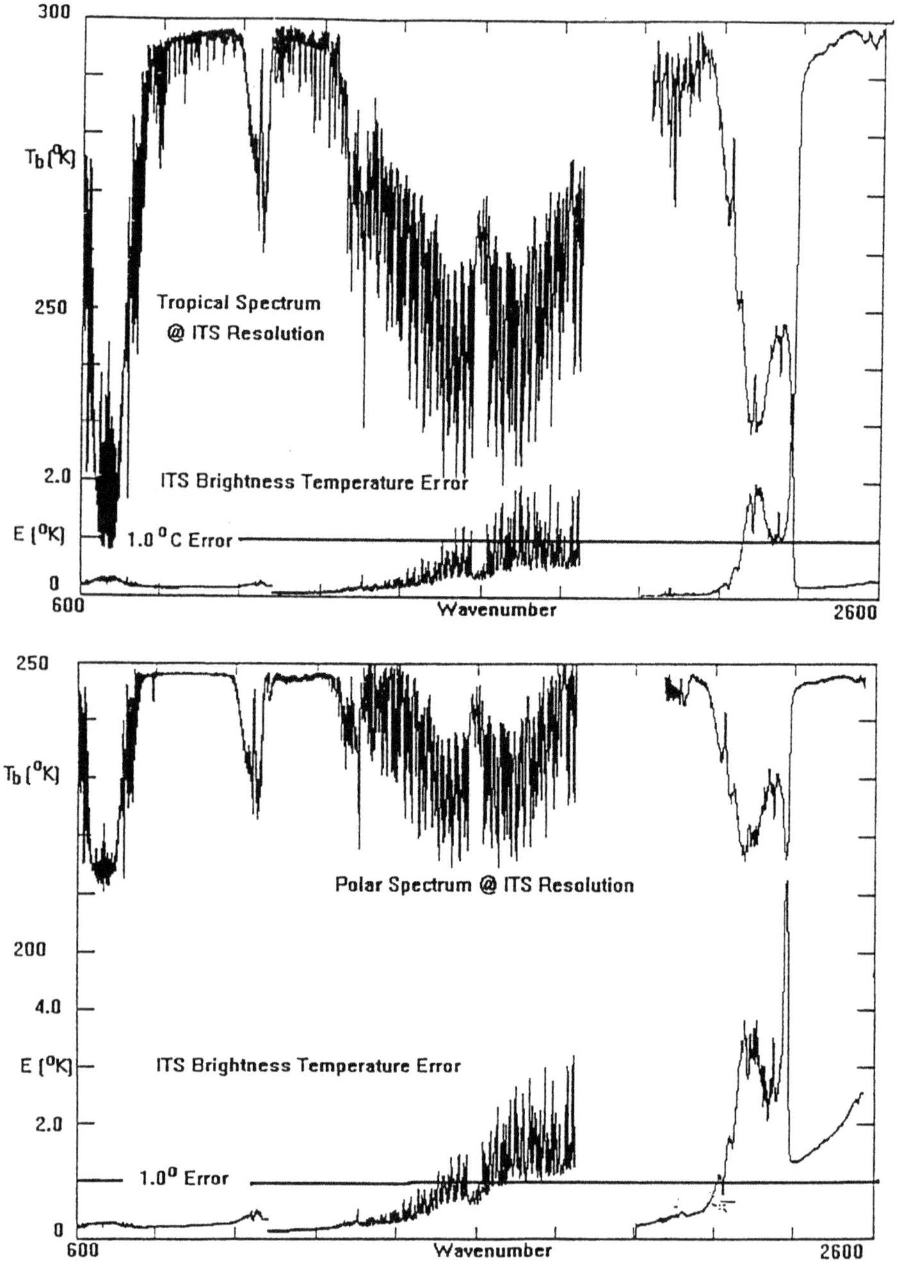

Figure 3: Brightness Temperature Error Spectra Associated with Brightness Temperature Observations for a Tropical and a Polar Atmosphere.

requirements of Table 1 into the spectral distribution of NEdN. The spectral distribution of Brightness Temperature and its error expected to be measured by the ITS for tropical and polar air mass conditions is shown in these illustrations. As can be seen, the NEdN performance requirement of Table 1 yields generally good NEdN performance within the longer wavelength regions of each band and at spectral wavelengths in between the water vapour absorption lines of band 2 where the brightness temperature is relatively large. It can also be seen that the single sample NEdN performance for the short wavelength band 3 is very poor for cold polar atmospheric conditions.

The absolute calibration of an ITS should be as good as practically possible, but in no case poorer than 1°K in terms of radiometric temperature. A 1°K calibration error is expected to induce a systematic error a 1°K for atmospheric temperature and a 10% error for atmospheric humidity.

6 FURTHER STUDIES REQUIRED FOR IMPLEMENTATION

6.1 Meteorological Requirements Definition

The ESSWG report focuses on the requirements for the operational temperature and humidity sounding missions. Further work is needed to assess the information content for the retrieval of the ozone mixing ratio profile at the Interferometer Thermal Sounder spectral resolution and to assess the need for higher spectral resolution. Similar analysis should be performed to evaluate the requirements for deriving concentration contents of other minor constituents and trace gases in order to allow more accurate quantification of planetary scale distribution of their sources and sinks.

6.2 Instrumental Requirements Definition

A method for assessing system performance and information content is through the comparison of analysis/retrieval error covariance with background error covariance. The ESSWG recommends this type of study for the advanced sounder as it is important, particularly with regard to instrumental trade-off options, to know the effect of a proposed change in the instrument specification in terms of the change in information content for an NWP system (i.e. the amount by which the error in the NWP system's representation of the atmospheric state can be reduced by the new measurements).

6.3 Processing Procedures

Atmospheric Transmittance Algorithms

Central to any system for utilizing sounding data of very high spectral resolution will be accurate calculations of atmospheric transmittance spectra. Studies should be continued and extended to develop in detail fast models for a high resolution sounder and to investigate their accuracy and speed. A related problem requiring attention will be how to perform empirical tuning of these models to correct for anticipated biases between measurement and theoretical calculation.

Use of Ancillary Data

Deriving geophysical parameters from instruments having the characteristics of MTS and MHS constitutes a novel work. After the launch of AMSU-A and AMSU-B some relevant experience will be drawn, and early efforts should be undertaken to be ready to make an appropriate use of the new sensors. Also, AVHRR data, when merged with TOVS, is still not used operationally even for cloud identification purposes and a large effort will hence be needed to enhance the use of such valuable data.

Retrieval Methodology

At present a considerable amount of research and development is under way at a number of NWP centres on improved methods for assimilating the information in satellite sounding radiances into NWP models. By the late 1990s, it is anticipated that such methods will be well-established and will have been used operationally with TOVS and ATOVS data. It can also be anticipated that NWP centres will be seeking to adapt these techniques to the new instruments for high spectral resolution. Some unresolved problems are already recognized and more will arise in the implementation of these methods. The ESSWG report addresses the following problems which require special study in the context of advanced sounders of high spectral resolution:

- Fast radiative transfer models: appropriate models will be required. Also needed will be their corresponding adjoint models or equivalently the Jacobians of the observations with respect to the thermodynamic parameters
- Observational error characteristics: the error structure of the measured interferogram and how this maps into errors in the retrieved spectrum needs to be investigated. A good understanding of the errors involved in simulating the spectrum from a given atmospheric profile and in the methods used for mapping between spectral and interferogram space will also be needed
- Treatment of clouds: improved understanding of the effects of cloud on infrared radiances will be required in particular with regard to spectral variances in cloud effects. New processing schemes will be needed to treat the effects of cloud:
    - either as a separate cloud detection/clearing step prior to inversion;
    - or as part of a simultaneous inversion.
- Data volumes: studies are required to determine to what extent the increase of data volume will create problems for data transmission and storage and for the computer power required to process and assimilate this data. If as expected, methods of data compression are required studies of the most appropriate methods will be needed

7. CONCLUSIONS

This paper gives a summary of the ESSWG report which contains the specifications for an operational interferometer sounder to satisfy the operational high vertical resolution temperature and moisture sounding mission and which indicates further studies and calculations required for implementation. The operational interferometer sounder statement of requirements is important as a point of reference for industrial studies, providing not only an initial specification for the instrument, but also a framework within which decisions on instrument options and trade-offs maybe considered.

8. MEMBERSHIP OF THE EUMETSAT SOUNDER SCIENCE WORKING GROUP

Marilena Perrone (EUMETSAT)
Paul Ingmann (ESTEC)
Alain Chedin (LMD)
John Eyre (ECMWF)
Rolando Rizzi (EUMETSAT/Univ. of Bologna)
William Smith (EUMETSAT/CIMSS)
Dietrich Spänkuch (German Met. Service)
Jan Svensson (Swedish Met. Service)

DEVELOPMENT OF A FTIR SOUNDER - IMG

Haruhisa SHIMODA, Toshihiro OGAWA[1)]
Tokai University Research and Information Center
2-28-4, Tomigaya, Shibuya-ku, Tokyo, Japan

1. INTRODUCTION

Global change problems are the largest problem for human-kind. Increase of global mean temperatures of several degrees and sea surface elevations of several tens of centimeters are expected on the middle of next century from recent studies. However, models used in these studies include many assumptions on parameters concerning atmosphere and ocean. A slight modification in these parameters may induce very different scenario on the future global climate.

In order to effectively counteract the problems related to the atmosphere, following items should be measured over the globe.

1) Concentrations of greenhouse effect gases.
2) Temperature and water vapor profiles.

In order to collect these data on a global basis, utilization of artificial satellites are inevitable. IMG(Interferometric Monitor for Greenhouse Gases) is now under development to measure distributions of atmospheric constituents.

---

[1]Department of Earth and Planetary Physics
Faculty of Science, University of Tokyo
7-3-1, Hongo, Bunkyo-ku, Tokyo

NATO ASI Series, Vol. I 9
High Spectral Resolution Infrared Remote Sensing
for Earth's Weather and Climate Studies
Edited by A. Chedin, M.T. Chahine and N.A. Scott

The primary objective of IMG is to continuously monitor spatial distribution(horizontal and vertical) of greenhouse effect gases such as CO_2, H_2O, CH_4, N_2O, etc. Especially, measurements of the horizontal distribution of greenhouse gases are more important, because sinks and sources of these gases are still unknown.

There are many methods to measure such physical parameters, i.e. occultation measurements, limb sounding, etc. The most appropriate method for the purpose described above is to accurately measure spectra of infrared radiation emitted by atmosphere below the satellite.

Distributions of gases can be retrieved from these spectra, but in order to obtain high accuracy for measurements, each absorption line of gases should be discriminated. In the troposphere, width of absorption lines are in the order of $0.1cm^{-1}$, hence spectral resolution of the sensor should be in this region. This high spectral resolution can also provide very high accuracy temperature and water vapor profiles which will not be obtained by the other methods.

Another point that should be considered is the wavelength region of the measurements. There exists many kinds of greenhouse gases, the absorption lines of which distributes over wide range of infrared spectrum. In order to measure broad region of infrared with this high spectral resolution, FTIR(Michelson-Fourier interferometric spectrometer) is the best choice. IMG is a nadir looking FTIR which covers 3.3-14μm region with $0.1cm^{-1}$ resolution.

This sensor will be installed on ADEOS(Advanced Earth Observation Satellite) which will be launched on Feb. 1996 by NASDA(National Space Development Agency of Japan). This is the first occasion that such a high spectral resolution spectrometer will be on-board a satellite.

IMG is developed by JAROS(Japan Resources Observation System

Organization) deputed by MITI (Ministry of International Trade and Industry of Japan). Further, data analysis of IMG is studied in Central Research Institute of Electric Power Industry. In this paper, the development of IMG is described.

2. The IMG System

2.1 ADEOS

weight	3.6 ton	
orbit	type	sun-synchronous
	altitude	800 km
	inclination	98.6°
	recurrent period	41 days
	period	101 min.

Table 1 Main characteristics of ADEOS.

Table 1 shows the main characteristics of ADEOS. ADEOS is a large Earth observation satellite with 8 sensors on board. These sensors are listed in Table 2. Five of the eight sensors (i.e. TOMS, POLDER, ILAS, RIS and IMG) are designed to measure atmospheric parameters. Another sensor called OCTS (Ocean Color and Temperature Scanner) can also be used to detect clouds.

TOMS, which is a well known sensor, collects total ozone over the globe. ILAS is an infrared occultation limb looking sounder, which measures ozone profiles over the high latitude regions. POLDER is a visible polarimetric multi-spectral scanner, and is intended to measure atmospheric aerosols as well as cloud

Sensor	Country	Organization	Objectives and Specifications
OCTS (Ocean Color and Temperature Scanner)	Japan	NASDA	Measurement of Ocean Color and Sea Surface Temperature Swath width:1400km, IFOV:700m Wavelength:Visible 6bands, NearIR:2 bands Middle IR: 1 band, Thermal IR: 3 bands
AVNIR (Advanced Visible and Near Infra-red Radiometer)	Japan	NASDA	Observation of Land and Coastal Zone Swath width:80km, IFOV:16m(multi),8m(pan) Wavelength:Visible 3 bands, Near IR: 1 band Panchromatic(visible): 1 band Stereo capability
NSCAT (NASA Scattero-meter)	USA	NASA/GSFC	Measurement of surface wind speed and direction over the global ocean Swath width:1200km, Frequency:14GHz, Accuracy;Wind speed:2m/s, direction:20°
TOMS (Total Ozone Mapping Spectro-Meter)	USA	NASA/JPL	Observation of total ozone changes, evaluation of changes in UV radiation and the observation of sulfur dioxide Swath width: 2795km, Wavelength:304.0,312.5 325.0,317.5,332.6,360.0 nm
POLDER (Polarization and Directiona-lity of the Earth's reflec-tance)	France	CNES/LERTS	Observation of bidirectionality and polarization of the solar radiation reflected by the atmosphere Swath width:1440x1920km, Wavelength:435,670,880 nm(3 polarization) 490,520,565,765,950 nm (no polarization)
IMG (Interferometric Monitor for Greenhouse Gases)	Japan	Ministry of International Trade and Industry	Measurement of temperature, water vapor and greenhouse effect gases IFOV:8km x 8km, Spectral resolution:0.1cm^{-1} Wavelength:3.3-14μm
ILAS (Improved Limb Atmospheric Spectrometer)	Japan	Environmental Agency	Observation of the limb atmospheric micro-ingredient over high latitude area Observation altitude:10-60km Wavelength:753-784 nm,4.1-6.9,7.3-11.8μm
RIS (Retroreflector in Space)	Japan	Environmental Agency	Measurement of ozone, fluorocarbon, carbon dioxide, etc. by laser beam absorption technique, transmitted from ground station and reflected by the retroreflector on ADEOS Corner cube reflector:ϕ50cm Wavelength:0.3-14μm(according to laser)

Table 2 Major Specifications of sensors on board ADEOS.

parameters. RIS is an on-board corner reflector, and acts as a reflector for ground based laser radars.

2.2 Configuration of IMG

Fig. 1 shows a block diagram of IMG. IMG is composed of a Michelson interferometer, a focusing mirror, a focal plane with detector assembly, a He-Ne laser for measuring the displacement of the scanning mirror, an Image Motion Compensator(IMC) mirror to compensate for the satellite motion, an on-board blackbody for calibration, and a Stirling-cycle cooler which cools the detector assembly. Table 3 shows the main characteristics of IMG.

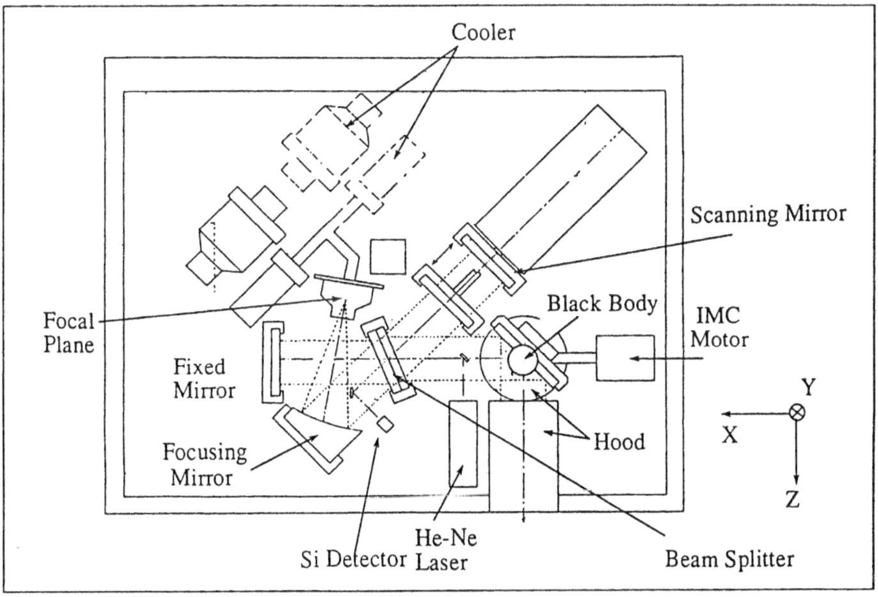

Figure 1 Block diagram of IMG.

The interferometer is composed of 1 fixed mirror, 1 scanning mirror and a 45 degree inclined beam splitter. The diameter of light beam is 10 cm. The diameters of the mirrors and beam splitter are 12-15 cm. The beam splitter is made of KBr coated with Ge. All the mirrors are made of Al coated by Au.

Fig. 2 shows a block diagram of the mirror scanning mechanism. The scanning mirror is supported by a magnetic bearings and scanned with the aid of a linear motor. The use of a magnetic bearing system eliminates the need for lubrication, assures a long lifetime, and is highly reliable. Any miss-alignment of the interferometer can be corrected by adjusting the orientation of the scanning mirror. High interferometric efficiency is achieved by controlling the orientation of the mirror with accuracy of 0.0001 degree over the entire scanning range.

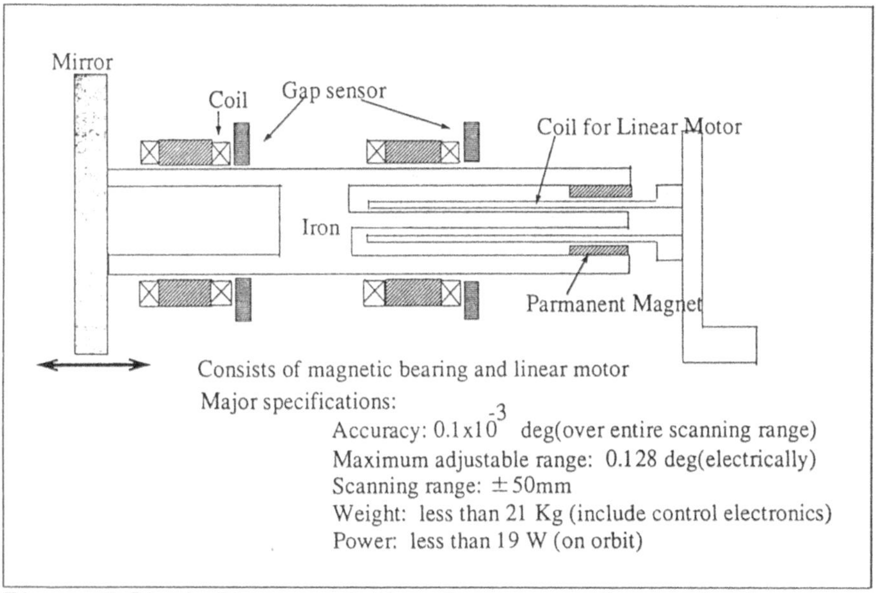

Figure 2 Block diagram of the mirror scanning mechanism.

The distance of the scanning mirror displacement is 10 cm in order to assure a 0.1cm^{-1} spectral resolution after apodization. This displacement is measured by a wavelength stabilized He-Ne laser.The wavelength of the laser is 633 nm, stabilized to $3x10^{-7}$, which meets the required 0.001 cm^{-1} wave number accuracy. This laser is also used to determine the sampling interval of IMG data(interferogram).

The incoming light to IMG is first reflected by the IMC mirror. The primary function of this mirror is to compensate for the motion of the spacecraft and for the effect of Earth's rotation in order to maintain a constant FOV throughout the duration of the scan. The IMC uses a dual axis gimbal mechanism. With this mechanism, the 2.6 degree movement which correspond to 10 seconds of spacecraft motion can be corrected with an accuracy of 0.01 degree.

The IMC is also used to view the deep space and the on-board blackbody for data calibration. The large changes in the orientation of the mirror required for this function are produced by a stepping motor added to the gimbal mechanism. This mechanism gives IMG the additional capability of taking off-nadir measurements.

Interfered light is focused to the focal plane by means of a parabolic mirror. At the focal plane, there are three detectors corresponding to three spectral channels. The configuration of the three detectors at the focal plane is shown in Fig.3. Two InSb detectors cover the shorter wavelength regions(3.3-4.3μm, 4-5μm; channel #1 and 2) and one HgCdTe detector covers the longer wavelength region(4-15μm; channel #3). The IFOV of each detector is 0.6 degree, which correspond to a footprint size of 8kmx8km.

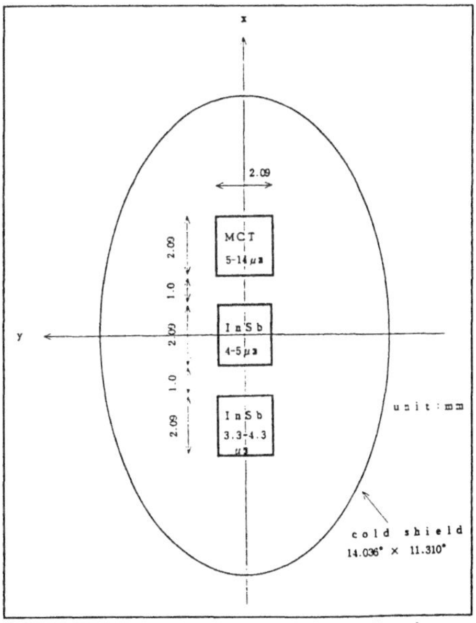

Figure 3 Configuration of three detectors at the focal plane of IMG.

The detectors, along with the cold slit, are cooled to temperatures below 80 K by a Stirling-cycle cooler. The actual temperature of the detectors depends upon the cooling capacity and power consumption of the proto-flight model(PFM) cooler.

The cooler is an Oxford type Stirling-cycle cooler with one compressor and one displacer. The vibrations caused by the compressor are absorbed with the aid of a suspension mechanism. The projected lifetime of the cooler is more than 25000 hours, which meets the lifetime requirement of 3 years.

2.3 Radiometric Performance of IMG

gas	accuracy	wavelength (μm)
CO_2	2~3%	14, 4.3
CH_4	5%	3.3
N_2O	1%	4.5
O_3	1%	10
H_2O	10%	7
CO	4.7%	4.7

Table 3 Measurements targets.

The radiometric performance specifications of IMG have been determined based on the measurement accuracy requirements of various atmospheric constituents. Table 4 shows the accuracy requirements for the primary constituents. From these requirements, signal-to-noise ratio specifications for IMG were determined. These data were interpreted to noise equivalent spectral radiation (NESR) for IMG hardware specifications. Table

Wavelength (μm)	Minimum detection sensitivity ($\mu W/cm^2 st cm^{-1}$)	Target Gas
14	0.030	CO_2
10	0.019	O_3
8	0.024	N_2O, CH_4
7	0.02	H_2O
4.7	0.004	CO
4.5	0.001	N_2O
4.3	0.0002	CO_2
3.3	0.00015	CH_4

Table 4 Required NESR.

5 shows the NESR IMG computed for each spectral region from the value of Table 4. The required performance and the estimated performance of the instrument are compared in Fig.4. As can be seen from this figure, the performance of IMG is within specifications in most of parts of the spectrum, except the shortest wavelength region.

The stability of the measurements will be better than 0.1 K over the entire spectral range. The temperature of the optical system, which influences the stability of the measurements, will be kept constant (close to ambient temperature).

Absolute radiance calibration is done by taking measurements of the deep space and the on-board blackbody. The temperature of the blackbody will be measured by platinum resisters. The absolute radiometric accuracy of IMG will be better than 1 K over the entire wavelength region and the design lifetime of the sensor(3 years).

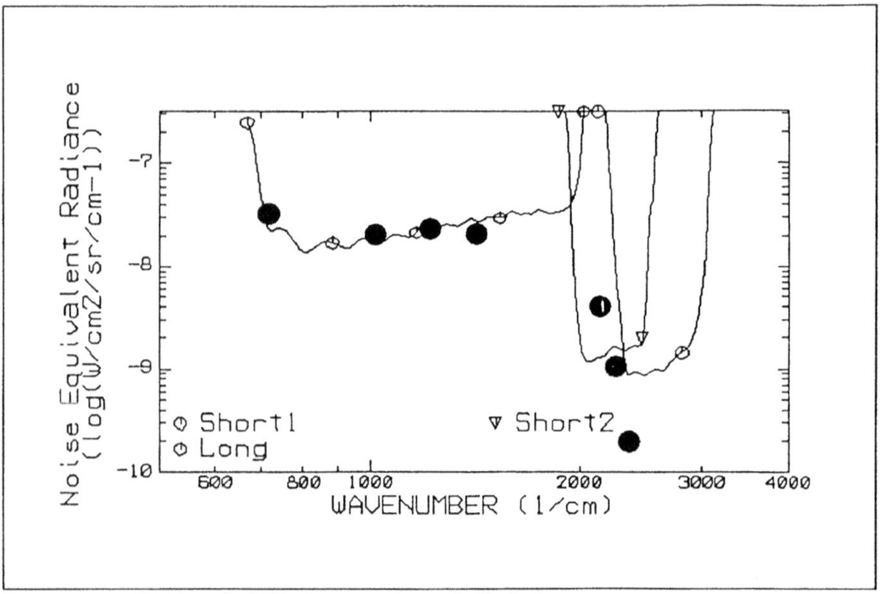

Figure 4 NESR of IMG Bread Board Model.
solid line shows NESR of BBM and marks shows specifications.

3. Preliminary Performance of IMG

3.1 Apparatus Function

Fig.5 shows a spectrum of a Ruby laser taken by the IMG Bread Board Model(BBM). It is an unapodized spectrum and the half height width of this spectrum is 0.04 cm^{-1}. This value is well within the specification of 0.1 cm^{-1} resolution after apodization.

In this figure, several side lobes other than the main lobe can be seen. These side lobes are primarily due to vibrations of the BBM. These vibrations will be eliminated in the Engineering Model(EM) of IMG.

3.2 Signal-to-Noise Ratio

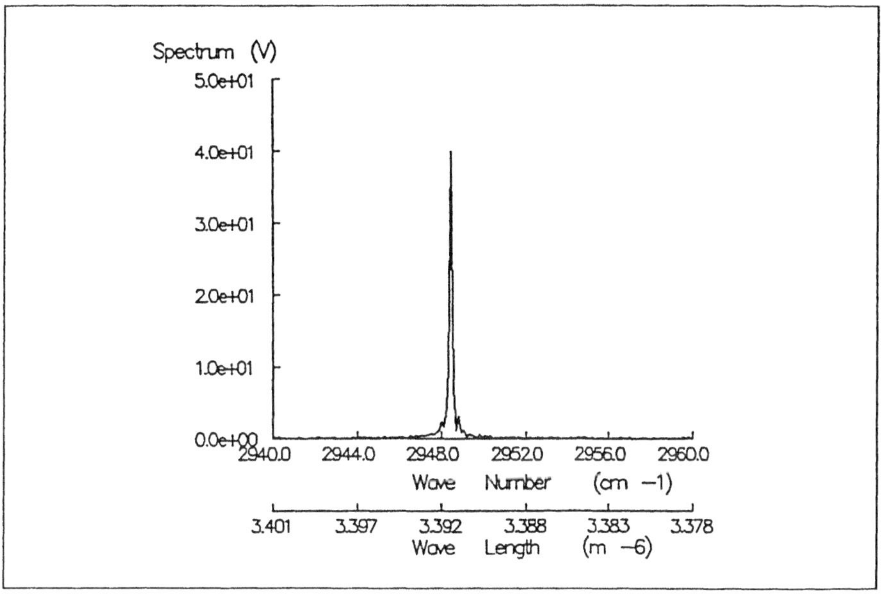

Figure 5 Laser spectrum taken by IMG Bread Board Model (BBM).

Fig.6 shows a typical spectrum of a blackbody taken by the IMG BBM. The NESR measured from this spectrum is shown in Table 5 with those values calculated from performance of each component. With both a performance increase in detectivity D^* and interference efficiency, NESR of EM will be 5.1, 3.9 and 6.3 times better than those of BBM for channels 1, 2 and 3, respectively. These results assure that the NESR of IMG will be within the specifications, except in the very short wavelength region.

4. Data Processing and Analysis

4.1 Data Processing

IMG will generate the raw data of three interferograms simultaneously every 13 seconds. An interferogram is sampled with a 16 bit A/D converter in 10 seconds at 1.05×10^5 points for the shorter wavelength channels, and at 1.57×10^5 points for the longer wavelength channel. The interferograms transmitted from the ADEOS satellite will be transformed into radiance spectra by

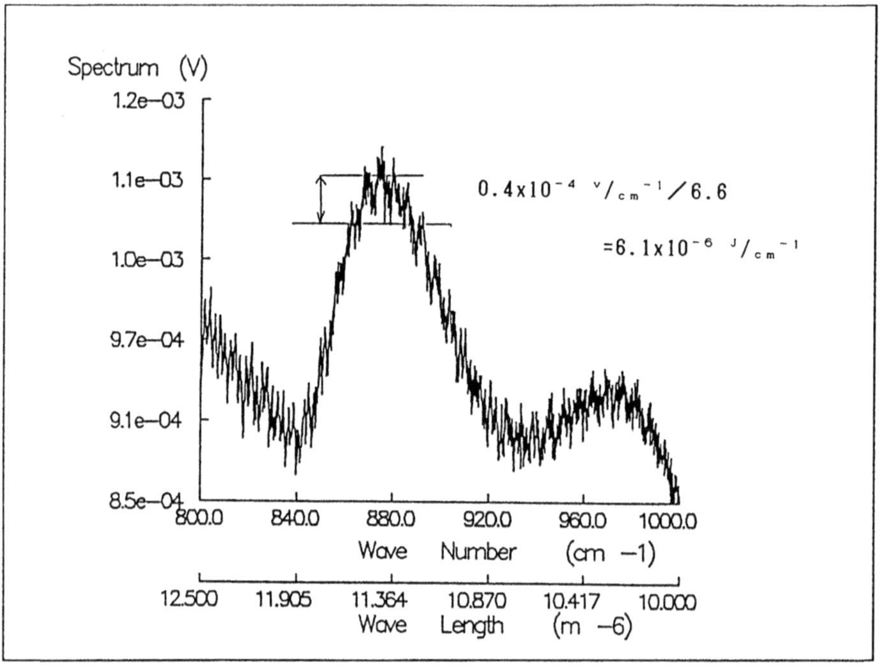

Figure 6 Blackbody spectrum taken by IMG Bread Board Model.

means of a Fast Fourier Transform (FFT). It is important to carefully select the apodization function for a precise reproduction of the spectra. Correction for the phase error of the inter ferograms will be made using the full range of the interferogram. Calibration is made in terms of the on-board blackbody and the deep space measurement once in every six measurements, at 107 second intervals. The calibration data are crucial to discriminate the incident thermal radiation from that produced by the interferometer optics. As mentioned before, IMG has three spectral channels corresponding short, medium and long wavelength region. In the retrieval process, however, these three channel data will be used simultaneously. In this respect, these three channel data should be inter-calibrated. The calibration data are also crucial to assure the precise measurement of spectral radiance to synthesize three spectra from three channels to a full range spectra in 3.3 - 14μm wavelengths.

The calibrated spectral data (so called "level 1" data) will be distributed to investigators. The spectral data are useful as such for Earth's radiation budget studies. The data can also be used to infer atmospheric and surface temperature, trace gas concentrations, optical properties of cirrus clouds, etc.

4.2 Retrieval

One of the principal objectives of the IMG observation is to infer the concentrations of greenhouse gases together with atmospheric temperature and humidity. Since IMG will provide thermal atmospheric radiation spectra with a spectral resolution of 0.1 cm^{-1}, it is expected that many of the rotational lines of various trace gases will be isolated from one another. This is

wave length μm	BBM estimated	BBM measured
14	0.522	0.179
11.5	0.079	0.067
10	0.105	0.067
8	0.156	0.125
7	0.161	0.196
4.8	0.0089	0.0073
4.7	0.0088	0.0072
4.5	0.0103	0.0082
4.3	0.0093	0.0104
4.2	0.0060	0.0056
3.3	0.0288	0.0105

Table 5 NESR of IMG Bread Board Model.

an advantage in the retrieval of trace gas concentrations since many spectral bins can be used.

In the IMG data retrieval, the linear simultaneous physical/statistical retrieval algorithm, proposed by Smith et al. (1991) will be applied. This procedure requires an initial guess and its covariance matrix for the atmospheric temperature and gas concentrations as well as an error covariance matrix for radiance measurement. Because the retrieval solution depends strongly on this a priori information, we have to choose their values carefully taking account of their seasonal and latitudinal variations. Thc data will be obtained either from numerical model calculations or from radiosonde observations.

It is important to quantify the effect of clouds in the IMG data analysis. We intend to use cloud image data obtained from co-located OCTS observations to provide estimates of the cloud cover in the IMG field of view.

5. Expected Results

5.1 Temperature and Humidity

The thermal atmospheric radiation spectrum measured by IMG is expected to obtain the profiles of atmospheric temperature and water vapor concentration highly accurately. A simulation has been made using the above mentioned algorithm and a hypothetical radiation spectrum calculated for the U.S. Standard Atmosphere (1976) with FASCODE 2 (Smith et al., 1978) and the HITRAN spectral data base (Rothman et al., 1987). The diagonal elements of the covariance matrix have been assumed to be 10 K^2 for atmospheric temperature and for effective radiation temperature of trace gases. Those for the error covariance matrix have been assumed to be 1 K^2.

Preliminary results indicate that the retrieval accuracy of atmospheric temperature is about 1 K in the troposphere and the accuracy of water vapor concentration is about 10 % and 20 % in

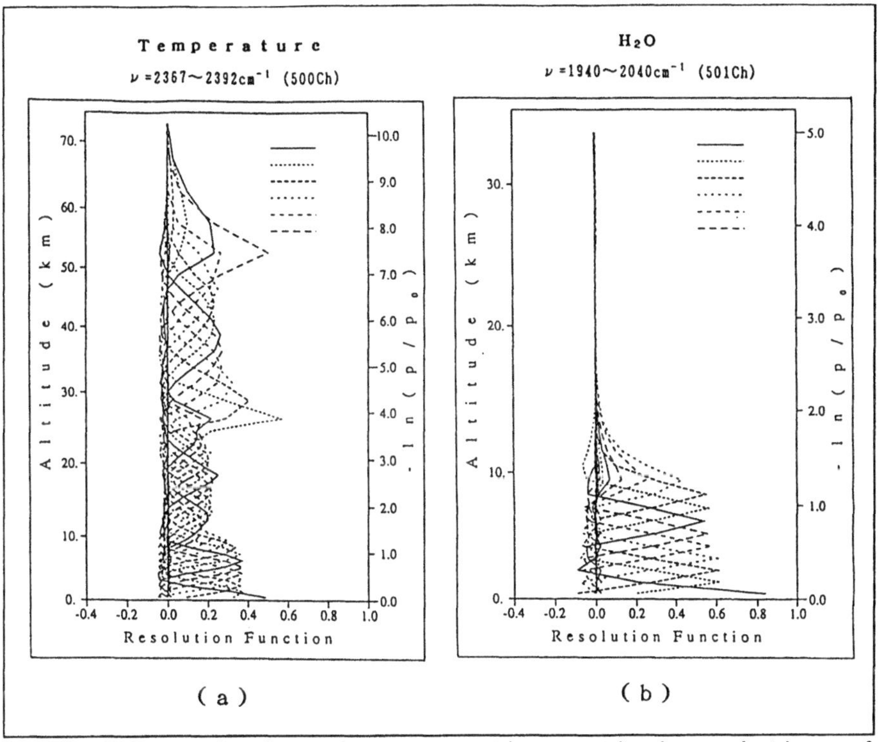

Figure 7 Averaging kernels representing vertical resolutions of the retrieval for temperature and water vapor.

the lower and upper troposphere, respectively. According to these results, current IMG will provide 2-3 times more accurate retrievals than operational satellite-borne radiometers used at the present.

The altitude resolution of the retrieval has been examined by evaluating the averaging kernels (Rodgers, 1990), and shown in Fig. 7 for temperature and water vapor. They have been calculated using 500 spectral bins with 0.1 cm^{-1} resolution in the CO_2 4.2 μm band and H_2O 5 ~μm band, respectively. Fig.7a indicates an altitude resolution of 1 - 2 km (full width at half maximum) for temperature. The altitude resolution of water vapor in Fig.7b seems better than that of temperature in the lower troposphere. This is because the weighting functions of water vapor are distributed over a narrower altitude range than those of temperature. These figures suggest that IMG has the capability of sensing the planetary boundary layer by separating the free troposphere, and that IMG will provide a much better altitude resolution than currently used operational satellite-borne radiometers.

5.2 Trace Gases

The IMG data is expected to provide spectral signatures of trace gases such as CH_4, N_2O, CO, O_3 and HNO_3 as well as H_2O and CO_2. The data may also be useful for the detection of volcanic effluent gases such as SO_2.

CO_2 mixing ratios will be inferred from the IMG data. A very high accuracy in radiometric measurement is required for this application, because the spatial variations of CO_2 mixing ratio are likely to be small in the free troposphere. It is hoped that IMG can be used to detect the horizontal inhomogeneities in the spatial distribution of CO_2; in particular, possible higher concentrations of CO_2 near emission sources such as urban

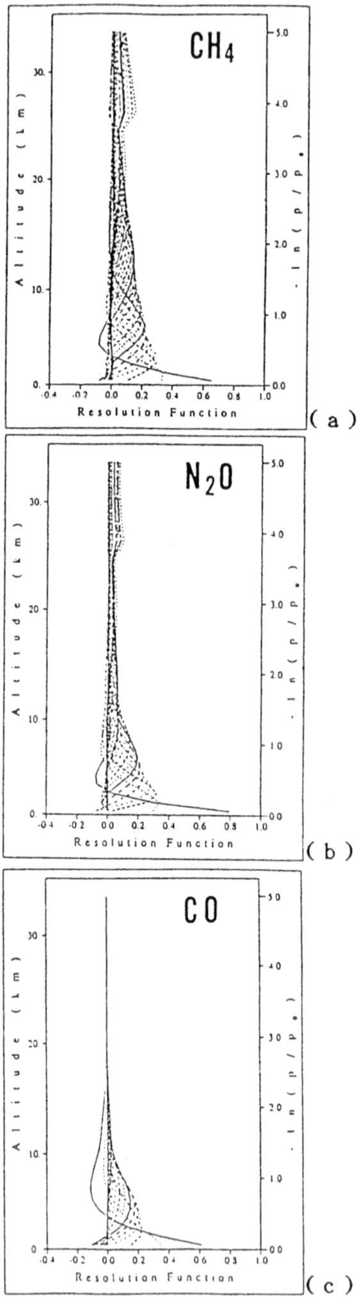

Figure 8 Vertical resolutions of the retrieval for CH_4, N_2O and CO.

areas, fires, volcanic eruptions, biomass burning, etc.

Figs. 8a, 8b and 8c show the vertical resolution functions for CH_4, N_2O and CO, respectively. They have been calculated using 555, 401 and 159 spectral bins with $0.1\ cm^{-1}$ resolution in the CH_4 7.8 μm band, N_2O 4.5 μm band and CO 4.7 μm band, respectively. These figures indicate an altitude resolution of about 5 km in the troposphere and of more than 10 km in the stratosphere. With respect to these three species, there is a possibility to separate the contribution of the planetary boundary layer from the free tropospheric contribution. It is also expected that the stratospheric contribution of CH_4 and N_2O can be isolated. The total atmospheric column content of O_3 and HNO_3 will be inferred from the IMG data. The high spectral resolution of IMG will enable the total ozone content to be inferred with better accuracy than conventional infrared atmospheric sounders. Therefore, it is expected that IMG can be used to provide an accurate mapping of the ozone distribution over high latitude regions during the polar nights.

6. MISSION OPERATION

6.1 Short-term operation plan

Fig. 9 shows the short-term mission operation plan of IMG. The mirror displacement of 10 cm requires 10 sec and it takes 3 sec for the mirror to return to its initial position. According to this sequence, an 8kmx8km area will be measured every 86 km along-track which correspond to 13 sec. of spacecraft motion. After six consecutive observations, IMG looks at the on-board blackbody and into deep space for calibration. The calibration phase takes 32 sec. Therefore, one complete cycle takes 110 sec, which corresponds to approximately 728 km on the ground.

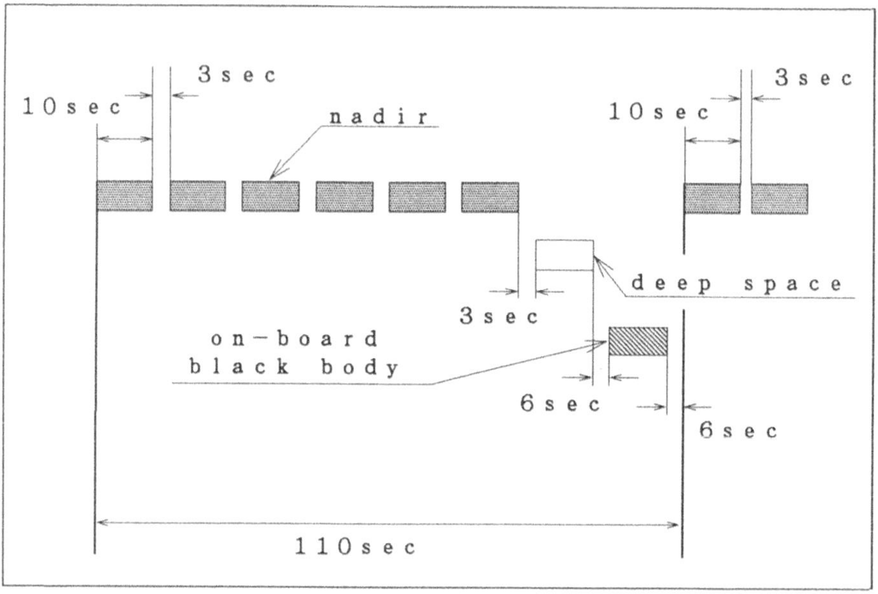

Figure 9 Short-term mission operation plan.

6.2 Long-term operation plan

IMG data will be recorded to on-board mission data recorder(MDR) and will be transmitted to ground receiving stations. Two stations, i.e. EOC (Hatoyama, Japan operated by NASDA) and Alaska (Fairbanks, USA operated by University of Alaska and US National Oceanic and Atmospheric Administration) will receive these MDR downlink data. As there are 7 sensors on-board ADEOS, these MDR operations cannot cover all the requirements from the on-board sensors. Hence, there is a possibility that IMG operations are restricted.

Three candidates have been retained for the IMG long-term mission operation plan. The first plan is to operate IMG continuously over the mission life time of 3 years. In this case, the lifetimes of the cooler and He-Ne laser will be the limiting factors.

According to the second plan, IMG is operated for 41 days

continuously and stops for the next 41 days. In this case, the duty factor is 0.5 and the only limiting factor will be the data transmission from the ADEOS on-board data recorder. As the recurrent cycle of ADEOS is 41 days, global data sets can be obtained within the shortest time frame.

According to the third plan, IMG is operated for 4 days continuously and stops for the next 10 days. This plan will be adopted only in case of severe limitations due to the data transmission.

6.3 Special events

As already mentioned in Sec. 3.2, IMG has the capability to point in off-nadir directions. Utilizing this function, IMG will be operated to observe some specific Earth's locations in the occurrence of some special event such as volcanic eruptions, biomass burning, in-situ measurements, etc.

6.4 Simultaneous operation with other sensors

As mentioned in Sec. 3.1, ADEOS will carry a total of seven sensors in addition to IMG. We are now requesting to operate OCTS simultaneously with IMG. The main reasons for this simultaneous operation are to use OCTS to detect small cloud fractions within the IMG field of view and to precisely determine the geometric location of the IMG observations. Again, the data recording and transmission capability of ADEOS will be the limiting factor of this simultaneous operation.

7. Data Products and Distribution

7.1 Standard Data Products

Table 6 shows a draft of IMG standard data products. Level 0 data correspond to the raw IMG sensor data. Level 1 data are spectra obtained from the level 0 data. Level 2 data correspond to retrieved physical quantities. Many types of data can be included in this last category. Table 7 shows a currently proposed list of physical quantities.

All the data obtained by IMG will be processed up to level 2 data. The retrieval algorithms may be modified after verifications of retrieval results. In this case, all the historical data will be processed again using the modified algorithms.

7.2 Data Distribution

The standard IMG data products will be publicly distributed in free of charge or at marginal cost.

Some problems are anticipated with the distribution of level 0 and level 1 data. Due to the size of these data sets (see Table 8). The choice of units as well as the type of storage media for the level 0 and level 1 data should be carefully considered.

level 0		
	0 A	raw data (interferogram) / TBD days with ephemeris data
	0 B	raw data (interferogram) / TBD days 0A + OCTS data
level 1		
	1 A	spectra / TBD days uncalibrated, after apodization with location data
	1 B	spectra / TBD days calibrated, after apodization with location data
	1 C	spectra / TBD days 1B + cloud information
level 2		
	a	temperature profile
	b	water vapor profile
	c	CO_2 profile
	d	total ozone
	e	CH_4 mixing ratio (troposphere)
	f	N_2O mixing ratio (troposphere)
	g	CO mixing ratio (troposphere)
level 3		
		global data set of level 2

Table 6 IMG standard data products (draft).

(a) Raw Data

882.4	kb/s	data aquisition
678.8	kb/s	mean data quantity
305.4	MB	data/1hour
7.33	GB	data/1day
29.3	GB	data/4days
300.6	GB	data/41days(global coverage)
2.68	TB	data/1year
8.0	TB	data/3years

(b) Spectra Data

55.75	MB	1revolution(101min)
795	MB	1day
3.18	GB	4days
32.6	GB	41days
290	GB	1year
870	GB	3years

Table 7 Data acquisition rate and size of IMG data.

8. CONCLUSION

In conclusion, the main characteristics of IMG are summarized :

1) IMG is the first wide range(3.3-14μm) high spectral resolution(0.1cm^{-1}) nadir looking FTIR in space.
2) Global data sets of
   the earth radiation budget
   temperature profiles
   water vapor profiles, and
   atmospheric constituents
   will be obtained from the IMG observations.
3) Use of a magnetic suspension and a linear motor system, will make IMG a highly reliable and accurate FTIR.

REFERENCES

Rodgers, C. D.,(1990) J. Geophys. Res., 95: 5587

Rothman, L. S., et al.,(1987) Appl. Opt., 26: 4058.

Smith, W. L., Woolf, H. M., and Revercomb, H. E.,(1991) Appl. Opt., 30: 1117.

Smith, H. J. P. et al.,(1978) AFGL=TR-0081

U.S. Standard Atmosphere, (1976) U.S. Government Printing Office

Chapter 2

Next generation IR sounders : temperature and moisture analysis

"The "Advanced-3I (Improved Initialization Inversion)" method for the processing of next generation sounder observations"

A. CHEDIN, N.A. SCOTT, F. CHERUY, B. TOURNIER - ARA/LMD, France

"Extraction of atmospheric signals from radiance measurements : some limitations"

M.J. UDDSTROM - New-Zealand Met. Service, New Zealand
L.M. McMILLIN

"GLA and LMD approaches to the processing of AIRS and IASI" observations"

F. CHERUY - ARA/LMD, France
J. SUSSKIND, N.A. SCOTT, A. CHEDIN, J. JOINER

"The impact of the initial guess on the accuracy of satellite retrievals"

L.M. McMILLIN - NOAA/NESDIS, USA
M. UDDSTROM

"Vertical sounding capabilities with high spectral resolution atmospheric radiation measurements. A demonstration with the High resolution Interferometer Sounder (HIS)"

W.L. SMITH, H.E. REVERCOMB, H.L. HUANG, R.O. KNUTESON
Univ. Wisconsin, USA

THE "ADVANCED-3I (Improved Initialization Inversion)" METHOD FOR THE PROCESSING OF NEXT GENERATION SOUNDER OBSERVATIONS

A. Chédin, N.A. Scott, F. Chéruy, B. Tournier
ARA / Laboratoire de Météorologie Dynamique du CNRS
Ecole Polytechnique, 91128 Palaiseau Cedex, France

ABSTRACT

The ARA (Atmospheric Radiation Analysis) group at LMD has developed an expertise in the retrieval of atmospheric temperature and moisture profiles with the "3I" algorithm using TOVS observations. However, for operational numerical weather prediction or climate research, the weaknesses of these observations are well recognized, namely insufficient vertical resolution and accuracy (= 2K rms in temperature and 20-30 % in water vapor profiles). High spectral resolution infrared sounders like AIRS (Atmospheric Infrared Sounder) or IASI (Improved Atmospheric Sounding in the Infrared) will provide a great improvement in the retrieval of atmospheric and surface properties. An increase in the spectral resolution of the observations must be accompanied by improvements in processing tools to be successful. Important developments include a good knowledge of spectroscopic line parameters, a fast and accurate radiative transfer model, a carefully selected set of channels throughout the infrared spectrum and a well elaborated retrieval/assimilation procedure. This paper presents an overview of these developments, with special emphasis being given to the latter.

1. INTRODUCTION

The weaknesses of the present generation NOAA/TOVS vertical sounding observations for operational numerical weather prediction and climate research are now well recognized, namely insufficient vertical resolution and limited accuracy (= 2K rms in temperature and 20-30 % in water vapor profiles). High spectral resolution infrared sounders like AIRS (Atmospheric Infrared Sounder) or IASI (Improved Atmospheric Sounding in the Infrared) will provide a great improvement in the retrieval of atmospheric and surface properties.

NATO ASI Series, Vol. I 9
High Spectral Resolution Infrared Remote Sensing
for Earth's Weather and Climate Studies
Edited by A. Chedin, M.T. Chahine and N.A. Scott

To meet the specifications issued by national and international agencies (i.e. temperature with an average rms error of 1K ; humidity with an average rms error of 10 % ; vertical resolution of 1 km, at least in the lower troposphere), an improvement of the spectral resolution by one order of magnitude is needed.

Next generation infrared sounders are designed to provide both new and more accurate data about the atmosphere, land and oceans for application to climate studies. Among the important observations that AIRS or IASI will contribute to the climate data set are day and night global measurements of :

- atmospheric temperature profiles with an average accuracy of 1K and 1 km vertical resolution,
- relative humidity profiles with an average accuracy of 10 % and a vertical resolution of about 2 km, and total precipitable water vapor,
- fractional cloud cover, cloud infrared spectral emissivity, and cloud-top pressure and temperature,
- total ozone burden of the atmosphere and possibly two layer total burden (troposphere and lower stratosphere),
- mapping of the distribution of minor atmospheric gases such as methane, carbon monoxide and nitrous oxide : total burden and possibly the resolution of about two layers,
- sea surface temperature with an average accuracy better than 1°C,
- land surface temperature and spectral infrared emissivity,
- surface albedo,
- snow and ice cover,
- outgoing longwave radiation,
- precipitation index (computed from a knowledge of the cloud height, surface temperature, and atmospheric thermal structure.

Such information is essential to the study of numerous interdisciplinary questions to be addressed in Earth science :

- cloud-radiation interaction,
- global energy and water cycles,
- measure of the increase of greenhouse gases,
- atmosphere-surface interaction,
- land properties,

and, last but not least,

- improvement of numerical weather forecasting.

The consideration of a much larger set of parameters describing the complexity of the atmosphere-ocean-land system obviously requires a multi-spectral approach based on a simultaneous analysis of a large set of measurement channels. An increase, by at least one order of magnitude, of the number of channels presently implemented on the operational weather satellites (NOAA series) is expected (from about 30 to a maximum of 3000). Specific signatures of radiatively active phenomena may then be easily isolated, interpreted and quantified.

2. PROCESSING OF OPERATIONAL WEATHER SATELLITE OBSERVATIONS FOR METEOROLOGY AND CLIMATOLOGY : PAST AND PRESENT EFFORTS AT LMD

During the past 15 years, the ARA group at LMD has developed an expertise in the field of direct or inverse atmospheric radiative transfer theory, covering the whole domain between molecular spectroscopy and the analysis of meteorological situations from space. Three main "tools" are now currently in use at LMD and in several other groups in the world :

- the "Gestion et Etude des Informations Spectroscopiques Atmosphériques" (GEISA") spectroscopic line parameters data bank, initiated in 1976 at LMD,
- the development of transmittance and radiance algorithms that have already been applied to numerous problems for earth and planetary atmospheres and, in particular, to the present NOAA weather satellite series,
- the development of the Improved Initialization Inversion (3I) retrieval algorithm for application to the NOAA series, now in use in about 20 groups (15 countries) for research or routine applications, from mesoscale to global scale.

2.1 A spectroscopic line parameters data bank : the GEISA catalog

Accurate modelling of radiative transfer phenomena in a planetary atmosphere requires very good knowledge of the parameters describing the radiation, absorption and emission, and properties of the optically active gases in the medium considered. The first compilation, oriented towards the Earth's atmosphere, was initiated at the Air Force Geophysics Laboratory (the so-called "AFGL-tape") by *McClatchey et al. (1973)*.

In 1976, a similar work was started at "Laboratoire de Météorologie Dynamique" with the development of the GEISA data bank which has been described by *Chedin et al. (1980)* and updated twice (*Chedin et al., 1981 ; Husson et al., 1982*). The major orientations of this work have been, first, to introduce molecules of interest not only for the Earth's atmosphere, but also for planetary atmospheres, in particular for the giant planets, and, second, to develop software for various kinds of extractions covering applications in the field of atmospheric physics and molecular spectroscopy.

The impact of the variation with temperature of the line halfwidth on the accuracy of radiative transfer computations led us to introduce, for the first time in such compilations, the value of the exponent, which summarizes this variation (*Chedin et al., 1984 : Husson et al., 1986*). Figure 1 shows the evolution of GEISA since 1978. The latest GEISA update has been documented by *Husson et al. (1992, in preparation).*

Many important problems still remain unsolved that could hamper very significantly the benefit expected from high spectral resolution atmospheric sounding from space. Resulting noises could even dominate all other sources of noise.

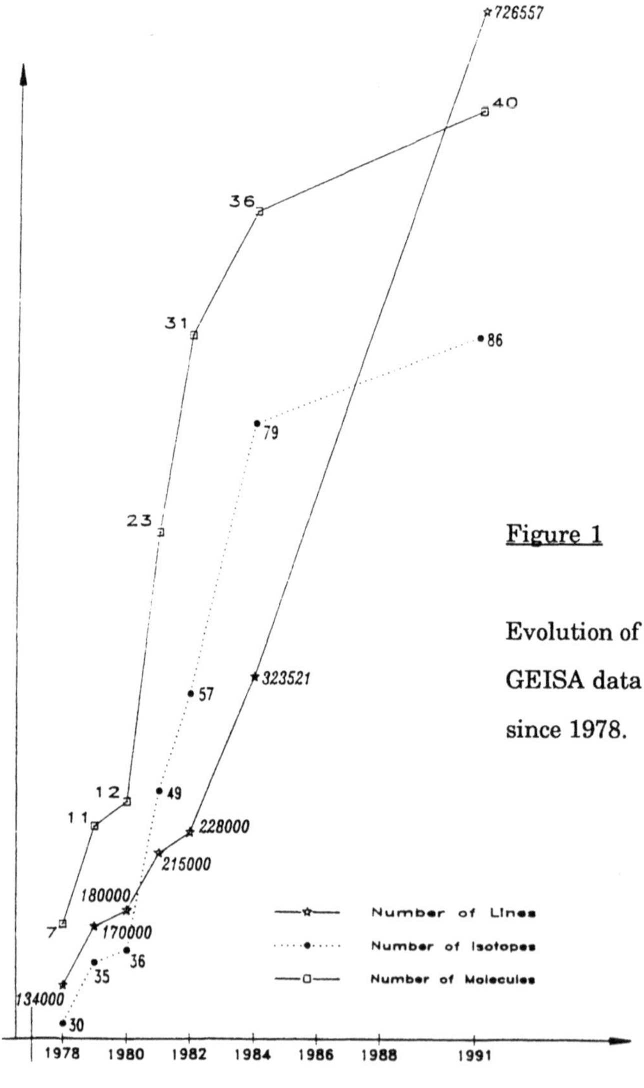

<u>Figure 1</u>

Evolution of the GEISA data bank since 1978.

2.2 Modelling of radiative transfer problems : two fast transmittance/radiance codes

In most cases, errors in retrieved meteorological or climatic parameters are due to errors in the computation of atmospheric transmittances which, in turn, are due to either spectroscopic or modelling problems. Moreover, it is well known that the processing of satellite data, covering the whole globe several times per day, requires fast models. From this analysis has followed the development of faster transmittance/radiance codes. Two such algorithms have been developed at LMD.

. *The Automatized Atmospheric Absorption Atlas (4A) model*

The approach developed several years ago by *Scott & Chedin (1981)* for obtaining a computationally fast and versatile line-by-line type method consisted, first, in using a standard extensive line-by-line code (*Scott, 1974*) (STRANSAC) to derive the monochromatic optical depths for the spectral region(s) of interest, through individual layers of a stratified atmosphere for a large number of plausible atmospheric conditions and, second, in developing a means of reconstructing the transmittance profile for any given atmospheric condition. This method, referred to as the Automatized Atmospheric Absorption Atlas (4A), resulted in a code running 30-40 times faster than the extensive line-by-line code, while accurately reproducing the latter's results. Furthermore, analysis of the deviations between satellite observations and results produced by 4A validates this approach to the inversion of the radiative transfer equation.

An example of 4A brightness temperature spectra, at a resolution of $\nu/\Delta\nu = 1200$ is illustrated in Figure 2 for 5 "standard" atmospheres : tropical, mid-latitude winter and summer, and sub-arctic winter and summer. In its revised version, the 4A model provides transmittance profiles, radiance and/or brightness temperatures (convolved, or not, by any apparatus function) as well as Jacobians (partial derivatives of satellite observations with respect to geophysical quantities like temperature, moisture, ozone, etc.). A spectrum covering the full spectral domain (15.5 - 3.0 μm) is obtained, from 4A, in 3 minutes (including Jacobian computations ; 1 minute without) with the SIEMENS-VP200 computer of CIRCE (CNRS). This model may consequently be used for studying in more detail such or such spectral domain and, in particular, its statistical behavior for large representative sets of atmospheric situations belonging to different atmospheric air mass types. Figures 3 (a-c) illustrate such statistics for the interval 710-725 cm^{-1} (14.1-13.8 μm respectively) for three sets of 137 tropical atmospheres, 545 mid-latitude atmospheres and 525 polar atmospheres. For each case, the mean brightness temperature value (green) and the curves deduced by adding or subtracting the standard deviation

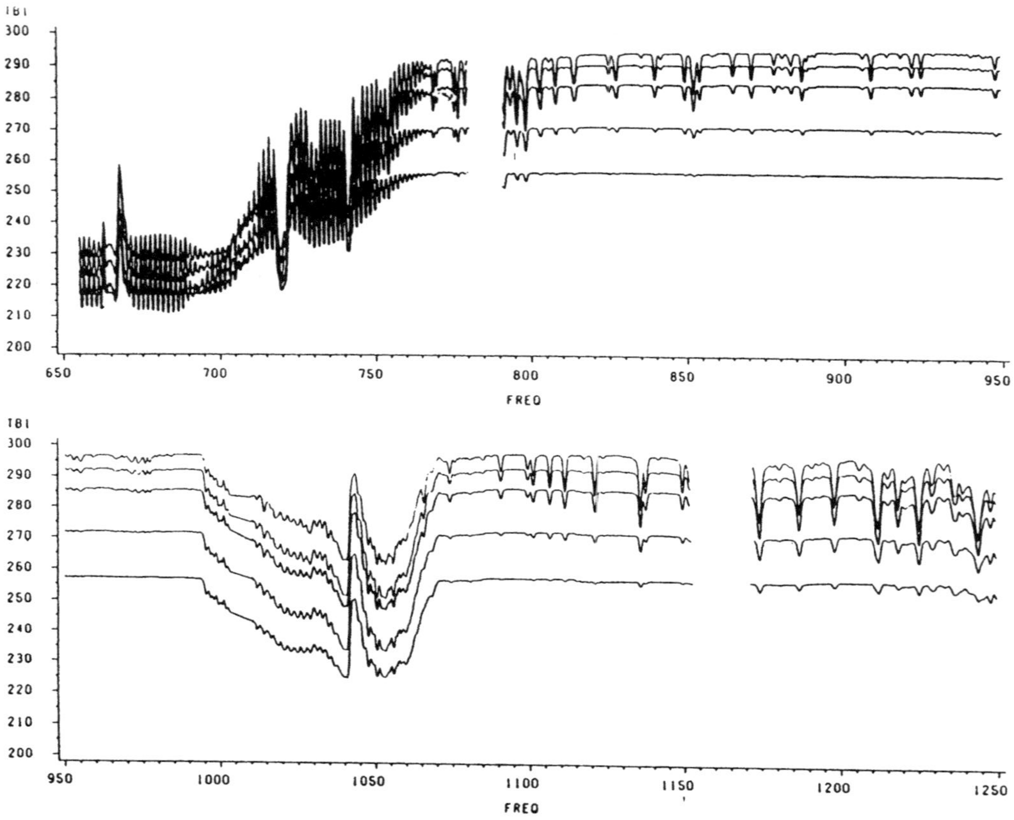

Figure 2

"4A" brightness temperature spectra at the AIRS resolution (1200) for 5 standard atmospheres. Red : tropical ; pink : mid-latitude summer ; cyan : mid-latitude winter ; green : polar summer ; dark blue : polar winter.

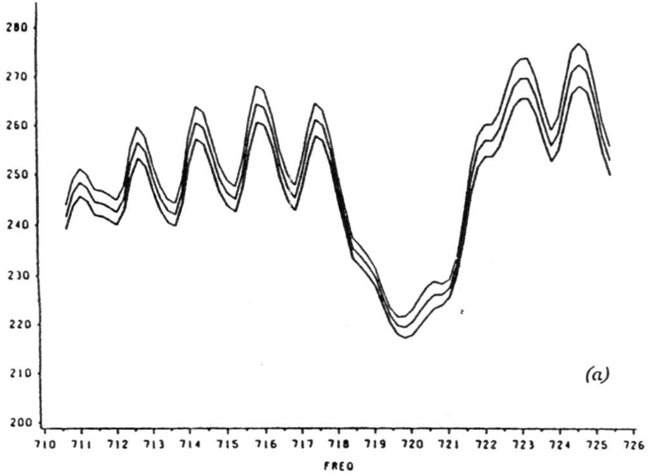

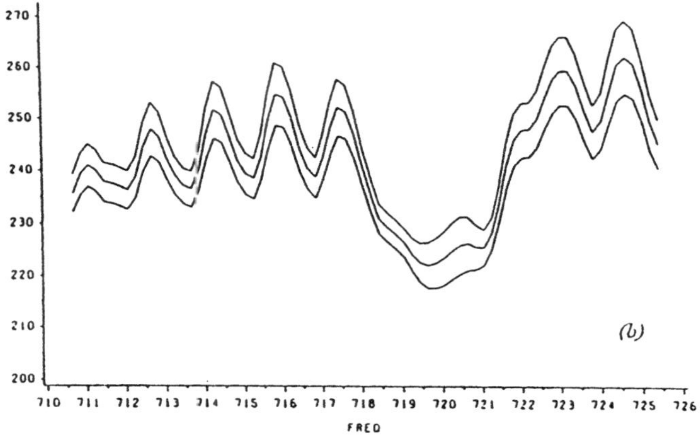

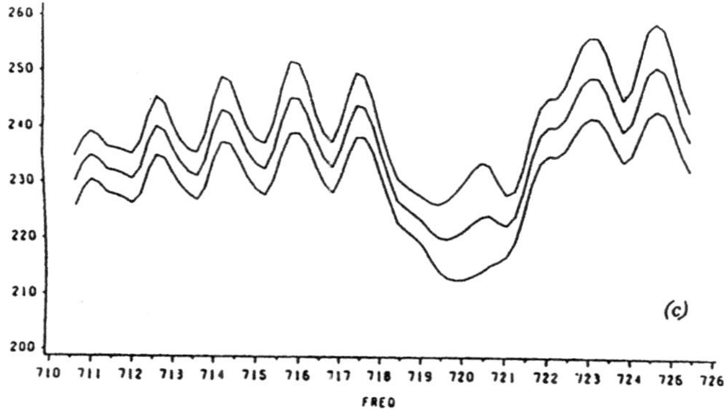

Figure 3

a) Mean brightness temperature and curves deduced by adding or subtracting one standard deviation for 137 tropical atmospheres from TIGR.
b) The same as for a), but for 545 mid-latitude atmospheres.
c) The same as for a), but for 525 polar atmospheres.

are drawn (at a spectral resolution of 0.3 cm^{-1}). Such results are essential to the optimization of a channel selection procedure.

The way to calculate temperature Jacobians is well known because they depend mainly on the Planck function variations with temperature. Calculation of the water vapor, ozone and minor gas Jacobians is not that simple. However, the structure of the 4A model allows for their accurate determination, as shown by *F. Cheruy et al. (1992)*. Computed, once and for all, for a representative set of atmospheric situations, they may be further stored.

Because of storage limitations, monochromatic Jacobians cannot be preserved for the full IR spectrum. Convolving them by the apparatus function, thus limiting the number of "points" (down to the limit of about 3,000 channels) is the way to solve this problem.

- *The Rapid Radiance Recognition (3R) model*

As seen above, a first attempt to increase the speed of the derivation of the transmittances and radiances has been made with the 4A model, relying on an atlas of precomputed line-by-line transmittances and the integration of the radiative transfer equation. Although incomparably faster than the classical line-by-line models, the 4A model is unable to produce radiances as quickly as required for particular applications, e.g. direct use in analyses and numerical weather prediction experiments.

This is the reason why a much faster model has been developed recently for computing transmittances and radiances (or brightness temperatures) from a given set of temperature and water vapor profiles for the NOAA TOVS (TIROS–N Operational Vertical Sounder) instruments, particularly for the 19 HIRS-2 infrared channels and the 4 MSU microwave channels.

This fast model, hereafter referred to as Rapid Radiance Recognition (3R) model, directly gives transmittances and radiances owing to a pattern recognition approach and a large library of precomputed radiances associated with carefully selected atmospheric situations like the TIGR (TOVS Initial Guess Retrieval) data set. TIGR was designed to improve the determination of the first guess profile in the 3I procedure of *Chedin et al. (1985)* for the retrieval of atmospheric parameters from the TOVS observations (see below).

On the same computer, 3R runs 5000 times faster than 4A. Figure 4 shows a global map of 3R–computed MSU-4 channel from ECMWF analyses (1/02/87 at 12 Z). One minute is required for the computation of such maps for all the TOVS channels.

In this scenario, it clearly appears that precomputing is the key for faster algorithms, i.e. the atlas required by the 4A model is obtained from the

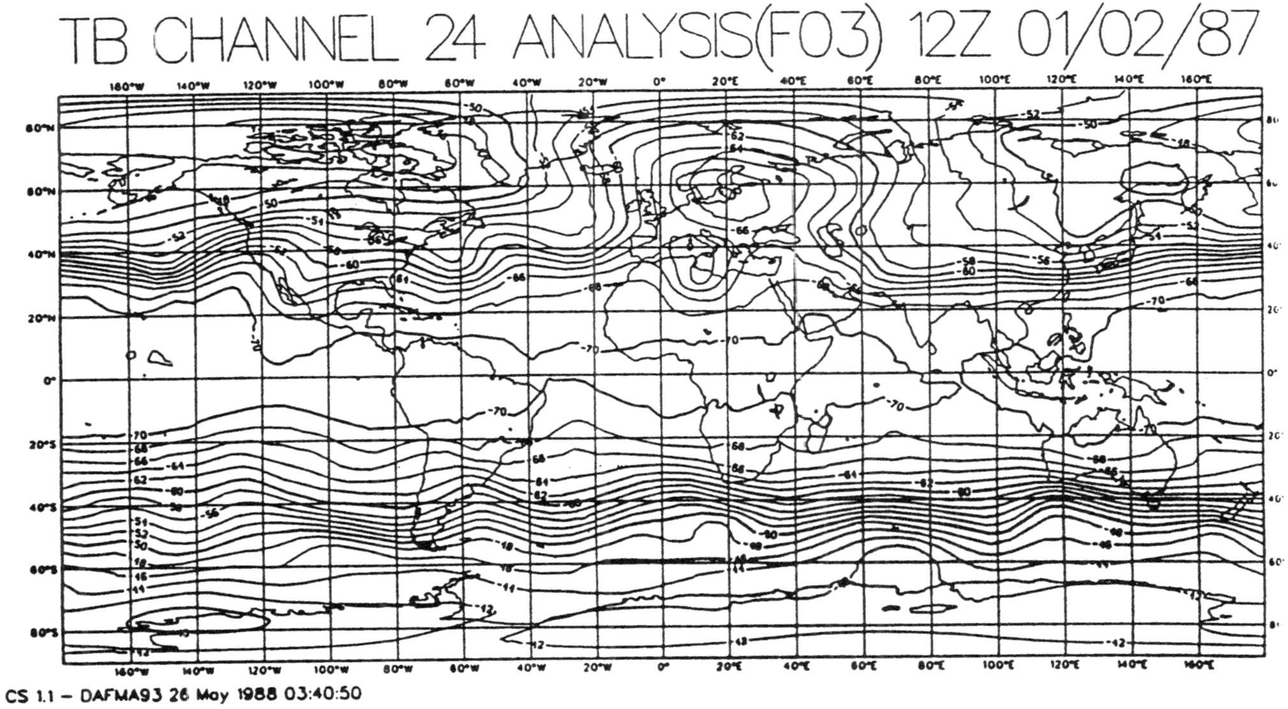

Figure 4 MSU-4 brightness temperature field computed with 3R from the 01/02/87 ECMWF analysis at 12Z.

classical line-by-line code, STRANSAC. The relative rapidity of 4A makes it possible to create the TIGR data set, which is at the core of the 3R model, itself founded upon a pattern recognition approach.

More details may be found in *Chedin et al. (1990)* and *Flobert et al. (1986)*. The 3R model is presently being adapted to the new Advanced TIGR (ATIGR) data base (see next section) and will be adapted to the AIRS or IASI channels.

2.3 The Improved Initialization Inversion (3I) retrieval system

The 3I system has been developed in the ARA group at LMD, for routinely processing NOAA/TOVS observations so as to retrieve the 3D thermodynamic structure of the atmosphere (temperature, moisture, surface properties, cloud parameters, etc.). This algorithm falls into the category of direct, non-iterative, physico-statistical methods, and gives results whose accuracy, as demonstrated at international conferences of specialists in the field (International TOVS Working Group / IAMAP), is among the best *(Le Marshall, 1985).*

The 3I approach has been widely described in the literature (*Chedin & Scott, 1983, 1985* ; *Chedin et al., 1985*). The main characteristics are given below. This physico-statistical procedure relies on a pattern recognition approach making maximal use of a priori information. Inversion of the radiative transfer equation is split into two steps, both of which take satellite measurements as their source of data. The first step "inverts" these measurements, according to the principle of "best initial profile" from among a vast selection of atmospheric situations archived in advance and classified in groups according to air mass type. This "library" of atmospheric situations, TIGR, consists, at present, of about 1800 situations carefully sampled by statistical analysis methods. The selected set of brightness temperatures serves to initialize the inversion algorithm in the second step.

Retrieval of water vapor quantities (dew point temperatures, relative humidities, integrated precipitable water vapor) is made for 3 layers delimited by the levels 1000, 800, 500 and 300 hPa. Following the temperature profile inversion, the brightness temperatures associated with the initial guess are corrected for deviations between the initial temperature profile and the final solution, giving rise to the initial guess for water vapor and surface temperature retrievals. The method used is based upon a simultaneous physical inversion of the water vapor amounts and of the surface temperature. Cloud parameters (mean top pressure and amounts) are also derived. The whole process requires about 20 minutes of a CRAY-2 at global scale.

Numerous applications have already been successfully carried out with 3I, either within the ARA group or in the groups working with it, from local mesoscale meteorological research (fronts, severe weather, polar lows, etc.)

and routine meteorological applications, to global scale impact studies in connection with medium range weather forecasting or climate research. See, for example, *Chedin et al. (1989), Chedin (1989) and Flobert et al. (1991).*

The present 3I system concept may easily be extended to a high spectral resolution infrared sounder like AIRS or IASI, in principle coupled with the Advanced Microwave Sounding Unit (AMSU).

3. TOWARDS THE A-3I (ADVANCED IMPROVED INITIALIZATION INVERSION) PROCESSING SYSTEM : ALGORITHM DESCRIPTION

3.1 The "Educated" ATIGR (Advanced-TIGR) data set : an off-line procedure

Our past experience with TIGR/version 1, which comprised 1207 atmospheric situations, revealed some "holes" in the sampling. This led us to reconstruct and enlarge TIGR to about 1800 atmospheric situations, extracted from a set of more than 150,000 radiosoundings, covering the whole globe and a large time period, using a statistical screening scheme based upon pattern recognition (see *Escobar et al., 1990*). This set (TIGR/version 2) is the basis for the ATIGR data set. Figure 5 illustrates the difference between TIGR-1 and TIGR-2. A principal component analysis of the temperature profiles archived in the latter has been carried out and all these profiles (≅ 1800) projected on the two first components giving Figure 5a. Figure 5b is the result of the same process for the situations archived in TIGR-1 (≅ 1200) which clearly appears as a subset of TIGR-2. Each situation is described by its temperature, moisture and ozone profiles at 40 levels between the surface and 0.05 hPa. The 4A, fast line-by-line, forward radiative transfer model is used to compute, on the tentative basis of 1000 AIRS or IASI channels, the transmittance profiles for a total of at least 10 viewing angles, sampling at best the whole scanning range. Resulting data, including terms of the radiative transfer equation, like $\int B dt\uparrow$ and $\int B dt\downarrow$, are stored, after data compression, on a fast access electronic device, giving rise to ATIGR. Figure 6 gives the flow chart for this off-line step. Due to the large number of channels considered, the size of ATIGR might be a problem. Three ways are presently being explored to compress the data , in particular the so-called wavelet packet transformation (cooperation M. Farge and M.V. Wickerhauser). Expected storage size should be of the order of 500 Mbytes.

This data set is then classified into air mass types (5 are considered at present : 2 polar, 2 mid-latitude and 1 tropical). The classification algorithm follows two steps :

- principal component analysis of the virtual temperature profiles archived in ATIGR,
- ascending hierarchical clustering of the sets of the 10 (at most) first principal components.

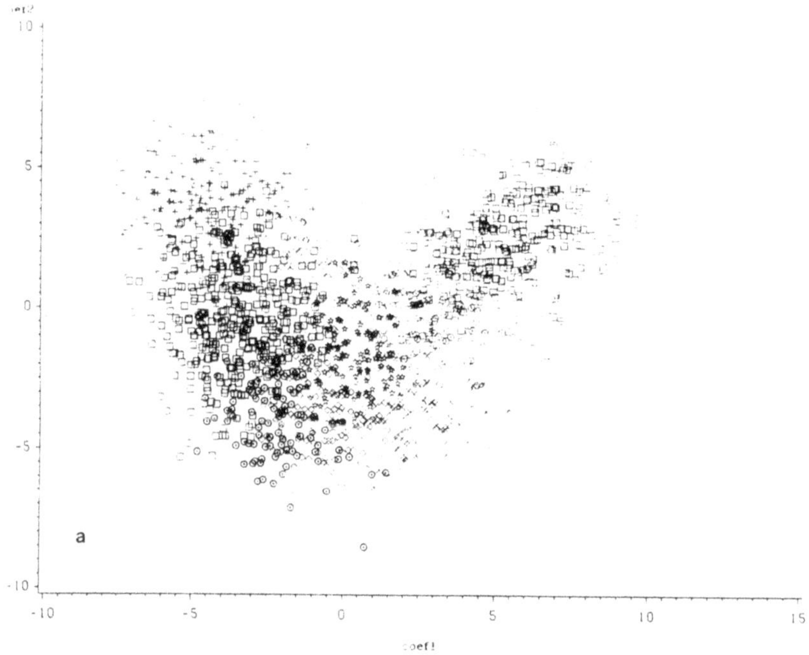

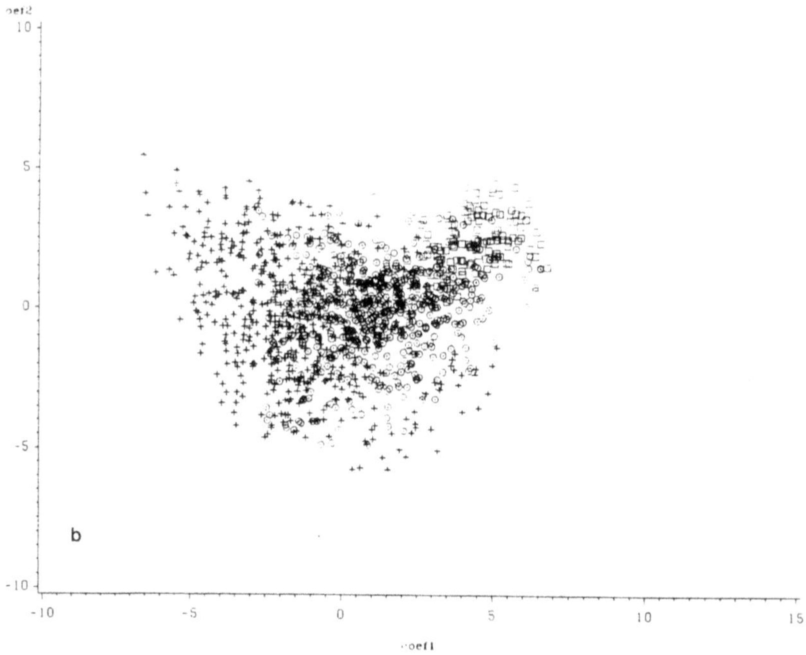

Figure 5 Projections of the situations archived in TIGR-2 (a) and TIGR-1 (b) on the two first eigenvectors of the virtual temperature variance-covariance matrix of TIGR-2 (V. Achard, Thesis, 1991). Five classes are here identified for TIGR-2 and three for TIGR-1. They range from polar/winter to tropical from left to right.

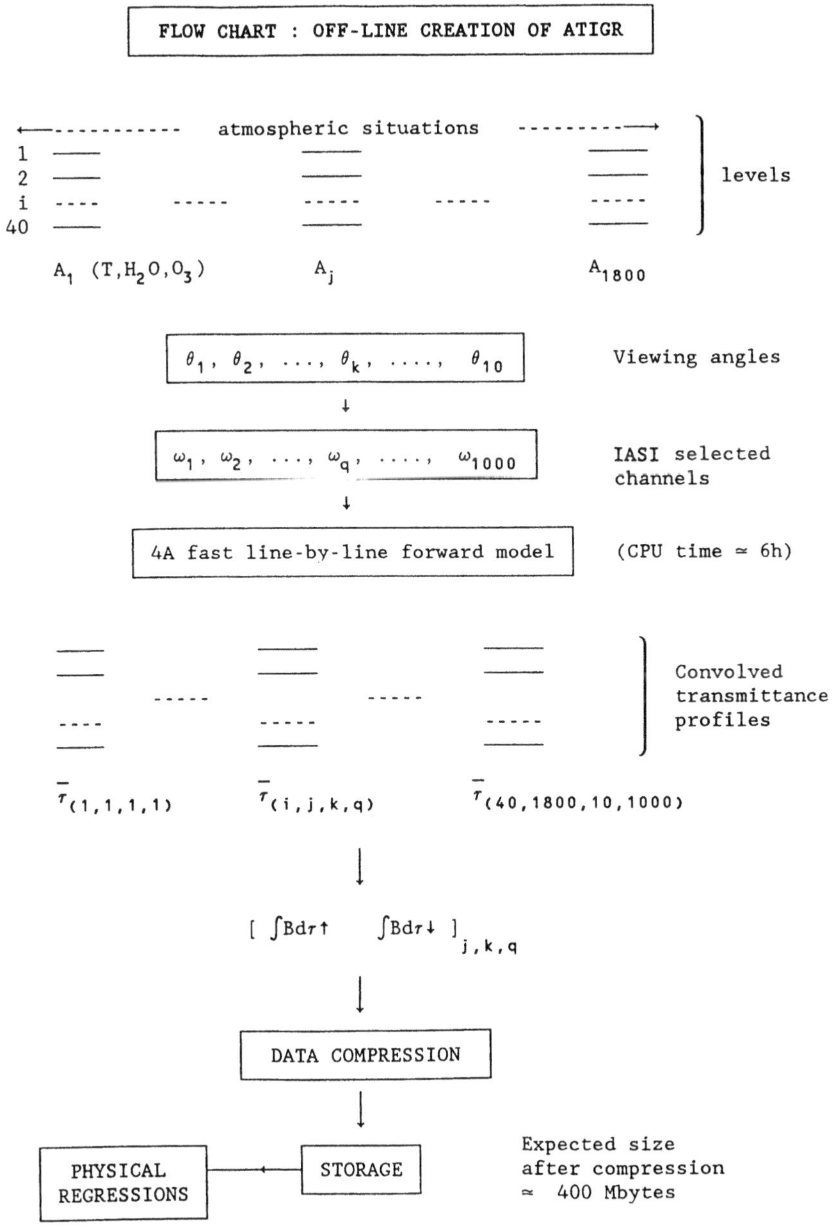

Figure 6 Flow chart of the off-line ATIGR creation.

For each cluster thus isolated, AMSU and IR channel brightness temperature gravity centers are determined for those channels not significantly affected by either the surface characteristics or the clouds (AMSU and stratospheric channels). Classification of real observations is made by reference to these gravity centers, plus some a priori information (variance-covariance brightness temperature matrices).

The ATIGR data set allows a number of physical regressions to be established, (e.g. inter-channel regressions : AMSU alone, IR alone, AMSU and IR mixed, etc.) which are used at different stages of the on-line processing. Any regression is determined separately for each observation condition and each air mass type.

3.2 The Advanced-3I (A3I) on-line system

The main features of the A3I processing system are briefly listed below and are illustrated by the flow charts of Figures 7 and 8.

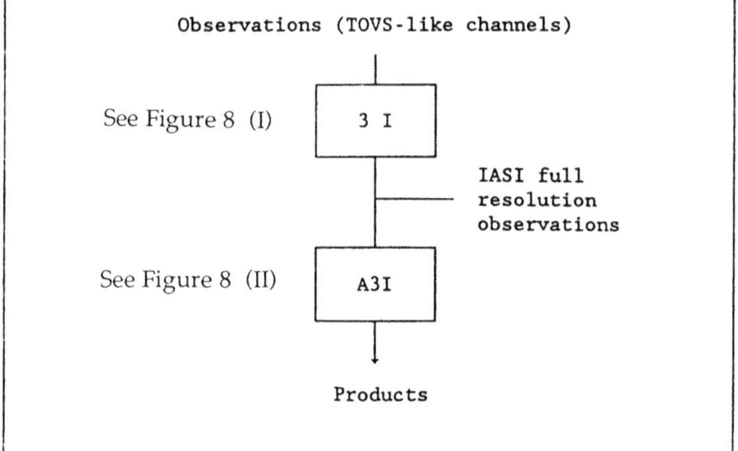

Figure 7 Synthetic A3I flow chart

1. AMSU tests are carried out to detect contamination due to thick clouds or to precipitation.
2. The 3I system, in its present configuration, is run on the basis of the AIRS or IASI HIRS-like and AMSU channels. Input data are the brightness temperatures of these channels and a priori information like the conditions

of observations (elevation, viewing angle, surface type, surface pressure from forecasts, operational SST values, etc.). Output data are, for each AIRS or IASI spot, the air mass type, the flag : "clear/cloudy", temperature and moisture profiles, cloud parameters (mean top pressure and amount), surface parameters (temperature, snow, ice, etc.) and an estimate of the total ozone content.

3. The Advanced-3R model (see 3.3) hyperfast forward radiative transfer model, analogous to the 3R model (see 2.2), is then used to reconstruct the spectrum from the results of step 2.

4. Comparison is made between the observed and the reconstructed spectrum and analysis conducted of the differences.

5. Refinements of the "cloud/clear" flags are made by looking at the differences between the peak and the window parts of several selected lines. The same step allows refinement of the cloud parameters. Several spectral regions are used along the whole spectrum for coherence check. Interchannel regressions (IR/AMSU) are also used.

6. IR cloud-cleared radiances are obtained in two ways :

   - a method similar to the so-called "ψ–method" (*Chedin, 1989* ; *Chedin et al., 1989*) which relies upon an initial guess solution and associated brightness temperatures (both in the infrared and microwave) and the ATIGR "educated" data set ;
   - a method relying upon the refined cloud parameters obtained in 5.

   Coherence checks are carried out to converge to the optimal solution.

7. Within several spectral domains, a set of clear or cloud-cleared channels is then used to construct the brightness temperature cost function :

$$\left(T_B - T_B^o\right)^t \, \Sigma^{-1} \, \left(T_B - T_B^o\right)$$

   where Σ is the observation error covariance matrix, T_B and T°_B are the observed and initialized brightness temperatures and t denotes transposition. Improvements in the initial profile are obtained through the minimization of this cost function in T(P), T_d(P) and T_s simultaneously. "Hyper" windows in each spectral domain will help refine the ground emissivity initial value.

8. Steps 3 to 7 are iterated once or twice.

9. Integrated contents of O_3, CH_4, N_2O and CO are finally determined from relevant spectral domains of each spectrum.

Figure 8 Flow chart : the A3I system

(I) THE CURRENT 3I PROCESSING

INPUT
- IASI TOVS-LIKE CHANNELS
- TOPOGRAPHY FILE
- SURFACE PRESSURE (forecast)
- SEA SURFACE TEMP. (operational gridded product)

↓ Proximity recognition

3I / TIGR DATA SET
FIRST INITIAL GUESS SOLUTION

↓

CLOUD CLEARING (ψ-method)

↓

(One iteration)

INITIAL GUESS

↓

INVERSION OF RTE IN TEMPERATURE

↓

CLOUD PARAMETERS

↓

CLOUD CLEARING OF H_20 SENSITIVE CHANNELS

↓

INVERSION IN T_d AND T_s

↓

END OF CURRENT 3I PROCESS

./..

Figure 8 (continued)

(II) THE A3I PROCESSING

INPUTS → OUTPUTS OF 3I

↓

HYPER FAST 3R MODEL

↓

RECONSTRUCTION OF IASI SPECTRUM FROM INPUTS

↓

COMPARISON BETWEEN OBS. AND CALC. SPECTRUM WITH
SEPARATE ANALYSIS OF ABSORBING AND WINDOW CHANNELS

↓

REFINEMENT OF CLEAR/CLOUDY FLAGS. REFINEMENT
OF CLOUD PARAMETERS. COHERENCE CHECK
FROM VARIOUS SPECTRAL DOMAINS

↓

One or two Iterations

IASI CHANNELS CLOUD CLEARING
BY TWO DIFFERENT METHODS. COHERENCE CHECK

↓

FOR SEVERAL SPECTRAL DOMAINS, A FEW SETS OF CHANNELS
ARE SELECTED AND COST FUNCTIONS ARE DETERMINED

↓

MINIMIZATION OF COST-FUNCTIONS
IN T(P), H_2O(P), T_S SIMULTANEOUSLY

↓

HYPER WINDOWS IN EACH SPECTRAL DOMAIN FOR
REFINEMENT OF SURFACE SPECTRAL EMISSIVITY

↓

INTEGRATED CONTENTS OF O_3, CH_4, N_2O AND CO
FROM RELEVANT SPECTRAL DOMAINS

↓

END

3.3 The Advanced-3R (A3R) very fast forward model

No matter what method is used to optimize the use of next generation sounding data (physical retrieval schemes or direct "variational" assimilation of radiances in a numerical model), there is a strong need for an accurate hyperfast, forward radiative transfer model and for "Jacobians" which are the partial derivatives of the brightness temperatures with respect to the geophysical parameters.

Based upon the 3R model (*Flobert et al., 1986* ; *Chedin et al., 1990*) developed for the TOVS data, we are designing a new model capable of handling AIRS or IASI like data : the Advanced–3R (A3R) model. This model will produce transmittances and brightness temperatures at the required spectral resolution, owing to a pattern recognition approach and a large library of precomputed brightness temperatures associated with carefully selected atmospheric situations (namely : ATIGR). The A3R basic principle is similar to the 3I one : initialization through the minimization of a cost function expressed in terms of atmospheric profiles (the input one and those archived in ATIGR),and refinement of the initial guess solution through a first order limited expansion using the Jacobian ($\partial T_B/\partial T$) stored in ATIGR.

In this approach, we assume that observed radiances depend on temperature, water vapor, ozone and minor gas profiles, as well as on surface characteristics, and that the computed value corresponding to the input pattern is obtained, for a channel k, by :

$$T_B^k = T_{B_c}^k + \Sigma_i \frac{\partial T_B^k}{\partial T_i}\left(T_{obs}^i - T_c^i\right) + \Sigma_l \, \Sigma_i \frac{\partial T_B^k}{\partial q_i^l}\left(q_i^l - q_{ic}^l\right)$$

Σ_i is the summation of i quantities corresponding to the layers ; Σ_l is the summation of l quantities corresponding to the absorbing gases and the subcript c denotes the closest situation.

$(\partial T^k_B/\partial T_i)(T=T^i_c)$ is the partial derivative of the brightness temperature with respect to the temperature in the layer i (temperature Jacobian).

$(\partial T^k_B/\partial q^l_i)(q^l_i=q^l_{ic})$ is the partial derivative of the brightness temperature with respect to the mixing ratios of water vapor, ozone and other minor gases in the layer i. The Jacobians are pre-computed and found in the ATIGR data set.

CONCLUSION

Since 1978, in spite of their limited technical capabilities, the TOVS instruments on board the present generation US operational weather satellites (NOAA) have demonstrated their usefulness for research or routine applications, from mesoscale, short range weather forecasting or analysis to global scale, medium range weather forecasting. The TOVS archives also represent an invaluable set of data for climate evolution studies. Research continues, aimed at increasing the impact of TOVS data on numerical weather prediction, mostly through a better and more coherent definition of the error functions (error covariance matrices), which must take into consideration the complexity of the satellite products and the physics involved in their understanding.

However, the limitations mentioned above forbid reaching accuracies comparable with that of the present generation of weather prediction models. A consensus exists to recognize that only an improvement, by at least one order of magnitude, of the spectral resolution of the infrared sounder may solve this problem. The high resolution concept is capable of meeting the key requirements concerning the vertical profiles of temperature (1 K rms ; 1 km vertical resolution) and humidity (10 % ; 2 km resolution).

In this document, we have presented the work presently under development in the Atmospheric Radiation Analysis group at LMD toward the design of routine processing algorithms. Many problems have to be addressed in doing so, from theoretical and experimental spectroscopy, atmospheric radiative transfer modelling and radiative transfer equation inversion, to assimi-lation of the satellite products in a model (classic or variational).

The present research has shown the capability of the pattern recognition initialization concept, which is at the basis of 3I and 3R, to deal with the large numbers of channels proposed by the planned IR instruments. Relying on the satellite observations, the initialization process should result in an initial guess solution bearing signatures of the real situation, like temperature inversions at any level, height of the tropopause, etc. Ongoing simulations will aim at quantifying these benefits, taking into account, as far as possible, what should be probably the main source of noise : our lack of knowledge in spectroscopy (mostly line shapes and intensities).

More work remains to be done to properly assess the information content and the vertical resolution resulting from high spectral resolution real (or realistic, if synthetic) observations for several instrument options.

REFERENCES

Achard V. (1991) : Trois problèmes clés de l'analyse 3D de la structure thermodynamique de l'atmosphère par satellites. Thesis. Paris 6.

Chedin A (1989) : The 3I retrieval method. Recent local and global applications. Proc. of the ECMWF Seminar on Data Assimilation and the Use of Satellite Data. 5-9 Sept. 1988, Reading, UK, pp. 181-123.

Chedin A, Husson N, Scott N A, Jobard I, Cohen-Hallaleh I, Berroir A (1980) : La banque de données GEISA. Description et logiciel d'utilisation. Laboratoire de Météorologie Dynamique du CNRS, Ecole Polytechnique, Palaiseau, France. Internal Note nb. 108, Oct. 1980.

Chedin A, Husson N, Scott N A, Cohen-Hallaleh I., Berroir A (1981) : La banque de données GEISA. Mise à jour n° 2. Laboratoire de Météorologie Dynamique du CNRS, Ecole Polytechnique, Palaiseau, France. Internal Note nb. 113, July 1981.

Chedin A, Husson N, Scott N A, Cohen-Hallaleh I, Berroir A (1984) : The GEISA data bank : 1984 version". Laboratoire de Météorologie Dynamique du CNRS, Ecole Polytechnique, Palaiseau, France. Internal Note nb. 127.

Chedin A, Scott N A (1983) : Improved Initialization Inversion ("3I"). Proceedings 1st Int. TOVS Study Conference, Igls, Austria, p. 14-79.

Chedin A, Scott N A (1985) : Initialization of the radiative transfer equation inversion problem from a pattern recognition type approach : application to the satellites of the TIROS-N series". In "Advances in Remote Sensing Retrieval Methods". A Deepak, H E Fleming & M T Chahine Eds., A Deepak Publishing

Chedin A, Scott N A, Wahiche C, Moulinier P (1985) : The Improved Initialization Inversion method : a high resolution physical method for temperature retrieval from the TIROS-N series. J Clim Appl Meteor 24, 124-143

Chedin A, Scott N A, Flobert J F, Andersson E (1989) : Recent developments and results of the 3I algorithm. Proc. of the Joint ECMWF-EUMETSAT Workshop on Use of Satellite Data in Operational Numerical Weather Prediction : 1989-1993. Reading, UK, p. 105-126

Chedin A, Scott N A, Achard V (1990) : Pattern recognition and assimilation of satellite data in numerical weather forecasting. Proc Int Symp on Assimilation of Observations in Meteorology and Ocenaography, Clermont-Ferrand, France, 7 pp

Cheruy F, Scott N A, Tournier B, Chedin A (1992) : Informativity content of high resolution spectra in the infrared from physically determined Jacobians. In preparation

Escobar J, Achard V, Chedin A, Scott N A (1990) : Statistical analysis and screening of a vast set of radiosonde data (more than 100,000) over the whole globe. Creation of a representative library of about 1800 situations". In preparation

Flobert J F, Scott N A, Chedin A (1986) : A fast model for TOVS radiances computation. Proc 5th Conf on Atmospheric Radiation, Williamsburg, USA, p. 186-189

Flobert J F, Andersson E, Chedin A, Hollingsworth A, Kelly G, Pailleux J, Scott N A (1991) : Global data assimilation and forecast experiments using the Improved Initialization Inversion method for satellite soundings. Mon. Wea. Rev., 119, 1881-1914

Husson N, Chedin A, Scott N A, Cohen-Hallaleh, Berroir A (1982) : La banque de données GEISA. Mise à jour n° 3. Laboratoire de Météorologie Dynamique du CNRS, Ecole Polytechnique, Palaiseau, France. Internal Note nb. 113, July 1991

Husson N, Chedin A, Scott N A, Bailey D, Graner G, Lacome N, Levy A, Rossetti C, Tarrago G, Camy-Peyret C, Flaud J M, Bauer A, Colmont J M, Monnanteuil N, Hilico J C, Pierre G, Loette M, Champion J P, Rothman L S, Brown L R, Orton G, Varanasi P, Rinsland C P, Smith M A H, Goldman A (1986) : Ann Geophysicae 4, 185-190

Husson N, Bonnet B, Scott NA, Chédin A (1992a) : Management and Study of Spectroscopic Information. To appear in JQRST, 48

Husson N, Bonnet B, Chédin A, Scott N A, Arié E, Ballard J, Bjoraker G, Brown L R, Camy-Peyret C, Deroche C, Fayt A, Flaud J M, Goldman A, Graner G, Hillman J F, Hilico J C, Lellouch E, Perrin A, Remedios J J, Rinsland C P, Rothman L S, Tiping R H (1992) : The GEISA spectroscopic line parameters data bank in 1992. Ann Geophysicae (in preparation)

Le Marshall J F (1985) : An intercomparison of temperature and moisture fields retrieved from TOVS. Proc. 2nd Int. TOVS Study Conf., Igls, Austria, p. 106-161.

McClatchey R A, Benedict W S, Clough S A, Burch D E, Calfee R F, Fox K, Rothman L S, Garing J S (1973) : Paper 434, AFCRL-TR-73-0096, Air Force Cambridge Research Labs.

Scott N A (1974) : A direct method of computation of the transmission function of an inhomogeneous gaseous medium. Description of the method. JQSRT 14, 691-704

Scott N A, Chedin A (1981) : A fast line-by-line method for atmospheric absorption computation : the Automatized Atmospheric Absorption Atlas, J Appl Meteor 20, 802-812

EXTRACTION OF ATMOSPHERIC SIGNALS FROM RADIANCE MEASUREMENTS: SOME LIMITATIONS

M. J. Uddstrom and L. M. McMillin[1]

National Institute of Water and Atmospheric Research
P.O. Box 3047, Wellington
New Zealand

ABSTRACT

The ability to extract useful information from measured radiances, whether for meteorological or climate purposes, is reliant upon a proper understanding of the error characteristics of the measurements and resulting products. Data error characteristics of the current generation of infrared and microwave sounders are discussed, and the resulting consequences for a physical retrieval algorithm demonstrated.

1. Introduction

All potentially iterative physical retrieval algorithms, whether explicit or implicit, have the following general form:-

$$\hat{T} - \overline{T} = D(r(T,q) - R(\overline{T},\overline{q}))$$

Where $\overline{T}$ and $\overline{q}$ indicate some assumed initial state of the atmosphere, R is the radiative transfer model operator and r is the satellite "measurement" operator on T and q, the true state of the atmosphere. D is the "retrieval" operator, and $\hat{T}$ the resultant estimate, or retrieval, of the true state of the atmosphere.

The most significant aspect of this equation is that the retrieval operator is applied to a difference signal which incorporates information from the satellite measurement process and a radiative transfer model. As a consequence, the retrieval operator is susceptible to a number of error sources, which combined, define the total system noise in the measurement and retrieval processes. Further, there is no a priori expectation that the mean bias $\overline{(r(T,q) - R(T,q))}$ will be zero for some sample of collocated satellite and radiosonde measurements.

In this paper, estimates of the total system noise for the High resolution Infrared Radiation Sounder (HIRS) and Microwave Sounding Unit (MSU) instruments of the TIROS Op-

[1] Satellite Experiment Laboratory NOAA/NESDIS/ORA, E/RA14 Room 810 NSC, Washington DC 20233, U.S.A.

NATO ASI Series, Vol. I 9
High Spectral Resolution Infrared Remote Sensing
for Earth's Weather and Climate Studies
Edited by A. Chedin, M.T. Chahine and N.A. Scott

erational Vertical Sounder (TOVS) will be discussed, and the consequential impact upon the accuracy of an explicit, optimal, physical retrieval algorithm specified.

2. Total System Noise

Estimates of the total system noise in the satellite measurement / retrieval process may be computed from samples of collocated radiosonde and satellite soundings.

2.1 Noise Components

There are four general error terms that can be identified as contributing to the "measurement" noise in the retrieval process:-

- Satellite measurement errors,
- Radiative transfer model errors,
- Validation data errors, and
- Non representativeness errors.

The first incorporates the fundamental errors arising from the instrument's optical, detector and electronic characteristics, together with errors introduced by ensuing processing. The latter includes contributions from sources such as calibration, cloud clearing and limb correction procedures.

Radiative transfer models are also known to suffer from errors arising from a fundamental lack of knowledge of molecular collision processes; e.g. the characteristics of absorption features in terms of their pressure and temperature dependence are not well understood (Chedin et al. 1988). Compounding these problems is the need to compute atmospheric transmittances in a timely manner. All practical (i.e. fast) transmittance models utilise parameterisations of one type or another (e.g. Weinreb et. al. 1981)

Implicit to validation experiments, where the noise characteristics of some measurement process are to be determined, is the assumption that the true state of the system can be estimated with appropriate accuracy and precision by independent means. Here, this assumption implies that all collocated radiosondes can be relied upon to provide unbiased, accurate and precise estimates of the temperature and water vapour structures of the atmosphere sounded by the satellite. A further implication is that the extrapolated temperature measurements are accurate to 0.1 hPa, the top level for the radiative transfer model used here. Unfortunately, neither of these assumptions is correct (Nash and Schmidlin (1987), Schmidlin and Luers, (1987), Uddstrom (1989), McMillin et. al. (1992)).

The last error source, non-representativeness, includes all those error terms that relate to the fundamental differences between satellite remotely sensed and in-situ measurements. Specifically, the differences between point (radiosonde) and areal (satellite) measurements, as well as time and distance mismatch considerations. For sufficiently large samples these errors might however be expected to contribute zero bias to the total system noise.

2.2 Total System Noise Estimates

2.2.1 Data sources

The data used in this experiment were extracted from the NOAA/NESDIS collocation archive (termed the DSD5), and consist of approximately 30000 NOAA10 collocations having the following characteristics:-

- Global distribution.
- Sampled from the period October to April (weighted toward January) and the years 1989 to 1991.
- Includes clear, nstar and cloudy retrieval path observations.
- Each radiance observation, whether clear, nstar or cloudy, is collocated with only the closest radiosonde, where the definition of closeness incorporates both time and distance considerations.
- Day, night, land and sea data are included.
- Radiosonde shortwave radiation errors have been corrected using the method of Finger and McInturff et al. (1979). For the U.S.A. VIZ radiosondes in the sample a long wave correction is applied above 20 hPa.
- Badly cloud contaminated nominally clear and nstar data have been deleted from the sample.
- The data have been divided into dependent (90%) and independent (10%) samples by random selection without replacement.

2.2.2 The forward model

With the exception of some additional enhancements, the radiative transfer model utilised is algorithmically identical to the operational NESDIS model (Weinreb et al. 1981). The enhancements affect the model in only two ways. All radiating layers above the physical surface at the location of the radiance observation contribute to the computed radiance, regardless of surface pressure, and an estimate of the radiance arising from the atmosphere above the topmost layer of the model is included. The surface skin temperature is estimated directly from the window channels and associated radiosonde profile (Uddstrom 1991). The "gamma" transmittance adjustment factors were set to unity.

2.2.3 Radiance temperature adjustments

The radiance temperature difference $\overline{(r(T,q)-R(T,q))}$ for the dependent sample, as a function of TOVS channel number (HIRS 1 to 20, and MSU 1 to 4 are TOVS channels 1 to 20, and 21 to 24 respectively) and airmass class is plotted in Fig. 1. The airmass classes were defined by Typical Shape Function (TSF) classification (Uddstrom and Wark, 1985) and ordered such that airmass class 1 essentially corresponds to the "most" polar atmospheric class, and airmass class 19 is the "most" tropical. Three points are evident:-

- There is a strong, channel dependent, bias in the raw retrieval operator signal, which if uncorrected will lead to spurious retrieval results. The bias in the shortwave infrared channels

(13 through 17) is significantly more variable than that in the longwave channels (1 through 7). Some channels, sounding similar regions of the atmosphere (e.g. 7 and 13) have biases of opposite sign. Note, by definition, the biases for channels 8, 9, 18, 19, 20 and 21 have been set to zero.

- There is some indication that the bias errors, both infrared and microwave are airmass dependent in some organised fashion.
- Outside the polar regions, the most opaque water vapour channels (i.e. 11 and 12) have characteristics altogether different from those of the other channels. The bias increases with increasing atmospheric water vapour.

Notwithstanding the error sources identified above, it is evident from Fig. 1 that at least in terms of the bias statistic, the radiative transfer models currently in use do not well simulate the satellite measurements. Since all retrieval algorithms are built on the implicit assumption that a zero radiance temperature signal will produce a zero perturbation to the first guess profile, either the measured or modelled radiance temperature measurements must be "adjusted" before use in a retrieval algorithm. Generally the measured radiance temperatures are adjusted to "equivalent" modelled radiance temperatures, since the measurement space apparently includes the modelled space.

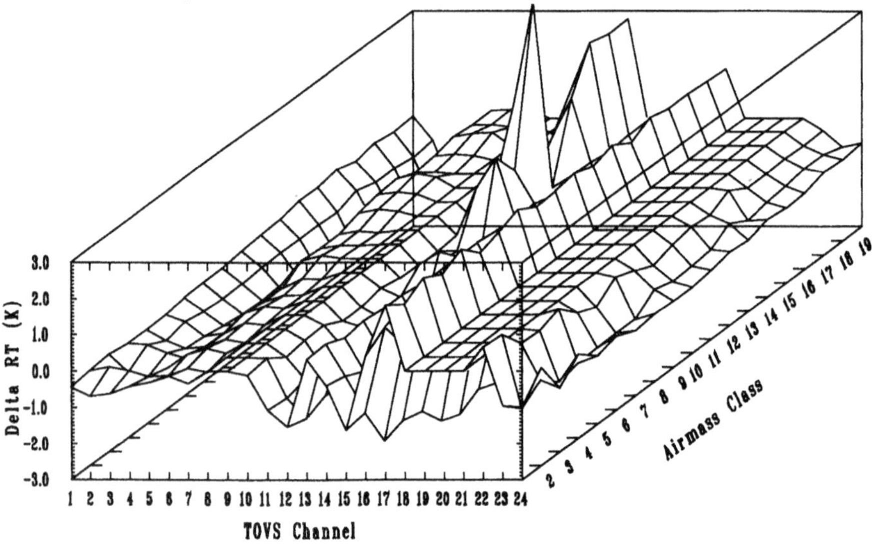

Fig. 1: Mean bias between NOAA10 measured, and modeled radiance temperatures (z axis), TOVS channel number (x axis) and airmass class (y axis), for the dependent sample.

The adjustment technique used here is the shrinkage estimator of Oman, (1982), a constrained regression method, i.e.:-

$$\hat{\mathbf{y}} = (\mathbf{S}_{yx} + \Upsilon \mathbf{C}_0^T)(\mathbf{S}_{xx} + \Upsilon \mathbf{I})^{-1}\mathbf{x} + \mathbf{C}$$

where $\mathbf{x}$ is the vector of predictors, which includes all the satellite measured channels to be corrected, together with two additional predictors; a land/sea indicator, and the cosine of the solar zenith angle. The predictands $\hat{\mathbf{y}}$, are the model (equivalent) radiances. A separate equa-

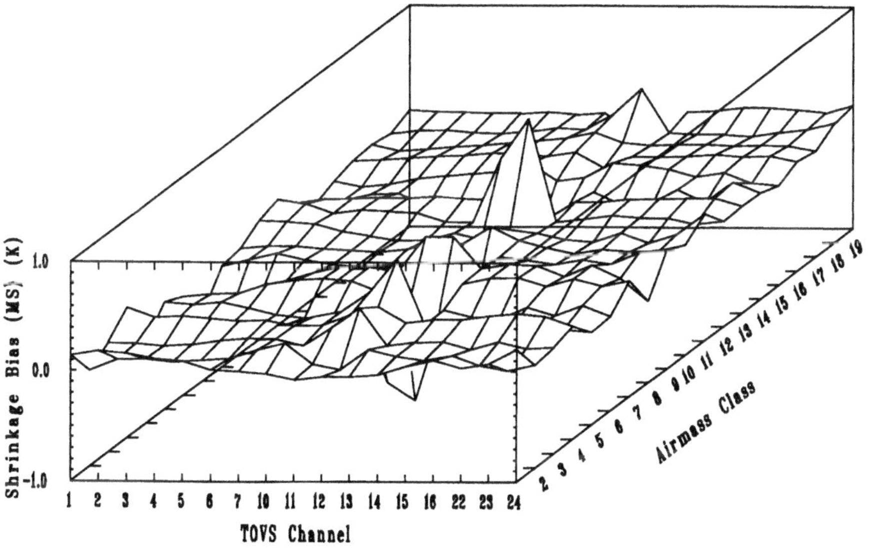

Fig. 2: Mean bias between NOAA10 measured, adjusted to model equivalents, and modeled radiance tem peratures for the independent samples.

tion is derived for each airmass class, and the constraint $\mathbf{C}_0$ is chosen so as to constrain the significant predictor channels to lie close, in the vertical sense, to the relevant predictand channels. By this means, for example, a near surface channel will not be used to correct a stratospheric channel, or vice versa. $\mathbf{S}$ is used to represent the covariance (subscript *xx)* and cross-covariance (subscript *xy)* matrices generated from the dependent sample, and $\mathbf{C}$ is the constant matrix. Υ is a ridge smoothing parameter.

Generating shrinkage adjustment equations for each atmospheric class, and applying these to the independent samples, the total system noise properties of the measurement process can be defined.

The adjusted (model equivalent) signal bias, $\overline{(\hat{r}(T,q) - R(T,q))}$, is demonstrated in Fig. 2. Evidently, the adjustment procedure has removed, at least on an ensemble basis, the signal bias that was evident in Fig. 1. None of these biases are significantly different from zero. In

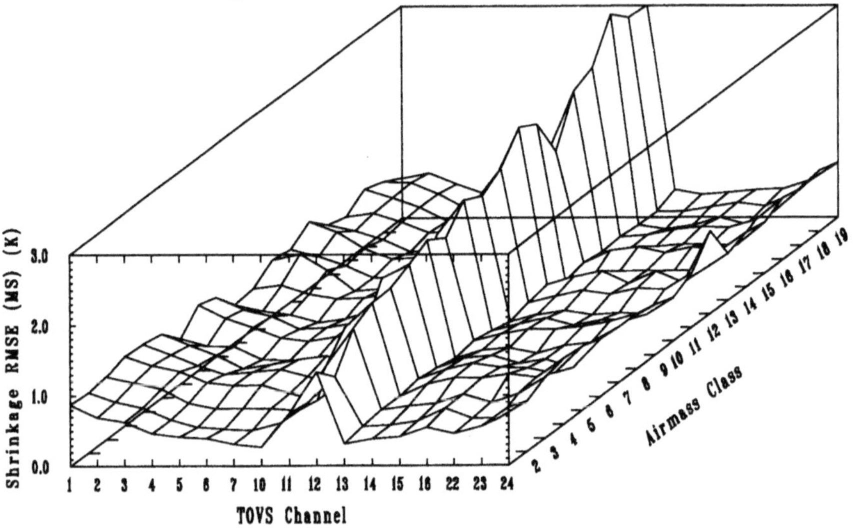

Fig. 3: RMS errors for NOAA10 measured, adjusted to model equivalents (by shrinkage estimation), and modeled radiance temperatures for the independent samples.

Fig. 3, the root mean square (rms) error in the adjusted signal radiance temperatures is plotted. It is these rms error that define the total system noise in the "measurements" provided to a retrieval operator, and this noise contains inseparable elements from the four components mentioned earlier. These values differ in a significant way from the NEΔT values specified in the NESS 107 (Rev. 1) document (or inferred from the NEΔN values given). Further, it is apparent, from Fig. 3 that the opaque water vapour channels (HIRS 11 and 12) are not well modelled, or else the validating atmospheric water vapour measurements do not describe well, the state of the atmosphere.

The correlation between the measured, adjusted to model equivalent radiance temperatures, $\hat{r}(T, q)$,, and modelled radiance temperatures is presented in Fig. 4. While the rms error in the adjusted satellite measurements (Fig. 3) is essentially flat, with respect to airmass, the information content in the measurements is substantially less in the tropics than in the extra tropics, since the atmospheric variance is less. Apart from the opaque water vapour channels, upper tropospheric channels in all three spectral regions are the most severely affected in this regard.

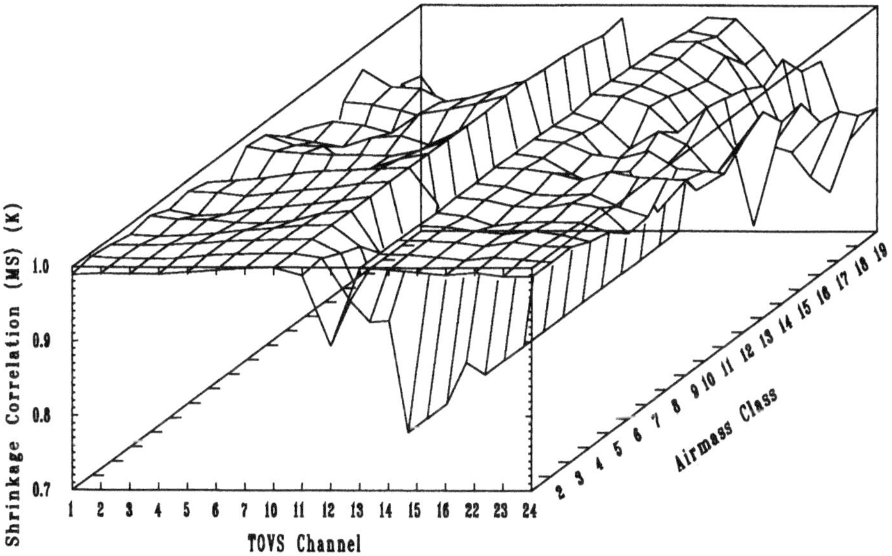

Fig. 4: Correlation between NOAA10 measured, adjusted to model equivalents (by shrinkage estimation), and modeled radiance temperatures for the independent samples.

3. Retrieval Experiments

3.1 The Retrieval Algorithm

In this study, explicit retrievals of mixing ratio, atmospheric and skin temperatures were computed using the Typical shape function Maximum a posteriori Sequential (TMS) simultaneous physical algorithm of Uddstrom (1988). This algorithm utilises pattern recognition in temperature or moisture space to define airmass dependent first guess constraints, and radiance adjustment equations as outlined above. Given a measurement vector, the atmospheric class of the observation and its probability of membership, as determined by Bayesian discrimination in the satellite measurement space, the measurements are adjusted to model equivalents according to the chosen airmass class. A retrieval is then performed, using an optimal, sequential, simultaneous, physical retrieval estimator. The retrieval portion may be iterated, and a Bonferroni radiance temperature convergence test (Neter et al., 1989) is applied as a final quality control measure. In these experiments, the channels listed in Table 1 were employed in the retrieval algorithm:-

Table 1: TMS estimator channels

Clear Path:	1, 2, 24, 23, 22, 15, 14, 13, 3, 4, 5, 6, 7, 8, 10
NStar Path:	1, 2, 24, 23, 22, 3, 4, 5, 6, 7, 10

Four experiments were conducted in order to estimate the impact of system measurement noise on the retrieval algorithm. In the first experiment, the radiances used in the retrieval step (but not the discrimination step) are noise free modelled radiance temperatures i.e. perfect radiance temperatures. Next, random single field of view NEΔT noise was added to the modelled values, and the retrievals re-computed. In the third experiment, random total system noise (as defined by Fig. 3) was added to the modelled radiance temperatures, and finally, retrievals were computed from the satellite measurements (adjusted to model equivalent values).

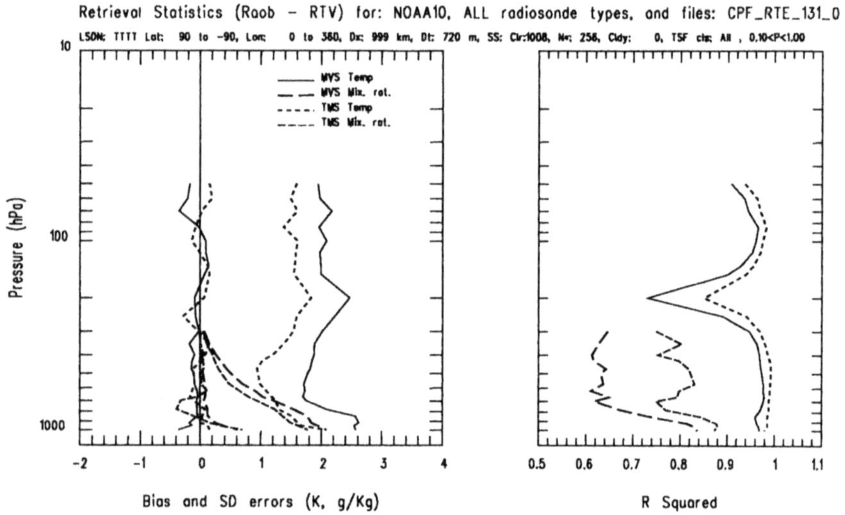

Fig. 5: Retrieval error statistics for a sample of 1264 perfect radiance temperatures. SD refers to the standard deviation error. Temperature errors are in K while mixing ratio errors are in g/kg.

3.2 Perfect Radiance Temperatures

Fig. 5 indicates the bias, standard deviation and R^2 retrieval error statistics for temperature and mixing ratio, when perfect radiance temperatures are utilised in the retrieval step. The sample size is 1264, with collocations from all latitudes, over both land and sea surfaces. Results from the NESDIS operational product are included for reference purposes only (labelled MVS). Evidently, the TOVS instruments, together with the TMS retrieval algorithm are, in principal, capable of resolving many atmospheric features, with resulting thickness errors (not shown) of not more than 12 m for any layer between 1000 hPa and any other standard level, up to 20 hPa.

3.3 NEΔT Contaminated Radiance Temperatures

When the radiance temperatures used in the retrieval step are model values, contaminated with Gaussian noise having standard deviations as given in Table 2 the accuracy of the resulting retrievals (Fig. 6) is substantially poorer than that for the perfect radiance case. The sample size is fractionally larger, at 1301, a difference that is directly attributable to the effect of

Table 2: Assumed Instantaneous Field of View NEΔT values (K)

Channel	NEΔT	Channel	NEΔT	Channel	NEΔT
1	0.97	8	0.06	16	0.27
2	0.76	10	0.15	17	0.32
3	0.56	11	0.43	18	0.05
4	0.30	12	0.87	19	0.04
5	0.18	13	0.08	22	0.30
6	0.18	14	0.06	23	0.30
7	0.14	15	0.18	24	0.30

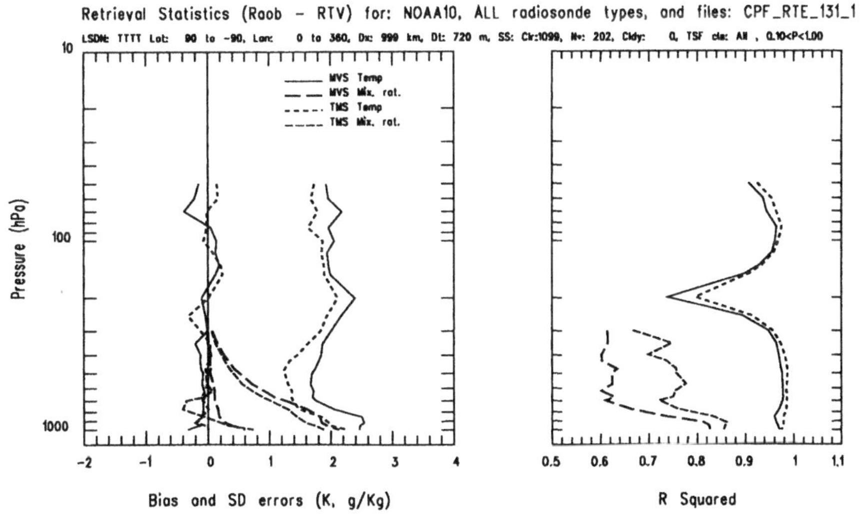

Fig. 6: Retrieval error statistics for a sample of 1301 NEΔT contaminated perfect radiance temperatures.

the Bonferroni test. Temperature retrieval accuracy at all levels has been reduced by approximately 0.3 C, since the bias error characteristics are essentially unchanged between Figs. 5 and 6.

3.4 Total System Noise Contaminated Radiance Temperatures

The retrieval accuracy error statistics, when Gaussian noise having standard deviations equal to the the total system noise is added to modelled radiances, are shown in Fig. 7. The sample size is slightly increased at 1346, and the accuracy of the retrievals is now much like that generated by the NESDIS operational processing, with temperature retrieval accuracy at all levels having been reduced by approximately 0.5 C from the perfect radiance temperature case.

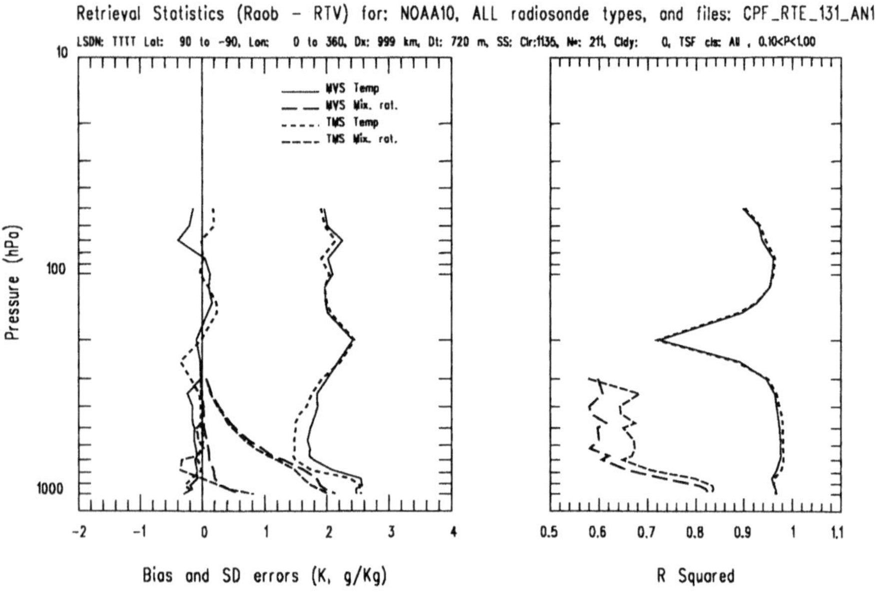

Fig. 7: Retrieval error statistics for a sample of 1346 total system noise contaminated perfect radiance temperatures.

Of particular note is the degradation in accuracy of the stratospheric retrievals, since the change does not arise from first guess errors, but is instead simply due to noise contamination of the stratospheric channels. Also, there has been a significant loss of mixing ratio retrieval accuracy.

3.5 Satellite Measured Radiance Temperatures

Finally, when the TMS algorithm is applied to the satellite measured radiance temperatures (adjusted to model equivalents), the TMS results (Fig. 8) can be seen to have properties similar to those of the previous case. Here the sample size is 1203, which is smaller than the above cases because only "clear" radiance temperature retrievals have been included in the statistics (although nstar soundings are not well represented in the DSD5 archive). Some types of cloud cleared radiance temperatures (e.g. over land) are known to have undesirable properties, which can cause a physical retrieval estimator to produce spurious error characteristics. In this case the TMS error statistics are directly comparable with the NESDIS operational (labelled MVS) results since both retrievals are computed from satellite measurements, and the samples are identical.

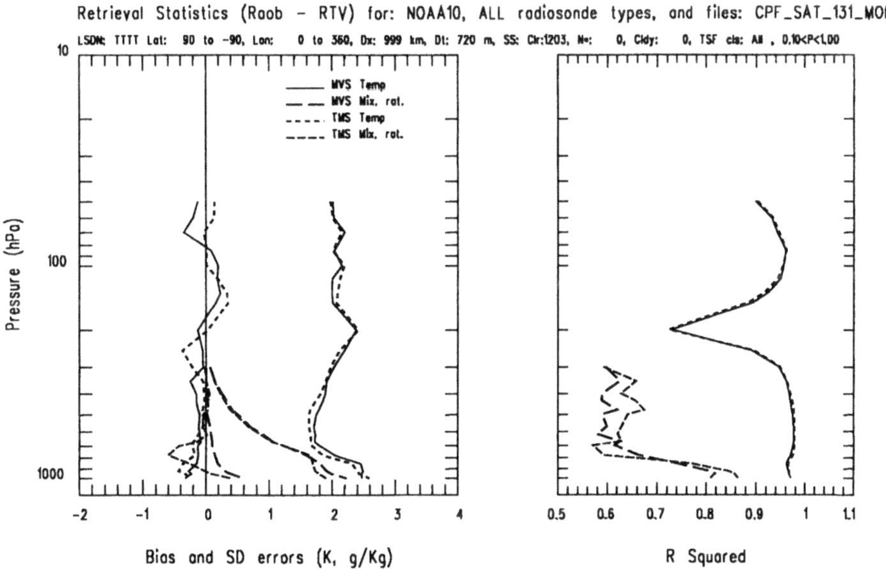

Fig. 8: Retrieval error statistics for a sample of 1203 satellite measured radiance temperature profiles, computed from the TMS and MVS (NESDIS Operational) algorithms.

Although not shown, the radiance temperature residual statistics (measured, adjusted to model equivalents, minus model values computed from retrieved profiles) for this sample indicate excellent convergence in radiance temperature space. Mean radiance temperature biases are not greater than 0.08 K for any channel (and are more typically near 0.02 K), and standard deviation errors are close to 0.1 K for the HIRS and 0.3 K for MSU channels. The source of the resulting bias in the retrieved temperatures is unclear.

4. Discussion

These experiments have been designed in order to specify the intrinsic capability of the TOVS sounding system, and to demonstrate the impact of total system "measurement" noise upon an optimal retrieval system. It is important to note that the differences evident in Figs. 5 to 8 do not arise from first guess dependencies, since in all cases the first guesses are identical, being determined from the spacecraft measurements, by the Bayesian procedure indicated earlier. Instead, these differences are driven entirely by the effect of noise in the measurement signal supplied to, in this case, an explicit retrieval operator, and by implication, to the noise term in the retrieval operator. This suggests that the information content in the TOVS data is strongly noise limited.

The accuracy of physical retrieval estimators would appear to be constrained by the accuracy of the radiance adjustment procedure, which is always necessary, and which includes components from simple measurement errors, both satellite and radiosonde, transmittance, and non-representativeness errors. In essence this noise limits the amount of information that can be extracted from radiances arising from significantly overlapping weighting functions. Fundamental to the concept of the Advanced InfraRed Sounder (AIRS) is the idea that significant vertical information may be extracted from measurements in channels having highly overlapping weighting functions. Under these conditions, given the TOVS experience outlined here, retrieval accuracy may be rather poorer than might otherwise be expected, since total system noise will severely limit the number of channels that may be effectively used in any retrieval or direct assimilation algorithm. Accordingly, these results indicate a rather pessimistic view for instruments having hundreds, or thousands of channels.

However, it is possible to point to a number of areas of endeavour that might provide a better understanding of, and perhaps lead to an improvement in the situation.

4.1 Satellite Measurement Errors

The following discussion reflects directly on the TOVS results presented here, but is indicative of areas where careful design will be required.

During the period for which data were collected for this experiment, NESDIS performed retrievals from instantaneous fields of view (ifov). Prior to May 1986 (L. McMillin, pers. comm.), retrievals were computed from an average of up to 9 x 7 ifovs. While the single field of view retrieval approach was introduced to improve retrieval spatial resolution, it has resulted in an up to 8 fold increase in the satellite measurement noise component of the total system noise, assuming no meteorological variance across the 9 x 7 array. Since the latter assumption will not generally be true, and because some of the component ifovs may be rejected for some reason (e.g. high terrain), averaged tropospheric channels are unlikely to see such a large reduction in measurement noise. However, important gains in noise reduction in upper tropospheric and stratospheric channels might be expected, and as the present results indicate, that would improve the accuracy of upper tropospheric and stratospheric retrievals. In this regard it is interesting to note from Table 2 and Fig. 3 that the total system noise for the the stratospheric channels is close to (or less than) the nominal NEΔT value, which would seem to indicate that the instrument is performing rather better than the NEΔT specifications might suggest, since for these channels, it is known that there will be a significant contribution to total system noise estimates from radiosonde radiation errors. Clearly a retrieval from an array of 9 x 7 ifovs is not desirable, but the noise penalty incurred from single field of view retrievals is not desirable either. Some averaging in the radiance domain before retrieval would reduce total system noise.

Another component of the direct satellite measurement noise arises from inappropriate calibration of the satellite radiances. Currently, the NESDIS TOVS calibration procedure results in calibration jumps halfway through each superswath (Wark, pers. comm.). The significance of this noise source is unknown. Further, the present approach to limb correction and cloud clearing at NESDIS, where the former is done before the latter, introduces what are almost certainly non zero bias errors.

If each of these error sources were removed, it might be expected that the direct measurement noise component of the total system noise could be reduced by a factor of perhaps two to three in the lower troposphere and by a greater amount in the stratosphere and upper troposphere.

4.2 Radiative Transfer Model Errors

According to Scott et al. (1986) current fast radiative transfer models match line by line model results with an accuracy which is better than the variations among the different line by line models (Spänkuch, 1988). However it is also evident that current models suffer deficiencies with respect to the way they handle the continuum, line mixing and the temperature dependence of the line widths (Chedin, 1988). Spänkuch (1988) also indicates that the differences between candidate line by line radiance integrations increases, with improving instrument spectral resolution.

It might be hoped that there would be much to gain from significant investment of effort in the forward model physics problem.

4.3 Validation Data Errors

Current short and long wave radiation error temperature correction algorithms are rather simplistic. Results from recent experiments (Schmidlin and Luers, 1987) and modelling efforts (Luers, 1990, and McMillin et al., 1992) suggest improved methods for defining and correcting such errors. In particular, infrared radiosonde errors, which may be as large as 2 K at 20 hPa, are now known to be airmass dependent and sensor specific. Shortwave errors are dependent upon both the solar elevation and the albedo of the under flight surface. These errors, which differ from one radiosonde type to another, should be corrected before the true noise in the retrieval estimator signal can be defined. Unfortunately, this remains a difficult task, since the error characteristics of many radiosonde sensors are unknown. Compounding this confusion is the increasing use of undocumented post flight correction procedures at observing stations. There remains a need for some ongoing BUAN (Baseline Upper Air Network) (Reale et al. 1991) type monitoring with a very select set of well defined upper air observing sites.

Radiosonde measurements of water vapour are becoming increasing suspect, especially above 850 hPa. Validation of radiances from those channels most susceptible to water vapour absorption will remain a challenging problem.

4.4 Non-Representativeness Errors

As long as satellite measurements are validated against radiosonde measured profiles of temperature and water vapour (or resulting derived variables, such as modelled radiances) non-representativeness errors will always result. Again, for the purposes of generating radiance temperature adjustment equations and validation of retrieved (or assimilated) quantities, a BUAN type observing system would seem to be a requirement. Selected observing sites, flying radiosondes with well understood error characteristics, very close to satellite overpass time, or perhaps multiple flights, bracketing overpass time might provide better estimates of this error's contribution to total system noise.

5. Conclusions

The process of retrieving information from radiance temperature observations, is dependent upon an understanding of the noise characteristics of the signal provided to an explicit or implicit retrieval estimator, where the total system noise includes components from direct measurement noise, forward model noise, and validating data errors.

The experiments reported here indicate that when realistic estimates of total system noise are added to modelled TOVS radiance temperatures, the resulting retrieval accuracy is similar to that derived from satellite measurements. In temperature space the accuracy of these retrievals is approximately 0.5 K poorer than for those computed from perfect radiance temperatures. In effect, the total system noise in the satellite sounding measurement and retrieval processes determines a lower bound to retrieval product accuracy, and strongly limits the amount of information that can extracted from highly dependent observations.

From the TOVS experience it is clear that validation of opaque water vapour channels is extremely difficult. Currently, these channels cannot be simply used in a physical retrieval algorithm. In the AIRS era, validation of radiative transfer models and radiance adjustment procedures, will be critical elements to enabling these data to yield their expected promise.

Acknowledgements: The work of M. J. Uddstrom was supported under the Resident Research Associateship Program of the National Research Council of the National Academy of Sciences.

References

Chedin, A., 1988: Editor; Report on the ITRA (Intercomparison of Transmittance and Radiance Algorithms) Campaign and Workshop. Pub. International Radiation Commission.

Luers, J. K., 1990: Estimating the temperature error of the radiosonde rod thermistor under different environments. *Journal of Atmospheric and Oceanic Technology*, **7**, 882-895.

McInturff, R.M., F.G. Finger, K.W. Johnson and J.D. Laver 1979: Day - night differences in radiosonde observations in the stratosphere and troposphere. NOAA Technical Memorandum NWS NMC **63**, 47 pp.

McMillin, L.M., M.J. Uddstrom and A Coletti, 1992: A procedure for correcting radiosonde reports for radiation errors. Submitted to *Journal of Atmospheric and Oceanic Technology*.

Nash, J. and F.J. Schmidlin, 1987: Final Report of the WMO International Radiosonde Comparison (U.K. 1984, U.S.A. 1985), Instruments and Observing Methods, Rep No. **30**, 103 pp.

NOAA/NESS 107 Rev 1, 1988: Data extraction and calibration of the TIROS-N/NOAA Radiometers. NOAA Technical Memorandum NESS 107 -- Rev. 1, 58 pp.

Neter, J., W. Wasserman and M.H. Kutner, 1989: Applied linear regression models, Second Edition. Pub. Irwin, Homewood, IL, U.S.A. 667 pp.

Oman, S.D., 1982: Shrinking towards subspaces in multiple linear regression. *Technometrics*, **24**, 307-311.

Reale, A.L., H E. Fleming, D.Q. Wark, C.S. Novak, F.S. Zbar, J.R. Neilon, M.E. Gelman and H.J. Bloom, 1990: Baseline Upper Air Network (BUAN) Final Report. NOAA Technical Report NESDIS **52**, 57 pp.

Schmidlin, F.J. and J.K. Luers, 1987: US Radiosonde temperature errors, methods of correction and evidence of data improvement. In Extended Abstracts to *6th AMS Symposium on Methods of Observation and Instrumentation*, New Orleans, LA, U.S.A. Pub American Meteorological Society, 463-466.

Scott N. 1986: Transmittance calculation improvements. In Report on the Third International TOVS Study Conference, Madison, WI.. Pub. CIMSS, University of Wisconsin, 40-41.

Spänkuch, D., 1988: In Report on the ITRA (Intercomparison of Transmittance and Radiance Algorithms) Campaign and Workshop. Pub. International Radiation Commission.

Uddstrom, M.J. and D.Q. Wark, 1985: A classification scheme for satellite temperature retrievals. *Journal of Climate and Applied Meteorology*, **24**, 16-29.

Uddstrom, M.J. 1988: Retrieval of atmospheric profiles from satellite radiance data by typical shape function maximum a posteriori simultaneous retrieval estimators. *Journal of Applied Meteorology*, **25**, 515-549.

Uddstrom, M.J. 1989: A comparison of Philips RS4 and Vaisala RS80 radiosonde data. *Journal of Atmospheric and Oceanic Technology*, **6**, 201-207.

Uddstrom, M.J. 1991: Forward model errors. In the technical *Proceedings of the sixth International TOVS Study Conference*, Airlie, VA, U.S.A., Pub CIMSS, University of Wisconsin, 501-516.

Weinreb, M.P., H.E. Fleming, L.M. McMillin and A.C. Neuendorffer, 1981: Transmittances for the TIROS operational vertical sounder. NOAA Technical Report NESS **85**, 60 pp.

GLA AND LMD APPROACHES TO THE PROCESSING OF AIRS AND IASI OBSERVATIONS

F. Cheruy (*), J. Susskind (**), N.A. Scott (*), A. Chedin (*), J. Joiner (**)

(*) Laboratoire de Météorologie Dynamique du CNRS(*)
Ecole Polytechnique, 91128 Palaiseau Cedex, FRANCE

(**) Goddard Laboratory for Atmospheres
GSFC, Greenbelt, 20771 Maryland, USA

INTRODUCTION

We assess components of the methods developed at GLA and LMD to retrieve temperature, humidity and ozone profiles from AIRS (Advanced Infrared Radiometric Sounder) and IASI (Improved Atmospheric Sounder in the Infrared) like data. Both methodologies involve use of a physical inversion scheme but differ among other things, by the initialization process and the use of a priori information.

GLA can utilize any guess but currently uses a climatology guess to initialize the inversion process so as to determine how much information can come from the radiance alone. The solution is stabilized by an appropriate choice of vertical degrees of freedom and a smoothness constraint. The LMD approach is based upon a pattern recognition scheme in an educated database (TIGR: TOVS Initial Guess Retrieval). In addition it uses a priori information from this data base to stabilize the solution. These aspects have been separately tested within the context of the general GLA analysis scheme to determine the degree that an improved initial guess will improve AIRS sounding capability. Both GLA and LMD approaches are under development and work is still going on for determining the number as well as the positions of the channels used for the retrievals. For this preliminary work we assume all radiances to be clear, and observed at nadir.

1. RETRIEVAL OF GEOPHYSICAL PARAMETERS FROM HIGH SPECTRAL RESOLUTION DATA : GENERAL CONSIDERATIONS

The radiative transfer equation (RTE) depends on geophysical parameters in a non linear way. To retrieve geophysical parameters from radiances measured at the top of the atmosphere, it is convenient to linearize the RTE and solve the problem iteratively. The linearization of the RTE requires knowledge of the partial derivatives of the brightness temperature with respect to the

NATO ASI Series, Vol. I 9
High Spectral Resolution Infrared Remote Sensing
for Earth's Weather and Climate Studies
Edited by A. Chedin, M.T. Chahine and N.A. Scott

geophysical parameters (temperature, water vapor, ozone...), that is to say the Jacobians. The process is then initialized using an appropriate first guess. The system to be inverted has to be stabilized because correlations in the observations cause the system to be under-determined even if more observations are present than numbers of parameters to be solved for. The choice of the stabilization constraint and also the initial guess may contribute to reduce the number of iterations needed. We will briefly describe the system used by GLA and LMD in this study.

1.1 Scheme for the retrieval

The process followed to retrieve the geophysical parameters from high spectral resolution data is briefly described on Figure 1. The solution is sequentially and iteratively deduced for each set of parameters. The retrieval algorithm consists of the following main steps :

(1) obtain an initial guess for the surface temperature, temperature, water vapor and ozone profiles,

(2) solve for the surface parameters (skin temperature, surface emissivity, surface reflectivity), using initial guess values,

(3) solve for the temperature profile, assuming that the surface values are the ones found at step (2),

(4) solve for the water vapor profile, taking benefit of the previously retrieved parameters.

A pattern recognition initial guess, devoted to the atmospheric temperature retrieval and the water vapor retrieval may be introduced before steps (3) and (4), as described in section 2.3. This must be done after the surface temperature retrieval to help account for effects of surface terms on the observed radiances. Since the results are obtained sequentially, changing the water vapor initial guess before the second pass in the temperature module may influence the temperature retrieval itself.

In the GLA approach, the method is designed to be relatively guess independent. The process is currently initialized with an initial guess coming from a climatological dataset and selected according to the season and the latitude. At each step of the retrieval, the inversion process is a weighted least square method stabilized with a damping factor (eq. 1).

$$\Delta P = (H^t W H + d)^{-1} \quad H^t W \ \Delta\theta \tag{1}$$

In this equation :

. ΔP is a vector of the differences between the value of the parameter at the iteration n-1 and the new value, the initial guess being the first estimation for the parameter. P has as many columns as the number of parameters solved for in the inversion.

. $\Delta\theta$ is the vector of the differences between observed brightness temperatures and those computed from n^{th} iteration observations. It has as many column as the number of channels used to retrieve the selected parameter.

. H is the matrix of Jacobians for n channels and m parameters, computed every iteration.

. W is a diagonal matrix with the weights of each channel linked to the noise in the observations and computations.

. d is the stabilization matrix.

The parameters to be solved in a given step are transformed so as to be eigenfunctions of H'wH with eigenvalue λ_i. The stabilization matrix d was then taken as a constant diagonal matrix with a value of .001 $\lambda_{i,max}$. This damps coefficients of poorly determined eigenfunctions (small λ_i) but not these of well determined eigenfunctions (large λ_i).

The complete sequence, from surface parameters to ozone profile retrieval, may be repeated to improve the results. In the first pass, the temperature sounding channels were whosen so as to be independent of moisture profile. In the second pass, some channels sensitive to water vapor distribution are added to the temperature sounding set because a better estimate of water vapor has been obtained.

Among the choices that may be utilized are : the number and the thickness of the layers for which quantities will be retrieved, the frequencies and the number of channels used to solve for each parameter as well as the stabilization process used. We do not address those questions in this paper. More details are given in Susskind et al., 1992, elsewhere in this volume. The results presented are based upon the state-of-the art at GLA at the time we made this work. We had 14 layers for the temperature retrieval (1000, 850, 700, 600, 500, 400, 300, 200, 140, 90, 50, 20, 10, 3, 1 hPa), 8 for the water vapor (1000, 850, 700, 600, 500, 400, 300, 200, 1 hPa) and 4 for the ozone (1000, 150, 70, 30, 1 hPa). The Jacobians are computed on line by taking finite differences using trapezoidal perturbating functions. The width of the trapezoid in pressure corresponds to the layers for which the inversion is performed.

2. LMD APPROACH FOR THE INITIALIZATION IN THE GLA INVERSION ENVIRONMENT

In this study, all aspects of the GLA inversion scheme were used as above, but the first guess profiles were used for temperature and moisture soundings obtained by the LMD approach.

Only modules of the LMD approach involved in the present study will be described. A more general description of the LMD method 3I ("Improved Initialization Inversion") may be found in the open literature (Chedin et al., 1985). In the 3I approach, the initial guess is obtained through a pattern recognition type approach in the educated database : TIGR.

2.1 Computing the brightness temperatures for the TIGR database

The TIGR database (TOVS initial guess retrieval) groups together 1800 atmospheric situations described by their temperature, water vapor and ozone profiles reconstituted from thousands of radio-sounding observations, which have been interpolated and extrapolated from 1013 to 0.05 hPa in 40 pressure levels. The surface temperature is assumed to be equal to the temperature of the 1013 hPa level. The database is splitted into 5 air mass classes corresponding to atmospheric characteristics from tropical to polar . We computed once and for all and stored the atmospheric contribution to the outgoing radiance as well as the total transmittance for each AIRS channels (about 3500, from 600 cm^{-1} to 2700 cm^{-1} with a resolution of 1200) and each TIGR-atmosphere with the rapid forward algorithm available at GLA. We could have computed directly the brightness temperature from the database, but to be consistent with the GLA approach, we decided to use the surface temperature coming from the first step of the retrieval instead of the one provided by TIGR. Nevertheless, for the statistics needed for our approach (see next paragraph), we have computed brightness temperature, assuming that the surface temperature is equal to the temperature of the 1013 hPa level. These brightness temperatures have been used to compute the mean brightness temperature and the standard deviation for each AIRS channel in each of the 5 atmospheric classes.

2.2 The pattern recognition module for selecting the initial guess

We defined a pattern recognition process to select as the first guess the "closest situation", in terms of radiances, to the observed situation among the 1800 situations from the TIGR database. This approach is directly derived from the one used at LMD for the processing of the TOVS data (Chedin et al. 1985).

A set of parameters describing the observation is made of the brightness temperatures observed for a selected set of channels (see 2.3). In the future, other extra parameters describing the local properties of the observation: latitude zone, viewing angle, surface pressure, percentage of water in the terrain observed could be added. From these parameters a classification proceeds through the hierarchy organizing the TIGR dataset and finally reaches the actual atmospheric conditions.

The brightness temperature for the initial guess, is computed using the retrieved surface properties and the precomputed total transmittance and atmospheric contribution to the radiance for each particular situation in the database. The idea of proximity of one set of brightness temperatures to another is based upon the normalized least-squares criterion (2) applied to the set of channels selected. The least square criterion is applied two times, once to select the appropriate air mass type, the other to isolate the actual atmospheric situation.

$$D_i = 1/n \ \Sigma \left(T^b_{obs}(k) - T^b_T(k) \right)^2 / S_k \qquad i = 1,m \quad \text{or} \quad i = 1,5 \qquad (2)$$

In the expression of this criterion, n is the number of channels which have been selected, the $T^b(k)$'s are brightness temperatures: T^b_{obs} are the observed one, $T^b_T(k)$ are either the value for the situation i among the m situations in the air-mass class (second pass, i=1,m) or the corresponding mean value of $T^b_T(k)$ over the sample of m situations (first pass, i=1,5). S_k is the variance of $T^b_T(k)$ for the channel k over the sample of m situations, that is to say over the atmospheric classes.

Once the air-mass class and the actual atmospheric situation found, the selection criterion is relaxed in order to avoid selecting one particular situation that may eventually bear signatures of too local phenomena. We look for all the element whose distance D to the observations is such that :

$$D < D_{min}(1+e) = d_R \qquad (3)$$

where D_{min} is the distance to the closest element and "e" has been set equal to 0.25 for this study. The initial guess solution is obtained by averaging the elements within the "distance circle" of radius d_R.

2.3 Initial guess and channel selection

With the present generation of vertical sounders, the problem of the channel selection is relatively easy to solve since no more than 24 channels are used (TOVS with HIRS2 and MSU). With the high spectral resolution sounders, the number of channels at our disposal is increased by two orders of magnitude. The present state-of-the art of our spectroscopic knowledge will first limit the usefullness of some of them, but the sensitivity to the geophysical parameters to be retrieved, as well as the vertical resolving power of these channels should be addressed, to optimize (quality of the results, computational time ...) the use of high spectral resolution data.

Geophysical parameters sensitivity and initial guess

Because the GLA retrieval approach is sequential, the location of the pattern recognition module in the inversion process (see Fig. 1) is significant. It should be after the ground temperature retrieval step because knowledge of ground temperature helps the classification method. We investigated several possibilities regarding the respective weights to be given to temperature and water vapor sounding channels in the selection criterion to provide a guess. With this intention, different experiments were done :

- The first one consisted in using the methodology described in paragraph 2.2 to find an initial temperature profile guess as close as possible to the true temperature profile to be used in the first step in the inversion after the retrieval of surface parameters. This procedure took as initial guess for the water vapor the corresponding water vapor profile associated with this temperature profile in the TIGR data base.
- For the second one, water vapor sensitivity was added in the selection of the initial guess, in order to try to take greater benefit of the TIGR initial guess for the water vapor retrieval.
- Finally, we also explored the possibility of introducing the pattern recognition module after the temperature retrieval to select a better initial guess in terms of water vapor, taking into account the benefit of the previous temperature retrieval.

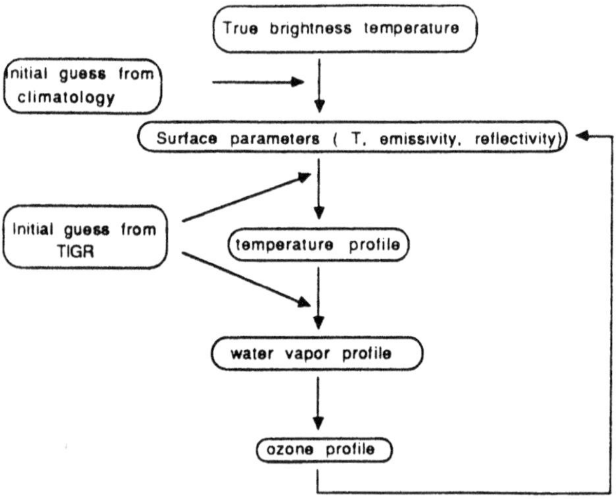

Fig. 1 Scheme for sequentially retrieving surface parameters (surface temperature, reflectivity and emissivity), temperature, water vapor and ozone profiles. In each box the retrieval is iteratively done. On the left part of the diagram, we indicated different initialization modules (climatology or TIGR through pattern recognition approach). The arrow on the right part indicates that the whole process may be iterated to improve the results, the initial guess being then the results coming from the previous retrieval.

Channel selection for the pattern recogniton process

i) Selecting channels principally sensitive to the temperature profile

We present here an attempt to select a set of channels principally sensitive to the temperature profile, and well representing all the atmospheric column. Those channels will be used for the pattern recognition. In addition, they could be involved in the direct inversion process, but were not in the current test.

Channels were chosen according to the following method: for each atmosphere in each air-mass class and for each channel we determine the pressure level where the transmittance reaches 0.37, which roughly represents the maximum of the corresponding channel weighting function. Looking for channels mostly sensitive to the temperature, we eliminated channels too sensitive to water vapor and ozone (i.e. channels for which the water vapor transmittance - respectively the ozone transmittance- is lower than 0.8 - respectively 0.7 -). Since the initial guess we are looking for is used for the retrieval of atmospheric parameters, we have partially eliminated those channels which could be too contaminated by the surface. Nevertheless, to correctly represent the bottom layers of the tropical atmospheres, and because of the high variability of the water vapor content close to the surface for this kind of atmospheres, we have declared acceptable channels whose transmission to the surface could be as high as 0.4 (≤). The number of temperature sounding channels selected for use in pattern recognition is shown in Table 1 for the 5 air mass classes. Since the use of this pattern recognition module is done after the surface parameters retrieval, the above constraint does not affect the quality of the results. Even if this process has to be refined in the future, it provides a set of channels correctly representing the whole atmospheric column. This is shown by looking at the histogram displaying the number of channels selected for which the transmittance is equal to 0.37 as a function of the pressure level (Fig. 2). When compared to the set of channels selected for the temperature retrieval at GLA it appears that most pressure level gaps have been filled. (Fig. 3).

air mass class	(650-790cm^{-1}) 15 μm channels	(2050-2400cm^{-1}) 4 μm channels
Tropical	162	117
Mid.lat.1	149	115
Mid.lat 2	138	113
Polar 1	133	111
Polar 2	134	111

Table 1 Number of channels selected for the pattern recognition for the selection of the initial guess temperature profile in the 15 and 4 micrometers bands for each class of the TIGR data base.

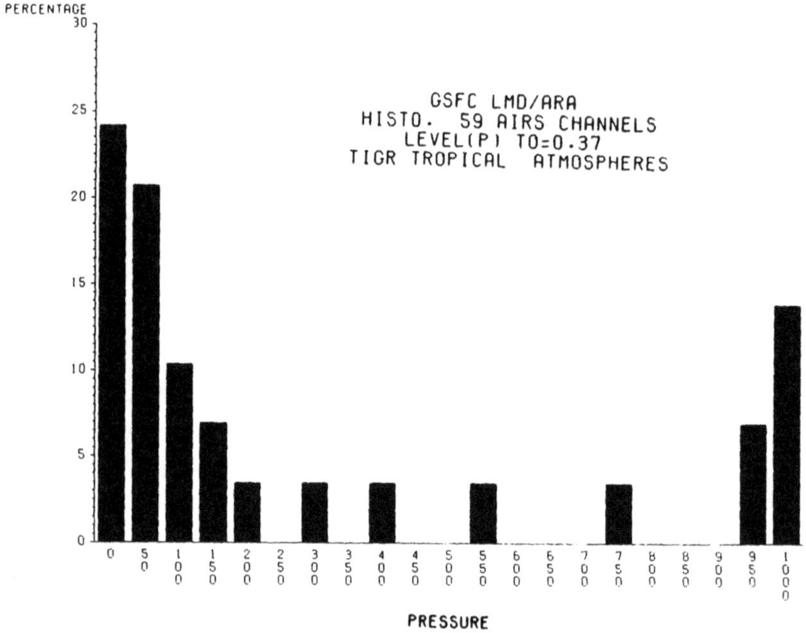

Fig. 2 Histogram displaying the number of channels selected for which the transmittance is equal to 0.37 as a function of the pressure level for the TIGR tropical air mass.The 59 channels are selected in the 4 μm and 15 μm bands for temperature profile retrieval in the GLA Scheme.

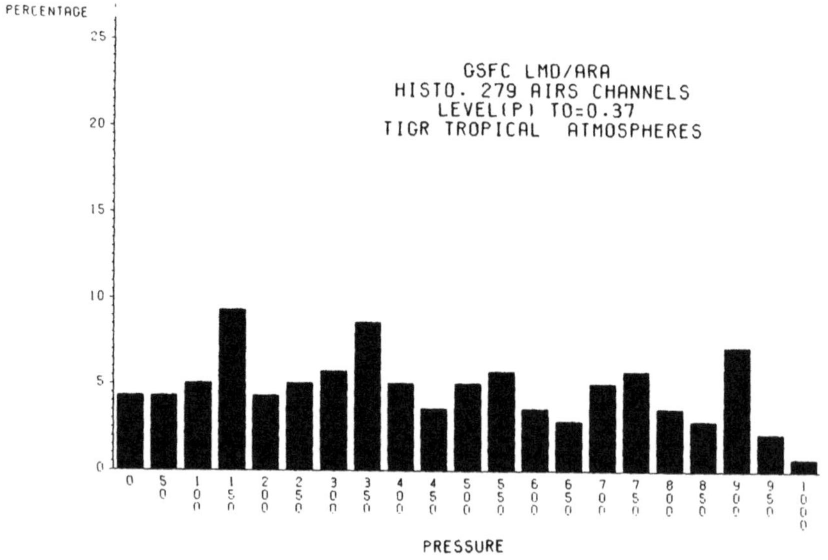

Fig. 3 Same as Fig. 2, but for the channels used for the temperature retrieval.

ii) Channels sensitive to water vapor and surface temperature

In this approach, we include moisture sounding channels in addition to the above temperature sounding channels to select the temperature and water initial guess. An analogous methodology (i) has not been yet applied to the selection of the water vapor sensitive channels for use in pattern recognition. At this time, we used the channels selected at GLA for the water vapor retrieval. These channels were selected in the 6.3 μm water vapor branch which are sensitive to humidity throughout the atmosphere. Also channels in the 11 μm band, sensitive to the water vapor continuum were used, in order to improve the sensitivity to the lower tropospheric humidity (Susskind, 1992, elsewere in this volume).

iii) Channels sensitive only to water vapor

In the third experiment, we tried to produce a new moisture firstguess after the temperature profile has been retrieved. This used only the moisture sounding channels mentioned above, but not the temperature sounding channels.

3. IMPACT OF THE FIRST GUESS ON THE QUALITY OF THE RETRIEVAL

To test our approach, we used a restricted set of 28 temperature, water vapor and ozone profiles from the so-called Phillips dataset constructed from radiosonde reports collected during five consecutive weeks in winter and in summer time at tropical and mid-latitude sites (Phillips et al., 1988). This selection is unambiguously independent from the TIGR data. AIRS observations were simulated for this dataset with the same GLA rapid algorithm that is used in the simulation in this study and was used to simulate the TIGR data base radiances.

The impact of first guess on temperature retrievals is shown in Fig. 4. All statistics represent layer mean temperature in 1 km layer for 1000-200 mb, 2 km for 200-100 mb, and 4 km above 100 mb. The guess coming from the TIGR dataset is significantly closer to the real profile than the guess coming from the climatology. Some degradation in the initial guess appears, due to the addition of water vapor sensitive channels on the pattern recognition (second experiment), but the quality of the retrieval is not significantly affected. However, we have not been able to observe any significant improvement in the retrieved profiles. A preliminary explanation could be that the high spectral resolution data should provide enough information and make the retrieval process powerful enough to correct the initial guess even when far from the solution. Nevertheless we do not consider this assumption as definitive and will look for more detailed explanation.

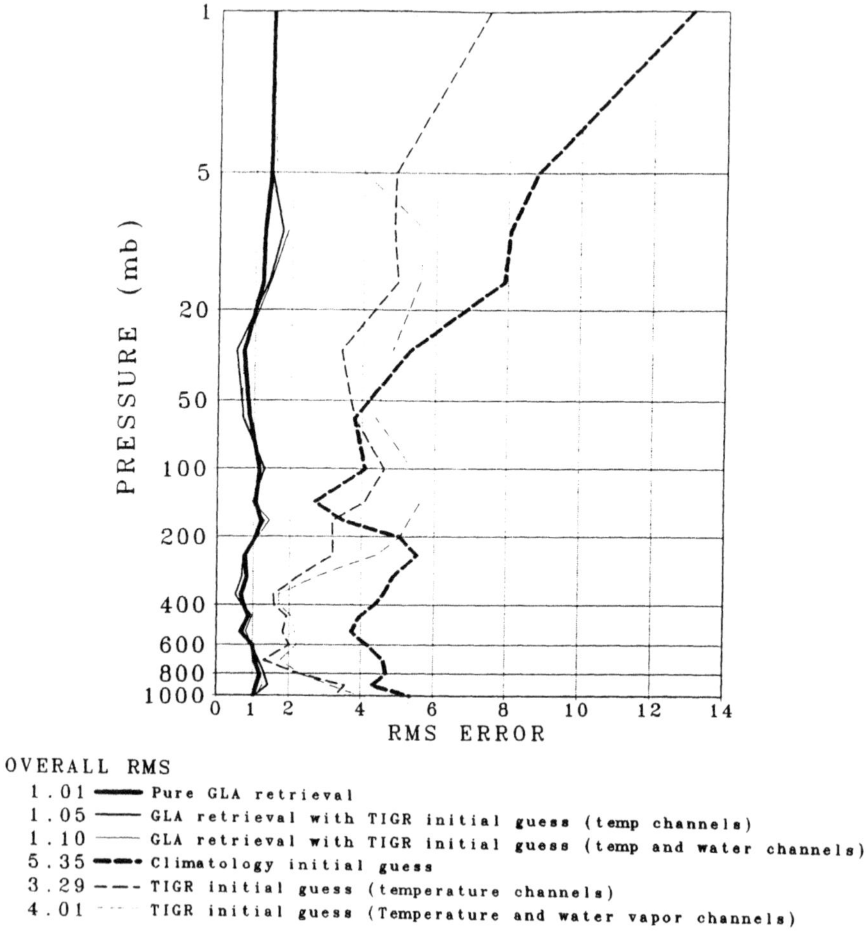

Fig. 4 Layer mean rms temperature errors for the initial guess for the 28 Phillips profiles. Layers are 1 km thick up to 200 mb, 2 km thick from 200 mb to 100 mb, and 4 km thick above 100 mb.

As already mentioned, we investigated three different choices for the initial guess, two being more particularly devoted to water vapor retrieval. The first attempt has been done by taking an initial guess selected with temperature only selection criterion (I-3-bi). That means that we do not pay attention to the quality of the first guess, in terms of water vapor profile. The second exercise uses water vapor channels in the pattern recognition in addition to the temperature channels already used. The last exercice introduces the pattern recognition module after temperature retrieval, using only water vapor channels.

In this case, we have to take into account the retrieved temperature profile which is not one of the 1800 from TIGR. This requires to recompute the brightness temperature using this retrieved temperature profile, or at least to correct the results in the TIGR database with Jacobians. Doing that, we lose part of the benefit of the precomputations, done in principle once and for all, in the TIGR dataset. Fig. 5 shows results for percent error in water profile. In this figure, errors to the total precipitable water are shown at 1000 mb and errors in precipitable water layers in 2 km are shown at lower pressure. In all cases, but especially when water vapor channels are used, the first guess is better than the one coming from the climatology.

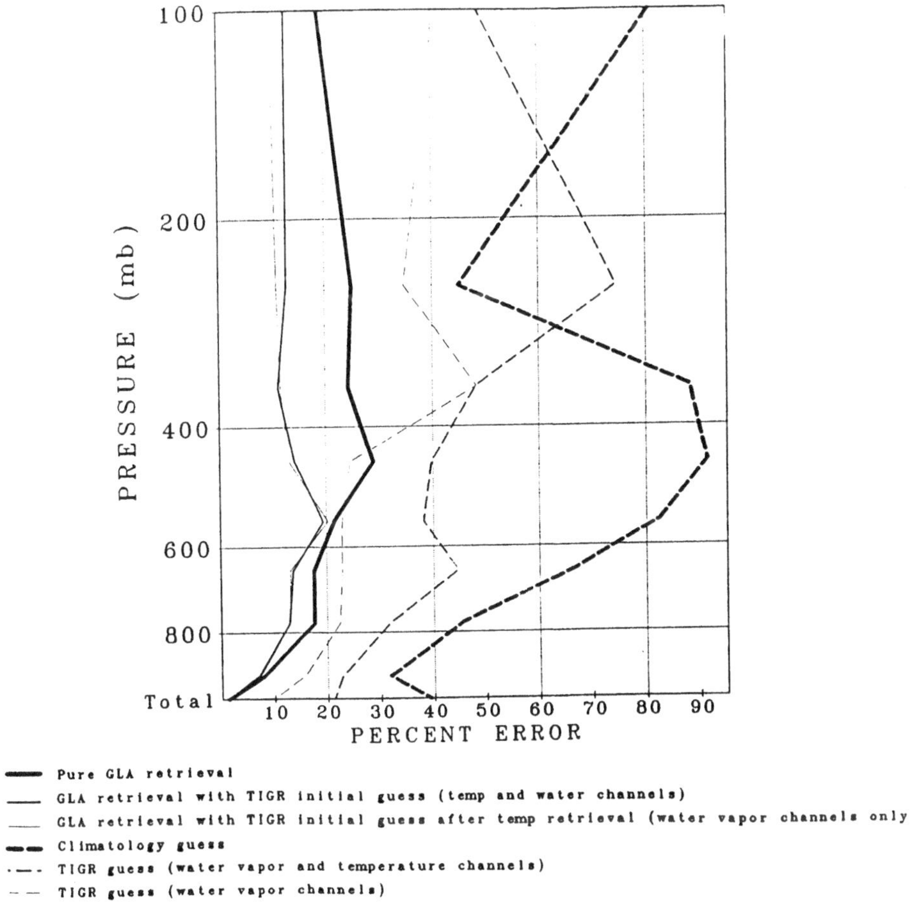

Fig. 5 2 km layer mean precipitable water vapor percent errors for 28 profiles. Total precipitable water errors are shown up to 1000 mb.

It induces a clear improvement in the retrieved water vapor profile. Introducing a special pattern recognition for the water vapor profile seems to not substantially improve the results when compared to those taking water vapor channels in addition the temperature channels in the first pattern recognition.

4. CONCLUSION AND FURTHER DEVELOPMENTS

We tested the role of the initialization on the retrieval of atmospheric temperature and water vapor profiles from AIRS data.

According to the limited test we conducted, we conclude that :
- the pattern recognition approach allows selection of a better initial guess than the direct selection from a climatology ;
- the impact on the quality of the temperature retrieval in the GLA environment is not obvious ;
- an improved first guess in terms of water vapor profile greatly improves the sounding capabilities of the high spectral resolution infrared sounder.

Concerning ozone retrieval, we may expect a behavior comparable to the one observed for the water vapor retrieval. The problem for this variable is to obtain a sufficiently consistent dataset based upon accurate measurements, to help obtain the initial guess. Such a dataset is not presently at our disposal.

These results are encouraging but the study has to be extended to a more complete test dataset, that could be the entire TIGR dataset. The role of the stabilization process will then be studied, paying particular attention to the number of iterations needed. The information content of AIRS like data, in conjunction with the thickness of pressure layers fixed in that inversion, has to be studied too. The case of cloudy situations also has to be considered. Since the TIGR initialization could come after a cloud clearing process such as that used by GLA (Susskind et al., 1992), the results should be correct in those cases as soon as the cloud clearing module is well defined and efficient. Work will be continuing at GLA and at LMD to give more complete answers to these questions.

REFERENCES

Chédin A, Scott NA, Wahiche C, Moulinier P (1985) The Improved Initialisation Inversion method: a high resolution physical method for temperature retrievals from the TIROS-N series J. Clim. Appl. Meteor., 24, 128-143

Phillips N, Susskind J, Mc Millin L (1988) Results of a joint NOAA/NASA sounder simulation study. J Atmos Ocean Tech 5: 44-56

Susskind J, Joiner J, Chahine MT (1992) Determination of temperature and moisture profiles in a cloudy atmosphere using AIRS/AMSU. This issue.

THE IMPACT OF THE INITIAL GUESS ON THE ACCURACY OF SATELLITE RETRIEVALS

Larry M. McMillin
Michael J. Uddstrom
National Oceanic and Atmospheric Administration
National Environmental Satellite, Data, and Information Service
Satellite research Laboratory
Washington DC 20233
Note M Uddstrom is an NRC/NOAA Senior Associate on leave from the New Zealand Meteorological Service

1. Introduction

High resolution sensors such as the Atmospheric Infrared Sounder (AIRS) will provide 100 times more measurements of a single profile than is being received by the current TIROS Operational Vertical Sounder (TOVS). One of the reasons such instruments are being considered is that numerical models used for weather prediction continue to improve, and as they improve they require increasingly more accurate observations. For example, the current TOVS series of satellites have been flying since 1978, and since that time the accuracy of the model used by one forecast center, the National Oceanic and Atmospheric Administration's (NOAA) National Meteorological Center (NMC), has increased by over 0.4 K at 500 hPa. The forecast model run by at least one center, the European Centre for Medium-Range Weather Forecasts (ECMWF), has reached the point where the positive impact TOVS soundings used to have on forecasts has disappeared in the northern hemisphere. TOVS soundings are still required to stabilize models in the data sparse southern hemisphere and new approaches are being developed to make better use of TOVS information. Still, the limitations of the current instruments are becoming a limitation on the use of the data. Although a more accurate instrument with higher vertical resolution will improve the accuracy, it should be recognized that other factors will have to be considered if the potential of these new instruments is to realized. For example, simulations with the current TOVS instruments show a 500 hPa accuracy of slightly less

NATO ASI Series, Vol. I 9
High Spectral Resolution Infrared Remote Sensing for Earth's Weather and Climate Studies
Edited by A. Chedin, M.T. Chahine and N.A. Scott

1.0 K. Yet the best accuracy obtained with measured radiances is 1.5 K due to errors in calibration, cloud clearing, the forward model, and the radiosondes which are used both for tuning and as a definition of truth. Unless sufficient effort is devoted to solving these problems, there will be a time when they dominate the retrieval error and limit additional improvements in instrument accuracy. The operational processing of high resolution data will require an approach that is both fast and accurate. A recent study of different retrieval approaches using different guesses provides some useful insight to a retrieval algorithm that will satisfy the needs for speed and accuracy. The retrieval algorithm is independent of a numerical forecast, a feature that is useful for some applications and one that enables retrievals to be made when a model forecast is not available.

Because of the spectral width of the interference filters and the limited number of channels on the TOVS, the vertical resolution of the sounding instruments is limited (Smith et al. 1979). To compensate for the thickness of the atmospheric layers represented by the measured radiances, retrieval procedures start with an initial guess of the atmospheric profile, then modify the initial temperature profile, moisture profile, or other parameter according to the differences between the radiance temperatures of the measurement and the guess. It is obvious that the final result of such an approach depends on both the guess and the measurements. More specifically, retrievals usually reflect the vertical fine structure in the guess but modify the large scale structure to force agreement with the measurements. This naturally raises questions about the interaction between the retrieval method and the initial guess.

In this paper the relationship between the guess and the measurement is examined using a fixed set of data. First the data sets are described; then the initial guess procedures are defined; the retrieval approaches are described; the combinations of available data are described; the results are presented; and finally some conclusions are drawn.

2. Data description

This study is based on a set of winter profiles collected for the months of October through April of 1989-1991. The profiles come from the collocations used to generate coefficients for the NESDIS operational TOVS processing (Dey et al. 1989, Reale et al. 1990) and consist of radiosonde measurements matched with the closest TOVS observation. The TOVS observations include the operational retrieval in addition to the cloud cleared measurement. The entire data set consists of about 30,000 soundings. However, satellite soundings are more useful over data sparse regions such as the oceans, and soundings from cloudy measurments are less accurate than those from clear ones. Accordingly, this study was limited to clear and N* soundings over ocean areas. This reduced the sample to between 3000-4000 profiles. A random number generator was used to select 10% of these profiles for an independent data set to be used for evaluation. The other 90% were used to generate coefficients. Although attempts were made to keep the data sets consistent, it was not always possible. Specifically, one of the initial guesses was provided by the operational procedures used at the time the data were collected. This guess was not included in the file until June 19, 1990. In addition, when it was included, only the channels used in the current operational retrievals were included. This consists of TOVS channels 2-7, and 13-15, as well as MSU channels 2-4 and SSU channels 1-3.

3. Initial Guesses

In this study, three major types of initial guesses were used. Although variations of the guesses expanded the number, only one is important enough to be included. The three guesses are the library search (Goldberg et al. 1988) that is used as the guess in the operational processing at NESDIS, the Typical Shape

Function approach used by (Uddstrom 1988), and the eigenvector classification used by McMillin (1991). The library search keeps a library of recent (1 - 2 months) soundings in a library of up to 1500 soundings. In a recent version, the library is divided into 5 latitude bands (60-90N, 30-60N, 30S-30N, 60S-30S, and 90S-60S), each containing up to 300 soundings. When a retrieval is made, the latitude band containing the measurement and any adjacent bands are scanned to find the 20 closest soundings using the procedure described by Dey et al. (1989) and Goldberg et al. (1988) to obtain the relationship

$$d(c) = [b - b(c)]^T B^{-1} [b - b(c)]. \tag{1}$$

In (1), d(c) is the distance for class c, b is the radiance, b(c) is the mean radiance for class c, $()^T$ denotes the transpose, B denotes the radiance covariance matrix, and $()^{-1}$ denotes the inverse of a matrix. These 20 are then averaged to provide the initial guess.

In the TSF approach, a set of soundings is classified by temperature (Uddstrom 1988). At some point, the radiative transfer equation

$$I = B_s \tau_s + \int_{\tau_s}^{1} B \, d\tau \tag{2}$$

(where I is the calculated radiance, B_s is the radiation emitted by the surface, τ_s is the transmittance of the atmosphere between the satellite and the surface, B is radiation emitted by the atmospheric layer $d\tau$, and $d\tau$ is the layer) is used to generate profiles of calculated radiances. Once the profiles have been assigned to a class, the mean profiles of radiance temperature, temperature and moisture are generated. In addition, a relationship based on a constrained regression relationship (Crone et al. 1989) is derived to make the measured radiances agree with calculated values. These steps are done once and are not part of the retrieval steps for a specific profile. Once a measurement is obtained, the radiance temperatures are adjusted to minimize any inconsistencies between measured and calculated radiance temperatures. Next, the measured radiances are compared to the radiances for the class means. Then the Bayesian probability that the radiance vector belongs to each class is generated. Finally, the class with the highest probability is selected, and the mean quantities for the class become the guess.

In the Eigenvector Classification (EC), eigenvectors are generated for the dependent sample. Then the principal component scores are generated for the dependent sample, and the limits of the scores for the four eigenvectors with the largest eigenvalues are determined. The resulting ranges are divided into 14 intervals for the first eigenvector, 4 for the second, and 3 each for the third and fourth. This results in 224 possible classes. However, only about half of the possible classes are populated. When a measurement is obtained, the four principal component scores are generated and used to select the class. An important part of this procedure is the use of a profile mean in the classification. It has been found, and verified in this paper, that classification works best if it is used to determine the shape of the profile as suggested by Uddstrom and Wark (1985) and Wark (1986). This is done by determining, for each profile, a mean tropospheric temperature and subtracting it from the measured radiance temperatures used to select the classes. This has the effect of placing profiles that differ by a constant amount over all levels in the same class. When the mean temperature is subtracted, the resulting eigenvectors become more sensitive to the profile shape and the covariance matrices become more diagonally dominant. In practice, the mean tropospheric temperature is obtained by averaging selected channels. It should also be noted that the initial guess that is produced for the retrieval step does not have the mean subtracted; the subtraction of a profile mean temperature is used only for the class selection. Because there was evidence that one the retrieval algorithms may be more sensitive to the closeness of the initial guess than to the shape, classifications were also performed with this step removed. The Eigenvector Classification

(EC) is explained more fully in McMillin (1991).

The guesses can be summarized in the following manner.

1. library search (Dey et al. 1989)
2. Typical Shape Function (TSF) (Uddstrom 1988)
3. eigenvector classification on shape (McMillin 1991)
4. eigenvector classification on distance (McMillin 1991)

4. Retrievals

Three retrievals are included in the comparison. These are the operational MVS (Dey et al. 1989), the Maximum A Priori MAP (Uddstrom 1988), and the stepwise regression (McMillin 1991).

The operational MVS retrieval (Dey et al. 1989) differs from many physical retrievals in that it uses an empirical first guess as explained in Fleming et al. (1986). That is, the first guess comes from a collection of collocations. Because of this feature, it avoids the use of calculated radiances and the need to force calculated and measured radiances to agree. At the same time, it is physical only in the sense that the operator is derived from physics. It differs from many physical retrievals that compare measured and calculated radiances after the retrieval to assure that the radiative transfer equation is satisfied. To keep the meaning of the word "physical" clear, three definitions of physical retrievals will be used for the remainder of the paper. A retrieval that is physical will be referenced as "physical O" if the operator is physical, physical G if the guess is physical, and physical RT if the radiative transfer equation is checked after the solution. A solution that employs all three will be referenced as a full physical solution.

The MAP retrieval (Uddstrom 1988) is a full physical retrieval that uses physics to determine both the initial guess and the retrieval operator. In addition, it is a sequential retrieval system. This results from the fact that a retrieval is done for one channel, the retrieval is used to calculate the guess radiance for next channel, the retrieval is done for the next channel, and the process is iterated until the measured and guess radiances converge. This differs from the usual procedure in which retrievals are made for all the channels before the radiative transfer equation is applied.

The Noise Guided Stepwise Regression NGSR (McMillin 1991) is a stepwise regression procedure in which an estimate of the noise is used to select the next channel in the stepwise procedure. This procedure tends to counteract the tendency of some regressions to produce large coefficients for upper atmospheric channels when predicting the surface temperature or vice versa. It is also unique in that the regression coefficients are generated from one covariance matrix. Covariances for all classes are averaged to form one large covariance matrix before it is inverted to produce the regression coefficients. This solves the instability problems that can result when the sample for a particular class is too small.

The retrievals can be summarized in the following manner.

1.	MVS	Dey et al. (1989)	physical O
2.	MAP	Uddstrom (1988)	full physical
3.	NGSR	McMillin (1991)	empirical

Results are not available for all possible combinations. Those combinations for which data are available are shown in Table 1.

Table 1. Combinations of guesses and retrievals for which data are available.

Retrieval	MVS	MAP	Regression
Guess			
1	x		x
2		x	x
3			x
4	x		x

5. Results

When comparing the results, it is important to remember that it was not possible to make everything equal. There are unavoidable differences in the sample sizes used for some of the comparisons. However, the operational result is available for all cases, and every test is matched profile to profile with the operation. Thus differences in accuracy due to sample differences are reflected in the change in the accuracy of the operational results that are shown for every comparison. When evaluating methods, the important variable is the improvement over the operational curve, not the absolute accuracy.

Fig. 1 compares the results from the TSF and the operational systems. That is guess 2 - retrieval 2 with guess 1 - retrieval 1. As can be seen, the TSF is a significant improvement over the operational method at most levels. In addition, the degree of improvement is a function of the probability. A part of the TSF classification is the calculation of the probability that a given profile belongs to the selected class. This has the capability to limit retrievals to those cases for which probability exceeds a given amount. To be fair, the other methods also have capabilities to change their selection procedures. Since these capabilities were not implemented for the other methods, and since that is not the topic of this paper, the evaluation of changes to the limits used in the profile selection procedures will be left to a later time. For this paper, the probability level that resulted in a sample size consistent with the other alternatives was selected in attempt to keep the comparisons as fair as possible. The data shown are for a probability of 10% or greater and produce a sample size of 347.

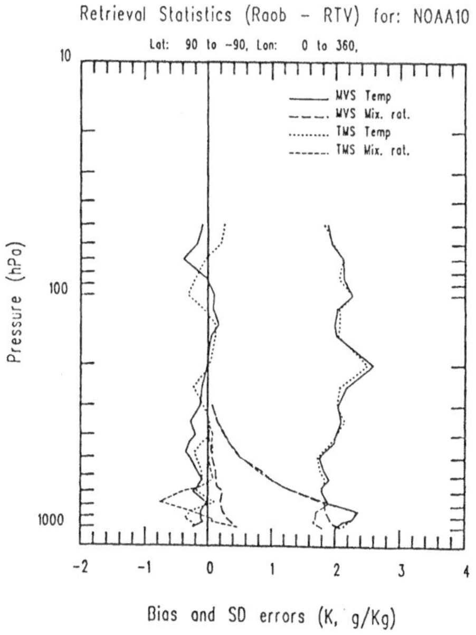

Fig. 1. Comparison of TSF guess - MAP retrieval and the operational methods for a TSF probability limit of 0.1. Sample size is 347 (71%).

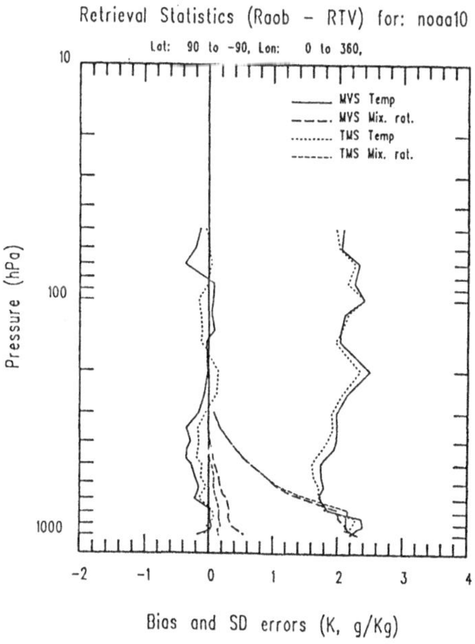

Fig. 2. Comparison of TSF guess - NGSR retrieval methods for all profiles. Sample size is 487 (100%).

Fig. 2 compares the results of the TSF guess and the regression retrieval with the operational system. That is guess 2 - retrieval 3 with guess 1 - retrieval 1. The sample size is 487, the full independent sample, and the improvement in the accuracy shown in Fig. 2 is slightly greater than that shown in Fig. 1. Fig. 3 compares the results of the eigenvector classification - regression retrieval with the operational system (guess 3 - retrieval 3 with guess 1 - retrieval 1). The sample size is 416 and the accuracy of both the temperature and water vapor are improved over most of the profile, particularly near the surface. This method produced the largest increase in accuracy over the operational procedures. The sample size is reduced because a retrieval is attempted only for classes that have a sample size of 20 or more in the dependent set.

At this point it is appropriate to discuss another advantage of the classification approaches. The operational results shown in Fig 3 shows a slight reduction in sample size over Fig. 2. However, the reduction is accompanied by a significant reduction in the operational retrieval error shown in Fig. 3. In fact, at 500 hPa, the increase in the accuracy of the operational results in Fig. 3 over the operational results shown min Fig. 2 is the same magnitude as the difference between the two curves shown in Fig. 3. The ability of the classification approaches to select the more accurate retrievals could be an important improvement to the current system. The selection used by the TSF in Fig. 1 makes a similar reduction in the retrieval error. However, it is more pronounced in the upper atmosphere while the EC method improves the accuracy in the lower atmosphere.

Fig 4. shows a comparison of the operational guess - regression retrieval with the operational system (guess 1 - retrieval 3 with guess 1 - retrieval 1). In this comparison, the MVS and regression operators are compared when they are given the same initial guess. The sample size is 250 and the regression shows a slight increase in accuracy. As mentioned earlier, the sample is reduced only because the operational guess is only available after June 19, 1990. The regression is slightly different from the others for a similar reason. The operational program uses only HIRS channels 2-7 and 13-16, along with MSU 2-4 and SSU 1-3 for the retrieval. Guess radiances are provided only for these channels so only these channels were available for the regression. The other regressions use HIRS channels 1-17 and MSU 2-4. An independent evaluation of the effect of the additional channels indicated a decrease in retrieval error of about 0.2 K near the surface. At 500 hPa, the effect was negligible.

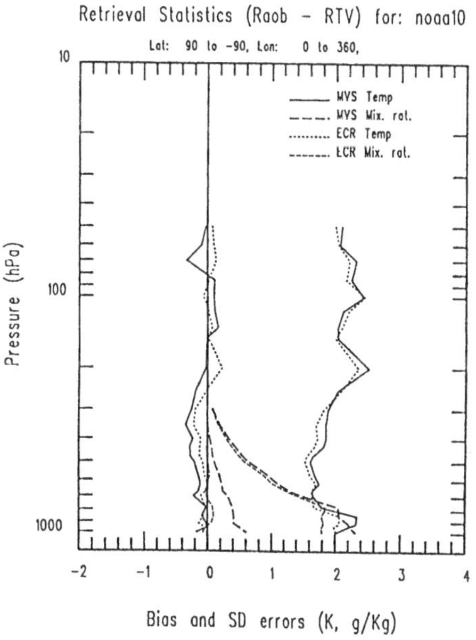

Fig. 3. Comparison of eigenvector classification - regression retrieval with the operational methods for class sizes greater than 20. Sample size is 416 (85%).

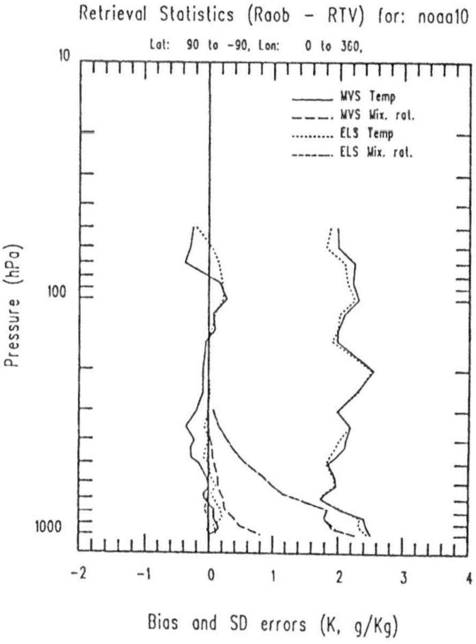

Fig. 4. Comparison of operational guess - regression retrieval with the operational methods for all profiles after June 19, 1990. Sample size is 250 (100% of possible). Only HIRS channels 2-7 & 13 -16 were used in the retrieval.

Fig. 5. compares the results for an eigenvector classification based on closeness rather than shape with the operational procedure (guess 4 - retrieval 3 with guess 1 -retrieval 1). It shows a sample size of 458 and an increase in retrieval accuracy that is generally similar to that shown in Fig. 3. However, there are slight differences, the main ones being that Fig. 5 shows a more accurate temperature near 100 Hpa., a less accurate temperature at the surface, and a significantly less accurate water vapor retrieval.

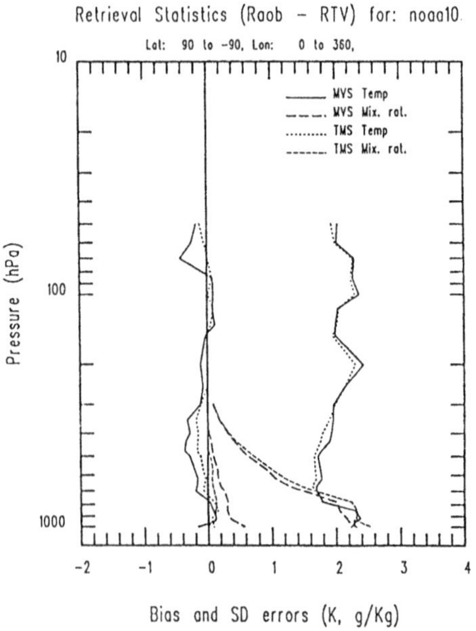

Fig. 5. Comparison of eigenvector classification based on closeness rather than shape and eigenvector regression with the operational methods. Sample size is 458 (94%).

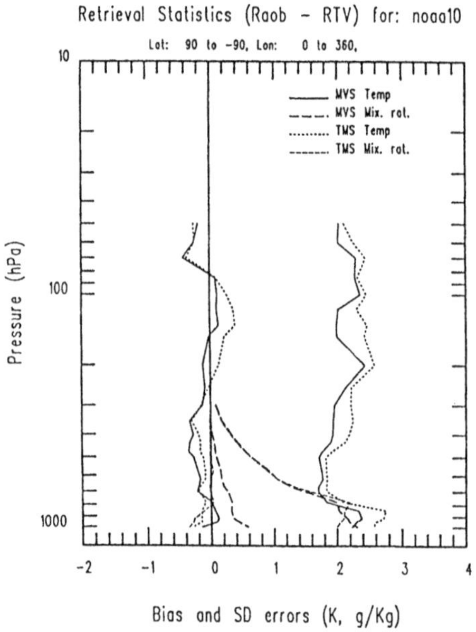

Fig. 6 Comparison of eigenvector classification based on shape and the MVS retrieval with the operational methods. Sample size is 460 (94%).

The remaining combinations are the classification guesses combined with the MVS retrieval. All produced results that are significantly worse than those produced by the current operation. Fig. 6 shows the result of the eigenvector classification based on shape with the MVS retrieval operator. The sample size is 460 and the retrieval accuracy is worse than that of the current operation at all levels for temperature and all but the surface for water vapor. This combination was shown because it is the best of the guesses that were combined with this retrieval operator. The others were less accurate.

6. Summary and Conclusions

The results of this study demonstrate several factors. The first is that the first guess is an important factor in the retrieval results. In addition, it is more important to have the correct shape in the guess than it is to have the guess radiances be close to the measured ones. This makes sense once it is realized that the guess is the only place for the retrieval to obtain the correct fine structure. Differences between the coarse structure of the guess and the true atmosphere are not relevant because retrieval operators can make very accurate adjustments to the coarse structure. Thus the selection of a good initial guess is more complex than just minimizing the radiance differences.

The second is that retrieval operators differ widely in their ability to use shape information. The MVS operator seems very sensitive to the closeness of the guess, while the other two operators can take a guess with the correct fine structure and make the corrections indicated by the radiance patterns. Both produce final results that are more accurate than the best that were produced by any combination involving the MVS. However, without further study, it should be remembered that this result applies only to the versions used in this study. It may or may not be characteristic of all MVS retrievals. Given a choice between the two physical retrieval operators, the MAP and the MVS, it is clear that the MAP provides the more accurate results.

The regression retrieval produced the most accurate results. This is surprising when one realizes the nature of the regression. One set of coefficients was applied for all latitudes in this study and all latitudes and seasons in McMillin (1991). In addition, McMillin (1991) demonstrated an important difference between this method and other regressions. Most previous implementations of a regression operator required constant monitoring and updating of retrieval coefficients. McMillin demonstrated that coefficients could be used for at least a year before any degradation in accuracy was noticed.

In the direct comparison between the MAP and regression retrievals, the regression produced a slightly more accurate result. This does not mean that a physical approach should not be used. For example, a comparison of the radiances calculated from the regression solution with the measured radiances to assure agreement with the radiative transfer equation would be a valuable addition, and agreement between measured and calculated radiances, not the method by which it was achieved, is the essence of a physical approach. If this step were implemented and retrievals that did not satisfy the relationship were discarded, the regression would become a physical RT retrieval. In addition, a full physical retrieval such as the MAP is more flexible in that it can account for features such as surface elevation in a more direct way. Certainly, efforts to improve the accuracy of the physical approaches should be continued because of the potential of the physical approaches to eventually become the more accurate alternative. In the meantime, the current accuracy advantages of the regression should be utilized.

One way of utilizing the advantages of the existing approaches is to use the regression retrieval as the first guess for a physical approach. This can be done with a modification to the physical approach. Many physical approaches are based on comparisons between measured and calculated radiances with all the problems associated with forcing agreement when the atmospheres are identical. The alternative that is currently being used at NESDIS is to use a measured radiance from collocations (Dey et al 1989, Fleming et al. 1986) for the guess. In this case measured radiances are compared with measured values and no radiance adjustment is required. However, this also precludes the ability to iterate. A logical solution is to use the measured radiances for the guess, but to also determine the calculated radiances for the initial guess. Then as the solution is iterated, use the radiative transfer equation to determine only the change in the guess radiance, not the entire quantity. In the case of the ECR retrieval, the first guess for the physical retrieval that follows would be the mean of the measured radiances for the class plus the difference between the radiance calculated for the retrieval after the regression minus the radiance calculated for the guess before the retrieval. As an equation this approach takes the form of

$$T^*(i) = T^*(\text{class},m) + T^*(i-1,c) - T^*(\text{class},c) \qquad (3)$$

where $T^*(i)$ is the guess radiance for iteration ,i, $T^*(\text{class})$ is the average of the measured radiances for the class, $T^*(i-1,c)$ is the radiance calculated by Eq. 2 for the retrieval from the (i - 1) iteration, and $T^*(\text{class},c)$ is the average of the calculated radiances for the class. If the regression does not produce a solution that satisfies the radiative transfer equation, then a physical retrieval can be applied.

There is currently an active discussion about the merits of an interactive retrieval system versus one that is independent of a forecast. Although the interactive approach is likely to be used in many centers, it is important to note that the increased accuracy of the ECR approach brings the accuracy to the levels being achieved by numerical models at most levels. This is regarded as the value required for a positive impact on the models. Since there has been no impact test for classification approach, its impact on a numerical model is not known and won't be, unless it is.

It is also important to note that both the regression and MAP retrievals have demonstrated that they can produce retrievals with an accuracy of better than 1.0 K given simulated data with the noise level of the instrument. However, when given the error in that is characteristic of the agreement between calculated and measured radiances, the simulated retrieval accuracy approaches observed levels (Uddstrom and McMillin 1992). This means that the accuracy is being limited by other factors such as a known calibration problem on the HIRS, errors produced by the current angle correction procedures, radiosonde errors, location and time differences in the collocations, and errors in the radiative transfer models. These results demonstrate that little additional improvement in retrieval accuracy can be expected until these errors are reduced. For example, a retrieval error of 1.0 K for the current instrument, and an error level of 1.5 K with these noise factors, means that the best that can be achieved by a perfect instrument is the square root of (1.5 squared minus 1.0 squared) which gives 1.1 K. While there is time before the new instruments are ready to reduce these errors given appropriate resources, there is a real danger that these errors will negate the expected benefits of the new instruments if these problems are ignored.

Although classification approaches have not been applied to a high resolution instrument, they are ideal for that use. The ECR method is extremely fast since the address of the class is calculated directly from the principal component scores of the eigenvectors and it was the most accurate method tested, producing retrieval accuracies that are equivalent to accuracies produced by current forecast models. In addition, high resolution data should contain information about surface inversions and fronts that can be recovered in the classification step once the number of classes is modified to reflect the additional information. The TMS method is entirely physical and it is also easy to make the ECR method more physical by adding a check of the radiance agreement after the retrieval. If the ECR method is used and the radiances don't agree, there is the option of doing a physical retrieval. This last consideration is an important one because a retrieval with the large number of channels is likely to be time consuming. If this step is limited to a small number of profiles, then more time can be spent retrieving these difficult but possibly important profiles.

Acknowledgments Dr. Uddstrom was supported under the Resident Research Associateship Program of the National Research Council of the National Academy of Sciences.

References

Crone LJ, Crosby DS, and McMillin LM (1989) A General Method of Using Prior Information in a Simultaneous Equation System. NOAA Technical Report NESDIS 47, U.S. Department of Commerce, National Technical Information Service, Sills Bldg., 5285 Port Royal Road, Springfield, VA 22161. p17

Dey CH, Peterson RA, Ballish BA, Caplan PM, Morone LL, Thiebaux HJ, Whiet GH, Fleming HF, Reale A, Gray DG, Goldberg MD, and Daniels JM (1989) An Evaluation of NESDIS TOVS Physical Retrievals using Data Impact Studies. NOAA Technical Report NWS NMC 69. U.S. Department of Commerce, National Technical Information Service, Sills Bldg., 5285 Port Royal Road, Springfield, VA 22161. p25

Goldberg MD, Daniels JM, and Fleming HE (1988) A Method for Obtaining an Improved Initial Approximation for the Temperature/Moisture Retrieval Problem. Third Conference on Satellite Meteorology and Oceanography, Anaheim, Amer Meteor Soc 20-23

Fleming HE, Crosby DS, and Neuendorffer AC (1986) Correction of Satellite Temperature Errors Due to Errors in Atmospheric transmittances. J Climate Appl Meteor, 25: 869-882

McMillin LM (1991) Evaluation of a Classification Method for Retrieving Atmospheric Temperatures from Satellite Measurements. 30,4: 432-446

Reale AL, Fleming HE, Wark DQ, Novak CS, Zbar FS, Neilon JR, Gelman ME, Bloom HJ (1990) Baseline Upper Air Network (BUAN) Final Report, NOAA Technical Report NESDIS 52., U.S. Department of Commerce, National Technical Information Service, Sills Bldg., 5285 Port Royal Road, Springfield, V A 22161. p57.

Smith WL, Hayden CM, Wark DQ, and McMillin LM (1979) The TIROS-N Vertical Sounder. Bull Amer Meteor Soc 60:1177-1187

Uddstrom MJ (1988) Retrieval of Atmospheric Paths from Satellite Radiance Data by Typical Shape Function Maximum a posteriori Simultaneous Retrieval Estimators. J Appl Meteor 27:515-549

Uddstrom MJ and Wark DQ (1985) A Classification Scheme for Satellite Temperature Retrievals. J Climate Appl Meteor 24:16-29

Uddstrom MJ and McMillin LM (1992) Extraction of Atmospheric Signals from Radiance Measurements. High Spectral Resolution Infrared Remote Sensing for Earth's Weather and Climate Studies. NATO Advanced Workshop. In preparation.

Wark DQ (1986) Temperature Profiles Derived from a Restricted Range of Solutions Using Satellite Measurements. Advances in Remote Sensing Retrieval Methods. Deepak A, Fleming H, and Chahine M, Eds. A. Deepak Publishing

Vertical Sounding Capabilities With High Spectral Resolution Atmospheric Radiation Measurements-A Demonstration with the High resolution Interferometer Sounder (HIS)

W.L. Smith, H.E. Revercomb, H.-L. Huang, and R.O. Knuteson
C I M S S
University of Wisconsin-Madison
1225 West Dayton Street
Madison, WI 53706

Abstract

This paper provides one example application of groundbased and aircraft Fourier Transform Spectrometer (FTS) measurements to the diagnosis of significant atmospheric structure. This capability is relevant to the development of future space and groundbased high vertical resolution sounding instruments.

Introduction

During February and March, 1992 both aircraft and groundbased High resolution Interferometer Sounder (HIS) instruments were operated as part of a field program called STORMFEST. A boundary layer subprogram was a key element of STORMFEST. The GroundBased HIS (GB-HIS) was operated continuously at SENECA, KS in an uplooking mode to monitor boundary layer thermodynamic structure. The Aircraft-HIS (AC-HIS) operated in a downlooking mode aboard the NASA-ER2 which flies at an altitude of 20 km. During the STORMFEST, both GB-HIS and AC-HIS aircraft data were achieved for a wide variety of meteorological conditions. In this paper we present results obtained during one boundary layer aircraft mission during the early morning hours of 1 March 1992. The results demonstrate the capabilities of both groundbased and aircraft Michelson Interferometers and satellite high resolution infrared spectrum measuring instruments such as the EOS Atmospheric InfraRed Sounder (AIRS) to monitor the evolution of the Planetary Boundary Layer (PBL) thermodynamic structure.

NATO ASI Series, Vol. I 9
High Spectral Resolution Infrared Remote Sensing for Earth's Weather and Climate Studies
Edited by A. Chedin, M.T. Chahine and N.A. Scott

Instrument Characteristics

The AC-HIS is a three detector instrument which observes the spectrum in three bands; (1) 600 - 1100 cm^{-1}, (2) 1100 - 1800 cm^{-1}, (3) 2000 - 2800 cm^{-1} (Revercomb et. al., 1988). The unapodized spectral resolutions and radiometric noise levels achieved in these bands are approximately 0.25 cm^{-1}, 0.5 cm^{-1}, 0.5 cm^{-1}, and 0.4 mw/m^2 - str - cm^{-1}, 0.2 mw/m^2 - str - cm^{-1}, 0.1 mw/m^2 - str - cm^{-1} respectively. From the ER-2 altitude of 20 km, the AC-HIS resolves a 2 km diameter circular scene. The Michelson mirror scan time and sampling rate is set so that independent spectra for all three bands are achieved approximately every kilometer along the flight track, (i.e. approximately two independent samples per instrument field of view are obtained). For atmospheric retrievals it is common practice to reduce the noise level of the data by either reducing the spatial resolution and/or the spectral resolution of the data. The data is spatially averaged to a ground resolution of 15 km and apodized to a spectral resolution of 0.5 cm^{-1}, 2.0 cm^{-1}, and 2.0 cm^{-1} for bands 1 - 3, respectively. The NEΔN's are about 0.1, 0.05, and 0.01 mw/m^2 - str - cm^{-1}, for channels 1 - 3, respectively. Absolute calibration of the AC-HIS is achieved by viewing two onboard reference blackbodies with a frequency of two minutes.

The GB-HIS is a two detector instrument which observes the spectrum in two bands; (1) 500 - 1600 cm^{-1}, and (2) 2000 -2800 cm^{-1}. The unapodized spectral resolution in both bands is 0.5 cm^{-1} and the NEΔN's are approximately 0.1 and 0.005 mw/m^2 - str - cm^{-1} for a three minute integration period. In practice, spectra are output in a ten minute cycle during which three minutes of the cycle are devoted to atmospheric scene viewing and the remaining seven minutes is devoted to views of three different calibration blackbody targets; two blackbodies controlled at temperature of 320°K and ambient (~280°K) and a tub of liquid nitrogen which radiates at a temperature of 77°K.

The Meteorological Condition and Observed Spectral Features

The morning of 1 March was clear over the central United States from the Gulf of Mexico to the Canadian Border, including the STORMFEST boundary layer experiment domain (Figure 1). There was strong radiative cooling of the surface causing a somewhat typical nocturnal inversion over a large portion of the region. The situation was also dynamic due to the existence of a strong low level jet stream produced by east-west thermal and pressure gradients produced by the sloping land surface across the Plains. As shown in figure 1, the winds at 850 mb were strong

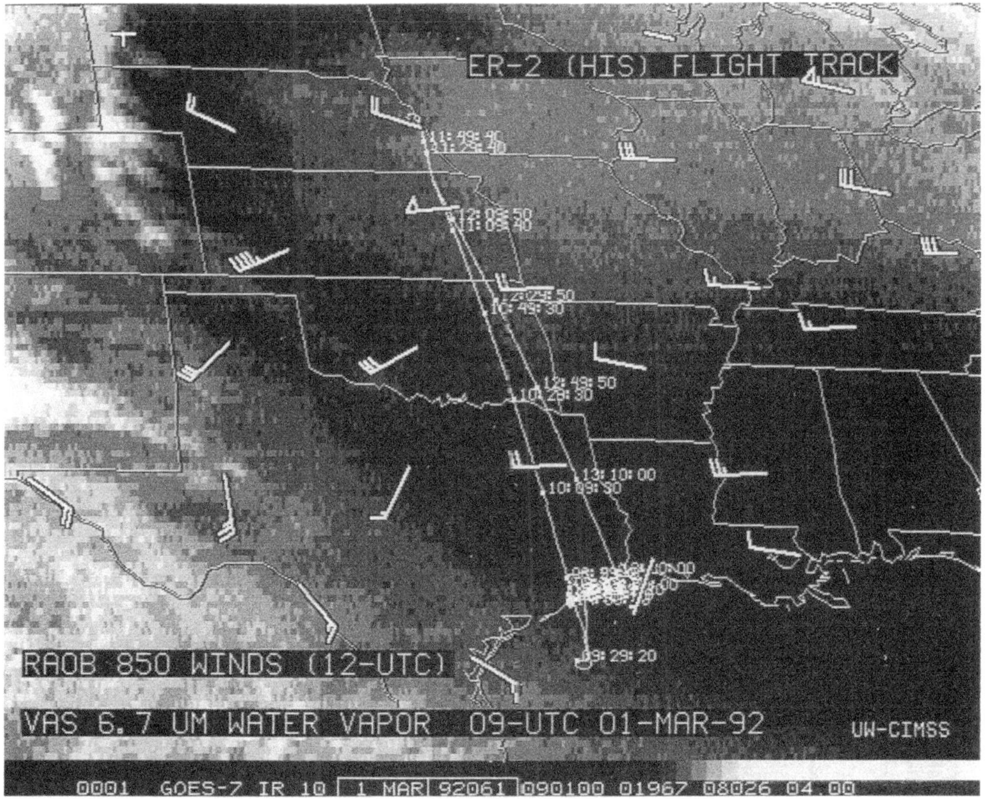

Figure 1: Flight Track of the NASA ER-2 aircraft and radiosonde 850 mb wind vectors superimposed on a GOES 6.7 μm water vapor image.

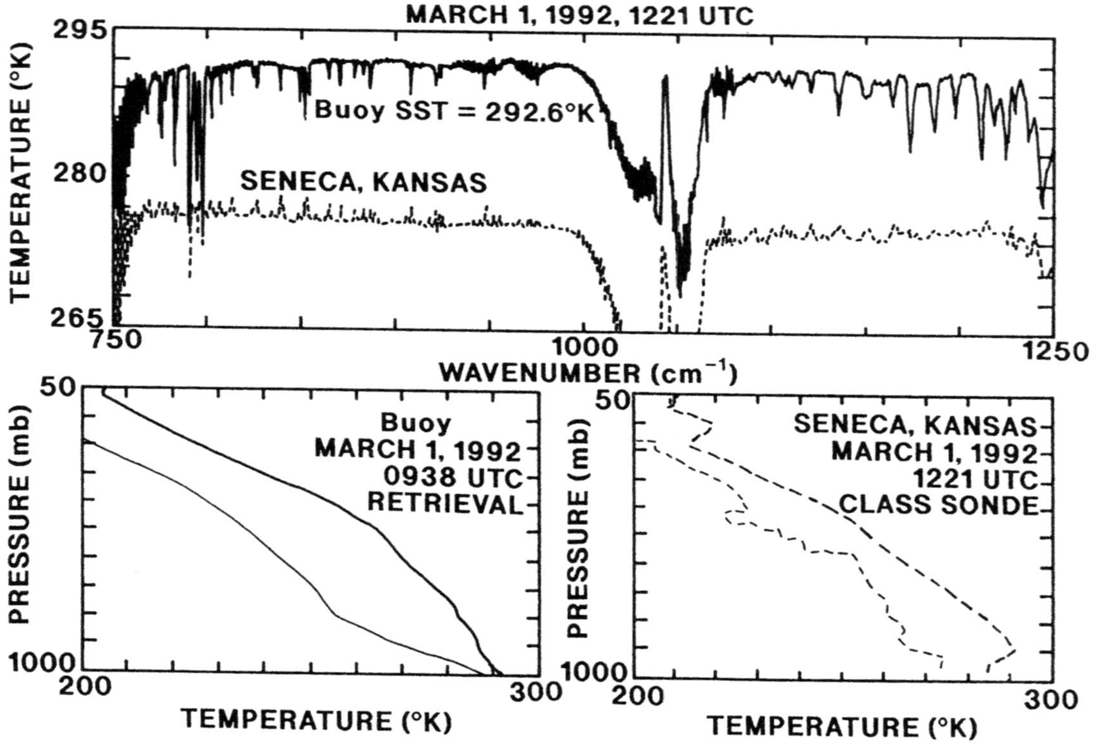

Figure 2: HIS observed brightness temperature spectra observed in the 8 - 13 μm "Window" region of the spectrum for two atmospheric conditions; (a) a warm oceanic temperature "Lapse" condition (Buoy) and, (b) a continental nocturnal temperature "inversion" situation (Seneca, KS).

along the Flight track of the ER-2. Because of the variability of the low level winds, the intensity of the low level inversion also varied as a result of non-uniform mixing.

Figure 2 shows two radiant Brightness Temperature Spectra observed within the 750 - 1250 cm^{-1} (8.0 - 13.3 μm) atmospheric "window" region of the spectrum revealing the two extreme meteorological conditions sampled during this flight. The "warm" spectrum was observed over an oceanographic buoy located in the Gulf of Mexico just south of Houston (see figure 1). The "cold" spectrum was observed over Seneca Kansas, near the Kansas-Nebraska border (see figure 1), which was the location of the upward viewing GB-HIS. The temperature and water vapor profiles retrieved from HIS spectra over the Buoy and observed with a Class-sonde over Seneca are shown to assist the interpretation of the differences in these Brightness Temperature spectra.

The spectrum observed over the Buoy is typical of that to be observed over ocean with a lapse rate of temperature and water vapor throughout the lower troposphere. One notes that the brightness temperature observed in the "Cleanest" portions of the window, in between weak absorption lines, is very close to the 292.6^{o}K water surface temperature observed from the Buoy. The absorption of the warm surface radiation caused by the water vapor, ozone, and carbon dioxide spectroscopic lines cause a decrease in brightness temperature due to the emission of radiation by the colder atmosphere.

The "window" region brightness temperature spectrum observed over Seneca Kansas is very different than that observed over the Gulf of Mexico. First of all the 277°K surface skin temperature is responsible for the much lower brightness temperatures over Seneca. Most interesting, however, is the "flipping" of the weak absorption line features due to the colder than air surface skin temperature and the lower level temperature inversion observed by the Class-sonde at the time of the HIS measurement. The weak absorption lines are flipped upwards because of the dominance of the emission by the warmer than surface atmosphere at these spectral locations. Within the narrow "windows", in between absorption lines, the colder surface skin is sensed with little atmospheric attenuation of the signal.

Further insight into the surface and atmospheric profile conditions over Seneca can be gained by comparing simultaneous brightness temperature spectra observed with the downward viewing aircraft HIS and the upward viewing groundbased HIS. As shown in figure 3, in general the upwelling radiance spectrum looks like a mirror image of the downwelling radiance spectrum, large absorption features producing low radiance in the upwelling spectrum and high radiance in the downwelling spectrum, respectively. These differences are due to the warm surface and cold space backgrounds of the atmosphere emission for the upwelling and downwelling spectra, respectively.

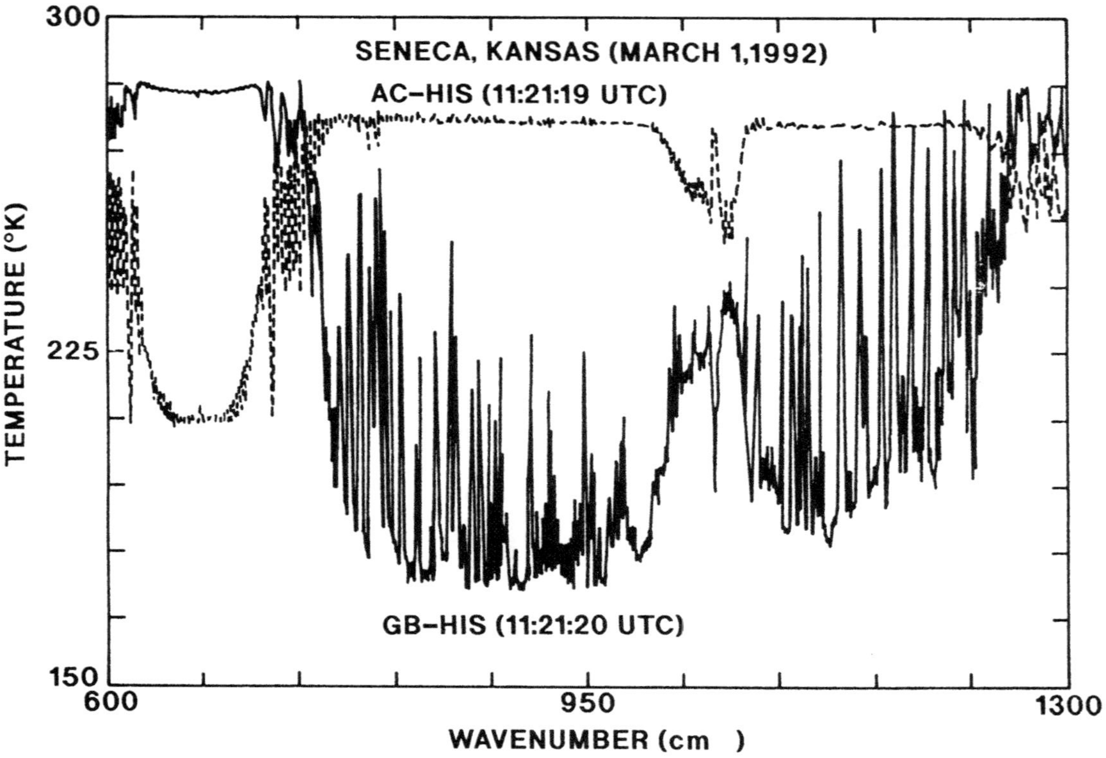

Figure 3: Brightness Temperature spectra observed at Seneca Kansas with the (a) upward looking Ground-Based HIS (GB-HIS), solid line, and (b) downward looking Aircraft HIS (AC-HIS) dashed line.

Details of these spectra reveal details of the atmospheric boundary layer thermal structure. Figure 4 shows the groundbased and aircraft spectra of downwelling and upwelling radiance brightness temperature at Seneca Kansas within the 600-900 cm^{-1} (11.1 - 16.7 μm) region. First, it can be seen that at the most opaque region of the downwelling GB-HIS spectrum (655-675 cm^{-1}) the surface atmospheric temperature is about 283.5^{o}K compared to the 276.5^{o}K surface skin temperature observed in the clean "windows" of the upwelling AC-HIS spectrum at 900 cm^{-1}. The fact that the surface skin is colder than the surface air temperature is also observed by the "flipped" lines of emission in the window region of the AC-HIS spectrum. Evidence of an inverted low level atmospheric temperature structure is revealed by the GB-HIS spectrum within the 630-710 cm^{-1} CO_2 absorption/emission region. Here we see the brightness temperature increase towards the weaker absorption wings of the band indicating an increasing temperature with altitude within the boundary layer. The brightness temperature decreases abruptly below 630 and above 710 cm^{-1} because at these wavelengths the transmittance for the entire depth of atmosphere becomes greater than zero so that the emission is influenced by the cold (4°K) cosmic background.

Temperature and Water Vapor Retrieval Methods

The AC-HIS and GB-HIS radiance spectra are transformed to atmospheric temperature and moisture profiles using established "inversion" techniques. For the aircraft data, a two step procedure is used. In the first step, a statistical regression algorithm relates the temperature and mixing ratio profile to a measurement vector, defined as the convolution of the weighting function matrix with the observed brightness temperature spectrum (Smith et al., 1991). This algorithm produces a good first approximation of the solution which in turn is used as a first guess to a "Maximum Likelihood" solution (Rodgers, 1987). The entire spectrum, 600 - 2500 cm^{-1}, is used in both steps of the retrieval process. The water vapor retrieval is in terms of level mixing ratio. For both steps, a climatology of temperature and water vapor profiles is used as statistical input for the solution of the retrieval coefficients.

For the upward viewing case, the weighting functions all peak at the surface such that the weighting function covariance matrix is highly ill-conditioned with respect to mathematical inversion. As a consequence a recursion retrieval method has been developed. The recursion initiates at the surface level, where the energy contribution peaks, and continues upward where the energy contribution rapidly diminishes, at a rate depending on the spectral absorption characteristics (Smith et al., 1992). The recursion technique utilizes a first guess profile which is initially a climatological profile or a prior radiosonde and subsequently a prior GB-HIS retrieval.

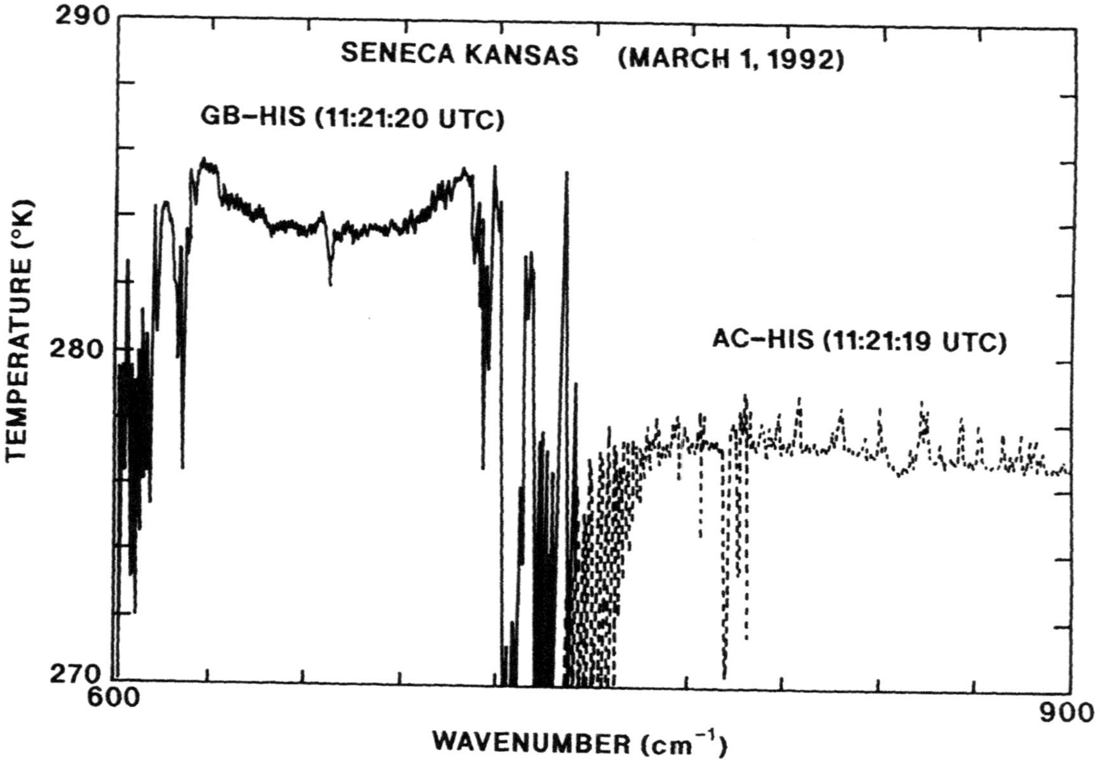

Figure 4: Surface and lower atmospheric brightness temperature spectra observed with the groundbased (solid line) and aircraft (dashed line) HIS instruments at Seneca Kansas.

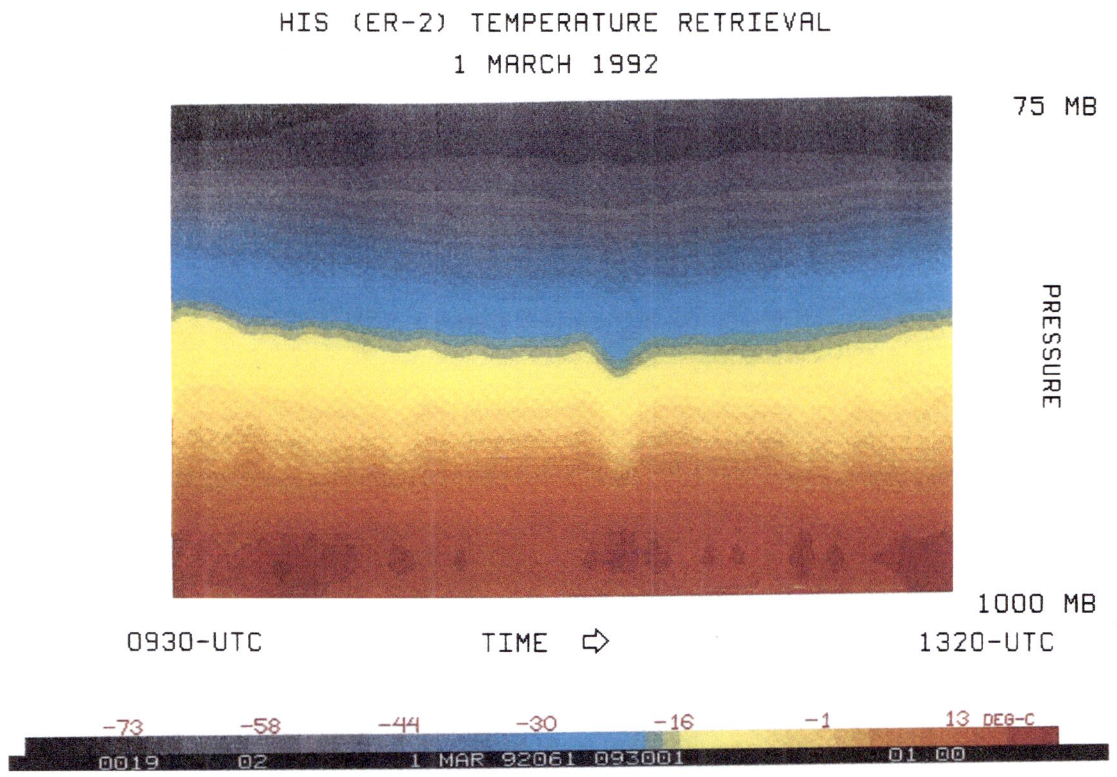

Figure 5: Vertical cross-section of retrieved atmospheric temperature along the Flight track of the ER-2.

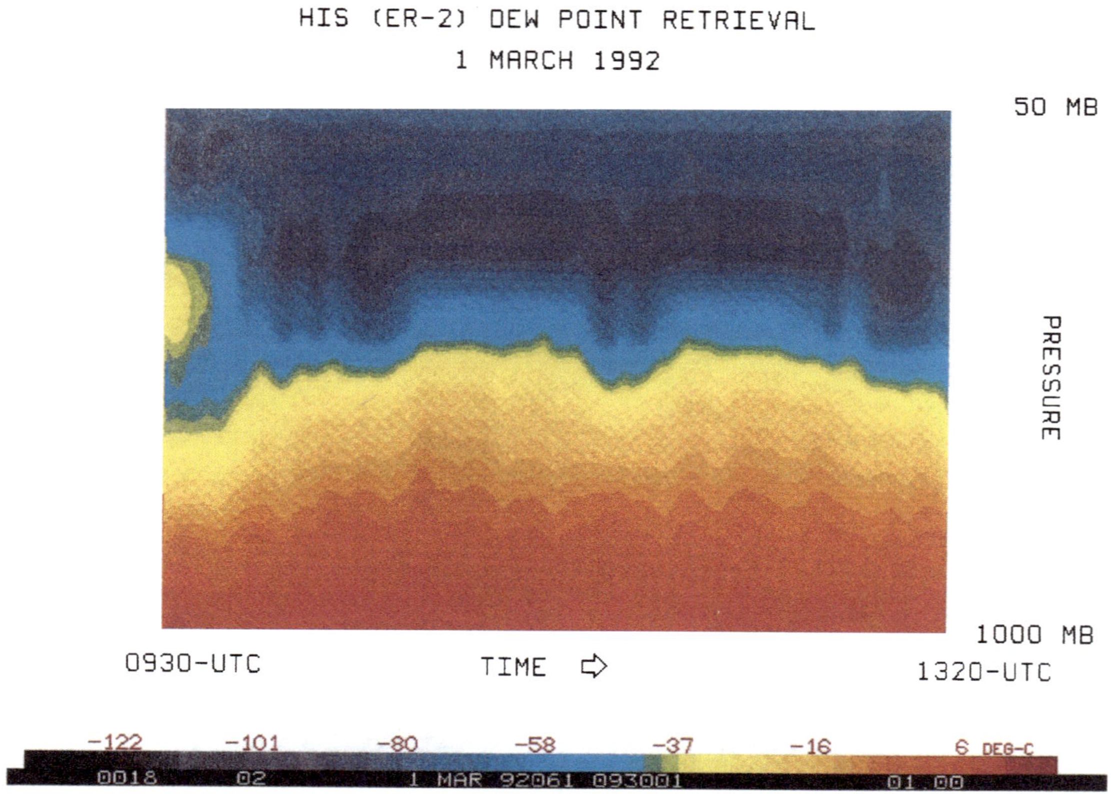

Figure 6: Vertical cross-section of retrieved atmospheric dewpoint temperature along the Flight track of the ER-2.

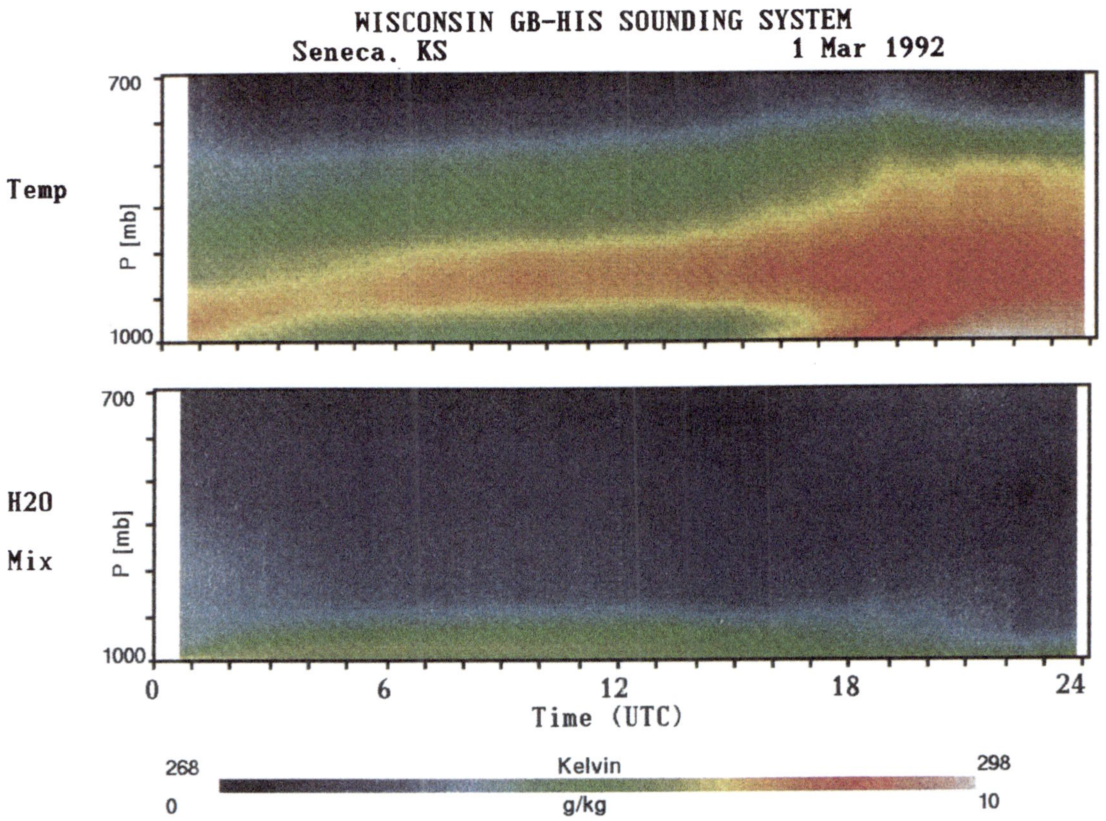

Figure 7: Time section of atmospheric temperature (top) and mixing ratio (bottom) retrieved from ten minute interval GB-HIS spectra at Seneca Kansas.

Since the GB-HIS is operated to produce time averaged spectra every ten minutes, the use of the previous retrieval as a first guess provides a close approximation to the current atmospheric state.

Atmospheric Retrieval Results

Figures 5 and 6 show time cross-sections of retrieved temperature and dewpoint temperature from the AC-HIS spectra. Most notable in the HIS retrievals is changing low level atmospheric boundary layer temperature structure (figure 5) from 9:30 UTC, when the aircraft was over the Gulf of Mexico, to 11:40 UTC when the aircraft was at its northern most extent over Omaha, Nebraska (see figure 1). During this portion of the flight one can see the changing stability of the boundary layer from the ocean to the land, with a low level inversion over land of varying strength along the flight track. The cold lower troposphere anomalies near 11:40 UTC 13:00 UTC are due to sharp banking of the aircraft which causes a raising of the weighting functions due to the increased optical path of the observed radiance.

The water vapor cross-section, figure 6, shows a moist upper troposphere over the Gulf at 9:30 UTC and over northern Kansas and Nebraska as observed near the middle portion of the flight. This result is in good qualitative agreement with the GOES-VAS 6.7 μm H_2O image shown in figure 1. The wave like structure in the lower troposphere is interesting and is possibly associated with the low level Jet across the flight path of the ER-2.

Figure 7 shows a time section of the GB-HIS retrievals over Seneca Kansas throughout the entire day of 1 March, 1992. Most interesting is the temperature section which clearly shows the development and decay of a strong nocturnal inversion between 0 and 18 UTC (6:00 PM - 12:00 PM local time). Note the increasing intensity of the inversion between 11:00 and 13:00 UTC, the period of the HIS overflights. As can be seen from the AC-HIS time-section, the strength of the inversion appears stronger over northern Kansas on the return leg which is in qualitative agreement with the time tendency shown in the upward viewing GB-HIS.

Finally figures 8 and 9 show AC-HIS and GB-HIS retrievals compared with the Class-sonde released at Seneca at the time of the ER-2 overpass. The first guess shown in figure 8 is the mean of the climatological radiosonde sample used to define the covariance matrices used in the regression and maximum likelihood solution. The ability to retrieve the low level temperature structure from the brightness temperature signature discussed earlier, is clearly evident. However, the absolute accuracy of the retrievals is deficient, presumably due to uncertainties in the atmospheric transmittances used to achieve the physical solutions. (The minimization of this source of error is the major topic of the Intercomparison of Transmittance and Radiance

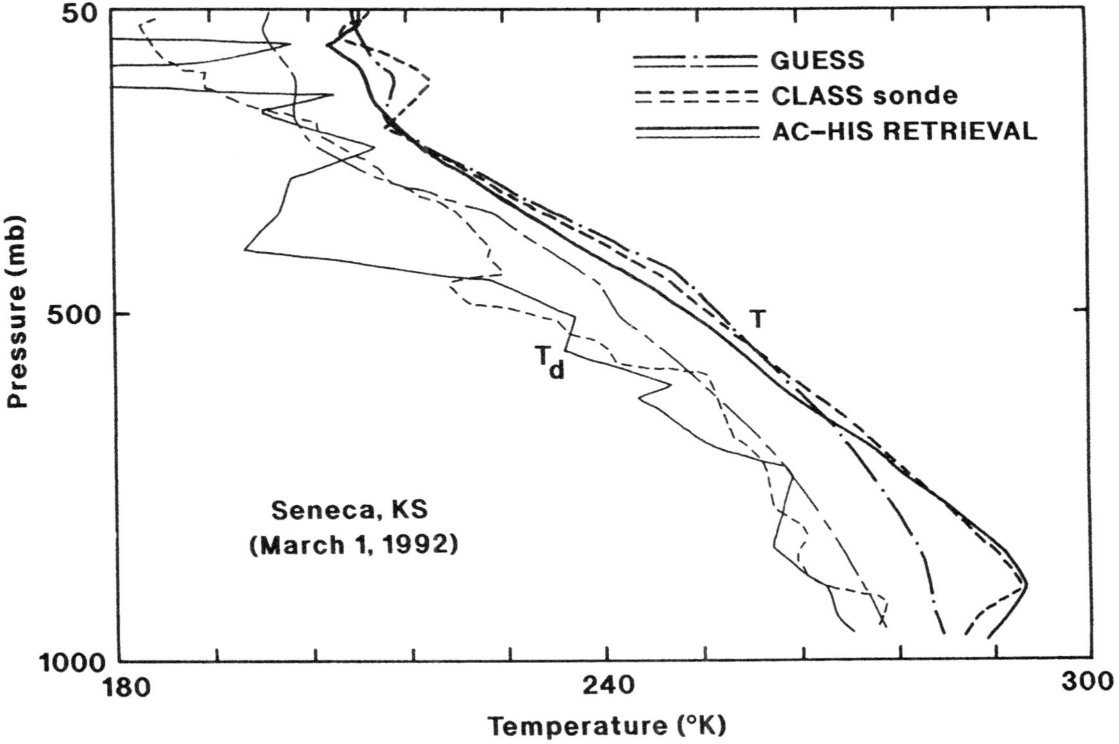

Figure 8: Comparison of AC-HIS temperature and dewpoint profile retrieval with a CLASS balloon sonde. The "first guess" for the retrieval is also shown.

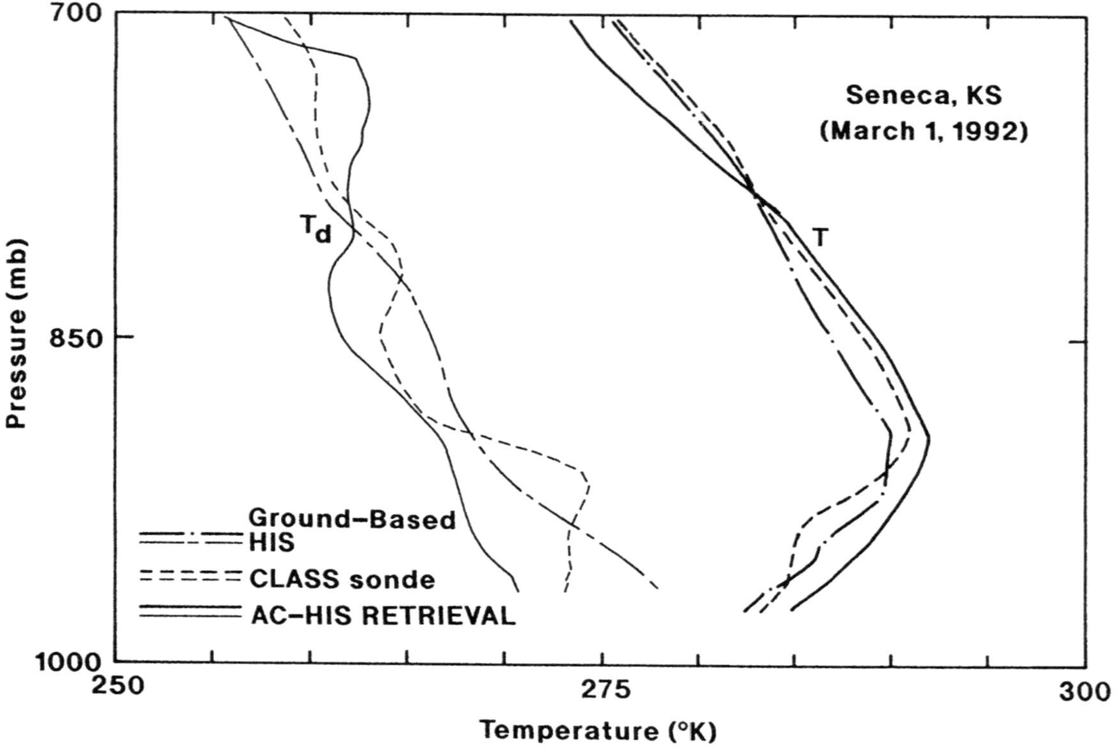

Figure 9: Comparison of lower tropospheric temperature and dewpoint profiles observed by AC-HIS, GB-HIS, and CLASS at 1220 GMT on March 1, 1992.

Algorithms (ITRA) working group of the International Radiation Commission). As expected, the groundbased system appears to provide somewhat higher absolute accuracy in the low troposphere. It appears that a combined satellite and groundbased high spectral resolution remote sounding system would optimize the result. For water vapor the conclusion is less definitive, with inconsistencies in fine scale structure being observed between all three sounding systems. In general, the groundbased HIS is superior to the aircraft HIS in the lower troposphere (figure 9). This result is expected because of the reduced sensitivity of upwelling radiance to water vapor in a near isothermal or inverted low level temperature environment.

Conclusion

High resolution infrared atmospheric radiance spectra observed from high flying aircraft and the ground resolve the thermal structure of the boundary layer nocturnal inversion. As expected, the groundbased spectral measurements provide more accurate profiles near the surface, whereas the aircraft spectra provide a horizontal distribution of profiles from the ground throughout the middle and upper troposphere where the groundbased system lacks profile sensitivity. The results shown here support the expectation of improved global sounding capabilities when higher spectral resolution instruments are implemented on Polar orbiting satellites (e.g. the EOS-AIRS, 1991). For a continental region such as the United States and Europe, optimal remote sensing profile results would be achieved from a combination of geostationary satellite and groundbased high spectral resolution measurement systems.

References

H. E. Revercomb, H. Buijs, H. B. Howell, D. D. LaPorte, W. L. Smith, and L. A. Sromovsky, "Radiometric Calibration of IR Fourier Transform Spectrometers: Solution to a Problem with the High-Resolution Interferometer Sounder," *Appl. Opt., 27,* 3210-3218 (1988).

C. D. Rodgers., 1987: A General Error Analysis for Profile Retrieval, *RSRM87, Advances in Remote Sensing Retrieval Methods*, Deepak, 285-297.

W. L. Smith, H. E. Revercomb, R. O. Knuteson, H. B. Howell, and F. A. Best, "Remote sensing of the Planetary Boundary Layer with a High resolution Interferometer Sounder (HIS)," Submitted to *J of Applied Meteorology.*

W. L. Smith, H. M. Woolf, H. E. Revercomb, 1991: A linear simultaneous solution for temperature and absorbing constituent profiles from radiance spectra. *Appl. Opt.*, **30**, 1117-1123.

M. T. Chahine, 1991: AIRS-Atmospheric Infrared Sounder Science and Measurement Results. JPL Document D6665, California Institute of Technology. 33 pp.

Chapter 3

Next generation IR sounders and cloud field analysis

"Determination of temperature and moisture profiles in a cloudy atmosphere using AIRS/AMSU"

J. SUSSKIND, M.T. CHAHINE, J. JOINER - NASA/JPL-GSFC, USA

"Topics in optimal inversion schemes applied to atmospheric structure retrieval"

V. CUOMO - Univ. of Basilicata/IAMEA, Italy
U. AMATO, R. RIZZI, C. SERIO, U. TRAMUTOLI

"Retrieval of cloud parameters by multiple high spectral resolution observations in the near infrared under conditions of varying solar illumination"

T.J. KLEESPIES - Phillips Lab./USAF, USA

"Spectral variability of light scattering by atmospheric ice crystals"

A. MACKE - GKSS, Germany

"Detection of Polar Stratospheric Clouds with next generation IR sounders"

R. MEERKOETTER - DLR, Germany

"Problems in cloud identification and clearing"

R. RIZZI, ECMWF, GB - C. SERIO, V. CUOMO

"The potential of medium resolution spectral infrared measurements for high cloud studies"

D. SPÄNKUCH - German Weather Serv., Germany

"Modelling reflecting and semi-transparent cloud for infrared sounding"

P.D. WATTS, A.J. BARAN - UK Met.O., GB

Determination of Temperature and Moisture Profiles in a Cloudy Atmosphere Using AIRS/AMSU

Joel Susskind
Joanna Joiner [1]
Goddard Space Flight Center
Greenbelt, MD 20771, USA

Moustafa T. Chahine
Jet Propulsion Laboratory
California Institute of Technology
Pasadena, CA 91109, USA

Abstract

High spectral resolution infrared satellite data can be used to retrieve accurate atmospheric temperature and water vapor profiles. In order to achieve the greatest retrieval accuracy, all the factors affecting the radiances, especially clouds, must be accurately accounted for. We have developed a methodology for retrieving atmospheric and surface parameters under cloudy conditions using the AIRS/AMSU instruments. Since all fields-of-view are contaminated to some degree by clouds, the task of eliminating clouds must be analytical, objective, and accurate. This is accomplished with great success by making use of observations in adjacent fields-of-view utilizing both the microwave and infrared bands. The treatment of clouds is carried out systematically and does not require any field-of-view to be necessarily clear. The only assumption about cloudiness is that clouds have identical spectral properties in both fields-of-view. The algorithm permits the determination of atmospheric temperature and water vapor profiles accurately under cloudy conditions and shows little degredation in their accuracy as a function of increased cloudiness.

1 Introduction

Microwave and infrared sounders flying aboard satellites provide a wealth of data which can be used to study global climate and weather. The current operational low earth orbiting meteorological satellite sounder consists of HIRS2, a 20 channel infrared sounder, MSU, a 4 channel microwave sounder, and SSU, a 3 channel infrared pressure modulated radiometer sensitive to radiance in the mid-upper stratosphere. These three instruments measure temperature and water profiles as well as land and sea surface skin temperatures, snow and ice cover, cloud heights and amounts, and total ozone (see *e.g.,* Susskind *et al.*, 1984).

The Atmospheric Infrared Sounder (AIRS) and the Advanced Microwave Sounding Unit (AMSU-A/B), which will fly on the EOS PM platform, have more channels and

[1] also at Universities Space Research Association

NATO ASI Series, Vol. I 9
High Spectral Resolution Infrared Remote Sensing
for Earth's Weather and Climate Studies
Edited by A. Chedin, M.T. Chahine and N.A. Scott

a higher spectral resolution than HIRS2/MSU/SSU. AIRS is a high-resolution grating spectrometer with approximately 3600 channels which cover wavelength regions from 3.4 to 15.4μm with spectral resolution $\nu/\Delta\nu = 1200$ and a spatial resolution of 15 km at nadir. AMSU-A is a 15 channel microwave radiometer which covers the frequency range from 23.8 to 89 GHz, with the majority of the channels centered around the 50 GHz oxygen cluster. AMSU-B has 5 channels ranging from 89 GHz to 183 GHz, with the majority of the channels centered around the 183 GHz water line. The spatial resolutions of AMSU-A and AMSU-B are 50 and 15 km, respectively. The AIRS/AMSU combination will provide superior rms accuracy and vertical resolution of temperature and constituent profiles compared with HIRS2/MSU/SSU.

An extremely important step in deriving the atmospheric temperature and water vapor profiles is the determination of the clear-column radiances (*i.e.*, the radiance emerging from the clear portions of the scene). The objective is to determine the clear-column temperature and water vapor profiles such that they agree well with both the microwave and derived clear-column infrared radiances or brightness temperatures. In this paper, we have simulated cloudy AIRS/AMSU radiances and developed a methodology, using observations in adjacent fields-of-view, for retrieving atmospheric and surface parameters under partially cloudy conditions. The only assumption about cloud spectral properties is that they are identical in each field-of-view. We have tested our algorithm using a set of temperature, water vapor, and ozone profiles derived from radiosondes. Our retrieved temperature and water vapor profiles do not show appreciable degredation due to clouds, even for up to 75% cloudiness.

2 Simulating Radiances: Forward Problem

The observed radiance, $\overline{R}_i$ for channel i in a scene with j different cloud types is given by

$$\overline{R}_i = (1 - \sum_j \alpha_j) R_{i,\mathrm{CLR}} + \sum_j \alpha_j R_{i,j,\mathrm{CLD}}, \tag{2.1}$$

where α_j is the fraction of the scene (as seen from above) covered by cloud type j, $R_{i,\mathrm{CLR}}$ is the clear-column radiance for channel i (*i.e.*, the radiance emerging from the clear portion of the scene), and $R_{i,j,\mathrm{CLD}}$ is the radiance for channel i emerging from the cloudy portion of the scene corresponding to cloud type j. A given relative combination of clouds in the scene is referred to as a cloud formation. For example, 10%, 20%, and 30% of three different cloud types in a scene is the same cloud formation as 15%, 30%, and 45% of the same respective cloud types in another scene. One can consider the total cloud fractions for the first and seconds scenes to be 60% and 90%, respectively, with both scenes having the same $R_{i,\mathrm{CLD}}$, the radiance which would be observed if the whole of each scene were covered by that cloud formation. If the second scene had 15%, 30%, and 50% of the respective cloud types, the extra 5% would be attributed to a second cloud formation.

The radiance of channel i with central frequency ν_i for a clear sky homogeneous scene, $R_{i,\mathrm{CLR}}$, can be computed from the radiative transfer equation. The radiative transfer equation gives the clear-column radiance in terms of four components: surface emission, atmospheric emission, reflected downward flux, and reflected solar radiation, respectively, according to

$$R_{i,\mathrm{CLR}} = \epsilon_{\nu_i} B_i(T_s)\tau_i(p_s) + \int_{p=p_s}^{p=0} B_i[T(p)]\Big(\frac{d\tau_i}{d\ln p}\Big) d\ln p + \rho_{\nu_i} R_i^d \tau_i(p_s) + \rho'_{\nu_i} H_i \tau'_i(p_s), \tag{2.2}$$

(Susskind *et al.*, 1984) where p is pressure, p_s is the surface pressure, $T(p)$ is the atmospheric temperature profile, T_s is the surface temperature ϵ_{ν_i} is the surface spectral emissivity, ρ_{ν_i} is the surface spectral reflectivity, R_i^d is an effective downwelling thermal flux averaged over all angles, H_i is the radiance of the sun (assumed to be a blackbody at 5600 K), ρ'_{ν_i} is the surface spectral bi-directional reflectance of solar radiation, $\tau'_i(p_s)$ is the two path transmittance from the sun to the surface to the satellite, and B_i is the planck function which gives the radiance emitted by a blackbody at channel frequency ν_i and temperature T. The channel average transmittance τ_i is the atmospheric transmittance from pressure p to the top of the atmosphere, given by

$$\tau_i(p) = \int_0^\infty f_i(\nu)\tau_\nu(p)d\nu, \tag{2.3}$$

where $f_i(\nu)$ is the response function for channel i, and $\tau_\nu(p)$ is the monochromatic transmittance which is a function of the temperature, moisture, and ozone distribution at pressure p. Channel average transmittances at pressure p depend on the distribution of temperature, moisture, and ozone in the entire profile above p (Susskind *et al.*, 1983). This dependence was parameterized along the lines of Susskind *et al.* (1983) for all AIRS and AMSU channels.

In our simulations, we neglected the small downwelling thermal radiation, R_i^d, for infrared channels. We also simplified the spectral dependence of the surface emissivity and bi-directional reflectance. The surface emissivity was treated as having two discrete values, one at low frequencies, ϵ_l, and one at high frequencies, ϵ_s, while the bi-directional reflectance had one value, ρ', independent of frequency. We computed the reflected downward flux term for the microwave channels only and assumed the surface reflectance at microwave frequencies to be specular so that $\rho'_{\nu_i} = 1 - \epsilon_{\nu_i}$. Since the Planck function is proportional to temperature at microwave frequencies, the microwave brightness temperature can be written as

$$\Theta_i = \epsilon_{\nu_i} T_s \tau_i(p_s) + \int_{p=p_s}^{p=0} T(p)\Big(\frac{d\tau_i}{d\ln p}\Big) d\ln p + (1-\epsilon_{\nu_i})\Theta_i^d \tau_i(p_s), \tag{2.4}$$

where Θ_i^d is the downwelling flux computed at the satellite zenith angle.

In reality, it is difficult to compute $R_{i,\mathrm{CLD}}$ for a given scene due to complex spectral absorption and reflection properties of clouds, as well as geometric shadowing factors. In the assumptions made in the simulations involving use of observations in adjacent fields-of-view with differing amounts of the same cloud formation, the details of the computation of $R_{i,\mathrm{CLD}}$ are irrelevant as long as cloud properties are identical in both fields-of-view. Therefore, we computed cloudy radiances, $R_{i,\mathrm{CLD}}$ by simulating a scene cloud formation as a single blackbody cloud ($\epsilon = 1$ and $\rho' = 0$) with cloud top pressure $p_{c,j}$. Equation 2.2 was used to compute $R_{i,\mathrm{CLD}}$, where we substituted p_c and T_c for p_s and T_s and assumed that $T_c = T(p = p_c)$.

3 Accounting for Cloud Effects on Observed Radiances

3.1 The Cloudiness Factor

A general parameter commonly called cloudiness is used to describe the effect of clouds. This raises a basic question: what are clouds? When we observe Earth from space we see certain horizontal inhomogeneities, clouds and aerosols, which we call cloudiness. This factor is quantitatively defined by measuring cloud horizontal extent or opacity relative to a specific background. In this paper, the effect of cloudiness is derived from infrared observations in the 15μm CO_2 band and the background is defined by microwave observations in the O_2 cluster at 50 GHz. Since the infrared observations are affected by clouds while the microwave observations are not affected by most types of clouds, the computed clear (no clouds) emission based on atmospheric temperature and humidity will satisfy the microwave observations but will not satisfy infrared observations.

There are three ways to deal with the effects of clouds on the radiances. The simplest way is to identify scenes that are completely clear and only perform retrievals in those cases, assuming the radiances are unaffected by clouds. This method suffers from two potential shortcomings. First, some cases identified as clear may have some clouds and erroneous retrievals will occur. Second, and more important, there will also be a significant sampling bias when only clear conditions are used which can give an erroneous picture of temperature and moisture distribution and trends.

The second method of handling cloud effects is to estimate or reconstruct the radiances which would have been observed by the satellite if the entire scene were identical to the clear portion of the observed scene. Chahine (1974, 1977, 1982) developed an approach which treats the cloud effects as short-period oscillations over the clear-column radiance. This approach requires radiance data from two spectral regions measured over multiple adjacent fields-of-view. It is assumed that the fields-of-view have the same clear scene channel radiances and the same cloudy scene channel radiances for each cloud formation, but differ only in the amount of formation cloud cover α_j. The uncoupling of the effects of clouds is carried out analytically without a priori information about the amounts, heights, and types of clouds in the fields-of-view. This is the approach we use, both in the analysis of HIRS2/MSU/SSU data (Susskind *et al.*, 1984) and in this analysis of simulated AIRS/AMSU data.

The third and most sophisticated method for accounting for cloud effects on IR radiances is to attempt to find geophysical parameters for both the clear and cloudy portions of the scene which enable the computation of $\overline{R}_i$ via the computation of $R_{i,\mathrm{CLR}}$, $R_{i,j,\mathrm{CLD}}$, and α_j as in Eq. 2.1. This approach does not need the assumption of uniformity of conditions in adjacent fields-of-view but does require the ability to accurately model cloud geometry and radiative transfer properties.

3.2 Important notations

To account for the effects of clouds on the infrared observations, we introduce three forms of the radiance R_i and brightness temperature Θ_i corresponding to a given channel i.

$\overline{R}_{i,k}$ and $\overline{\Theta}_{i,k}$ are the measured or observed channel i radiance and brightness temperature, respectively, corresponding to field-of-view k (not necessarily cloud-free).

$R^n_{i,\mathrm{CLR}}$ and $\Theta^n_{i,\mathrm{CLR}}$ are the clear-column radiance and brightness temperature *computed* from the radiative transfer equation (Eq. 2.2) using the nth iterative parameters (*i.e.*, temperature, water vapor, and ozone profiles, surface temperature, etc.)

$\hat{R}^n_{i,\mathrm{CLR}}$ and $\hat{\Theta}^n_{i,\mathrm{CLR}}$ are the nth iterative estimated clear-column radiance and brightness temperature *reconstructed* from observations over adjacent fields-of-view.

$\Delta\Theta_i^n$ is the nth iteration residual defined as the difference between the reconstructed and computed clear-column brightness temperatures (*i.e.*, $\Delta\Theta_i^n = \hat{\Theta}_{i,\mathrm{CLR}}^n - \Theta_{i,\mathrm{CLR}}^n$). The set of $\Delta\Theta_i^n$ are the key quantities used to obtain the $n+1$ estimates of parameters and also to check for final convergence of the solution.

3.3 Treatment of radiances in cloudy atmospheres

Consider observations made over k adjacent fields-of-view, $\overline{R}_{i,k}$, which are thought to be homogeneous except for the amount of cloud cover of different cloud types in each field-of-view. If the clear-column radiances in each field-of-view are assumed to be equal and the observed radiances in each field-of-view are different, the difference in the observed radiances is attributed to the difference in cloudiness between the fields-of-view.

Chahine (1977) has shown that the reconstructed clear-column radiance can be written as a linear combination of the measured radiances in k fields-of-view according to

$$\hat{R}_{i,\mathrm{CLR}} = \overline{R}_{i,1} + \eta_1[\overline{R}_1 - \overline{R}_2] + \eta_2[\overline{R}_1 - \overline{R}_3] + \ldots + \eta_{k-1}[\overline{R}_1 - \overline{R}_k], \tag{3.3.1}$$

where $\eta_1 \ldots \eta_{k-1}$ are unknown constants and k fields-of-view are needed to solve for $k-1$ cloud formations. Susskind *et al.* (1984) successfully used this approach with two fields-of-view assuming one cloud formation in the analysis of HIRS2/MSU data. Susskind and Reuter (1985) have performed simulations with two cloud formations and three fields-of-view. In this simulation study, we considered only one cloud formation and two fields-of view. It is important to bear in mind that one cloud formation can contain multiple cloud layers.

For one cloud formation and two fields-of-view, the reconstructed clear-column radiance for channel i can be written as

$$\hat{R}_{i,\mathrm{CLR}} = \overline{R}_{i,1} + \eta\left(\overline{R}_{i,1} - \overline{R}_{i,2}\right). \tag{3.3.2}$$

In deriving Eq. 3.3.2, the only assumptions made are that $R_{i,\mathrm{CLR}}$ is the same in each field-of-view and $R_{i,\mathrm{CLD}}$ is also the same in each field-of-view (Chahine, 1977). For example, the clouds in the scene have equivalent spectral emissivities and spectral bi-directional reflectances in each field-of-view. Moreover, with these assumptions, the value of η is the same value for all channels, and η can be written in terms of α_1 and α_2, the cloud fractions in the two fields-of-view, according to

$$\eta = \frac{\alpha_1}{\alpha_2 - \alpha_1}. \tag{3.3.3}$$

It is not necessary to know α_1 or α_2 to determine η. With a good estimate of the computed clear-column radiance, $R_{i,\mathrm{CLR}}^n$ obtained from use of Eq. 2.2 with n^{th} estimate parameters, η_i^n corresponding to channel i and iteration n can be estimated using

$$\eta_i^n = \frac{R_{i,\mathrm{CLR}}^n - \overline{R}_{i,1}}{\overline{R}_{i,1} - \overline{R}_{i,2}}. \tag{3.3.4}$$

If $R_{i,\mathrm{CLR}}^n$ is correct for a given channel, then η_i is computed correctly (up to noise effects) for all channels. If $R_{i,\mathrm{CLR}}^n$ is too large however (corresponding to an nth estimated temperature profile which is too warm averaged over the atmospheric response of channel i), then η_i^n will be too large and $\hat{R}_{i,\mathrm{CLR}}$ will in general be too large for all channels. This will result in turn with the next estimate of the temperature profile still being too warm. Chahine (1974) and Halem *et al.* (1978) have shown that this methodology for determination of η will converge correctly with a reasonable first guess, provided 15μm channels are used to compute η, and reconstructed clear column radiances for 4μm channels are used for lower tropospheric temperature sounding. However, under some conditions, using only IR channels for temperature sounding can lead to a divergent solution (Susskind *et al.*, 1984). Susskind *et al.* (1984) showed that biases in $R_{i,\mathrm{CLR}}$ arising from a first guess which is biased over a coarse tropospheric layer can be removed by the incorporation of a single microwave (MSU) channel in the cloud filtering procedure. We can further improve on this approach by using AMSU-A channels to retrieve a temperature profile which is consistent with all AMSU-A channel radiances. This temperature profile, which is used to compute initial IR radiances $R^0_{i,\mathrm{CLR}}$, is expected to be relatively unbiased over coarse tropospheric layers. Therefore $R^0_{i,\mathrm{CLR}}$ is expected to be accurate provided $R^0_{i,\mathrm{CLR}}$ is relatively insensitive to the surface skin temperature, which is not determined accurately from AMSU retrievals.

Using different channels in Eq. 3.3.4 will result in different computed values of η_i^n due to local errors in the temperature profile and noise effects. Therefore, we take a weighted average of η_i^n over a set of 15μm cloud filtering channels using

$$\eta^n = \frac{\sum_i \eta_i^n W_i}{\sum_i W_i} \tag{3.3.5}$$

to get a single value of η^n. The channels are weighted according to

$$W_i = [1 - \tau_i(p_s)]^2 [\overline{\Theta}_{i,1} - \overline{\Theta}_{i,2}]^2. \tag{3.3.6}$$

If the denominator in Eq. 3.3.4 is small, errors in the numerator will be amplified. Therefore, the second factor in Eq. 3.3.6 gives channels with a larger contrast in brightness temperature between the 2 fields-of-view a higher weight. The first factor in Eq. 3.3.6 gives lower weights to channels which are affected by surface emission in order to reduce the effects of errors in the estimated surface temperature used to compute radiances.

Once a value of η^n is computed, the clear-column radiances for all channels can be reconstructed using Eq. 3.3.2. The reconstructed clear-column radiances are used in the subsequent retrieval of atmospheric and surface parameters.

4 Retrieval of Atmospheric and Surface Parameters in Cloudy Atmospheres

4.1 Overview of the Retrieval Algorithm

Our AIRS/AMSU retrieval algorithm for cloudy conditions consists of the following main steps: (0) Obtain an initial guess for the temperature, moisture, and ozone profiles. (1) Perform an AMSU-A temperature profile retrieval to get a relatively unbiased temperature profile for use in obtaining an initial value of η^0. In the initial pass AMSU-A retrieval, we assume the surface temperature to be equal to the initial guess surface air temperature. In addition, in solving for the atmospheric temperature, we also solve for the product of the microwave emissivity and the surface temperature ($\epsilon_{\nu_i} T_s$). Microwave emissivities were simulated independently from the surface temperature. There is not enough information in the radiances to solve for ϵ_{ν_i} independent of T_s, so we set $(1 - \epsilon_{\nu_i}) = (1 - \epsilon_{\nu_i} T_s / T_s^o)$ in the last term of Eq. 2.4. The AMSU-A temperature profile retrieval is necessary to get an initial estimate of η^0 and $\hat{R}^0_{i,\mathrm{CLR}}$ before AIRS channels can be used. In subsequent iterations, after a value of the surface temperature has been retrieved, we solve for the microwave ϵ_{ν_i}, using the retrieved value of T_s. (2) Compute the nth iterative clear-column radiances $R^n_{i,\mathrm{CLR}}$ from Eq. 2.2 using the nth iterative AMSU-A retrieved temperature profile (assumed to be unbiased over coarse layers of the atmosphere) and other n^{th} iteration parameter estimates. (3) Determine η^n from Eqs. 3.3.4-3.3.6. (4) Reconstruct the clear-column radiances, $\hat{R}^n_{i,\mathrm{CLR}}$, for all channels using Eq. 3.3.2. (5) In the initial pass, retrieve an improved estimate of the ground temperature using the reconstructed clear-column IR radiances, $\hat{R}^n_{i,\mathrm{CLR}}$, from (4) and then repeat (1)-(4), this time using the improved estimate of the surface temperature obtained in (5), solving for ϵ_{ν_i} instead of $\epsilon_{\nu_i} T_s$ in the AMSU temperature retrieval. This provides an improved estimate of η^0 and $\hat{R}^0_{i,\mathrm{CLR}}$. In the initial pass, this represents the end of the basic start up procedure. (6) Solve sequentially for the surface parameters, temperature profile, water vapor profile, and ozone profile using the nth iteration reconstructed IR clear-column radiances, $\hat{R}^n_{i,\mathrm{CLR}}$. (7) Compare residuals of reconstructed and computed brightness temperatures to check for convergence. If not converged, then repeat (1)-(4) using the new parameter estimates as a first guess for the AMSU temperature profile retrieval in (1) to generate improved estimates of η^{n+1} and $\hat{R}^{n+1}_{i,\mathrm{CLR}}$ as in steps (2)-(4). Repeat step (6) solving for the surface parameters, temperature, moisture, and ozone profiles using the improved reconstructed clear-column radiances. In general, the iterative procedure to determine η^n converges after two iterations.

The general approach to solve for parameters within step (6) is also iterative and in the form of constrained least squares solutions, one for each set of variables to be solved for. The method uses the full physics of the problem to compute the dependence of the brightness temperatures as a function of all variables in each iteration, as well as their partial derivatives to all variables being solved for. The success of the methodology depends on accurate computation of the brightness temperature as a function of geophysical parameters. Even more critical is the choice of variables to be solved for and the accurate computation of the dependence of the brightness temperatures on changes in these variables. This dependence is done empirically and iteratively for all channels used in the retrieval process for a set of variables. The details of the retrieval methodology are beyond the scope of this paper and will be given in a forthcoming paper.

4.2 Channel and Variable Selection

When solving for a set of geophysical parameters, it is important to choose an appropriate set of parameters and channels which are both sensitive to those parameters and relatively insensitive to other parameters. However, some channels will be affected by more than one parameter. For example, channels sensitive to the water vapor and ozone distributions are also sensitive to the temperature profile and surface skin temperature. Our approach is to solve sequentially for the surface parameters, temperature profile, water vapor profile, and ozone profile in that order. In this approach, variables already solved for, used in conjuction with first guess variables, are kept fixed when solving for the next set of variables. Table 1 shows the variables solved for and the channels used in each step. Channels selected for a given step are relatively insensitive to variables to be solved for subsequently. A total of 257 AIRS channels and 12 AMSU channels were selected for use in the retrieval algorithm.

TABLE 1 - VARIABLES AND CHANNELS

Ground temperature		
$T_S, \varepsilon_L, \varepsilon_S, \rho', W, T_{AIR}$	53 channels	758.1 → 1234.5 cm^{-1}
	40 channels	1974.3 → 2684.0 cm^{-1}
93 ground temperature channels		
Temperature profile		
14 layer temperature functions		
	33 channels	651.2 → 691.6 cm^{-1}
	26 channels	2178.8 → 2400.4 cm^{-1}
59 AIRS temperature channels + 8 AMSU channels		54.9 → 57.3 GHz
AMSU only - 10 layer temperature functions +εT		
12 AMSU channels		50.3 → 57.29 GHz
Moisture profile		
8 layer column density functions	70 channels	1593.0 → 1971.8 cm^{-1}
70 moisture profile channels + 25 ground channels		
Ozone profile		
4 layer column density functions	18 channels	1001.0 → 1059.0 cm^{-1}
Cloud clearing		
η	17 channels	709.6 → 745.9 cm^{-1}
	257 AIRS channels + 12 AMSU channels	

In the ground temperature retrieval, we selected window channels from two frequency bands generally avoiding even weak absorption lines. For window channels, the transmittance at the surface, $\tau(p_s)$, is close to 1. However, there is absorption due to the water vapor continuum and the nitrogen continuum. Therefore, the radiances in window regions depend not only on the surface temperature T_s, longwave emissivity ϵ_l, shortwave emissivity ϵ_s, and surface bi-directional reflectance ρ', but also on the temperature and moisture in the boundary layer. To account for these dependencies, we solve for these four variables and also two additional variables dealing with the total precipitable water (W) and the mean boundary layer air temperature (T_{AIR}). These variables are refined subsequently in the temperature and moisture retrievals. The temperature retrieval uses the retrieved values of the ground temperature parameters T_s, ϵ_l, ϵ_s, and ρ' in the radiative transfer equation to compute the temperature sounding channel radiances.

In the temperature retrieval, we selected channels which are relatively insensitive to the ozone and water vapor distributions. Along the lines of Kaplan *et al.* (1977), we selected channels in the CO_2 Q-branch near 666 cm^{-1} to sound the mid to upper stratosphere and

15 μm channels in between CO_2 absorption lines to sound through the upper troposphere. Channels in the CO_2 R-branch near 2380 cm^{-1} are used to sound the mid to lower the troposphere. We also included microwave channels near 50 GHz in the temperature retrieval. A total of 14 independent pieces of information about the temperature profile were solved for. The 4.2μm IR channels are highly sensitive to solar radiation reflected off the ground during the day. Our results indicate that this factor is well accounted for to the extent that ρ' is independent of frequency, as assumed in the simulation. Solar radiation reflected off clouds will be accounted for in the cloud correction to the extent that clouds in both fields-of-view reflect solar radiation in the same way.

For constituent retrievals, the selected channels should have transmittances which depend primarily on the retrieved constituent. In the moisture retrieval, we selected channels in the 6.3μm water vapor band which are sensitive to humidity throughout the troposphere. We also selected channels in the 11μm window which are sensitive to the water vapor continuum in order to improve sounding of the lower tropospheric humidity. We selected channels in the 9.6μm ozone band for the ozone retrieval. The moisture and ozone profile retrievals solved for 8 and 4 pieces of information, respectively.

Chahine (1974) showed that it is necessary to use cloud filtering channels in one spectral region and tropospheric temperature sounding channels in a different spectral region. Therefore, as described in section 3, we selected channels in the 15μm CO_2 band which sound the mid to lower troposphere for cloud filtering. These channels are not used for temperature sounding both because of cloud effects, as well as the relatively large sensitivity of 15μm mid-lower tropospheric sounding channels to the water vapor distribution. In order to get sharp weighting functions which result in high vertical resolution, we used channels generally away from even weak absorption lines for cloud filtering as well as in the temperature, moisture, and ozone retrievals (Kaplan *et al.*, 1977).

5 Simulation and Results

We tested our algorithm on an ensemble of 28 temperature, water vapor, and ozone profiles which were derived from radiosondes (Phillips *et al.*, 1988). These profiles also contained a surface skin temperature with reasonable values of skin-surface air temperature differences. In order to have a meaningful sample for our simulation, profiles were selected from different latitudes, seasons, and surfaces (land or ocean). Using these profiles, appropriately computed channel transmittances, and several assumptions described below, we computed radiances and corresponding brightness temperatures for all channels using the radiative transfer equation (Eq. 2.2). For simplicity, all observations were taken at nadir and the sun was assumed to be at zenith. Noise was added to the radiances, where the noise had a gaussian distribution and a standard deviation equal to the equivalent noise temperature for that channel. The equivalent noise temperature ($NE\Delta T_i$) was assumed to be 0.2K for $\nu_i < 770$ cm^{-1}, 0.1K for 770 cm^{-1} $< \nu_i <$ 820 cm^{-1}, and 0.05K for $\nu_i >$ 820 cm^{-1}. These noise values are generally representative of expectations for AIRS channels. For the AMSU channels, we used instrument specified $NE\Delta T_i$ values ranging from 0.3 to 1.2 K. The simulation also contains computational errors to the extent that the computed transmittances depend on the iterative estimates of temperature, moisture, and ozone used in their computation and therefore differ from the true transmittaces. Other sources of computational error, such as spectroscopic uncertainty in absorption characteristics or instrument calibration errors, were not included in this simulation. Experience with analysis of HIRS2/MSU data indicates that these sources of noise tend to have a substantial systematic component which can be identified and accounted for.

In simulating observed radiances using Eq. 2.2, we assumed one constant value of the infrared emissivity at long wavelengths ($\nu < 1600$) and a different constant at short wavelengths ($\nu > 1600$). We used different constants for the long and short wave emissivities for each of the different profiles. The long wave emissivity was allowed to vary between 0.9 and 1.0 and the short wave emissivity between 0.8 and 0.9 over the different profiles. The true solar bi-directional reflectance (ρ'_ν) was constant for all infrared frequencies and was allowed to vary between 0.03 and 0.08 over the different profiles. For the microwave channels, we assumed ϵ_ν be constant with frequency (channels covered the range 50.3-57.3 GHz) and allowed it to vary over the profiles between 0.9 and 1.0 over land and between 0.6 and 0.7 over ocean.

We simulated radiances for the 28 profile ensemble using four different cloud conditions. We used average cloud fractions of 0% (clear), 25%, 50%, and 75%. The difference in cloud fraction between the two fields-of-view was 30% (*i.e.*, $\alpha_2 - \alpha_1 = 0.3$) and $p_c = 300$ mb for all four cases. These cases correspond to $\eta = 0.0$, 0.33, 1.17, and 2.0, respectively. In the cloud filtering procedure, η was treated as an unknown and solved for iteratively as described earlier. Note from Eq. 3.5.2 that η multiplies a difference of observed radiances in two fields-of-view. Therefore, the larger the value of η, the more sensitive one is to noise in the observations.

The success of the retrieval algorithm depends to some degree on the initial or first guess of the geophysical parameters. In order to determine how much information is contained in the radiances, we used climatology as the initial guess for atmospheric profiles and the guess surface air temperature was taken as the guess ground temperature. When performing retrievals, the long and short wave emissivities and reflectivities were treated as unknowns. We used the mean values of 0.95 and 0.85 as the first guess for the long and short wave emissivities, respectively. The first guess infrared bi-directional reflectance was 0.04. We used 0.95 and 0.65 as the first guess microwave emissivity over land and ocean, respectively.

Figures 1 and 2 show results of our simulations giving RMS temperature and moisture profile errors for different average cloud fractions. In Figure 1, the RMS error of the surface skin temperature is shown as the lowest point and other values indicate 1 km layer mean temperature errors up to 200 mb, 2 km layer mean errors between 200 mb and 100 mb, and 4 km errors above 100 mb. Errors of the climatology first guess are also shown. The RMS temperature profile error for the 20 layer profile is indicated in the legend varying from 1.03K for clear conditions to 5.35K for the first guess. Under clear conditions, the errors are on the order of 1K and are consistent with the sounding goals of AIRS/AMSU. Errors do not appear to degrade significantly up to 75% average cloud cover ($\eta = 2$). We did not attempt to simulate retrievals under cloudier conditions or with instrumental noise larger than those of the instrument specifications.

Similar results are shown for the moisture profile in Figure 2. Here the percent error (RMS layer precipitable water error divided by layer mean precipitable water amount) of water vapor is shown for 2 km layers between 1000 mb and 240 mb. Statistics for total water vapor above 240 mb are plotted at 120 mb and for total precipitable water are plotted at 1000 mb. Other values are plotted at the mid-layer point. Total precipitable water is determined to the order of 1% accuracy and 2 km layer mean precipitable water

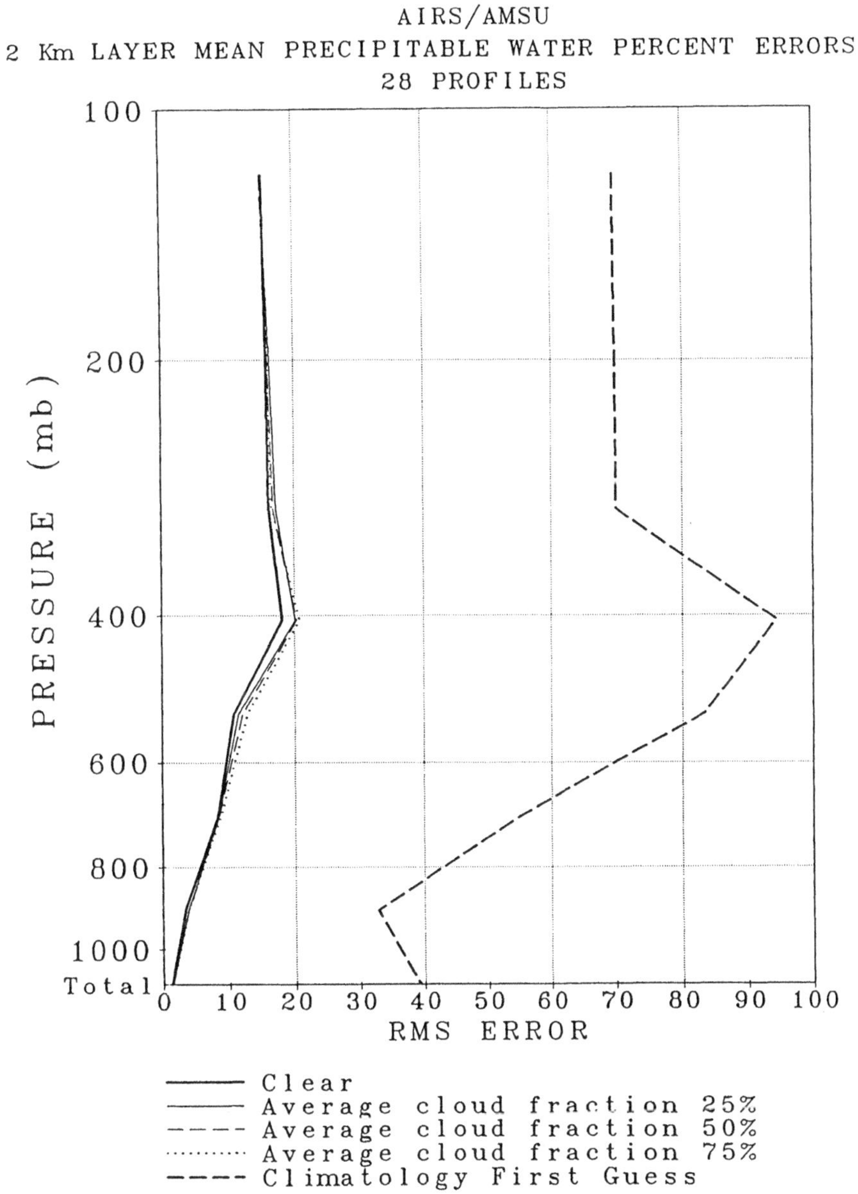
AIRS/AMSU
2 Km LAYER MEAN PRECIPITABLE WATER PERCENT ERRORS
28 PROFILES
PRESSURE (mb)
100
200
400
600
800
1000
Total
0 10 20 30 40 50 60 70 80 90 100
RMS ERROR
Clear
Average cloud fraction 25%
Average cloud fraction 50%
Average cloud fraction 75%
Climatology First Guess

Figure 2

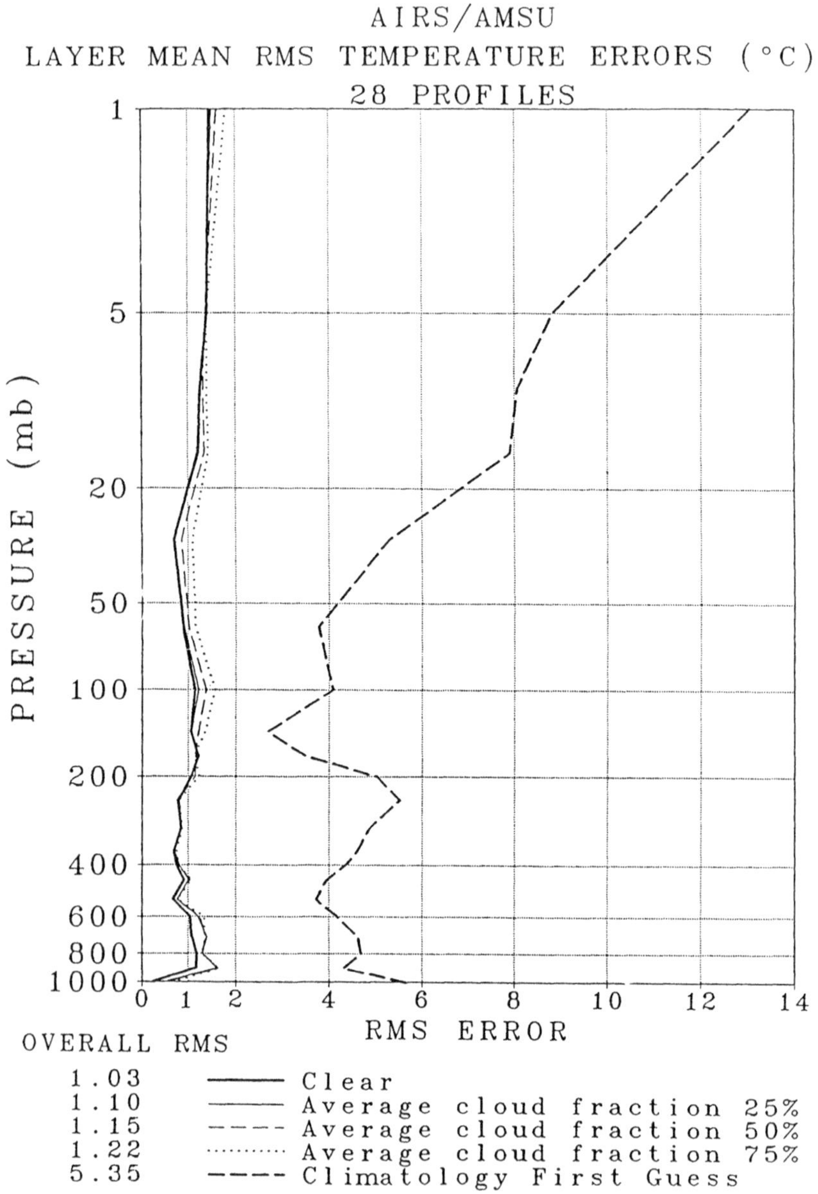
AIRS/AMSU
LAYER MEAN RMS TEMPERATURE ERRORS (°C)
28 PROFILES
PRESSURE (mb)
1
5
20
50
100
200
400
600
800
1000
0 1 2 4 6 8 10 12 14
RMS ERROR
OVERALL RMS
1.03 Clear
1.10 Average cloud fraction 25%
1.15 Average cloud fraction 50%
1.22 Average cloud fraction 75%
5.35 Climatology First Guess

Figure 1

is determined to better than 10% beneath 500 mb. Again the results do not appear to degrade appreciably with increasing cloud cover.

6 Conclusions and Future Work

We have simulated AIRS/AMSU observations and retrievals of temperature and water vapor profiles with a single cloud formation. The results show little degradation in the retrievals with cloudiness up to 75%. These simulations imply that the sounding goals of AIRS, to provide 1K RMS temperature accuracy in 1 km layers in the troposphere, and layer precipitable water RMS errors of 10% in 2 km layers in the troposphere, may be achievable even under extensive cloud cover. While the simulations contain many levels of sophistication, there were also simplifying assumptions. Further simulations will be conducted under more complex conditions, including simulations with multiple cloud formations, non-homogeneous surface characteristics within a field-of-view, a more detailed frequency dependence of the surface emissivity and bi-directional reflectance, and uncertainties in the molecular absorption characteristics. The algorithm will also be used to test the effects of different instrument noise characteristics.

7 References

Chahine MT (1974) Remote sounding cloudy atmospheres. I. The single cloud layer. J Atmos Sci 31: 233-243

Chahine MT (1977) Remote sounding cloudy atmospheres. II. Multiple cloud formations. J Atmos Sci 34: 744-757

Chahine MT (1982) Remote sounding of cloud parameters. J Atmos Sci 39: 159-170

Halem, M, Ghil M, Atlas R, Susskind J, Quirk W (1978) The GISS sounding temperature impact test. NASA Tech Memo 78063: 2.9-2.82

Kaplan LD, Chahine MT, Susskind J, Searl JE (1977) Spectral band passes for a high precision satellite sounder. Appl Opt 16: 322-325

Phillips N, Susskind J, McMillin L (1988) Results of a joint NOAA/NASA sounder simulation study. J Atmos Ocean Tech 5: 44-56

Susskind J, Rosenfield J, and Reuter D (1983) An accurate radiative transfer model for use in the direct physical inversion of HIRS and MSU temperature sounding data. J Geophys Res 88: 8550-8568

Susskind J, Rosenfield J, Reuter D, Chahine MT (1984) Remote sensing of weather and climate parameters from HIRS2/MSU on TIROS-N. J Geophys Res 89: 4677-4697

Susskind J, Reuter D (1985) Intercomparison of physical and statistical retrievals from simulated HIRS2 and AMTS data. In: Deepak A, Fleming HE, Chahine MT (eds) Advances in remote sensing retrieval methods. A Deepak, Hampton VA, p 641

TOPICS IN OPTIMAL INVERSION SCHEMES APPLIED TO ATMOSPHERIC STRUCTURE RETRIEVAL

V. Cuomo, U. Amato[1], R. Rizzi[2], C. Serio[3] and V. Tramutoli
Istituto di Metodologie Avanzate di Analisi Ambientale CNR
Tito Scalo, Potenza
Italy

1. Introduction

Spatial and spectral resolution are the two main points which must be accounted for in developing advanced infrared sounders aimed at improving water vapour and temperature retrievals. Ongoing research at IMAAA center about this topic will be discussed in this paper.

This paper is organized in three sections. Section 2 deals with the cloud clearing problem; both filtering techniques and cloud detection schemes will be addressed. In Section 3 the extent to which spatial resolution could improve the whole cloud clearing step will be discussed. Section 4 is mainly devoted to the analysis of some topics in the retrieval schemes; the characteristics of the mathematical problem will be discussed with particular attention to the possible improvements consequent to higher resolution infrared radiances.

2. The cloud clearing

[1]U. Amato is with Istituto per Applicazioni della Matematica CNR, Napoli, Italy
[2]R. Rizzi is with ECMWF, Reading, UK and EUMETSAT, Darmstat, Germany
[3]C. Serio is with Dipartimento Scienze Fisiche, Università di Napoli Federico II, Italy

NATO ASI Series, Vol. I 9
High Spectral Resolution Infrared Remote Sensing
for Earth's Weather and Climate Studies
Edited by A. Chedin, M.T. Chahine and N.A. Scott

Successful cloud clearing schemes rely on the previous identification of clear infrared radiances. This does not imply that we need a close or deterministic *definition* of clear sky which would be rather platonic for the problem at the hand. Indeed, a stochastic definition is consistent with the current physical-statistical inversion schemes aimed at retrieving the atmosphere structure. In practice, this means that the identification of clear radiances implies the introduction of suitable thresholds on atmospheric transparency. Such a task may be accomplished by using only HIRS/2 and MSU observations and may greatly improve by incorporating AVHRR data in the cloud-clearing processing.

A recent intercomparison exercise (Rizzi et al 1992) of various cloud clearing techniques showed that some approaches may introduce errors even higher than 10 degrees in terms of brightness temperature. The main cause relies on the lack of suitable cloud detection schemes, the current one being based on HIRS and MSU data.

The intercomparison showed that our method, MR59, provided the best results. However, in the light of the experience gained performing the intercomparison exercise, further improvements can be introduced, which now we shall discuss. For this purpose we briefly summarize the scheme MR59 (Cuomo et al 1992) and its two recent evolutions: MR59+ and MR60, respectively.

Basically, the MR59 procedure consists of five steps: a) clear Field-of-View (FOV) identification using the method currently running at UK meteo-office (UKMO) (Eyre and Watts 1987; McMillin and Dean 1982); b) an interpolation scheme to obtain preliminary clear-column estimates at cloudy FOVs by the nearest neighbour clear data; c) a HIRS-MSU regression relation to fill gaps left at the step b); d) removal of local biases eventually produced at step c); filtering of the radiance field to produce a final smooth version of the field.

Kalman filter, which is intrinsecally 1-dimensional, was the basic filtering tool adopted in MR59. A 2-dimensional filtering approach, based on radial basis functions, is the basic inprovement which leads to MR59+. More marked differences separate MR59 and MR60. In MR60, the cloud detection scheme (step a)) is based on AVHRR data; step b) proceeds as in MR59 whereas step c) and d) are replaced from the following procedure. First, FOVs detected as clear at the step a) are used to obtain the regression coefficients of the linear correlation relation (*in place correlation*)

$$T_k(H) = a_{0k} + \sum a_{ik} T_i(M) \tag{1}$$

where $T_k(H)$ and $T_i(M)$ denote respectively HIRS brightness temperature in channel k and MSU brightness temperature in channel i. Then (1) is used to obtain clear column HIRS brightness temperatures at cloudy FOVs. Finally, a 2-dimensional smoothing is considered as in MR59+.

The three cloud-clearing schemes were intercompared and validated using he procedure described in Rizzi et al (1992). The validation procedure consists of comparing clear products in HIRS channel 8 against clear column brightness temperature estimates obtained from AVHRR data.

Let us define $\Delta T = \hat{T}(8) - \hat{T}_c(8)$ where $\hat{T}(8)$ indicates the clear-column brightness temperature in HIRS/2 channel 8 computed according to the one of the three schemes above and $\hat{T}_c(8)$ the same quantity estimated from AVHRR data. Values of ΔT were computed for two satellite passes (Tab. 1) at all FOVs fulfilling the following conditions: a) the FOV is located over the sea; b) the *clearness index* C is greater than 5%.

PASS	scan lines	Clear FOVs (%)
A	102	15
D	62	25

Table. 1: Relevant information for the analyzed passes.

The clearness index is computed from AVHRR data co-located at HIRS/2 spots. It takes values from 0 to 100, the latter indicating that the HIRS/2 FOV is completely clear. Intermediate values give the fraction of area of the HIRS/2 FOV which is cloud-free, e.g. C=5% means that 5% of the HIRS/2 spot area is clear.

Tab. 2 summarizes the results in terms of the variance of ΔT.

Table 2: Variance of ΔT for MR59, MR59+ and MR60.

PASS	σ^2		
	MR59	MR59+	MR60
A (all pixels)	1.60	1.47	0.93
D (all pixels)	1.76	1.54	1.50
A (cloudy pixels)	2.50	2.11	1.30
D (cloudy pixels)	2.81	2.50	2.39

Note that MR60 gives better results than MR59+ which, in turn, works better than MR59. It is also interesting to note that the improvement in the filtering procedure accounts for much of the reduction of the variance (passage from MR59 to MR59+) while the improvement in the in place correlation (passage from MR59+ to MR60) seems to be less significant in the case of pass D. This effect is, in part, understandable in terms of the different air mass types which characterize clear FOVs. Such differences are not take into account in the in place correlation.

Fig. 1 compares the mesh-surface of the cloud-cleared field obtained by MR59 to the one obtained by MR60. The dynamic range on the T-scale in such a figure is about 15 degrees.

It is quite evident that the new version removes spurious micro-structures. Indeed, such micro-structures are due to a consistent bias which migrates through the field as a consequence of the interpolation step taken at the step d) of MR59. Based on our results the following preliminary conclusions may be taken.

a. The cloud mask, i.e., the preliminary identification of clear FOVs constitutes the most crucial and important point of any cloud-clearing scheme. Its failure accounts for most of the errors introduced in the retrieved fields.

b. The HIRS/MSU regression on the basis of pre-computed regression coefficients (forward modeling error) is the second important source of error.

As far as item a) is concerned, the cloud mask can be greatly improved by adding information from AVHRR data. Thus, the ability of parallel processing of HIRS/2 and AVHRR data can greatly improve the results of cloud-clearing. Furthermore, having a good cloud mask, the HIRS/MSU regression relation based on data bank profiles, can be replaced by an in place HIRS/MSU regression relation, that is the coefficients can

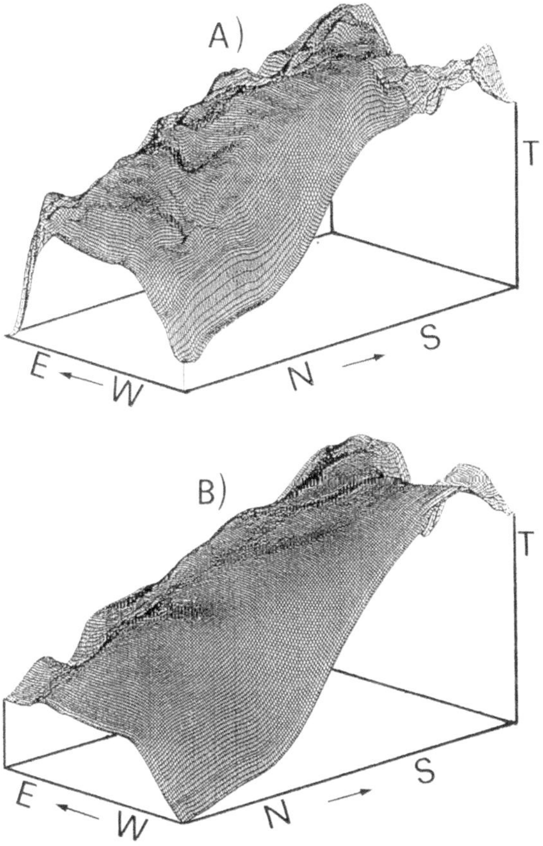

Fig. 1: Cloud-cleared field obtained by MR59 (a) and MR60 (b).

be computed for the given pass of interest on the basis of the clear observation. The effect is to have a regression that does not produce biased results.

Finally, the overall analysis shows that radial basis functions can be effectively used in the design of cloud-clearing schemes.

3. The role of spatial resolution

For the cloud-clearing problem, the more clear FOVs we have, the better. The number of clear FOVs crucially depends on the spatial resolution. In order to investigate such a point it is quite useful to define a cluster index which takes into account the average distance of cloudy pixels from clear pixels.

If (n_o, m_o) is the position of a cloudy pixel on a grid with grid-spacing $L\Delta x$ (here L is a rational number which allows us to define various spatial resolutions simply by changing its values: $L = \frac{1}{2}, \frac{1}{4}$ and so on), we search the position of the nearest neighbour clear node, say (n_c, m_c), and compute the distance d as

$$d(k) = \left[\left(\frac{n_o - n_c}{L}\right)^2 + \left(\frac{m_o - m_c}{L}\right)^2\right]^{\frac{1}{2}} \qquad k = 1, \ldots, N_m \tag{2}$$

where N_m is the number of cloudy FOVs. Then, the average minimum distance, d_m, is defined as: $d_m = \sum_k d(k)/N_m$. We also compute the random average minimum distance d_r, which is the value of d_m as it would be observed if the N_c clear FOVs were uniformly distributed across the grid. Performance of interpolating schemes crucially depends on the difference $d_m - d_r$, the smaller, the better. Change in spatial resolution affects the value of such a difference, therefore in this section we present some preliminary results about the dependence of $d_m - d_r$ on the spatial resolution. Only the case in which spatial resolution is doubled is discussed. This means that if linear dimension of present FOVs is $L = 1$, the one of "new" high resolution FOVs is $L = \frac{1}{2}$. Consequently four new FOVs correspond to each present FOV and if the present HIRS field contains M pixels the higher resolution field will contain $4M$ FOVs. AVHRR data co-located at HIRS/2 spots were used in order to identify either clear or cloudy FOVs in the two different patterns ($L = 1$ and $L = \frac{1}{2}$). In Fig. 2 we present the clear HIRS mask for $L = 1$ and $L = \frac{1}{2}$ (Pass A).

The improvement is quite evident since now clear FOVs appear in areas which are detected as cloudy according to the current HIRS/2 spatial resolution. In Figs 3 and 4 we present some results regarding pass A.

Fig. 3 shows shows a histogram of the clearness index C as defined in section 3. Cases $L = 1$ and $L = \frac{1}{2}$ are compared. The bin $C = 100$ corresponds to FOVs that are completely clear whereas the bin $C = 0$ to FOVs that are completely cloudy.

Fig. 4 gives the number of cloudy FOVs as a function of the distance (in terms of the grid spacing) of the nearest completely clear spot. It is quite evident from such a

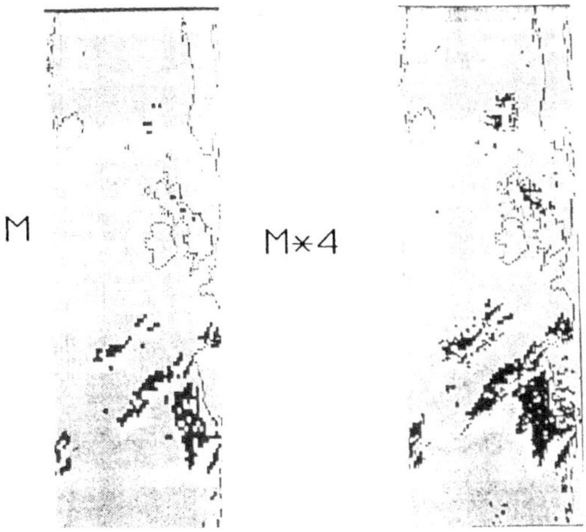

Fig. 2: Clear HIRS mask for $L = 1$ and $L = \frac{1}{2}$ (Pass A).

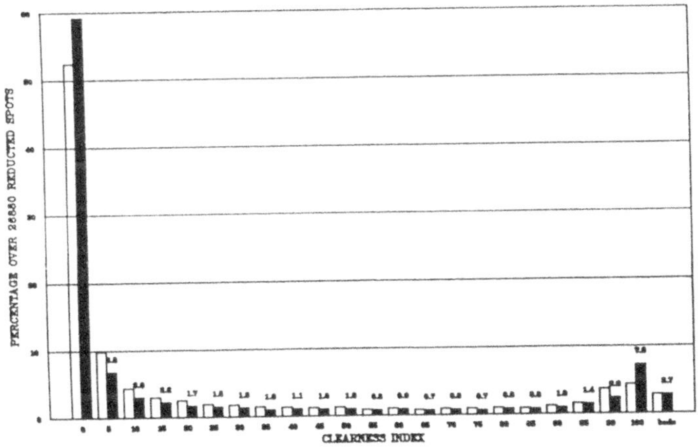

Fig. 3: Histogram of clearness index C for $L = 1$ (white bars) and $L = \frac{1}{2}$ (black bars).

figure that when $L = \frac{1}{2}$, cloudy FOVs are more likely to have closer clear FOVs. As a result, the increase in the spatial resolution should allow one to obtain fields with higher number of clear FOVs and lower d_m.

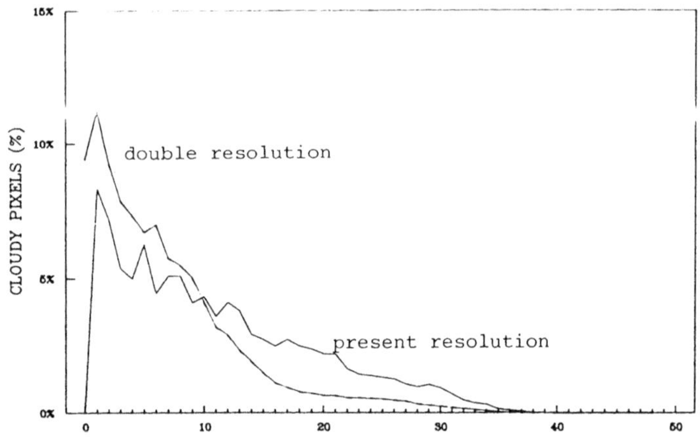

Fig. 4: Number of cloudy FOVs vs. distance of the nearest clear spot.

4. The temperature profile retrieval

The equation for calculating radiance from an atmospheric temperature profile is

$$R_\nu(z,\theta,\phi) = \varepsilon_o B_\nu(T_o(\theta,\phi))\tau(z_o,z) + \int B_\nu(T(\theta,\phi,z)\frac{d\tau_\nu(z,z_1)}{d\log p}d\log p \qquad (3)$$

where R_ν is the radiance at the top of the atmosphere measured by the channel ν (i.e., peaked at the frequency ν), z is altitude, z_o and z_1 represent ground and top atmosphere, and $d\tau_\nu/d\log p$ is the weight function of the channel ν.

Eq. (3) is a first kind Fredholm integral equation and such equations are ill-posed problems. For this reason two techniques are used to get reasonable solutions: regularization and introduction of available physical information into the problem.
Regularization is accomplished by means of a smoothing functional (usually based on the derivative of the solution) and a regularization parameter. Regularization represents an optimal technique form the mathematical point of view for solving inverse problems and actually it proves to be effective in many different applications. However, the power of the method dramatically depends on the choice of the regularization parameter. In fact, the solution of the regularized problem is strongly affected by it. If the aim is to devise an overall technique that is as objective as possible, it is clear that a crucial point is the development of some objective criterion for the choice of the regularization

parameter. The only criterion working when the variance of the experimental error is unknown is the Generalized Cross Validation (GCV) criterion (Wahba 1974). It holds under the hypotheses that experimental error is a white noise and asymptotical (i.e., for increasing number of measurements).

A great help in solving the inverse problem can come from properly exploiting available physical information. However, it is important to introduce such information into the problem in an objective way. The common practice is to obtain a preliminary solution ("first guess") and to use it in order to regularize the problem. Physical information is available through a large data base of temperature atmospheric profiles and corresponding radiances calculated by a forward model. One of the most successfull objective methods consists in choosing the profile of the data base whose corresponding radiances are "nearest" to the experimental ones in some sense (Chedin et al. 1985). It is clear that the first guess always belongs to the data base.

In our approach radiances and temperature profiles are linked through the following linear relationship:

$$R_\nu = \sum a_{\nu k} T(z_k) + c_\nu \tag{4}$$

where $T(z_k)$ is temperature at level z_k and $a_{\nu k}$ and c_ν are coefficients to be evaluated.

In order to obtain the first guess temperature profile, the following linear system of equations is solved:

$$R_\nu^{\mathrm{exp}} = \sum a_{\nu k} T(z_k) + c_\nu \tag{5}$$

The estimate of the coefficients $a_{\nu k}$ and c_ν is made basing on the profiles of the data base and corresponding radiances by linear regression. To this purpose, 1200 different profiles were considered. Note that $T(z)$ does not belong to the data base; the only role of the data base is to be wide enough in order to compute correctly the $a_{\nu k}$ and c_ν coefficients taking into account the different conditions of air masses and atmospheric structure.

Eq. (5) turns to be again an inverse problem. Then regularization tools can be resorted again, so that the solution is obtained by solving the following minimum problem:

$$\min_T \sum_\nu \left\{ R_\nu^{\mathrm{exp}} - \left[\sum_k a_{\nu k} T(z_k) + c_\nu \right] \right\}^2 + \lambda \mathcal{O}(T) \tag{6}$$

where $\mathcal{O}$ is a differential functional of T and λ is a regularization parameter.

As an example we have calculated the solution of pb. (6) in correspondence with three test temperature profiles. Three different values of the regularization parameter were considered: $\lambda = 0$ (that is, usual least squares solution); $\lambda = \lambda_{GCV}$ (that is, regularization parameter as obtained by GCV criterion); $\lambda = \lambda_{opt}$ (that is, the value of the regularization parameter which minimize the difference between the experimental and predicted radiances in the l_2 norm). Results are shown in Figs. 5a, 5b and 5c for $\lambda = 0$ and λ_{GCV} together with the true profile.

It is clear that the least squares method reproduces the true solution only approximately, while the optimal choice of λ always allows one to fit well the true solution. GCV criterion proves to be completely unsatisfactory in estimating the optimal value of the regularization parameter. This can be explained since GCV is an asymptotic method that requires a much higher number of measuring channels.

Summarizing we can affirm that higher resolution istrumentation is twice important, because temperature is much more resolved in its vertical behaviour and also because it becomes possible to use optimal mathematical methods.

5. Conclusions

1) Radiances of clear FOVs play an important role in retrieval processes. This requires that cloud detection schemes must be very well based. Up to now the best schemes seem to be based on AVHRR data. The analysis of new instrumentation infrared data should be accompanied by on line processing of an imager such as AVHRR.
2) Some cloud clearing procedures seem to be quite satisfactory even if further improvements can and must be pursued.
3) A doubling of spatial resolution increases the number of clear FOVs and the decrease of the cluster index of clouds. Consequently the increase of spatial resolution significantly affects retrieval processes. A more detailed assessment of the improvements due to higher spatial resolution should be based on a wide analysis of AVHRR radiance fields and on a sensitivity analysis of retrieval processes on cluster indexes of clouds.

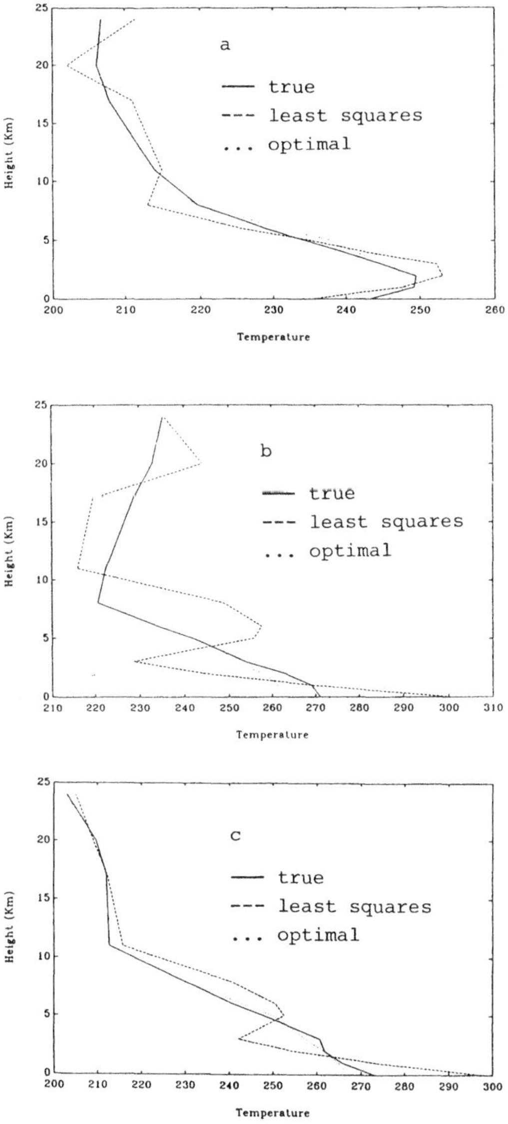

Fig. 5: True and retrieved first guess temperature profiles for 3 test cases (a,b,c).

4) Well based inversion techniques exist now (for example methods with regularization and choice of the regularization parameter by GCV). On the other hand GCV

holds only asymptotically (i.e., large number of measuring channels) and cannot be used at present. From this point of view an increase in vertical resolution is very important not only because it increases the quality of input data but also because it allows the use of "optimal" mathematical tools.

Acknowledgements: Work supported by CNR, *Progetto Strategico Clima Ambiente e Territorio del Mezzogiorno.*

References

Chedin, A., N.A. Scott, C. Wahicke and P. Mouliner, 1985: The Improved Initialization Inversion Method: a High Resolution Physical Method for Temperature Retrievals for Satellites of the TIROS-N Series. *J. Climate and Appl. Met.* **24**, 128–146.

Eyre, J.R. and P.D. Watts, 1987: A Sequential Estimation Approach to Cloud-Clearing for Satellite Temperature Sounding. *Quart. J. Roy. Met. Soc.* **113**, 1349–1376.

McMillin, L.M. and C. Dean, 1982: Evaluation of a New Operational Technique for Producing Clear Radiances. *J. Appl. Met.* **21**, 1005–1014.

Cuomo, V., R. Rizzi and C. Serio, 1992: A new Objective and Optimal Cloud-Clearing Scheme for Infrared Sounders. *Int. J. Remote Sens.*, in press.

Rizzi, R., C. Serio, G. Kelly, V. Tramutoli, A. McNally and V. Cuomo, 1992: The Cloud-Clearing Comparison Exercise. *Techn. Mem. ECMWF No. 185.*

Wahba, G., 1977: Practical approximate solutions to linear operators equations when the data are noisy. *SIAM J. Numer.Anal.*, **14**, 820-831.

Retrieval of Cloud Parameters by Multiple High Spectral Resolution Observations in the Near-Infrared Under Conditions of Varying Solar Illumination

Thomas J. Kleespies
Phillips Laboratory, Geophysics Directorate
PL/GPAS
Hanscom, AFB MA 01731
USA

1. Introduction

The near-infrared region of the upwelling earth's spectrum is important because of the combination of thermal and solar radiation. Clouds modulate both the incoming solar radiation by scattering, and the outgoing thermal radiation by emission. Unfortunately, this combination of thermal and solar radiation has made it difficult to use observations in this region, and numerous investigators have avoided using daytime 3.7 μm observations (e.g. Strong and McClain, 1984, d'Entremont, 1986, Saunders and Kriebel, 1989). Clouds have the effect of cooling the atmosphere above their tops, and warming of the atmosphere below their bases (c.f. Liou, 1980). Since stratus and stratocumulus (S,SC) are generally low in the atmosphere, their net effect is to cool the atmosphere as a whole. The annual frequency of occurrence of S,SC over the world's oceans is about 45%, while the annual S,SC cloud amount is about 35% (Warren, et al., 1988). The corresponding figures over land are about 25% and 18% (Warren et al., 1986). These low level water clouds have an effect on both the thermal and solar radiation due to a significant departure from unity of their effective emissivity (Hunt, 1973, Twomey and Cocks, 1982). This is because the cloud droplet size approaches the wavelength of the interacting light. For example, Hunt (1973) computes the emissivity at 3.7 μm for optically thick water clouds to be 0.74 at a mode radius of 4 μm, but 0.90 at a mode radius of 10 μm. In comparison, the emissivity at 11 μm was computed to be 0.997 and 0.998 for the respective mode radii.

The fact that radiation in the 3.7 μm spectral region is composed of both solar and thermal radiation is certainly a complication to radiative transfer. Most standard texts treat solar and thermal radiation as separable and pose independent solutions to the two problems (e.g. Liou, 1980; Goody, 1964; Chandrasekhar, 1960). Nevertheless, the formal statement of the equation of transfer is sufficiently general to allow for both solar and thermal radiation.

NATO ASI Series, Vol. I 9
High Spectral Resolution Infrared Remote Sensing
for Earth's Weather and Climate Studies
Edited by A. Chedin, M.T. Chahine and N.A. Scott

A number of investigators have used remotely sensed solar radiation to determine stratus cloud parameters. Nakajima and King (1990) use reflectance at 0.75 and 2.16 μm measured by the multichannel radiometer (MCR), flown on the NASA ER-2 during the FIRE marine stratocumulus field project in July 1981, to estimate the optical thickness and effective particle radius. Curran and Wu (1982) use channels centered at 0.83, 1.16, and 2.15 μm on an instrument flown on Skylab to determine cloud optical thickness, thermodynamic phase and effective particle size of orographic clouds. Rawlins and Foot (1990) use four channels (0.55, 1.04, 1.24 and 1.33 μm) on a similar MCR flown on a C130 aircraft during FIRE also to determine optical thickness and effective radius of marine stratiform clouds. Whereas the results of these and similar studies can be quite good, they make use of experimental instruments, and thus are useful only for case studies.

An advantage of using the 3.7 μm region is that there is a long operational archive of these data. The AVHRR-2 first flew on TIROS-N which was launched in October 1978 (Kidwell, 1991), and the VAS first flew on GOES 4 in September 1980 (Smith et al., 1981). The data from these satellites are readily available from NESDIS for regional, global and time series studies. The major reason that the 3.7 μm channel has been eschewed in daytime is the difficulty in separating the solar and thermal components. If there was *a priori* knowledge of the atmospheric state and composition, as well as of the cloud and surface characteristics, then the solar and thermal effects could be separated by radiative transfer calculations. However, in most remote sensing problems atmospheric and surface state and composition are among those parameters to be retrieved.

The first part of this paper will propose a method of removing the effects of thermal radiation from daytime observations in the 3.7 μm region, and from the resulting solar radiation, estimate cloud properties. The discussion will focus on the application of the technique to persistent opaque marine stratiform clouds. The second part of this paper will examine the use of high spectral resolution measurements to remove the constraint that the clouds be opaque.

2. Proposed Technique

The method proposed here makes use of the viewing geometry from geosynchronous orbit (Figure 1). The satellite views a cloud feature twice over a short period of time during which the thermal structure, location and radiative characteristics of the cloud change little, and the structure and character of the atmosphere above the cloud also change little. During this time the sun's position changes in the sky, thus the amount of radiation impinging on the cloud is different for the two viewing times. As the solar elevation changes, the path length from space to the cloud changes, which modulates the amount of sunlight reaching the cloud. With

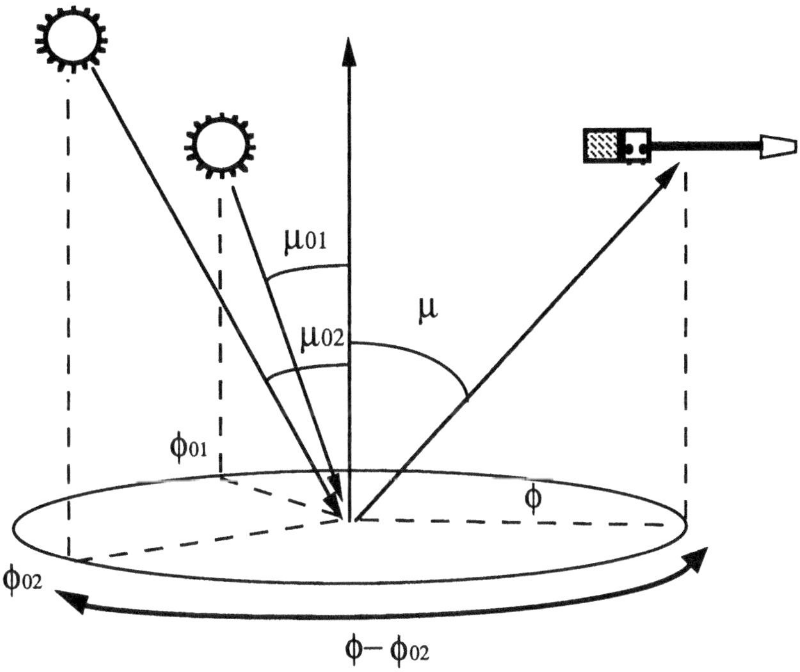

Figure 1. Viewing geometry from geosynchronous orbit. μ is cosine of the polar angle, φ is the azimuth angle.

these assumptions, the difference in the radiance measurements cancels the thermal component, and the result is the difference in the solar component.

The inversion of the radiance difference into cloud properties, such as emissivity, is not a straightforward problem. There is no closed mathematical form. However, it is possible to deduce these properties by making offline computations, and comparing the observations with the computations.

By assuming a particular cloud drop size distribution, say a modified gamma function, and using Mie theory to compute the scattering parameters, expected radiances can be computed for a range of angles and mode radii. Whereas these computations are tedious, they are straightforward. Then with the known measurements and observational angles, a table lookup can then be performed to deduce the effective mode radius which led to the reflected radiance.

Presently there is only one instrument in geosynchronous orbit with a near-infrared channel in the 3.7 μm region, the Visible-Infrared Spin Scan Radiometer (VISSR) Atmospheric Sounder (VAS). Although both the imager and sounder on GOES-NEXT will have channels in the 3.7 μm region, programmatic delays make the VAS the instrument of choice for a proof of concept study.

3. Application to the VISSR Atmospheric Sounder

The VAS has twelve infrared channels and one visible channel. It operates in either an imaging mode or a sounding mode. Visible and one, two, three or four infrared channels are available in the imaging mode. All twelve channels are available in the sounding mode. The near-infrared window channel is centered near 3.9 μm.

Sounding data were taken from 18 July 1991. The data were available every 90 minutes and the nominal spatial resolution was 32x16 km. The region of interest was an area of persistent marine stratus near the coast of California. Figure 2 gives the brightness temperatures of the 3.9 and 11 μm window channels as a function of time for a single location about 100 km west of San Francisco.

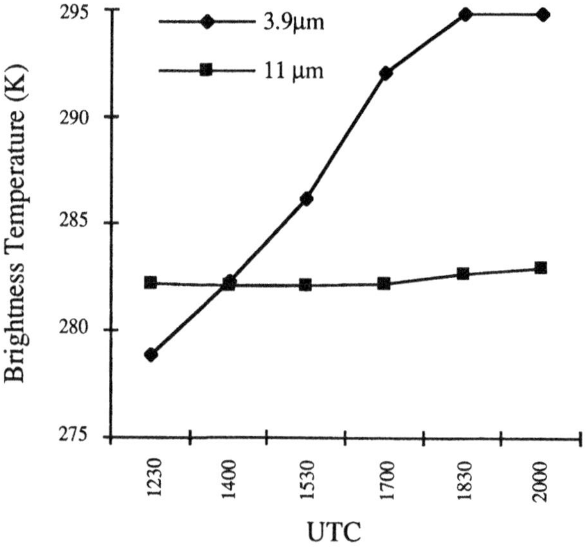

Figure 2. Brightness temperature as a function of time for VAS near- and mid-infrared window channels.

While the 11 μm brightness temperature changes only by a few tenths of a degree, the 3.9 μm brightness temperature changes from approximately 278 K before local sunrise (1230 UTC) to 294 K near local noon (1830 UTC) due to the influence of solar radiation. The fact that the 11 μm brightness temperature changes very little suggests that the assumption of thermal invariance of the cloud and the atmosphere was met. The ensuing discussion will concentrate on the interval from 1530 to 1700 UTC, when the greatest 3.9 μm brightness temperature change occurred.

Forward radiative transfer calculations were made in order to estimate the cloud properties that generated the measured radiance difference. The clouds were presumed to have a modified gamma size distribution.[1] Scattering parameters (extinction, single scattering albedo, phase function) were calculated from a Mie code (Wiscombe, 1988) for different polydispersions generated using a range of droplet mode radii, and some of which are presented in figure 3.

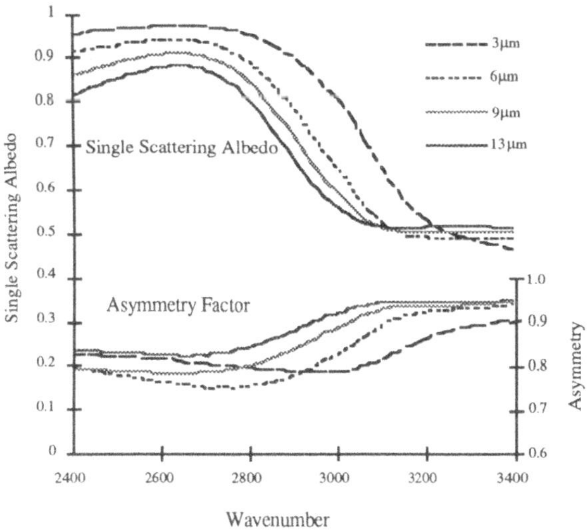

Figure 3. Single scattering albedo and asymmetry factor as a function for wavenumber for polydispersions with a variety of mode radii.

The complex indices of refraction used in producing Figure 3 were those tabulated by Downing and Williams (1975). These scattering parameters were in turn used in a multiple scattering code described by Stamnes et al (1988). This model is a stable, general purpose discrete ordinate model which can utilize a large variety of sources and boundary conditions. Sixty-four

1 Hansen and Travis (1974) note that the scattering properties of a polydispersion are largely independent of the size distribution, depending primarily upon an effective radius, which can be related to the mode radius through an effective variance. See Hansen and Travis for details.

streams were used in the calculations presented here. The solar forcing was extracted from the MODTRAN data base (Berk et al, 1989). The cloud was considered to be a single layer of optical depth 10 in contact with a lambertian surface with an albedo of 0.05. Absorption and scattering above the cloud were neglected. The observed radiance difference was compared with the computed radiance difference. Where they agreed, the mode radius that generated the matching computed radiance difference was the solution. The results are presented in figure 4.

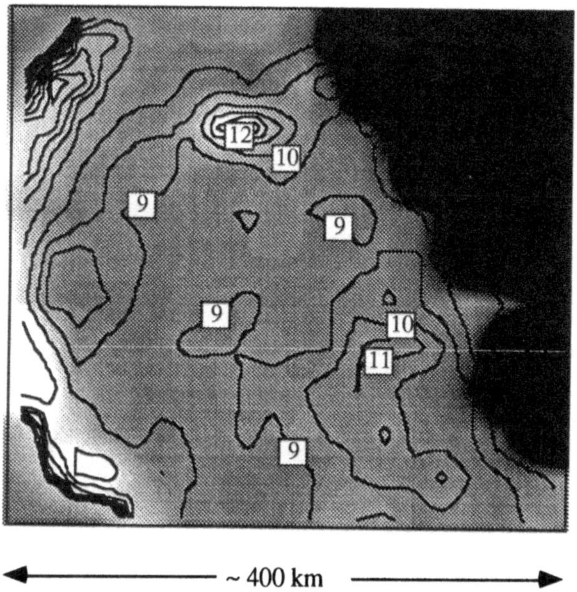

Figure 4. Contours of retrieved stratus mode radius in μm. Background is 11 μm brightness temperature difference from 1530 to 1700 UTC. Neutral gray indicates no 11μm brightness temperature difference.

The retrieved mode radius is depicted as a contour in figure 4. The gray shade background in figure 4 is the brightness temperature difference from 1530-1700 UTC and is used as a quality control measure to judge the validity of the retrieval. The dark area to the right is the California land mass heating dramatically over the period. A depiction of the entrance to San Francisco Bay can be seen right center, and Monterey Bay is in the lower right corner. The mode radius retrieval can be considered to be unreliable along the coastline because of land influence. On the left side of figure 4 is a bright feature. This is due to stratus advecting over a region of clear ocean. Since the cloud field was not continuous in this region for both viewing times, the retrieval on the left side of the figure is not valid. In the top center of figure 4 is a bright spot which is due to changes in the cloud field. The resulting "bullseye" in the contour analysis of mode radius is an artifact of this change and is not a true feature. The rest of the image is predominantly a neutral gray, which indicates the 11 μm brightness temperature difference is not greater than a few tenths of a degree. The lack of 11 μm brightness temperature variation gives confidence that the assumption of thermal invariance was met over the larger part of the region, and that the retrieved mode radii are valid.

Over the region where the retrieval is valid, the mode radius is about 9 μm. However, just to the west and south of San Francisco Bay is a region where the predominant cloud particle size is larger than the surroundings. One might speculate that this may be due to the effect of anthropogenic cloud condensation nuclei, but the mechanism is not clear at this time.

4. Application of High Spectral Resolution Data

The problem with the method presented in the previous section is that it presumes the clouds to be opaque. In reality, the radiance reflected by a cloud is a function not only of the mode radius, but of the optical depth as well. The radiance difference is therefore also a function of these two variables. The radiance difference was computed for a broad range of mode radius and optical depth and is presented in figure 5 to illustrate this relationship.

The computations done in the previous section were at an optical depth of 10. The observed radiance differences were in the range of 0.12 to 0.16 mW/(m^2 ster cm^{-1}). However figure 5 indicates that there is no unique mode radius that will generate a given radiance difference. Whereas this relationship changes slowly for mode radius larger than 6 μm and optical depths greater than five, it can be significant for clouds that fall outside this regime.

Suppose high spectral resolution measurements could be made from geosynchronous orbit under the conditions described in section two. To simulate this scenario, the radiance computations performed in section two were extended from a discrete wavenumber to the range 2400-3400 cm^{-1}, at an interval of 10 cm^{-1}. The computations were repeated at a number of optical depths and mode radii. Prior to examining the radiance difference, it would be profitable to first examine the radiance itself so as to better understand the physics of the problem. Figure 6 is the radiance for a 3 μm mode radius with various optical depths as a function of wavenumber, for a scattering angle of 160 degrees, which corresponds to the 1700 UTC case discussed in section three. Since it is desired to examine only the properties of cloud scattering, there is no thermal emission in these computations.

The first point to note about figure 6 is that the radiance for a given optical depth is a smooth function of wavenumber, with a pronounced maximum. The reason for this maximum lies in the scattering phase function. Figure 7 gives the scattering phase function

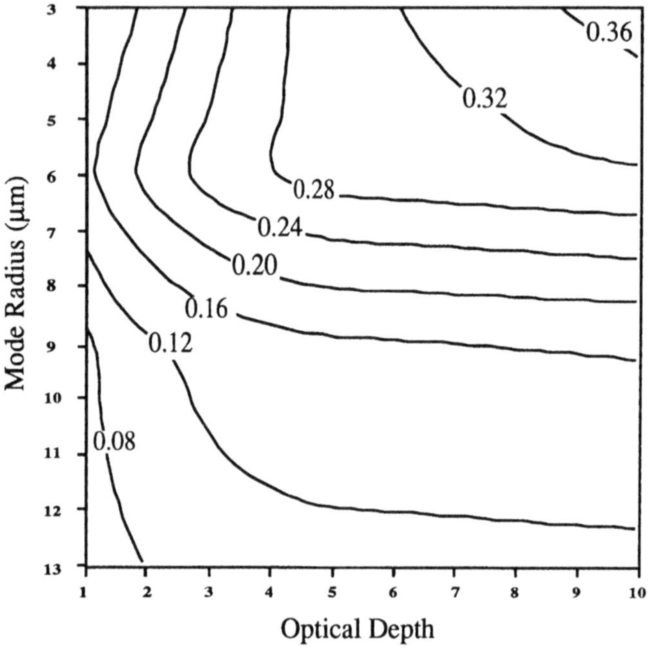

Figure 5. Radiance difference as a function of mode radius and optical depth. Units are mW/(m^2 ster cm^{-1}).

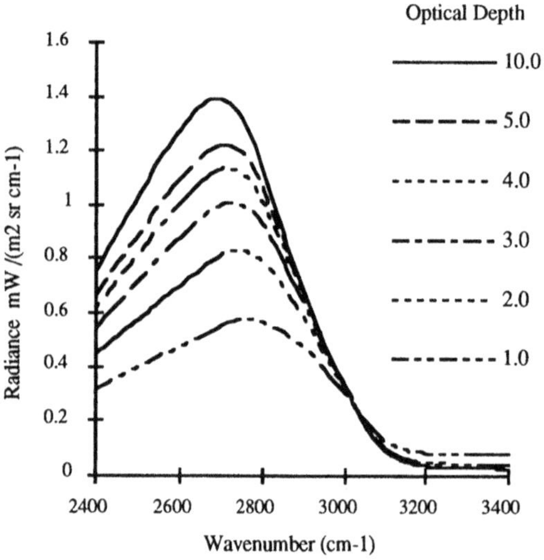

Figure 6. Radiance as a function of wavenumber and optical depth for a 3 μm mode radius and a scattering angle of 160 degrees.

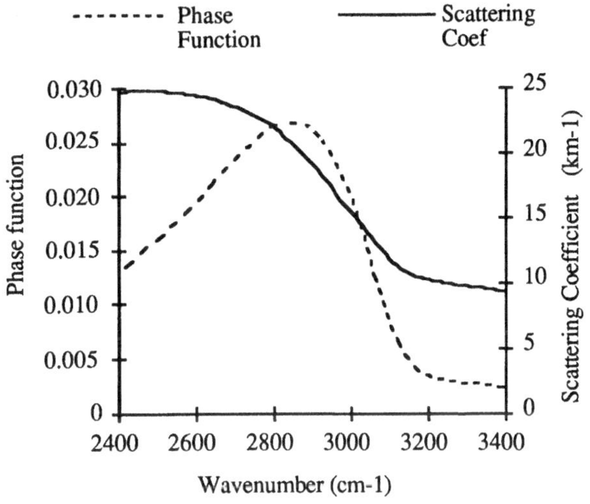

Figure 7. Scattering phase function for 160 degrees and scattering coefficient as a function of wavenumber for a mode radius of 3 μm and an optical depth of 5.

and scattering coefficient for an optical depth of five. The pronounced peak in the phase function value is what gives rise to the similar peak in the reflected radiance from the cloud. Since the scattering coefficient monotonically decreases with wavenumber, the product of the scattering coefficient and the phase function will cause the peak reflected radiance to be at a somewhat lower wavenumber than the peak of the phase function itself. In this case of the optical depth of five the peak of the phase function is at 2840 cm^{-1} while the peak of the radiance is at 2740 cm^{-1}.

The second fact worth mentioning about figure 6 is that as the optical depth increases, the reflected radiance increases as well. This is to be expected, since the scattering coefficient increases linearly with optical depth.

The third and probably the most significant fact to note about figure 6 is that the wavelength of the maximum radiance decreases with increasing optical depth, from nearly 2800 cm^{-1} at optical depth of 1 to 2700 cm^{-1} at optical depth of 10. The reason for this is not immediately obvious since the scattering parameters (phase function and single scattering albedo) do not change with optical depth. This phenomena can be understood by examining the absorption and scattering coefficients in figures 8 and 9.

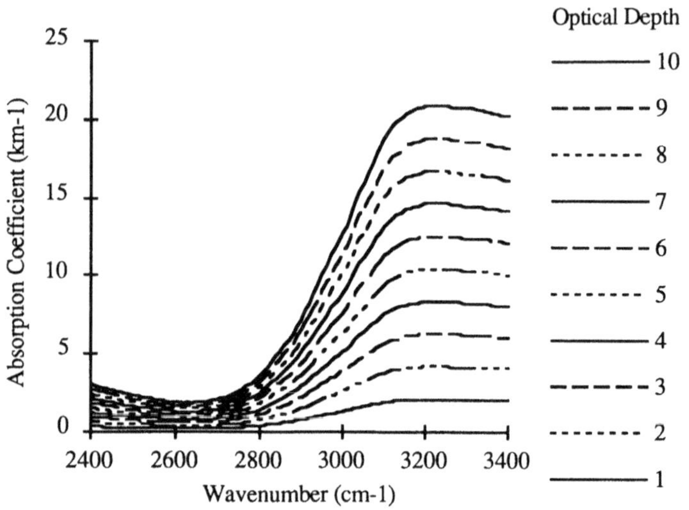

Figure 8. Absorption coefficient as a function of wavenumber and optical depth for 3 μm mode radius.

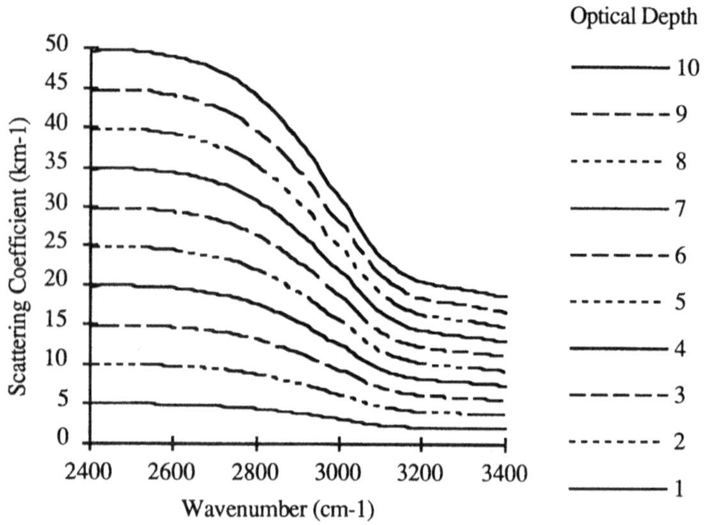

Figure 9. Scattering coefficient as a function of wavenumber and optical depth for 3 μm mode radius.

It is clear from figures 8 and 9 that there is a systematic difference in the absorption and scattering coefficients between 2800 and 2700 cm^{-1}. This difference increases with increasing optical depth. The interpretation is that for a given optical depth, there is more absorption at 2800 cm^{-1} than at 2700 cm^{-1}, and that this effect increases with optical depth. Correspondingly there is more scattering at 2700 cm^{-1} than at 2800 cm^{-1}, and that this effect also increases with optical depth. These two effects combine in such a manner that the maximum of the scattered radiation shifts from 2800 cm^{-1} to 2700 cm^{-1} with increasing optical depth.

The radiances for 145 degree scattering angle (corresponding to 1530 UTC) exhibit very similar behavior to those of the 160 degree scattering angle. The difference between these radiances exhibits properties similar to the radiances themselves, as evidenced in figures 10 and 11. The two most important properties are that as the optical depth increases, 1) the radiance difference below 3000 cm^{-1} increases, and 2) the wavenumber of the maximum difference decreases. Although results are presented here for the two mode radii (3 and 13 μm), calculation at intermediate values verify that as the mode radius increases, the radiance difference decreases.

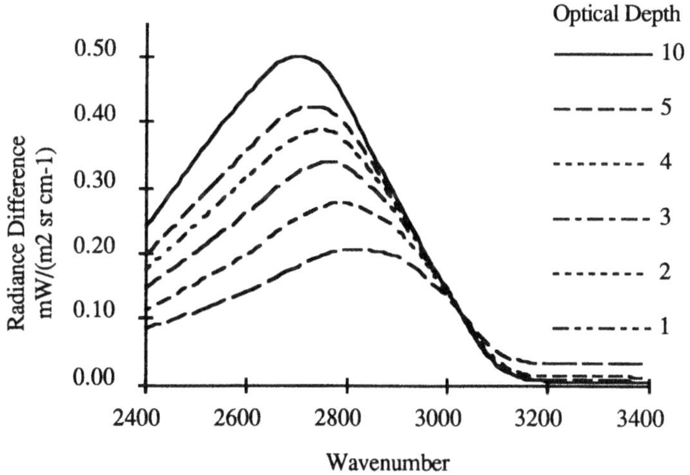

Figure 10. Radiance difference as a function of wavenumber and optical depth for a 3 μm mode radius.

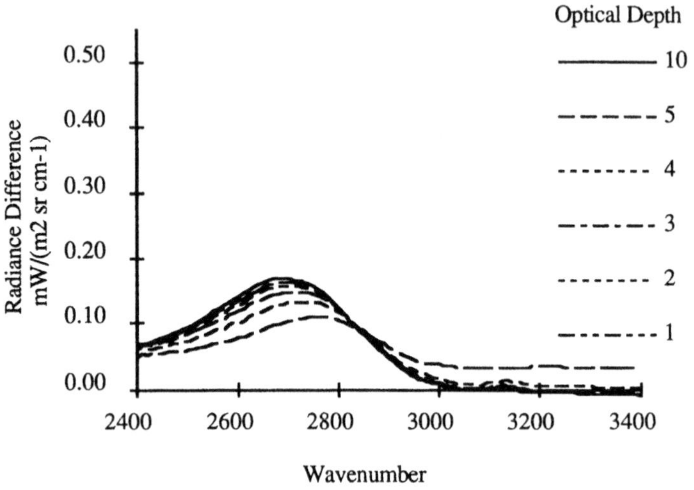

Figure 11: Same as figure 10 except for 13 μm mode radius.

These figures suggest that the reflectance spectrum may be reduced to two parameters with which to solve for two unknowns. The parameters are the maximum value of the radiance difference, and the wavenumber of the maximum value. The unknowns are the optical depth and the mode radius. Figure 12 gives an example of a simultaneous retrieval of both mode radius and optical depth.

It is understood from figure 5 that there is a continuum of optical depths - mode radius pairs that will produce the same radiance difference. However, in the interest of clarity, consider only discrete values. In this hypothetical example, the maximum radiance difference is 0.15 mW/(m^2 ster cm^{-1}) and the wavenumber of this maximum is at 2765 cm^{-1}. From the top part of figure 12 it can be seen that 0.15 mW/(m^2 ster cm^{-1}) can have two possible mode radii, 9 μm and 13 μm. However, by applying the additional constraint that the wavenumber of the maximum value was observed at 2765 cm^{-1} the ambiguity is resolved, and 9 μm is the recovered value. Furthermore, it is determined that the optical depth for this case is about 1.7.

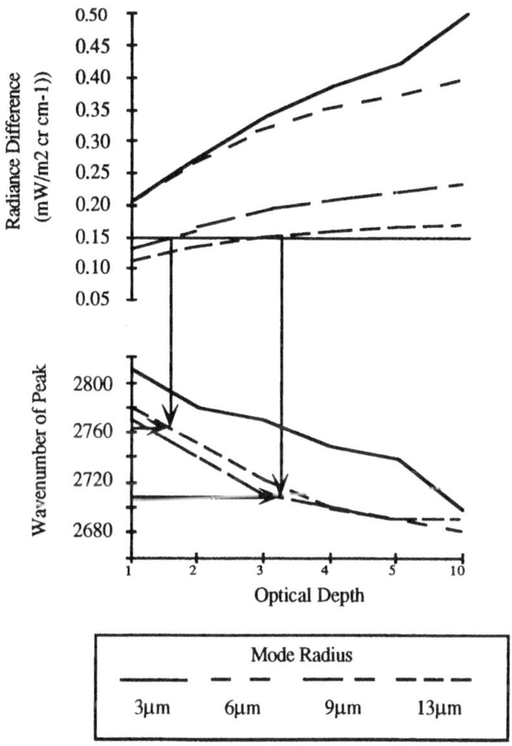

Figure 12: Example of simultaneous optical depth and mode radius retrieval. For this example, the maximum radiance difference is 0.15 mW/(m^2 ster cm^{-1}) and the wavenumber of the maximum value is at 2765 cm^{-1}.

5. Summary and Further Considerations

This paper presented a simple method for removing the thermal component from near-infrared window observations of water clouds that change little over a short time interval. The resulting radiance difference was then used to determine the cloud mode radius for persistent coastal marine stratus. A theoretical study was performed in which high spectral resolution radiances were used to remove the constraint that the clouds be optically thick which was placed on the single channel retrieval.

The calculations in this study did not consider absorption or scattering by aerosols above the cloud, nor absorption of water vapor in the cloud, which could be a significant factor. They also did not include any contribution by the atmosphere below the cloud, which might be improtant for some of the optically thin cases. Another area of concern is the vertical structure of the dropsize distribution below the cloud top. Future work will consider all of these effects.

6. References

Berk A, Bernstein L, Robertson D (1989) MODTRAN: A moderate resolution model for LOWTRAN 7. Air Force Geophysics Laboratory GL-TR-89-0122

Chandrasekhar S (1960) Radiative Transfer. Dover Publications.

Curran RJ, Wu MC (1982) Skylab near-infrared observations of clouds indicating supercooled liquid water droplets. J Atmos Sci 39:636-647

d'Entremont RP (1986) Low- and mid-level cloud analysis using nighttime multispectral imagery. J Climate and App Met 25:1853-1869

Downing HD, D Williams (1975) Optical constants in the infrared. J. Geoph. Resch. 19:1505-1509.

Goody RM (1964) Atmospheric Radiation, I, Theoretical Basis, Oxford Press

Jursa AS (ed) (1985) Handbook of geophysics and the space environment. Air Force Geophysics Laboratory

Hansen JE, LD Travis (1974) Light scattering in planetary atmospheres. Space Sci. Rev. 16:527-610.

Hunt GE (1973) Radiative properties of terrestrial clouds at visible and infra-red thermal window wavelengths. Quart J Roy Meteor Soc 99:346-369

Kidwell KB (1991) NOAA Polar orbiter data users guide. NOAA/NESDIS, Washington DC

Liou K-N(1980) An introduction to atmospheric radiation. Academic Press

Nakajima T, King MD (1990) Determination of the optical thickness and effective particle radius from reflected solar radiation measurements. Part I: Theory. J Atmos Sci 47:1878-1892

Rawlins F, Foot JS, (1990): Remotely sensed measurements of stratocumulus properties during FIRE using the C130 aircraft multi-channel radiometer. J Atmos Sci 47:2488-2503

Saunders RW, Kriebel KT (1988) An improved method for detecting clear sky and cloudy radiances from AVHRR data. Int J Remote Sensing 9:123-150

Smith WL, Suomi VE, Menzel WP, Woolf HM, Sromovsky LA, Revercomb HE, Hayden CM, Erickson DN, Mosher FR (1981) First sounding results from VAS-D. Bull Am Met Soc 62:232-236

Stamnes K, Tsay S, Wiscombe W, Jayaweera K (1988): Numerically stable algorithm for discrete-ordinate-method radiative transfer in multiple scattering and emitting layered media. Appl Optics 27:2502-2509

Strong AE, McClain EP (1984) Improved ocean surface temperatures from space-comparisons with drifting buoys. Bull Am Met Soc 65:138-142

Twomey S, Cocks T (1982) Spectral reflectance of clouds in the near-infrared: Comparisons of measurements and calculations. J Met Soc Japan 60:583-592

Warren SG, Hahn CJ, London J, Chervin RM, Jenne RL (1986) Global distribution of total cloud cover and cloud type amounts over land. DOE/ER/60085-H1, NCAR/TN-273+STR

Warren SG, Hahn CJ, London J, Chervin RM, Jenne RL (1988) Global distribution of total cloud cover and cloud type amounts over the ocean. DOE/ER/-0406, NCAR/TN-317+STR

Wiscombe, WJ (1980) Improved Mie scattering algorithms. Appl. Optics. 19:1505-1509.

SPECTRAL VARIABILITY OF LIGHT SCATTERING BY ATMOSPHERIC ICE CRYSTALS

ANDREAS MACKE
GKSS FORSCHUNGSZENTRUM
INSTITUT FÜR PHYSIK
POSTFACH 1160
W - 2054 GEESTHACHT
GERMANY

ABSTRACT

The scattering phase functions of arbitrary shaped ice particles are calculated by means of geometrical optics and diffraction theory. Their orientation is assumed to be random in space. Particle shapes are taken both from ice crystal classifications and from in-situ measurements made during the International Cirrus Experiment ICE, 1989. The spectral variability is discussed for hexagonal ice crystal columns in the range from 0.2 μm to 16.0 μm with medium spectral resolution ($\triangle\lambda \sim 0.5\mu m$), and from $2.5\mu m$ to $3\mu m$ with high spectral resolution ($\triangle\lambda \sim 0.02\mu m$). In principal, this variability reflects the different absorption mechanisms within the individual regions. The comparison of different particle types shows that backscattering is a sensitive indicator for ice crystal type identification. Aggregated particles like bullet rosettes show basically the scattering characteristics of their individual components. A qualitative estimate of the scattering behaviour of highly complex structured particles is demonstrated by scattering calculations at two dimensional fractal particles.

1 Introduction

The description of high altitude cirrus clouds, their microphysical, thermodynamical and radiative properties have been identified as one of the major unsolved problems in climate research (Liou 1986, Stephens et al. 1990). Several present projects like the First ISCCP Regional Experiment (FIRE) and the International Cirrus Experiment (ICE) cover the physics of ice clouds both from the experimental and the theoretical point of view (FIRE 1990, Raschke et al. 1990).

NATO ASI Series, Vol. I 9
High Spectral Resolution Infrared Remote Sensing
for Earth's Weather and Climate Studies
Edited by A. Chedin, M.T. Chahine and N.A. Scott

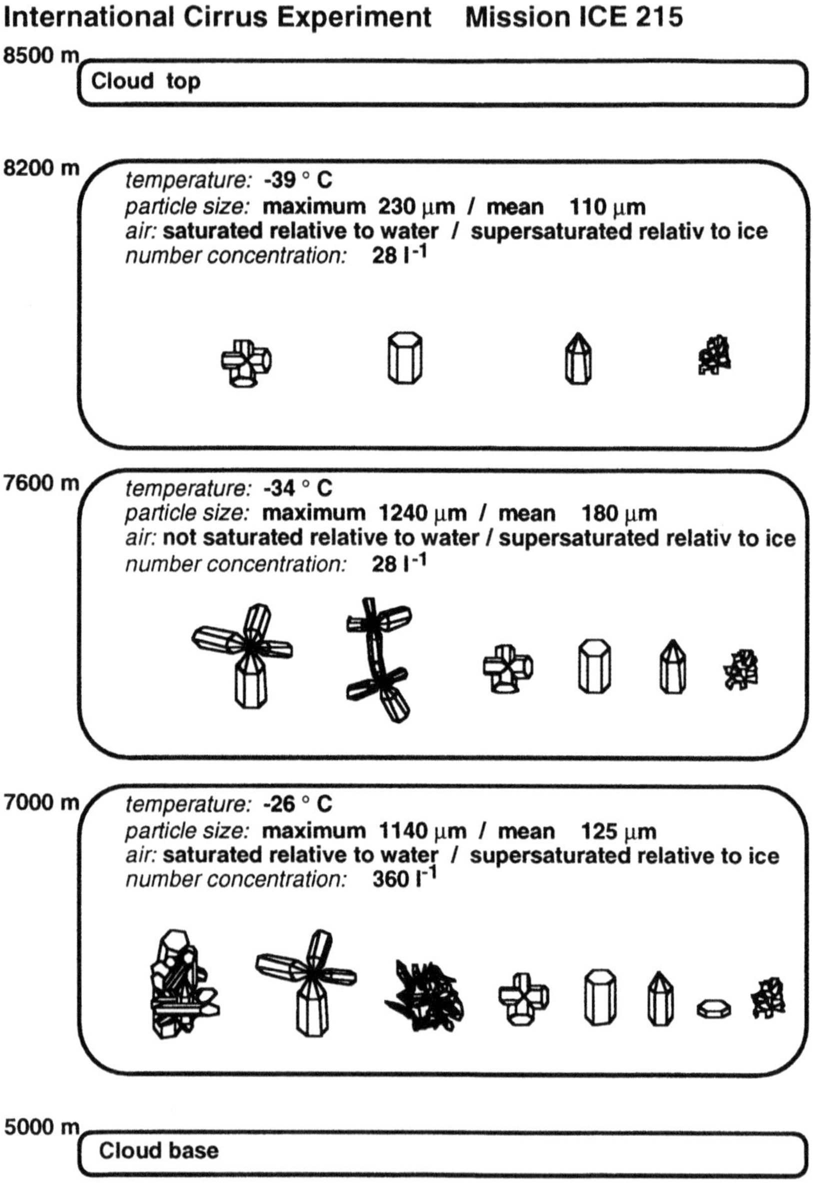

Fig. 1 Typical ice particle sizes and shapes observed during ICE IFO '89. The data were taken from an airborne holographic measurement system, operated by the Metereological Office Research Flight, UK

The overall effect of cirrus clouds on the energy balance of the Earth-atmosphere system depends on the global distribution of the greenhouse-versus-albedo effect of this cloud type. A detailed knowledge of the degree and extent of the cirrus cloud interaction with terrestrial and solar radiation is therefore essential for climate research. Global determination of cloud cover, type and optical properties from satellite radiance measurements requires improved radiative transfer computations for cloudy atmospheres at numerous wavelengths.

Recent work on the direct and indirect observation of ice crystal sizes and shapes has again emphasized the irregular nature of these particles (H vmsfield et al. 1990). The molecular structure of the solid phase of ice suggest e macroscopic form of ice crystals should be that of hexagonal columns or plates, assuming homogeneous growth conditions. Since the growth of ice crystals under real atmospheric conditions is not always homogeneous, this form can only be considered as an idealized one. Fluctuations of ambient temperature, pressure and wind velocity and aggregation and splitting of particles, can lead to any degree of irregularity. Recent analyses of holograms taken from aircraft in cirrus clouds during ICE show a large variety of crystal structures (Krupp 1991). A selection of these results is given in Fig. 1, where the smallest crystals are found at low temperatures ($\sim$ -39°C), while complex and quite large crystals occured in the lower and warmer layers of that particular cloud.

In what follows, the spectral variability of the single scattering phase function will be discussed for hexagonal ice crystal columns in the range from 0.2 μm to 16.0 μm. A spectrally more detailed look is provided around the vibrational mode at $3\mu m$. Subsequently the influence of different major ice particle types on the scattering characteristics is shown. To estimate the scattering properties of highly complex aggregated ice particles, ray tracing results at two dimensional fractal particles, so-called triadic Koch-islands, are discussed at the end of this paper.

2 The model

The choice of an appropriate single scattering method depends on particle shape, the size parameter (ratio of particle size and wavelength), and on both the real and imaginary part of the refractive index of the material (see Schuerman, 1979 for an overview of different single scattering models).

For large size parameters, the geometric optic approximation is known to describe the scattering properties of irregularly shaped particles with sufficient accuracy. This condition holds for almost all ice particles when visible irradiation is considered. Geometric Optics may also be applied to the near ($\lambda < 4\mu m$) and middle ($\lambda \sim 10\mu m$) infrared region when ice crystals are extremely large, as in lower layers of cirrus clouds (Fig. 1). The numerical counterpart of the geometric optics approximation is to trace a sufficient large number of light rays at a given particle. Direction and number of the incoming rays depend on the desired particle orientation and on the convergence speed of the final result, respectively. Rockwitz (1989) developed such a ray tracing technique to calculate the single scattering properties of non-absorbing, convex ice particles. This technique has been extended to consider both absorption (Macke and Raschke 1991) and concave particle shapes (Macke 1992).

The ray tracing calculations result in the single scattering phase function $P(\theta, \psi)$, where θ, ψ are the zenith and azimuth scattering angle, with respect to forward scattering, i.e., the direction of the incoming radiation. Since all particles presented here are assumed to have random orientation, the scattering phase function does not change in azimuth, and $P(\theta, \psi)$ reduces to $P(\theta)$.

3 Spectral variability

The spectral variation of refraction and absorption of ice can be expressed by the real and imaginary part of the complex index of refraction, respectively (Fig. 2).

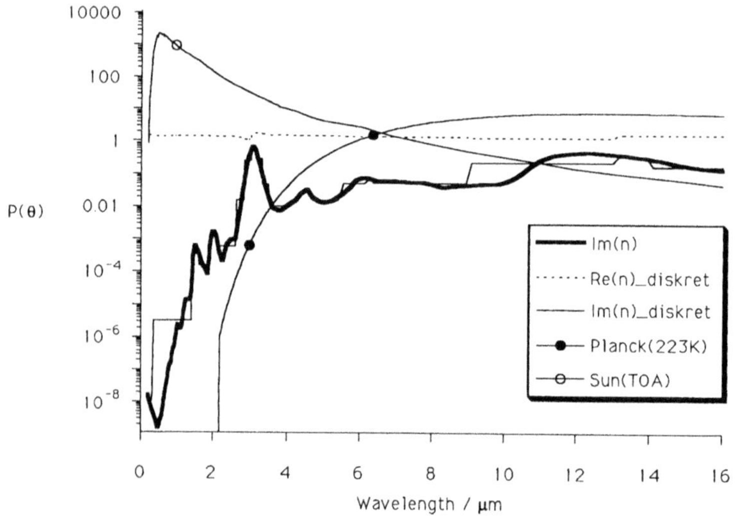

Fig. 2 The imaginary part of the refractive index of ice, in comparison with the thermal radiation at 6000K (sun) and 223K (upper troposphere). The discretised complex refractive index of ice is also included in this plot.

The data were taken from the compilation of Warren (1984). The changes in the absorption strength are caused by the different kinds of initiation mechanisms. These are electronic transitions in the UV region, vibrational modes between $3\mu m$ and $6\mu m$, a rotational oscillation at $12\mu m$, lattice vibrations in the far IR and overtones of all these mechanisms in the VIS and near IR. In order to select a finite number of spectral intervals which characterize the spectral variability within the considered spectral range, those spectral regions were choosen where both the absorptance of ice and the intensity of the incoming solar and/or thermal irradiation are high (Fig. 2). This selection leads to 22 wavelength intervals in the spectral region from $0.2\mu m$ to $16\mu m$. Fig. 3 shows the single scattering phase functions for a hexagonal random 3D oriented ice crystal columns at these intervals. The length to radius ratio of this crystal is $120\mu m$ to $30\mu m$.

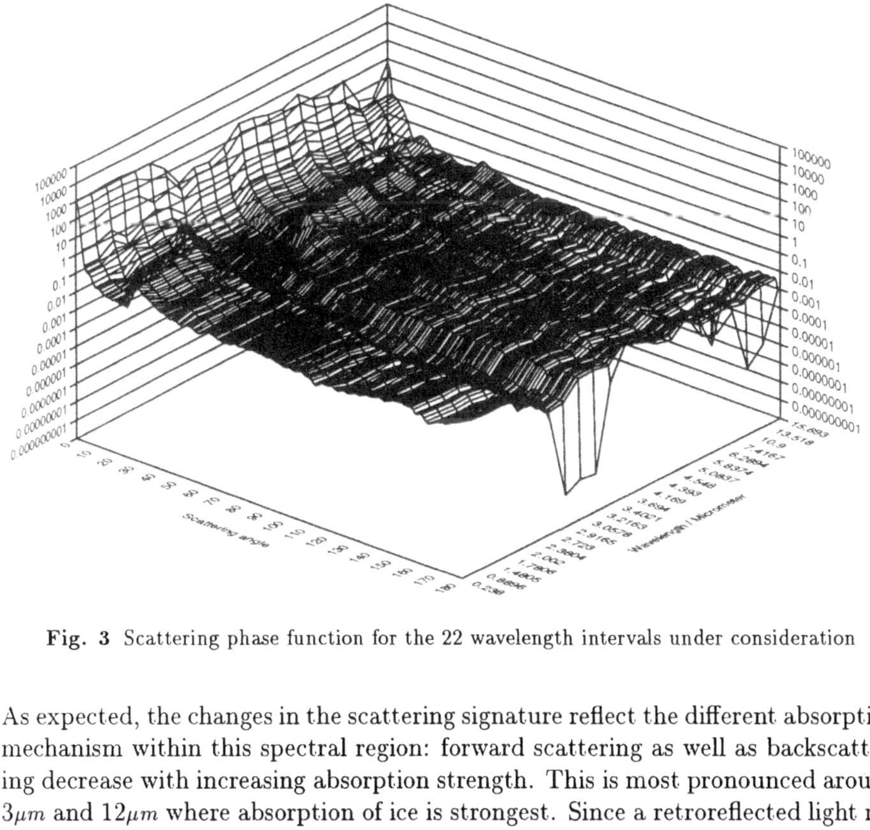

Fig. 3 Scattering phase function for the 22 wavelength intervals under consideration

As expected, the changes in the scattering signature reflect the different absorption mechanism within this spectral region: forward scattering as well as backscattering decrease with increasing absorption strength. This is most pronounced around $3\mu m$ and $12\mu m$ where absorption of ice is strongest. Since a retroreflected light ray requires at least one internal reflection within the crystal, backscattering senses absorption more than double that sensed in forward scattering. Hence, backscattering at 180° in Fig. 3 decreases more strongly with increasing absorption than forward scattering. Another feature of the two main absorption regions is the more or less constant scattering behaviour for almost all scattering angles; just a small forward scattering interval as well as backscattering at 180 ° provide an exception to the otherwise even scattering phase function. This is, besides the increased attenuation of a light ray within the particle, also caused by the fact that the angle of total reflection increases with increasing absorption strength (Macke and Raschke, 1991), increasing the probability that a ray is trapped inside the crystal. This will in turn disable transmission of light through the different ice prisms of which the crystal consists. Typical transmission features caused by minimum deviation at ice prisms (halos) will therefore vanish for high absorption. Thus, the scattering signature is determined only by diffraction and external reflections. The former provides a forward scattering peak and the latter the smooth side scattering behaviour.

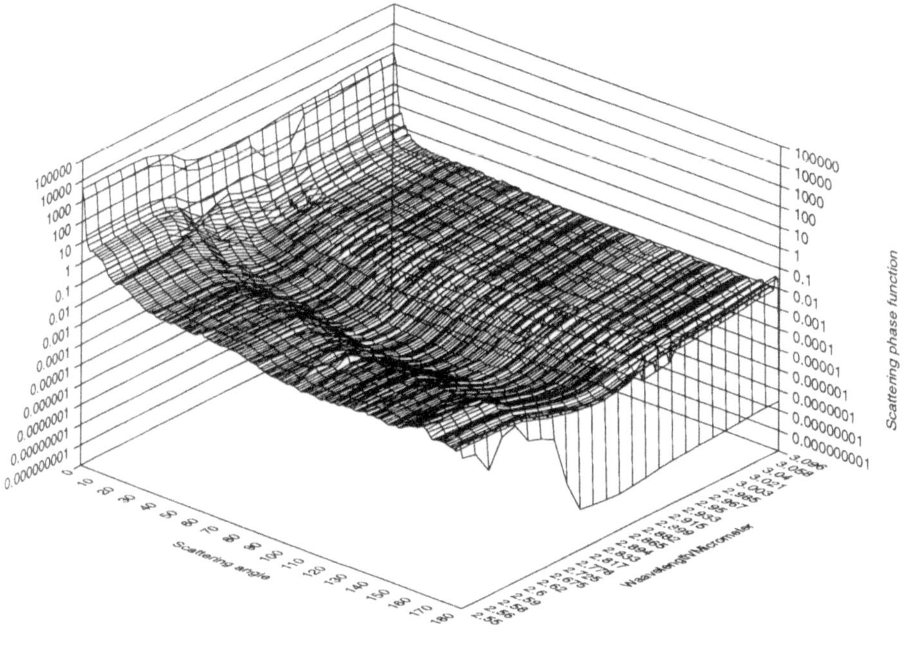

Fig. 4 Same as Fig. 3, but spectral interval $2.55\mu m$ to $3.096\mu m$ and enhanced to show high spectral resolution.

Fig. 4 shows the region of Fig. 3 between $2.55\mu m$ and $3.096\mu m$ at a spectral resolution of about $20nm$. This region covers the most intensive absorption band of ice at $3\mu m$. It is caused by stretching of the O-H covalent bond of the water molecules. The changes of the scattering signature with increasing wavelength (increasing absorption in this case) are in principal the same as shown in Fig. 3: below a distinct absorption strength, a halo peak caused by minimum deviation at 60° prisms as well as enhanced backscattering at 150° and retroreflection at 180° are visible. The halo peak moves to shorter scattering angles since the real part of refraction comes closer to the value of 1 (Fig. 2). Although absorption increase the magnitude of the halo peak does not decrease. This can be explained by Fresnels formulas: the lower the scattering angle for the transmitted light rays, the higher is the transmitted intensity. Again, with a further increase of absorption strength, the scattering signature becomes more or less flat. The only exceptions are the still relatively high forward scattering at 0° (diffraction) and very little backscattering at 180° (lack of retroreflections).

The spectral changes of the shape of the scattering phase function may be summarized as follows: for medium absorption strength, changing the real part of the complex refractive index causes a corresponding shift in the angular location of halo peaks; changing the imaginary part causes corresponding changes in the magnitude of the scattering phase function. These two features provide an effective method for a wavelength dependent parameterisation of the scattering behaviour. The developement of this parameterization will be done in the near future.

4 Different ice particle shapes

With the above mentioned wavelength-parametrization scheme in mind, the influence of different major ice particles on the scattering signature will be discussed briefly. Most of the particles presented here have actually been observed in nature. Following the ice crystal classification of Magano and Lee (1966), these crystals are the *pyramid* (C1a), the *cup* (C1b), the *solid bullet* (C1c), the *hollow bullet* (C1d), the *hollow column* (C1f), the *capped column* (CP1a), and the *bullet rosette* (C2a). From 78 classified ice crystals, these are the only particle shapes which satisfy the following two conditions: they are bound by plane surfaces, and there still exists a well defined recipe for their geometrical construction. Additionally, one example of a holographically measured particle is considered in this study. Figs. 5 and 6 summarizes the scattering phase functions of these particles. Particle sizes are of the order of $100\mu m$. All calculations were made at a single wavelength $\lambda = 0.55\mu m$. The corresponding refractive index of clear ice is $\hat{n} = 1.31 + i3.110^{-9}$. Particle orientation is assumed to be random.

The number of halo peaks in the forward scattering region increases with increasing complexity of the particle shape. This is caused by the increasing number of prism configurations. The combination of basic particle types such as the capped column and the bullet rosettes show basically the scattering characteristics of their individual components: multiple scattering between these components doesn't effect the total scattering behaviour seriously. The dotted lines in Fig. 6 a) shows the scattering characteristics of an in-situ holographic measured capped column. Due to its slightly irregular shape, the scattering signatures is smoother than its regular counterpart. A more detailed interpretation of the scattering properties of convex and concave polyhedric ice crystals can be found in Macke (1992).

The intercomparison of the individual scattering functions in Figs. 5 and 6 shows that their differences are greatest in the backscattering region around 180° scattering angle. Lidar measurements of cirrus cloud backscattering are therefore most suitable for the remote sensing of different particle types.

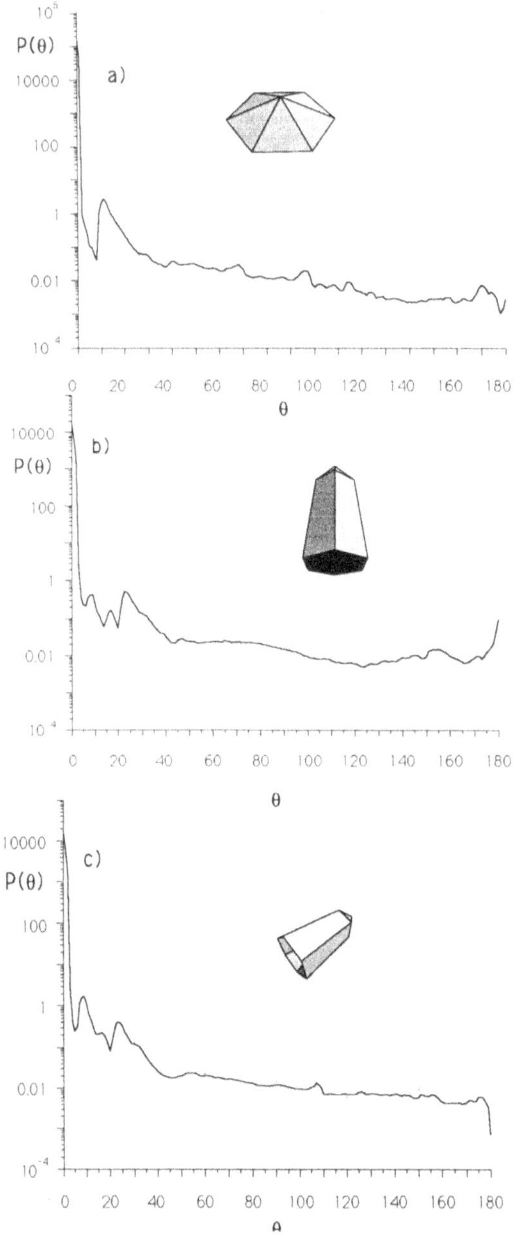

Fig. 5 Scattering phase function of the following ice crystal types: a) pyramid, b) solid bullet, c) hollow bullet

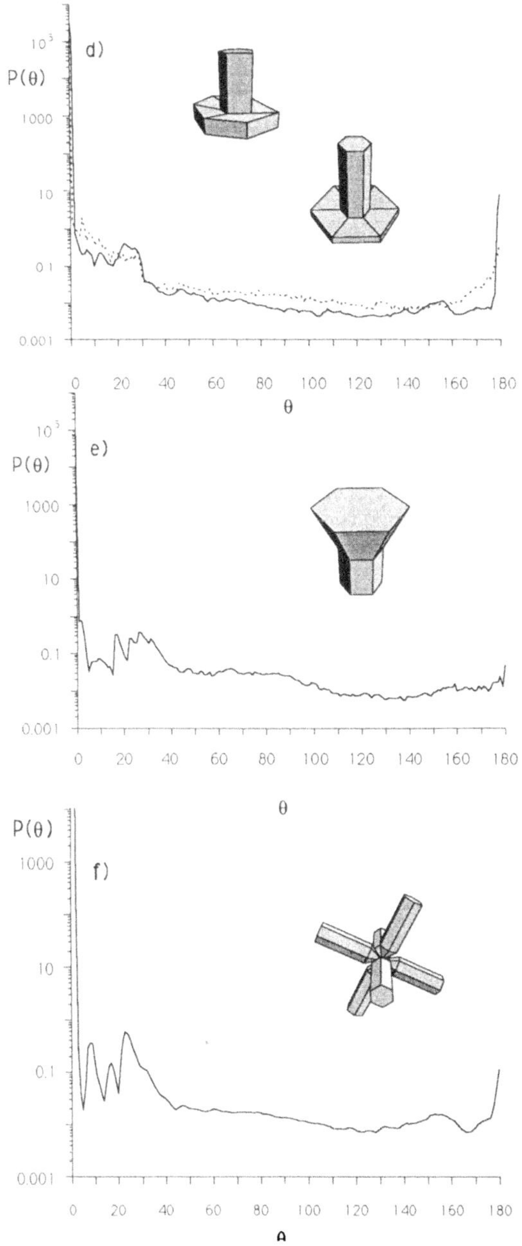

Fig. 6 Scattering phase function of the following ice crystal types: a) capped column (dotted: measured particle shape), b) cup, c) bullet rosettes.

5 Fractal particles

Fig. 7 shows a time record of in-situ measured ice particle shadow images (Albers 1992). These two dimensional projections of real ice particles outline the complex geometrical structure of ice particle shapes. Such graupellike structures result from particle growth due to aggregation of small broken crystals and aggregation has been shown to produce fractal structures. (Witten and Sander, 1981)

To estimate at least qualitatively the scattering behaviour of these kinds of particles, a ray tracing technique for two dimensional particles with self similiar surface structures has been developed. The most famous member of this particle family is the so called triadic Koch-island, first introduced by Koch (1904) and made popular by Mandelbrot (1977). It is constructed by replacing each line of a triangular *initiator* with a triangular *generator*. Repeating this procedure at the new line segments n times provides self-similarity within n times enlargements (about a factor 3 in the case of the regular triadic Koch-island). Another reason for taking this geometric shape lies in its hexagonal, i.e. snow-flake-like symmetry.

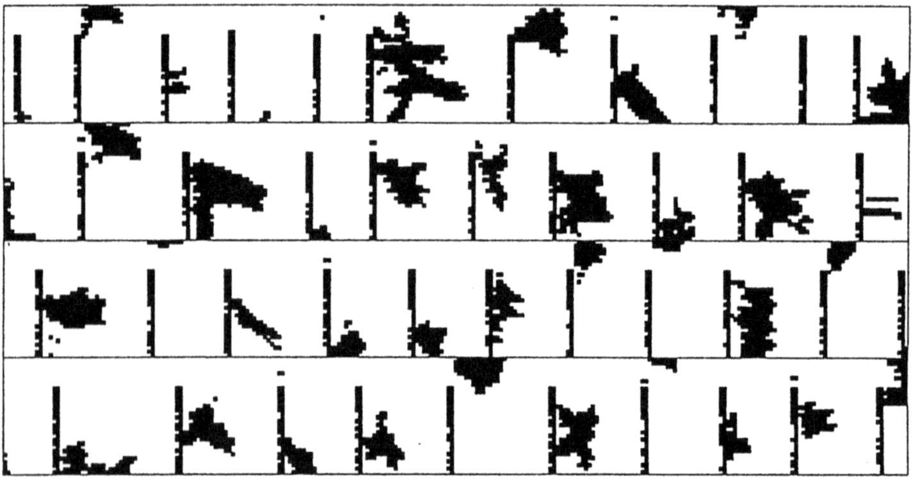

Fig. 7 Optical Array Probe measurement obtained during ICE IFO '89. Flight altitude ~ 8300m, temperature ~ -39°C (Albers, 1992)

Fig. 8 shows a triadic Koch-island after 4 of these replacements and its scattering phase function obtained by the ray tracing procedure. The outer particle radius and the wavelength of the incoming radiation are $1000\mu m$ and $0.55\mu m$, respectively. This corresponds to an outer size parameter (2π times outer radius/wavelength) of 11424, and an inner size parameter (2π times segment length/wavelength) of 63.

The scattering signature of this regular example of a fractal particle is extremely complex. There is a well pronounced 22° halo due to transmission through 60° prisms, and a 120° peak due to external reflections at 120° edges. Most of the other peaks can be explained by the so-called multi-order halos, which are caused by successive transmissions through different numbers of neighbouring prisms (Macke, 1992). The angular locations of these scattering peaks is typical for the regular triadic Koch-island; they do not change with increasing number of growth generation. The latter leads only to a slight decrease in their magnitude, and to a corresponding increase of the scattering backround. In mathematical terms it may be summerized that a) there exits a limit for the scattering signature in increasing generation-number n, and b) the convergence speed to obtain this limit is relatively large.

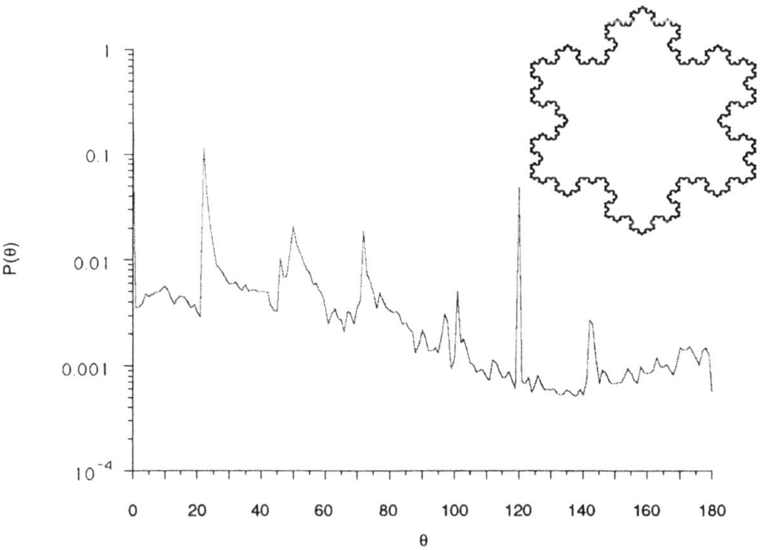

Fig. 8 Scattering phase function of a triadic Koch-island after 4 generations. The outer size parameter (outer radius)/wavelength is about 1818. The inner size parameter (segment length)/wavelength is about 10

At this point it has to be mentioned that Koch-islands of 6th and higher generations consist of segments with sizes comparable to or smaller as the wavelength of the incoming radiation. Applying Geometric Optics to those structures makes therefore no sense. However, if the wavelength can not resolve additional inner geometrical structures, the scattering behaviour will not change dramatically. On the other hand, real fractal particles consist of only a finite number of inner scales. A limit consideration for $n \to \infty$ is therefore not necessary.

This regular Koch-island is of course not very representative for real ice particles. One step closer to reality may be achieved by adding some disorder to the above described growth mechanism. Fig. 9 shows the resulting non-regular Koch-islands together with their scattering phase functions.

With increasing irregularity the individual scattering peaks decrease in magnitude and their angular location is broadened. Even slight disorder provide a considerable smoothing-effect. Above a distinct disorder, all individual scattering peaks vanish, and the whole scattering signatures come close to an exponential (i.e. linear in the semi-logarithmic presentation) shape. From the scattering behaviour of other fractal particles not presented here, a functional relationship between this constant slope of the scattering phase function and the fractal dimension of those particles is expected.

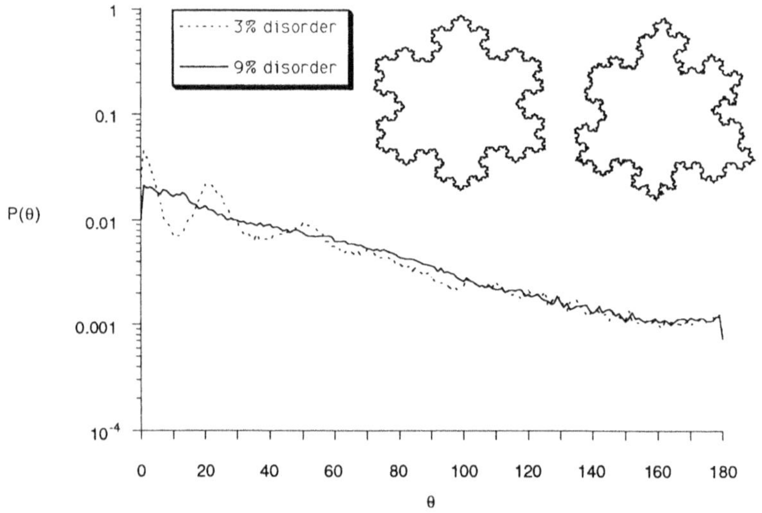

Fig. 9 Scattering phase functions of non-regular triadic Koch-island after 4 generations. Outer particle radius and wavelength as in Fig. 8.

6 Summary and Conclusion

The spectral variability of the scattering phase function of ice crystals may be divided into two regimes: for low and medium absorption, i.e. out of absorption bands, the phase function consists of several particle shape dependent scattering peaks. Angular location as well as the magnitude of these local maxima change with respect to the spectral change of the real and imaginary part of the complex refractive index, respectively. Backscattering is the most sensitive scattering region for changes in the absorption strength. Thus, radiometric measurements of the reflected solar irradiation along an absorption wing of ice (e.g., at $3\mu m$) may provide vertically resolved information on cirrus cloud microphysical properties, such as mean particle radius or ice content. For strong absorption, the signature of the scattering phase function is relatively flat. The only exceptions are the forward scattering peak due to diffraction and a fall-off at 180° scattering angle due to the lack of internal retroreflections.

The influence of different particle shapes on the scattering properties may be summarized as follows: symmetric crystals like hexagonal columns, pyramids or bullets show a relatively complex scattering signature. The dominance of several well defined ice prism configurations provide formation of strong pronounced halo peaks. Aggregated particles like capped columns or bullet rosettes show basically the scattering characteristic of their individual components: multiple scattering between these components doesn't effect the total scattering behaviour seriously. Particle asymmetry causes a smoothing of the scattering phase function. The same arguments hold for the two-dimensional fractal Koch-islands: symmetry in particle shape corresponds to a complex scattering signature, and asymmetric shapes provide smoother phase functions.

With regard to the geometry of in-situ measured ice crystals in cirrus clouds, single (hexagonal) symmetric ice particles seem to be the exception. More frequent are aggregates of these hexagonal components (bullet rosettes) as well as aggregates of small broken crystal parts (graupel). From the optical point of view, scattering at the former type can be replaced by independent scattering at its components, i.e. hexagonal plates, columns and bullets. Graupellike particles may be characterized by a more or less exponentional scattering phase function, where its exponent may be determined by the fractal dimension of the particle.

References:

Albers F (1992) Results from microphysical measurements during ICE mission 212. *Proceedings of the 5'th ICE-EUCREX Workshop in Clermont-Ferrand, France*

First ISCCP Regional Experiment (FIRE) - Issue (1990) . *Mon. Wea. Rev.* **118**

Heymsfield AJ, Miller KM, Spinhirne JD (1990) The 27-28 October 1986 FIRE IFO Cirrus Case Study: Cloud Microstructure. *Mon. Wea. Rev.* **118**:2313-2328

von Koch H (1904) Sur une courbe continue sans tangente, obtenue par une construction géométrique élémentaire. *Arkiv för Matematik, Astronomi och Fysik* **1**:681-704

Krupp C (1991) Holographische Messungen an Eiskristallen in Cirruswolken während ICE. *Master Thesis: Institute for Physics, University to Cologne, Germany.*

Liou KN (1986) Review: Influence of cirrus clouds on weather and climate processes: A global perspective. *Amer. Meteor. Soc.* **114**:1167-1199

Macke A, Raschke E (1991) Spectral variability of light scattering by hexagonal ice crystals. *Proceeding of the ICO Topical Meeting on Atmospheric, Volume and Surface Scattering and Propagation, Florence, Italy, August 1991*

Macke A (1992) Scattering of light by polyhedral ice crystals. Submitted to *Appl. Optics*

Magano C, Lee C (1966) Meteorological classification of natural snow crystals. *J. Fac. Sci. Hokkaido Univ., Ser. VII* **2**:321-335

Mandelbrot BB (1977) The fractal geometry of nature. W.H. Freeman and Company, New York

Stephens GL, Si-Chee Tsay, Stackhouse PW, Flatau PJ (1990) The relevance of the microphysikal and radiative properties of cirrus clouds to climate and climatic feedback. *J. Atmos. Sci.* **47**:1742-1753

Raschke E, Schmetz J, Heintzenberg J, Kandel R, Saunders R (1990) The International Cirrus Experiment (ICE) - A Joint European Effort. *ESA Journal* **14**:193-199.

Rockwitz K-D (1989) Scattering properties of horizontally oriented ice crystal columms in cirrus clouds, Part 1. *Appl. Optics* **28**:4103-4110

Schuerman DW (1980) Light scattering by irregularly shaped particles, 1st edn. Plenum Press, New York and London

Warren SG (1984) Optical constants of ice from the ultraviolet to the microwave. *Appl. Optics* **23**:1206-1225

Witten T, Sander M (1981) *Phys. Rev. Lett.* **47**:1400

DETECTION OF POLAR STRATOSPHERIC CLOUDS WITH NEXT GENERATION IR SOUNDERS

R. Meerkoetter
German Aerospace Research Establishment, Atmospheric Physics
8031 Oberpfaffenhofen
Federal Republic Germany

Abstract

A bispectral approach to temperature determination of semitransparent clouds is taken to identify areas covered by Polar Stratospheric Clouds (PSCs). The method uses information from two satellite pixels and two spectral channels in the infrared. PSCs form and emit in very cold stratospheric environments (<195 K), and solutions of the bispectral method yield low temperatures in a range that clearly differ from those derived for a cold but cloudfree stratosphere. Application of the method to HIRS data of the NOAA-satellite gained during the major PSC event on January 31 1989 identified large areas covered by dense PSCs which were in very good agreement with independent airborne measurements. Simulations demonstrate the potental of the method with high spectral resolution IR sounder information.

Introduction

Polar Stratospheric Clouds (PSCs) are a phenomenon of both Arctic and Antarctic regions and form as a consequence of dramatic stratospheric cooling during the polar winter. Three main types of PSCs are generally defined : a common type I consisting of nitric acid trihydrate particles, a rarer type II consisting of ice particles, and a type III occuring in relation to mountain lee waves. The formation process of PSCs is accompanied by dehydration and denitrification which influences ozone chemistry leading to the development of the ozone hole. Furthermore, optically thick PSCs of type II and III are said to force stratospheric dynamics via radiative heating or cooling.

NATO ASI Series, Vol. I 9
High Spectral Resolution Infrared Remote Sensing
for Earth's Weather and Climate Studies
Edited by A. Chedin, M.T. Chahine and N.A. Scott

Therefore, identification of PSC events and determination of the PSC temperature with data from satellites will be very important for monitoring regions of potential ozone depletion as well as regions where PSCs may have an impact on the polar stratospheric radiation budget.

In the following, a bispectral method which gives temperatures of optically thin clouds is proposed for the detection of PSCs. Model results and application to real satellite data demonstrate the ability of the method to identify PSC covered areas by deriving temperatures in a range that clearly differs from results achieved under conditions of a cloudfree stratosphere. Simulations further illustrate the potential of improvement when the method is applied to data of higher spectral resolution.

Method

The concept of the bispectral method for the determination of cirrus cloud temperature and altitude is developed and described in detail by Pollinger and Wendling (1984). It was shown that cloud temperature can be retrieved with information from two channels in the terrestrial spectral range (4-100μm) and two adjacent satellite pixels within a cirrus field, provided that the cirrus optical depth is different in each pixel but the cloud temperature is the same. The method needs spectral information from two channels whose weighting functions peak at different altitudes. Numerical simulations show, for example, a combination of channel 4 and channel 7 of the HIRS/2 instrument, corresponding to central wavelengths at 14.2μm and 13.4μm inside the CO_2 -absorption band, reacts to PSCs between 15-25 km and can be used to derive temperatures indicating their existence. The following equation enables the derivation of the PSC temperature from satellite measured radiances :

$$B_2(T) = \frac{L_1^1 L_2^2 - L_1^2 L_2^1}{L_1^1 L_1^2 - f(L_2^1 - L_2^2)} \qquad (1)$$

L_j^i are the radiances with pixel and wavelength denoted by i and j, respectively. $B_2(T)$ is the blackbody radiation at cloud temperature T. Since f, which is the ratio of the two spectral Planck functions $B_1(T)$ and $B_2(T)$, depends on the cloud temperature T itself, Equ. (1) is transcendent and has therefore to be solved iteratively with a first guess at the cloud temperature. In the first iteration step the blackbody radiance on the left side of Equ. (1) is calculated and the inverted temperature then serves to determine the factor f in the next step. When convergence of the iteration process leads to temperature differences between two consecutive steps below a given threshold, for example 0.1 K, a first estimate of the cloud temperature is obtained.

However this first estimate usually deviates from the actual cloud temperature because of atmospheric effects on the measured radiances from the layer above the cloud. This was eliminated, using radiative transfer calculations based upon a knowledge of the temperature profile above the cloud. An improved cloud temperature and height can then be determined during a further iteration that is started with corrected radiances.

Simulation Results

In this section the bispectral method is demonstrated based on simulations of HIRS observations of PSCs and clear stratospheres. Errors which can be expected from the bispectral method are demonstrated using theoretical simulations.

Spectral radiances of HIRS/2-channels 4 and 7 are calculated with a 1-D radiative transfer code that is based on matrix-operator-theory (Plass et al., 1973). An atmospheric profile typical for formation of dense type II PSCs which (Hanson and Mauersberger (1988)) condense below 190 K is constructed using the ECMWF analysis from January 31 1989 over Northern Europe. Following observations of the geometrical PSC depth during the Airborne Arctic Stratospheric Expedition (AASE) in 1989, a PSC embedded between the 18 and 21 km model layers adequately represents the cloud location. Temperatures at cloud top and base are indicated in Figure 1. To account for characteristic PSC II size distributions, a lognormal distribution of ice spheres with a mode radius r=5μm is assumed. Spectral optical parameters such as asymmetry factor, single scattering albedo, and volume extinction coefficients are derived by Mie-calculations with complex refractive indices for ice after Warren (1984). As mentioned above, the bispectral method needs information from two satellite pixels. Therefore, PSC optical depths, which have to appear differently in both satellite pixels, are simulated by varying total particle number densities at constant geometrical cloud depth.

Atmospheric correction above the cloud is reduced to one radiative transfer calculation for one fixed atmospheric layer from 25-50 km. An altitude of 25 km is a reasonable assumption for the lower boundary of the masking atmosphere because most PSCs are observed to form below this level.

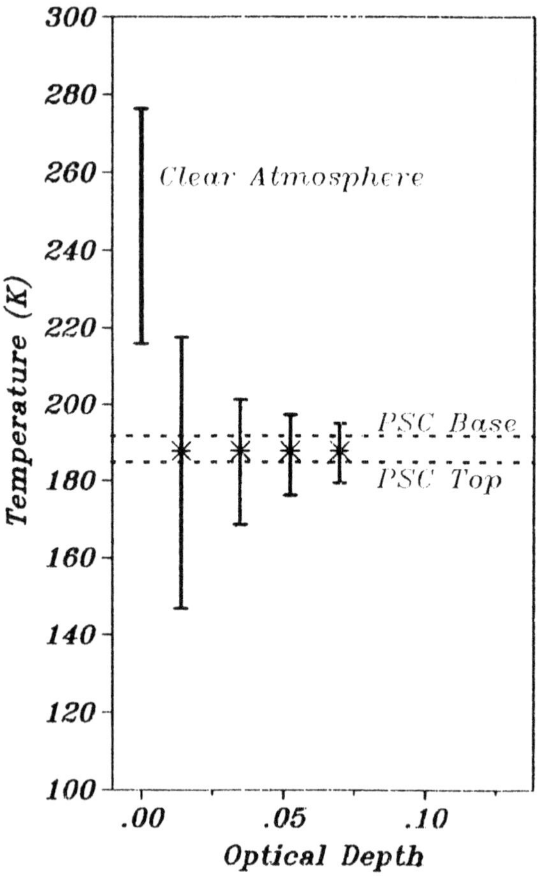

Figure 1 :
Solutions of the bispectral method for a clear atmosphere and for a single PSC layer as a function of cloud optical depth in HIRS/2 channel 4.

As a typical result, temperatures retrieved from simulated radiances and plotted as a function of PSC optical depth in channel 4 are shown in Figure 1. The optical depths correspond to particle number densities ranging from 0.02-0.1 cm^{-3} and concentrations differ by a factor of two in both HIRS/2 pixels to account for the necessary cloud inhomogeneities. Error bars illustrate the variety of solutions which have to be expected when noise equivalent differential radiances (NEΔN) of both HIRS/2 channels are taken into account. From Figure 1 it is obvious that the bispectral method tends to give PSC temperatures with increasing accuracy as the signal to noise ratios increase, i.e. when absolute differences of optical depth increase. Radiometric noise dominates the solution when the cloud appears optically thin and almost homogeneous in both satellite pixels. To clearly demonstrate how the algorithm reacts to emitting PSCs in a cold stratospheric environment in comparison to the same but cloudfree stratosphere, Figure 1 also exhibits solutions for the clear case. The temperature interval representing the clear case is a consequence of superimposed radiometric noise only; no solution would be obtained for noisefree clear column radiances of equal value in both pixels. Although error bars become rather broad for optically thin PSCs, all solutions for the PSC are significantly different from the clear case. This is due to the fact that increased optical depth between 18-21 km caused by PSCs significantly increases the contribution of layers emitting at very low temperatures. Clear case solutions are not expected to broadly overlap with temperatures derived for PSC covered pixels since temperatures above 220 K are calculated even though the cloudfree model stratosphere reaches values below 190 K.

Application to Satellite Data

Application to satellite data concentrates on the PSC event on January 31 1989, when large areas over the Norwegian Sea and Scandinavia were identified as covered with optically thick PSCs. For example, McCormick et al. (1990) report upon large scale PSCs of type I and II between 16-26 km (most frequent near 20 km) and between 60° -73°*N*, observed during an aircraft mission from 6:00-14:00 UTC. The bispectral method is applied to NOAA-10 data obtained within the same time interval and for the same region over Northern Europe. The NOAA-10 overpass is analyzed between 8:05 and 8:08 UTC. The corresponding overpass section located within the visibility circle of the data receiver station at the German Aerospace Research Establishment at Oberpfaffenhofen is shown in Figure 2. Temperatures are derived for adjacent pixel pairs in cross track direction. Atmospheric correction of measured radiances is performed appropriately for the zenith angle of observation for an atmospheric layer from 25-50 km. The temperature profile used to account for atmospheric masking is constructed following the ECMWF temperature analysis at 30 and 10 hPa (23 and 30 km) for January 31 1989. Above 30 km the temperature profile is adapted to the subarctic winter standard atmosphere from McClatchey (1972). Due to reduced gaseous absorption at these altitudes, uncertainties in the standard profile are not important.

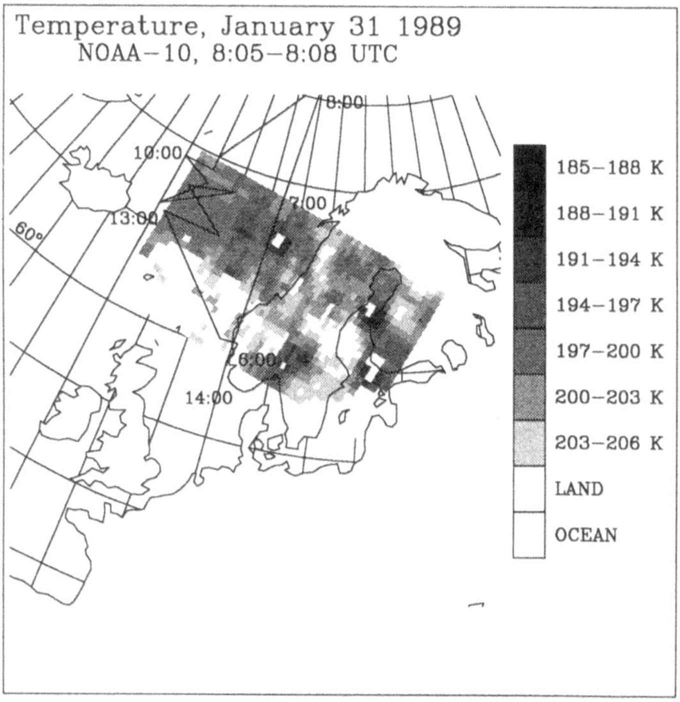

Figure 2 :
Temperature distribution indicating areas covered by PSCs on January 31 1989. Shown are results derived from NOAA-10 overpass data. For comparison the flight track of the research aircraft as given by McCormick et al. (1990) is reproduced.

Figure 2 displays the horizontal distribution of temperatures derived from the NOAA overpass radiances. Values shown are averaged over 4 · 4 solutions in order to smooth out the influence of radiometric noise. As can be seen in Figure 2, the solutions result in a horizontal temperature distribution exhibiting two cold centers, one over the Norwegian Sea and one over Southern Scandinavia.

The conclusion that these areas correspond to regions covered by PSCs is not only based on simulation results but also supported by independent measurements made during AASE. McCormick et al. (1990) report upon airborne measurements on January 31 1989 that show PSCs very similarly distributed compared to temperatures derived from the bispectral method. Along the flight track of the research aircraft marked in Figure 2, both types of PSCs are detected with an airborne lidar by analyzing characteristic scattering and depolarization ratios. PSCs of type II especially are observed in a region of very cold stratospheric air along the leg from Sola airfield (Southern Norway, 59°*N*, 5°*E*) to about 73°*N* from 6:00-7:30 UTC which nearly coincides in time with the NOAA-10 overpass at 8:00 UTC. Results of the bispectral method in fact confirm the existence of type II PSCs for areas crossing the flight legs between 66 and 68°*N* over the Norwegian Sea, especially since averaging over $4 \cdot 4$ solutions yields typical temperatures around 190 K with a standard deviation of ±15 K.

Potential of High Spectral Resolution and Concluding Remarks

It has been shown that PSC identification depends on the signal to noise behaviour with the signal mainly governed by the horizontal structure of cloud optical depth. As a consequence, the method succeeds for relatively thick PSCs that appear to be inhomogeneous between satellite pixels. Optically thin PSCs lead to solutions more dominated by radiometric noise. In this case PSC covered regions in an enlarged scale can be isolated by horizontal averaging within the scene of the derived temperatures.

However, the influence of radiometric noise, even on a smaller scale, may also be reduced by averaging over a set of solutions obtained from different channel combinations which would be available from high spectral resolution data. The potential of high spectral resolution can be illustrated already on the basis of simulated channels with a slightly enhanced resolution of $\Delta\nu=2.5cm^{-1}$ compared to the half power bandwidth of $\Delta\nu=16cm^{-1}$ of both HIRS/2 channels. The bispectral method is therefore applied to a set of six channel pairs selected in the range from 700-750cm^{-1} exhibiting absorption characteristics similar to those of HIRS/2 channels 4 and 7. Corresponding temperatures derived for the same atmosphere described before including a PSC layer with particle concentrations of N=0.01 and N=0.05cm^{-3} in the satellite pixels are presented in Figure 3. Error bars due to the influence of radiometric noise are not calculated.

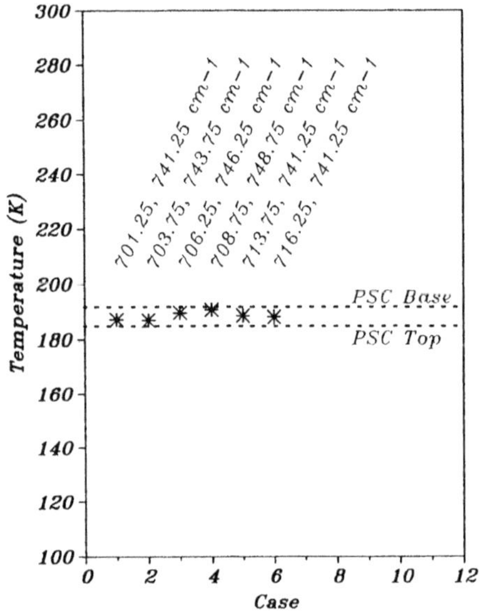

Figure 3 :
Solutions of the bispectral method for a spectral resolution of $\Delta\nu$=2.5 cm^{-1}. Indicated are the central wavenumbers of different channel combinations.

Figure 3 clearly illustrates that even on the basis of this relatively coarse spectral resolution different channel pairs giving the PSC temperature, as for example those denoted by case 1 and 2, are found. Averaging over such solutions in principle would allow to reduce the error introduced by radiometric noise. Since future radiometer conceptions will rather measure with a finer resolution of $\Delta\nu$=1 cm^{-1}, a much better accuracy in PSC temperature determination can be expected by the use of an enhanced number of appropriate channel combinations. Thus, the bispectral method will no longer be able only to indicate the existence of PSCs, rather a discrimination between different PSC types, which as already known condense at different temperatures, should be possible.

Acknowledgements. I would like to thank P. Wendling for many helpful comments and discussions, I am obliged to R. Buell for data preprocessing and for kind assistance in data handling, I wish to thank K. Meier for preparing most of the graphics.

References

Hanson D, Mauersberger K (1988)
Laboratory Studies of the Nitric Acid Trihydrate : Implications for Antarctic Ozone Depletion. Geoph Res Lett 17 : 855-858

McClatchey R A et al. (1972)
Optical Properties of the Atmosphere. AFCRL Rep. No. 72-0497, Air Force Cambridge Res. Lab., Bedford, Mass., p 108

McCormick M P et al. (1990)
Arctic Polar Stratospheric Clouds Observations by Airborne Lidar. Geoph Res Lett 17 : 365-368

Plass G N et al. (1973)
Matrix-Operator-Theory of Radiative Transfer. Appl Opt 12 : 314-329

Pollinger W, Wendling P (1984)
A Bispectral Method for the Height Determination of Optically Thin Ice Clouds. Beitr Phys Atm 57 : 269-281

Warren S G (1984)
Optical Constants of Ice from the Ultraviolett to the Microwave. Appl Opt 23 : 1206-1225

PROBLEMS IN CLOUD IDENTIFICATION AND CLEARING

R. Rizzi, C. Serio[1], V. Cuomo[2]
ECMWF
Shinfield Park
Reading RG2 9AX
UK

EUMETSAT
Am Elfengrund 45
D-6100 DArmstadt-Eberstadt
FRG

1. Introduction

The scientific community has devoted much attention to processing of clear-sky radiances. Over the years more physics has been introduced in the inversion algorithms with a steady reduction of the differences among various retrieval methods. Differences still arise depending on whether the first guess is based on forecast fields or on climatological databases.

Although the quality of retrieved products, and their usefulness in Numerical Weather Prediction (NWP) is fundamentally limited by the quality of the cloud cleared BTs, less attention has been paid to the cloud detection and clearing step, aimed at producing clear column radiances from measurements that can be cloud contaminated. The matter was for long considered a necessary preprocessing step with little scientific content.

Schemes have been developed to handle cloudy radiances, either by iterating between the cloud clearing and the inversion steps (Susskind et al., 1984) or by eliminating the need for a separate clearing stage. Huang and Smith (1985) perform a simultaneous retrieval that includes atmospheric profiles, surface parameters together with cloud top pressure and amount. Eyre (1989) applies the principles of nonlinear optimal estimation to extract the

---

[1]Dipartimento di Scienze Fisiche, Napoli

[2]Dipartimento di Ingegneria e Fisica dell'Ambiente, Potenza

NATO ASI Series, Vol. I 9
High Spectral Resolution Infrared Remote Sensing
for Earth's Weather and Climate Studies
Edited by A. Chedin, M.T. Chahine and N.A. Scott

same parameters for use in NWP. Although these techniques could in principle provide the needed profiles and the cloud parameters, no impact study has been done so far.

Despite the growing awareness of the relevance of cloud clearing to the retrieval problem, comparison among different techniques and validation studies are still lacking. A study was done at ECMWF (Rizzi et al., 1992) to compare three different cloud-clearing schemes: the NESDIS scheme, as implemented operationally in February 1989 (Reale et al., 1986), the operational scheme at the UK Meteorological Office (UKMO), based on the technique developed by Eyre and Watts (1987) and the method described in Andretta et al. (1989) and Amato et al. (1991), which will be referred to as MR59. Cloud identification products generated by LMD (Laboratoire de Meteorologie Dynamique) were also compared. Great attention was put in analysing NESDIS cloud identification flag and cloud cleared radiances, since these are used operationally in all NWP centres and their quality has an important impact not only in the daily operations of these centres but also in the way satellite data will be used.

The present paper will present some results contained in Rizzi et al. (1992) and some new material, focusing the attention on the importance of the cloud detection step for any cloud filtering procedure. Some implications will also be discussed pertinent to high spectral resolution infrared sounders and to the problems involved in the generation of databases for climate studies.

2. The comparison data set

The data set used for the comparison presented in this paper consists of three NOAA 11 nighttime orbits of February 11 (pass A) and 12 (pass C and D) 1989, covering the European and North Atlantic areas. Limb corrected simulated radiances were computed using the ECMWF analyses produced operationally, quadratically interpolated to the time of the passes. No bias correction was applied to the simulated BTs which therefore constitute only a reference field for the products obtained with the various techniques. Since ITPP and NESDIS processing software produce a HIRS/2 channel 8 BT "corrected" for water vapour attenuation, all references to BT in channel 8 identify a corrected value. The simulated BT in channel 8 is the averaged monthly values of the blended SST analysis produced by NMC (Reynolds, 1988).

AVHRR HRPT data and products were also available for passes A and D, provided by CMS (Centre de Meteorologie Spatiale) in Lannion, France. The products used were cloud masks (Derrien et al., 1990) and skin temperatures (Castagne at al., 1986) collocated at HIRS/2 FOVs. These products allowed accurate estimates of sea surface temperature (SST) to be obtained at HIRS/2 spots contaminated by clouds, thus permitting the validation of different cloud-clearing schemes in the most transparent channels. Furthermore, cloud-masks derived from AVHRR facilitated the intercomparison and validation of the different cloud-detection schemes based solely on TOVS data.

Cloud-detection from AVHRR enables the derivation of the amount of cloud coverage at every HIRS/2 spot, expressed as a number, C_c, which takes values from 0 for cloudy to 100 for clear. Intermediate values give the fraction of area of the HIRS/2 FOV which is cloud-free. Based on the HIRS/2 spots that were completely clear over the sea (67 FOVs in pass A and 129 in pass D), a linear regression relation was derived between the BT in HIRS/2 channel 8 and AVHRR SST with a root mean square error of 0.03 K. The regression relation has been used to estimate BTs in channel 8 at partially cloudy HIRS/2 FOVs, for which AVHRR data still allows the determination of skin temperatures (C_c>5%), thus providing a validation of the cloud-cleared products obtained with the different techniques.

3. Results

The first step in any cloud-clearing scheme is cloud detection. This is a very important task because the quality of the clear-column BT estimates at partially cloudy FOVs depends critically on the accuracy of cloud detection.

Figs. 1 to 3 show the result of the clear FOVs identification according respectively to NESDIS, UKMO and LMD, on top of channel 4 of AVHRR. They can be compared to the C_c =75 maps obtained from AVHRR, shown in Fig. 4. The comparison points out that the overall number of NESDIS clear FOVs is very small, that some of the clear FOVs appear quite severely cloud contaminated and that the processing does not identify most of the clear spots. The UKMO cloud-detection scheme seems to provide an operational cloud-masks similar to the AVHRR one in severe overcast conditions. LMD processing produces a more regular array of clear locations, some of which lie to the west of the front where at least 75% of the AVHRR FOVs, within each HIRS FOVs, are cloudy according to AVHRR products derived by CMS. The figures illustrate the range of different cloud masks that can be produced, starting with the same radiometric data and aiming at the same objective.

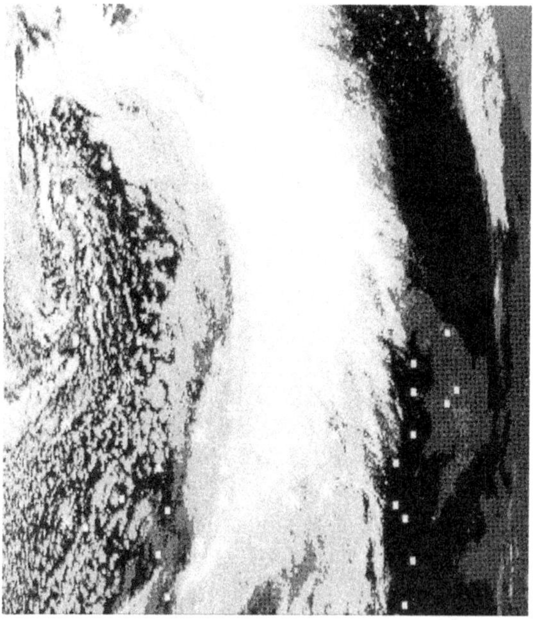

Fig. 1 Location of NESDIS clear (denoted by grey square) and NESDIS N* FOVs (marked by *) for the mid portion of pass A on top of AVHRR channel 4.

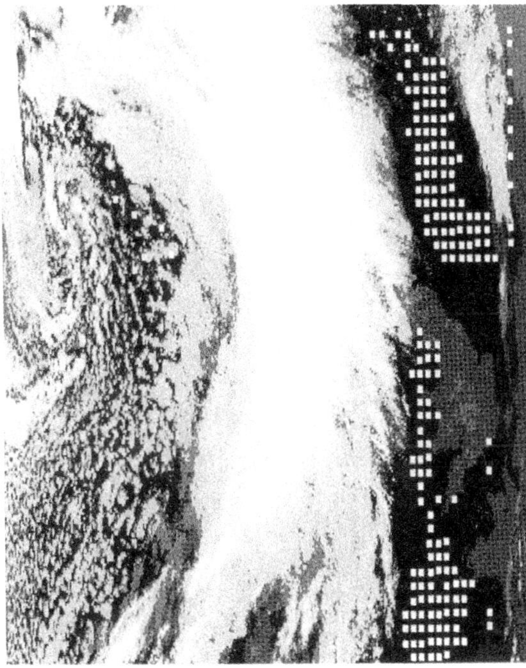

Fig. 2 Location of UKMO clear FOVs (denoted by grey square), for the mid portion of pass A on top of AVHRR channel 4.

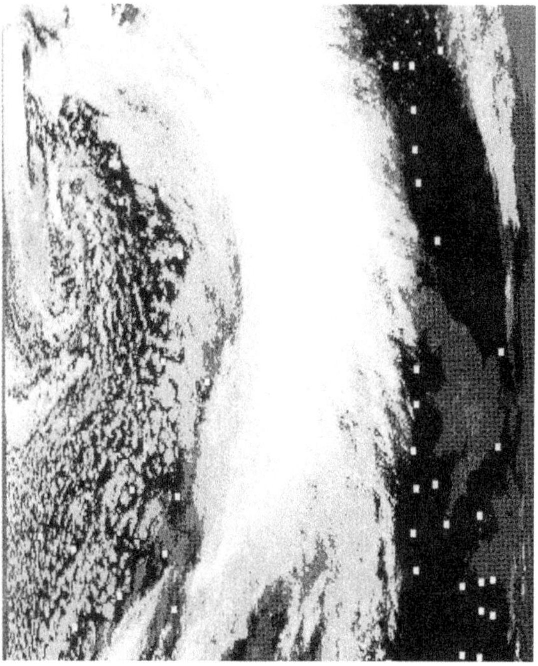

Fig. 3 Location of III clear FOVs (denoted by grey square), for the mid portion of pass A on top of AVHRR channel 4.

Fig. 4 Cloud-mask derived from AVHRR data for the mid portion of pass A on top of AVHRR channel 4. Grey squares indicates FOVs whose C_c > 75%.

The radiometric effect of assuming clear FOVs can also be evaluated quantitatively by computing, in all FOVs over sea, the differences between the cleared BT in HIRS/2 channel 8 and the value of the SST, estimated from AVHRR data as explained in the previous section. Figs. 5a and 5b show histograms of such differences for pass A and D in the case of NESDIS. It is possible to see that on average the cloud masks introduce a cold bias of about 2 K. The effect is less serious for UKMO and MR59 (Figs. 6a and 6b); MR59 produces the lowest bias and variance. The full analysis of the results shows that although a test based on AVHRR data is quite selective in pointing out the FOVs in which clouds are present, the amount of contamination, and its relevance for the subsequent inversion step, can best be judged in terms of BT differences.

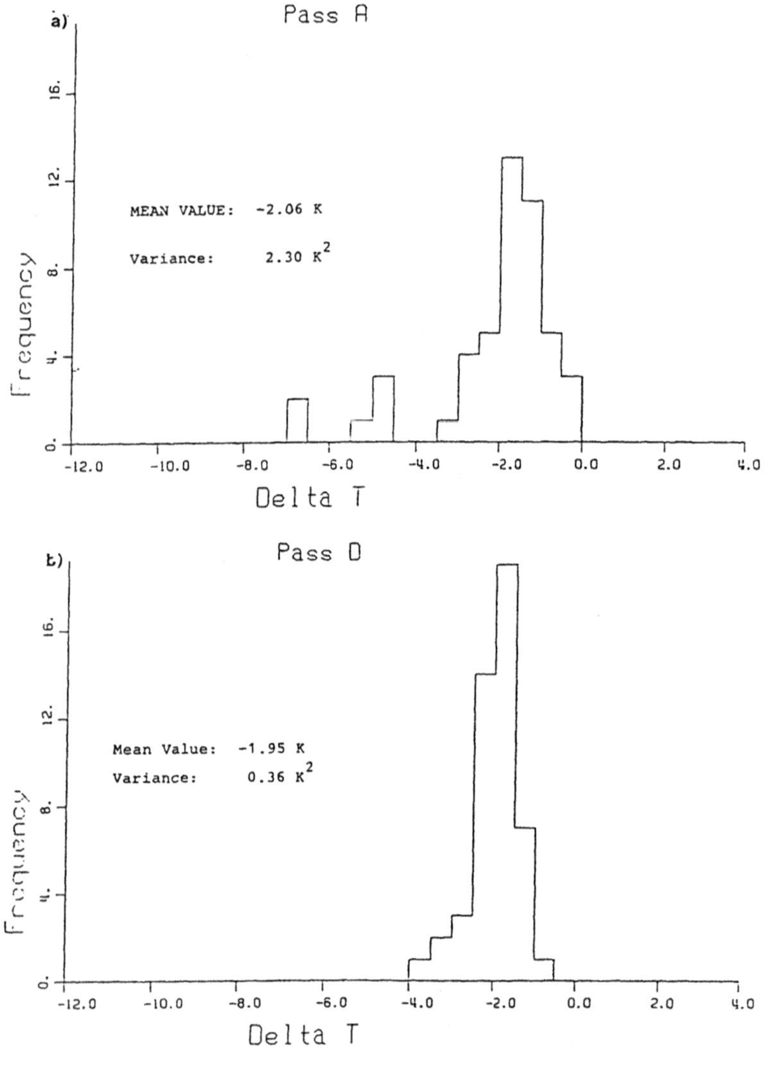

Fig. 5 Histogram of the differences between NESDIS cloud cleared BTs in HIRS channel 8 and SST estimated from AVHRR. Only clear FOVs according to the cloud-mask produced by NESDIS are used. a) pass A; b) pass D.

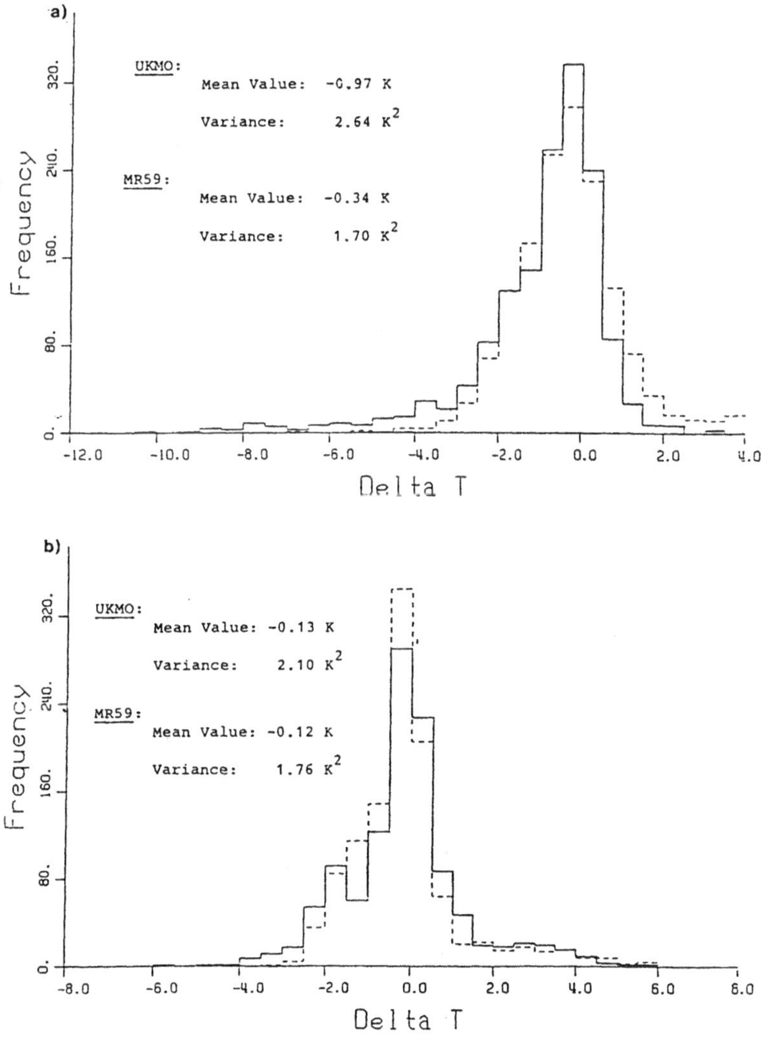

Fig. 6 Histogram of the differences between cloud cleared BTs in HIRS channel 8 and SST estimated from AVHRR. UKMO is solid line and MR59 is dashed line. All the FOVs for which a SST was available were used. a) - pass A; b) - pass D.

A better understanding of the underlying reasons for NESDIS' performance can be obtained by examining the location of the clear and N^* FOVs for pass A (respectively gray squares and stars in Fig. 1) on top of AVHRR channel 4 imagery. It is immediately evident that a large fraction of the FOVs in which the N^* technique is applied lies on top of a midlatitude frontal system, in regions where there seems to be no portion of the FOV seeing the surface of the sea and clouds appear as thick enough to be totally opaque. Clear FOVs are also identified westward of the front where the AVHRR cloud detection indicates that FOVs are at most 25% clear. Figs. 7a and 7b are enhanced views of the eastward orbit portion, the squares representing FOVs 75% clear or more according to the AVHRR analysis, while the stars are now NESDIS clear FOVs only. There are HIRS/2 FOVs eastward to the

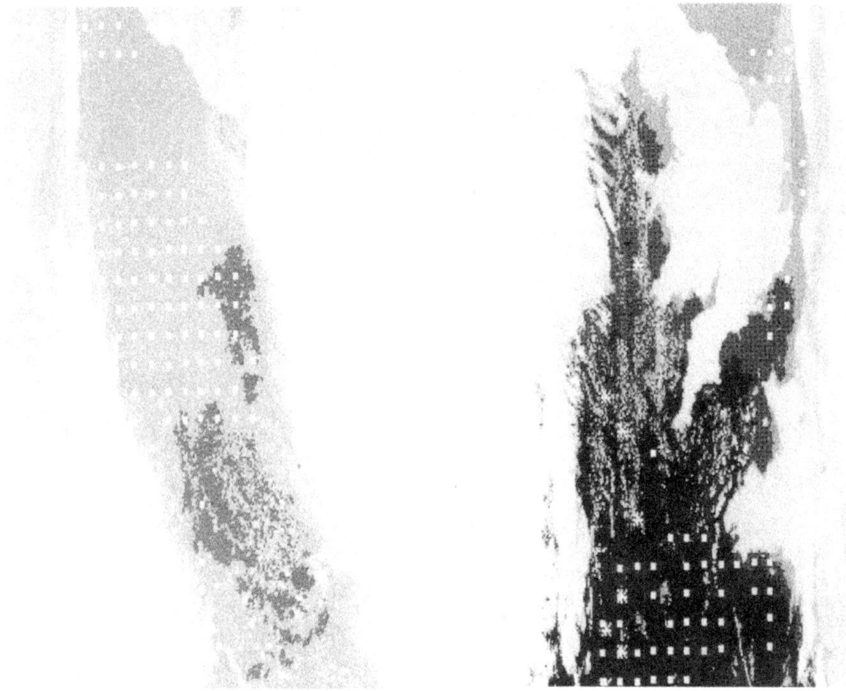

Fig. 7 Enhanced view of AVHRR channel 4 (pass A) to show location of AVHRR FOVs for which $C_c > 75$ (bright squares) and NESDIS clear FOVs (*) over the sea; a)- northern portion of pass A; b)- southern portion of pass A.

one selected by NESDIS, which are clear according to CMS but have not been selected altogether. North of 58^0 N, corresponding roughly to the lower limit in Fig. 7a, a selection criterium causes the extreme "boxes" (as defined in NESDIS processing) to be rejected on the grounds that the same region has been scanned during the previous orbit with a smaller scan angle. This can possibly explain why no FOVs are detected/selected clear in the orbit portion shown in Fig. 7a. In any event the selected clear FOVs to the east of the pass are not the clearest to be found and are certainly contaminated.

4. Conclusions

Cloud clearing techniques, aimed at producing clear column radiances from satellite data that can be cloud contaminated, are still needed in the processing of satellite data for NWP. In general not enough attention has been paid to cloud detection and clearing processing, the matter being considered as a necessary preprocessing with little scientific content. However the quality of retrieved products and their usefulness, both for NWP and Climate studies, is fundamentally limited by the quality of the cloud cleared radiances and the ability to specify this quality.

This paper presents some results of a comparison among different cloud-clearing schemes, based on data from three orbit sections covering the European and North Atlantic areas in Febrary 1989. Cloud masks produced by NESDIS, UKMO and LMD were compared to cloud masks obtained from AVHRR data. Clear-column BTs produced by NESDIS, UKMO and MR59 have been compared against AVHRR products. Great attention was put in analysing NESDIS cloud identification flag and cloud cleared radiances, since these are used operationally in all NWP centres and their quality has an important impact not only in the daily operations of these centres but also in the way satellite data will be used.

One of the clear results of the comparison is that the overall number of NESDIS clear FOVs was very small, some of the clear

FOVs were quite severely cloud contaminated and that the NESDIS processing did not identify most of the clear spots. As a consequence the quality of clear-column products was poor. The UKMO cloud-detection scheme seems to provide an operational cloud-masks similar to the AVHRR one in severe overcast conditions. LMD processing produces a more regular array of clear locations, some of which lie to the west of the front where at least 75% of the AVHRR FOVs, within each HIRS FOVs, are cloudy.

UKMO cloud-clearing scheme tends to give cold biased values when applied to the lowest HIRS/2 sounding channels (7 and 13 in our case). MR59 products compare best with both ECMWF forecasts and AVHRR products. The ability of the method to tune the amount of smoothing has proved to be quite useful in producing data with the needed smoothness and consistency.

A limitation to the present study is the amount of data that has been analysed; two or three orbit portions do not permit to draw many firm conclusions. This limitation may appear particularly relevant when dealing with global processing of satellite data. However there are no grounds to expect that problems arisen in the data examined would not show in different parts of the globe and at any time before March 89.

Clear radiances certainly have an important role in climate research and monitoring, besides the application to numerical weather prediction; it is however important to recognize that they are not a satellite measurement but a product. Since the implementation of cloud clearing schemes have changed over the years, the quality of the products has also changed considerably.

The difficulties that are encountered nowadays in handling clouds will not be reduced when higher spectral resolution data will be available, essentially because the size of the FOVs will remain practically unaltered with respect to today's operational sounders. Although the data will contain potentially many more information than today's sounders about the temperature and humidity profiles and the radiative properties of the lower boundary, most of this information will still be retrievable by modelling the differential spectral properties of the atmosphere and of the lower surface; these signatures may well be hidden by slight inaccuracies in the identification of the FOV content, such as the areal extent and height of clouds. The quality of cloud detection and clearing schemes will certainly need to be improved with respect to today's processing, to avoid a significant reduction of the advantage of oversampling caused by the introduction of correlated errors.

It appears therefore essential that some imager products, for each sounder FOV, be available to users. This will require a ground segment that addresses the problem of common sounder and imager processing. The availability of some imaging information measured directly by the sounder hardware will be a necessary tool to improve the collocation process between the sounder and the companion imager, but should not be considered an alternate solution, basically for the limited spectral coverage attainable by the sounder's imaging subsystem.

References

Amato U, V Cuomo, G Pavese, R Rizzi, C Serio and V Tramutoli, (1991) Cloud-clearing with Radial Basis Functions. Proceedings of the 6th International TOVS Conference, Airlie, VA, USA (CIMSS, Madison)

Andretta A, MM Bonzagni, V Cuomo, R Rizzi and C Serio (1990) A three-step Cloud Clearing Procedure for Infrared Sounder Measurements. International Journal of Remote Sensing, Vol. 11, 8: 1349-1368

Castagne N, P Le Borgne, J Le Vourch and JP Olry, (1986) Operational Measurements of Sea Surface Temperatures at CMS Lannion from NOAA-7 AVHRR Data. International Journal of Remote Sensing, Vol. 7, 8: 953-984

Derrien M, H Legleau, L Lavanant, L Harang, A Noyalet, D Pochic, (1990) Detection Automatique des Nuages avec l'Imageur AVHRR du Satellite defilant NOAA-11. EERM n. 4

Eyre JR (1989) Inversion of Cloudy Satellite Soundings Radiances by Nonlinear Optimal Estimation: Application to TOVS Data. Quarterly Journal of the Royal Meteorological Society, 115: 1027-1037

Eyre JR and PD Watts (1987) A Sequential Estimation Approach to Cloud-Clearing for Satellite Temperature Sounding. Quarterly Journal of the Royal Meteorological Society, 113: 1349-1376

Huang H-LA and WL Smith (1986) An Extension of the Simultaneous TOVS Retrieval Algorithm with the inclusion of cloud. Tech.Proc. 3rd Int. TOVS Study Conference, Madison Wisconsin 13-19 August 1986. CIMSS Report, University of Wisconsin

Reale AL, DG Gray, MW Chalfant, A Swaroop, A Nappi (1986) Higher Resolution Operational Satellite Retrievals. Proc. Second Conference on Satellite Meteorology-Remote Sensing and Applications, Am.Met.Soc.

Reynolds, RW, 1988: A Real-Time Global Sea Surface Temperature Analysis. Journal of Climate, 1: 75-86

Rizzi R, C Serio, G Kelly, V Tramutoli, A McNally, V Cuomo (1992) The cloud clearing comparison exercise. ECMWF Technical Memorandum n. 185

Susskind J, J Rosenfield, D Reuter, and MT Chahine, (1984) Remote Sensing of Weather and Climate Parameters from HIRS-2/MSU on TIROS-N. J.Geophys.Res., 89: 4677-4697

THE POTENTIAL OF MEDIUM-RESOLUTION SPECTRAL INFRARED MEASUREMENTS FOR HIGH CLOUD STUDIES

D. Spänkuch
German Weather Service
Meteorological Observatory Potsdam
Telegraphenberg
O-1561 Potsdam
Germany

I. Introduction

Cirrus clouds are challenging research objects. Neither our knowledge on their microphysical nor on their radiative properties and on the interrelationship between both is adequate to fully understand their role in the climate system which they affect through their impact on the atmospheric radiation field as well as on the hydrological cycle (e.g. Liou 1986, Dowling and Radke 1990, Stephens et al. 1990). Even the simplest climatology of cirrus clouds, their occurence, needs substantial improvement. On a global scale, only satellite remote sensing techniques are capable of estimating their relevant characteristics to some extent. But high clouds are not only remote sensing objects per se. It is a well known fact that the quality of remotely sensed tropospheric and surface parameters is severely detoriated by high cloudiness (e.g. Güldner 1980, Masuda and Takashima 1990).
Hence, appropriate remote sensing of cirrus clouds is a crucial component of remote sensing as well as climate monitoring.

II. Spectral radiances in the atmospheric window

In the field of passive remote sensing, measurements in the middle infrared ($\lambda \geq 5$ µm) and beyond are superior to measurements in other spectral regions from the aspect of round-the-clock monitoring. In the middle infrared, high clouds have the greatest effect on upwelling radiance in the 8-12 µm atmospheric window. Figure 1 presents two brightness temperature spectra (BT (ν)) for a cloudless and a cirrus-overcast sky in the tropics as measured with a Fourier spectrometer of 5 cm^{-1} spectral resolution aboard the Soviet satellite Meteor 28 (Spänkuch and Döhler, 1985).

NATO ASI Series, Vol. I 9
High Spectral Resolution Infrared Remote Sensing for Earth's Weather and Climate Studies
Edited by A. Chedin, M.T. Chahine and N.A. Scott

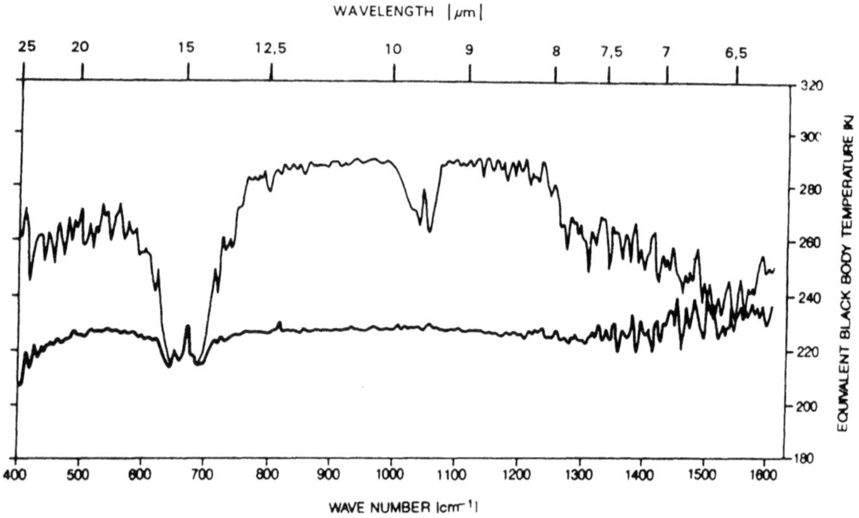

Fig. 1: Tropical brightness temperature spectra above a cirrus shield (lower heavy line) and for a clear atmosphere (upper line) measured with a Fourier spectrometer on 5 July 1977 aboard the satellite Meteor 28 (Spänkuch and Döhler, 1985)

The BT (ν) difference is about 60 K in the window region, about 30 K in the water vapour rotational band ($\lambda > 18$ μm) and 20 K and less in the water vapour rotational-vibrational band centered at 6.3 μm. This figure is also a good example of the significant impact of opaque high clouds on the Earth's radiation balance. Opaque high clouds in the tropics give the emitted radiance spectrum an appearance very much like that of the polar regions. While the absolute BT (ν) in the window is a good indication of the presence or absence of cirrus clouds within the field of view, brightness temperature differences $BTD(\nu_1,\nu_2) = BT(\nu_1) - BT(\nu_2)$ with $(\nu_1 \neq \nu_2)$ in the atmospheric window give additional information. Figure 2 derived from the Meteor 28 measurements shows that the spectral variation of BT across the window is different for cirrus than for all other types of emitting surfaces. In particular, BTD is higher between 8 and 9.5 μm (1250 to 960 cm^{-1}) and around 10 μm (1000 cm^{-1}), than at 11 μm (900 cm^{-1}). It is worthwhile noting that the BTD (ν) above cirrus clouds is also significantly different from the BTD (ν) of polar regions, where it is difficult to retrieve clouds from satellites. We note that the classification of the spectra of Fig. 2 was done by conventional classification algorithm using a couple of spectral channels in the 15 μm region and in the 10 μm window without the BTD information.

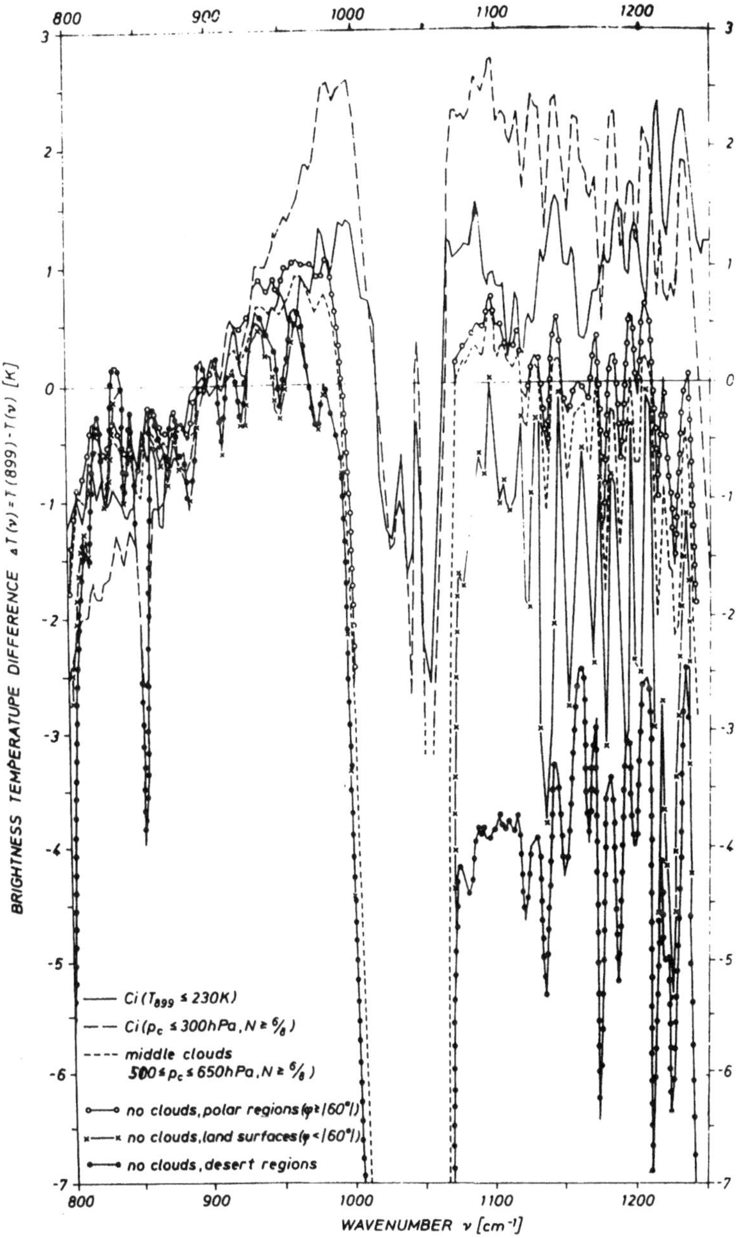

Fig. 2: Averaged spectra of brightness temperature differences BTD (ν) = BT (899 cm^{-1}) - BT (ν) in the atmospheric window for different conditions (Spänkuch and Döhler, 1985)

Due to the lack of supporting information in classifying high clouds, two classes of cirrus clouds were defined. The first class was defined when the brightness temperature in the window at 899 cm^{-1} was equal or less than 230K, the second class, when the cloud retrieval estimated a cloud amount of equal or larger than 6 octas and a cloud top pressure of less or equal 300 hPa. The use of the BTD-information in classifying high clouds was theoretically founded and demonstrated by Inoue (1985, 1987), Wu (1987), and Prabhakara et al (1988). Lee (1989) demonstrated the potential of the BTD method in identifying jet contrails even above dense cirrus clouds. Ackerman et al. (1990) showed that' a channel cloud detection technique using spectral measurements near 8, 11, and 12 μm poses the greatest potential for separating clear regions, cirrus clouds, and liquid water clouds' (Fig. 3).

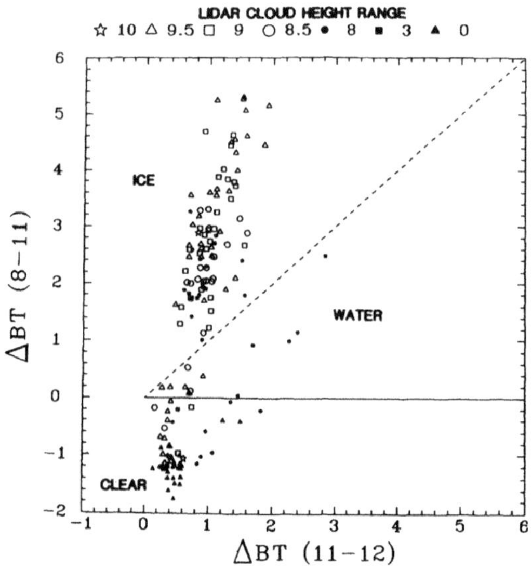

Fig. 3: Scatter diagram of BT (8 μm) - BT (12 μm) versus BT (11 μm) - BT (12 μm). The symbols represent cloud height ranges detected from simultaneous lidar data (Ackerman et al., 1990)

BTD (8-11) is generally positive for ice clouds, but negative for water clouds as stronger water vapour absorption at 8 μm compared to 11 μm results in negative BTD. Additionally, Ackerman et al. (1990), and Parol et al. (1991) show by theoretical calculations that the magnitude of the spectral variation in brightness temperature in the window region is related to the effective radius of the size distribution of spherical ice particles (r_{eff}), a parameter of highly informative character for parameterizations of cloud radiative properties. Figure 4

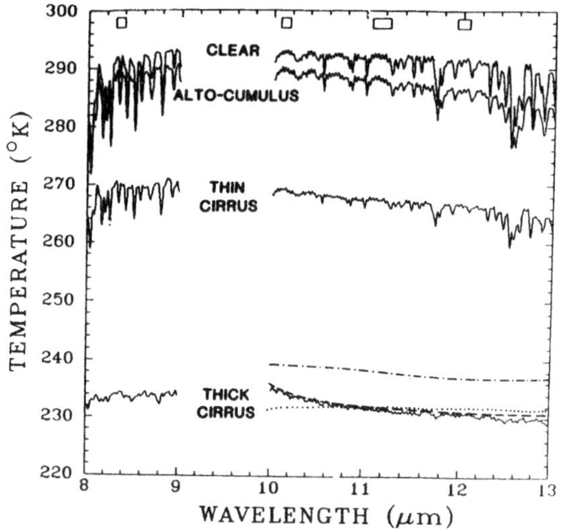

Fig. 4: HIS spectra for clear, altocumulus, thin and thick cirrus from 8 to 13 μm observed during the 28 October FIRE case-study day. The theoretical calculations for the thin cirrus case were done for clouds with an ice water content of 70 gm^{-2} and gamma size distributions with a variance of 0.25. The dotted, dashed, and dotdashed lines are for water spheres with r = 13 μm, ice spheres (13 μm), and ice spheres (25 μm), respectively (Ackerman et al., 1990)

shows a thick cirrus BT spectrum measured by the High-Spectral resolution Interferometer Sounder (HIS), (Smith et al. 1986), well approximated by a model cloud with an effective radius of ice spheres of 13 μm. Clouds with large ice particles display less spectral variance in BT, and water clouds with r_{eff} = 13 μm show no increase of BT from 11 to 10 μm. The results indicate that larger brightness temperature differences are associated with smaller effective radii of ice particles.

The BTD is not independent of the particle shape (Parol et al. (1991)). BTD is enhanced for nonspherical particles by the effects of scattering. Correspondingly the estimation of the both shape and effective size of the cloud particles is, in general, not unambiguously possible. Nevertheless, some constraints on particle shape and size can be derived.
The estimation of the effective radius of the cloud particles is done by the mass absorption coefficient ratio

$$K_1 \;/\; K_2 = \ln\,(1 - \varepsilon^b_{eff,1}) \;/\; \ln\,(1 - \varepsilon^b_{eff,2}) \tag{1}$$

derived from the effective beam emittance (Platt and Stephens 1980) at two wavelengths (i = 1,2) in the window

$$\varepsilon^{b}_{eff,i} = (L^{m}_{i} - L^{\uparrow}_{Cb,i}) \; / \; (B_{i}\,[T_{ct}] - L^{\uparrow}_{Cb,i}) \tag{2}$$

where L^{m}_{i} is the upwelling radiance measured by a satellite instrument, $L^{\uparrow}_{Cb,i}$ the upwelling radiance at cloud bottom and $B_{i}\,[T_{ct}]$ the blackbody radiation at cloud-top temperature T_{ct}. Equation (2) assumes no contribution to the measured radiance from the atmosphere above the cloud, which is rather well met in the window. Nor does Eq. (2) take not into account reflection of the upwelling radiance at cloud base, which is of some significance in the tropical atmosphere (Platt and Stephens 1980). However, Eq. (2) considers the findings of Platt and Stephens (1980) that the cloud-top temperature is an acceptable approximation when upward emittances are estimated.

The absorption coefficient ratio is particularly sensitive to size for $r_{eff} \lesssim 20$ to $30\mu m$ (Ackerman et al. 1990, Parol et al. 1991) and approaches an asymptotic value, dependent on particle shape and orientation. According to Fig. 2 this sensitivity is more pronounced when more distant wavenumber pairs between about 980 cm^{-1} (10.20 μm) and 833 cm^{-1} (12μm) are used for the absorption coefficient ratio.

Knowing r_{eff}, a first guess of other optical parameters such as single scattering albedo ($\tilde{\omega}_o$) and asymmetry factor (g) can be derived (Ackerman et al. 1990, Parol et al. 1991, Stone et al. 1990). One problem arises however, when window measurements are used to determine temperatures or altitudes of high clouds. Cirrus cloud top temperatures are generally biased to higher temperatures, and consequently, cirrus cloud top altitudes are biased to lower heights. Satellite derived upper tropospheric wind vectors, using high clouds as tracers, are, therefore, generally placed at too high pressure levels (Whitney 1983). This is due to two reasons. First, cirrus clouds are mostly semitransparent with emittances generally less than unity (Platt and Dilley 1981, Platt et al. 1987).

The weighting and the contribution function of a window channel have, therefore, two peaks in case of semitransparent clouds, one within the cloud layer and one beneath the cloud as shown in Fig. 5 for a one-level cloud layer (Spänkuch 1985). Second, scattering by cloud particles can not be neglected in general (Platt and Stephens 1980, Parol et al. 1991). There is hardly any chance to take its contribution operationally into account even if the corresponding microphysical parameters of the ice crystals (number density, distribution function, crystal shape and orientation) are known. As a rule of thumb, cirrus cloud top heights are underestimated by 1 to 1.4 km from window channel measurements (Kriebel et al. 1983, Spänkuch 1985).

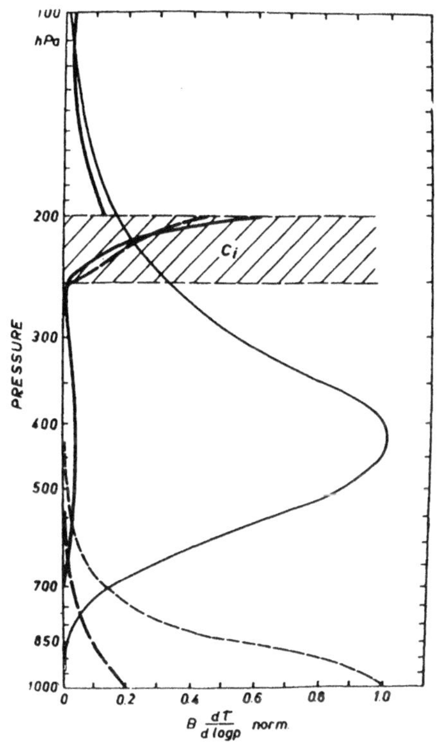

Fig. 5: Contribution functions for a water vapour channel at 488 cm^{-1} (full lines) and a window channel at 899 cm^{-1} (dashed lines). The thin curves are for a cloudless midlatitude summer atmosphere. The thick lines are for the same model but with a cirrus cloud between 200 and 250 hPa of a beam transmittance of 0.054 and 0.134 for ν = 1489 cm^{-1} (TOVS channel 12) and ν = 899 cm^{-1} (CH 8), located between 200 and 250 hPa, thin lines for the clear atmosphere respectively. In the cirrus cloud case the contribution function is normalized to the contribution of the whole cloud layer (Spänkuch 1985).

III. Cirrus cloud top estimates using spectral channels outside the atmospheric window

There are two approaches to estimate cirrus cloud top heights from middle IR measurements outside the atmospheric window. The first one is the use of appropriate channels at the wing of the 15 μm CO_2 band. The "CO_2 slicing" method introduced by Smith and Platt (1978), and developed by Menzel et al. (1983) uses two spatially independent radiance measurements of different cloud amount and/or cloud emittance assuming the cloud emittance to be independent of the spectral channels used. The accuracy of this method with the present generation medium-resolution spectrometers is of the same order as the window channel technique (Grund and Wylie 1989, quoted after Dowling and Radke 1990). Smith and Frey (1990) demonstrated a substantial improvement in estimated cloud top pressures with high spectral resolution measurement (0.5 cm^{-1}). Fig. 6 taken from their paper shows which portion of the spectral interval is particularly sensitive to variations in cloud top pressures. The most sensitive regions for high clouds are around 710 cm^{-1} and near 740 cm^{-1}.

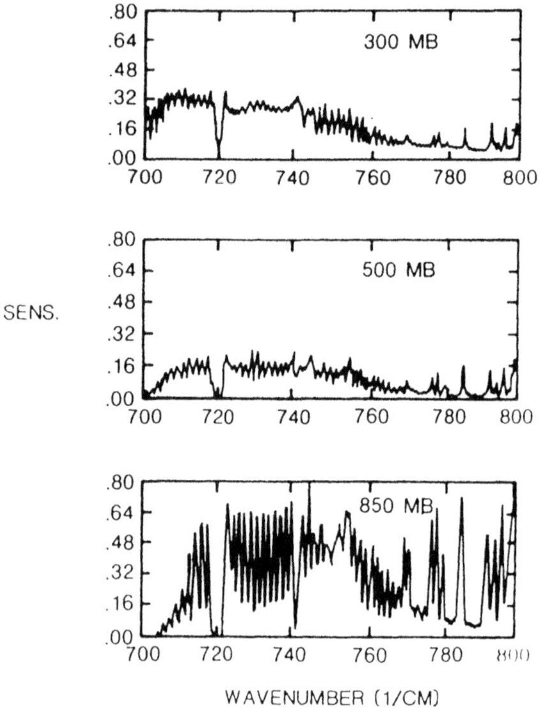

Fig. 6: Sensitivity of spectral radiance to variations of cloud top pressure (Smith and Frey 1990)

The second approach proposed and investigated to some extent by Szejwach (1982), Saunders and Hunt (1983), Pollinger and Wendling (1984), Spänkuch (1985),and Liou et al. (1990) uses measurements in the water vapour bands or a combination of window and water vapour channel measurements. The maximum contribution to the upwelling radiance originates in the water vapour channels in the cloud layer (see Fig. 5). The contributions from the adjacent atmospheric layers above and below the cloud depend on the spectral interval chosen. They are considerably less than in window channels where the contributions outside the cloud originates from the ground and from the lower troposphere which are significantly warmer than the cloud layer. The optimum spectral intervals depend on the height and effective emittance of the cloud. More opaque intervals are more appropriate for subtropopause cirrus clouds but could be saturated already for lower level cirrus. In the latter case less opaque intervals are more appropriate.

Table 1: Spectral radiances L_ν^m (mW/m² sr cm⁻¹) measured with a Fourier spectrometer of 5 cm^{-1} spectral resolution across a mid European low pressure system and some derived high cloud parameters. The L_ν^m of spectrum 09 behind the frontal system with an estimated cloud top pressure of 400 hPa is taken as $L_{Cb,\nu}^{1}$ (equ. 2). The cloud top temperature was estimated from measured spectral brightness temperatures of opaque intervals of the rotational water vapour band. More details about the synoptic pattern are given in Spänkuch and Döhler (1985).

Spectrum No. ν (cm^{-1})	05 833	05 988	06 833	06 988	07 833	07 988	08 833	08 988	09 833	09 988
L_ν^m (mW/ m²sr cm^{-1})	37.29	24.54	33.65	21.47	30.24	19.05	32.55	21.57	53.86	37.85
T (K)	222		216		212.6		215		239	
$\varepsilon_{eff,\nu}^{b}$	0.734	0.708	0.750	0.748	0.812	0.800	0.771	0.727		
K_{10}/K_{12}	1.16		1.20		1.19		1.12			
r_{eff} (μm)	~60		>60		>60		55			
$\tilde{\omega}_{ov}$	0.53	0.5	0.53	0.5	0.53	0.5	0.53	0.5		

Spänkuch and Döhler (1985) discussed a case study where Fourier spectrometer measurements were made above a mid-European low pressure system and where the tropopause temperature, measured by a nearby radiosonde, was measured as the spectral brightness temperature in an opaque region of the rotational water vapour band around 488.6 cm^{-1} (20.5 μm), too. Hence, using the expression for the apparent blackbody radiation at the top of a black cloud (e.g. Low 1978)

$$L_\nu^C \approx B_\nu\,[T(Z_t)] \approx (1 - \tilde{\omega}_{o\nu})\,B_\nu\,[T_c] + S_c \qquad (3)$$

with B the blackbody radiation at cloud top temperature $T(Z_t)$ and at a representative cloud temperature T_c, resp., S_c the scattered component and $\tilde{\omega}_{ov}$ the single scattering albedo of the system 'cloud plus absorbing gas', we find $\tilde{\omega}_{ov} = 0$ and $S_c = 0$ at this wavenumber. The scattered component S_c can be estimated for wavenumber intervals of nearly the same water line absorption not too far away from the reference spectral interval since there the assumptions of $T_c = T_{Z_t}$ and $\tilde{\omega}_{ov} = 0$ are approximately met. Nearly the same line absorption contribution as at 488.6 cm^{-1} are given at 502.7 cm^{-1} (19.9 μm) and 517.5 cm^{-1} (19.3 μm). With measured values of $L_{502.5} = 54$ mW/m² sr cm^{-1} and $L_{517.5} = 53.2$ mW/m² sr cm^{-1} instead of B (212.6 K) = 52.16 mW/m² sr cm^{-1} and 51.28 mW/m² sr cm^{-1}, resp., S_c is estimated to 1.84 and 1.92 mW/m² sr cm^{-1} or about 3.5 % of the emitted radiation. The effective beam emittance would consequently be estimated at these wavenumbers to 0.965 and the cloud top temperature overestimation would be 2.8 and 2.2 K, resp. . Unfortunately, the quality of our measurements did not allow applying the same procedure in the vibrational rotational band (6.3 μm) due to the reduced signal to noise ratio. We expect, however, a larger S_c there due to the three times larger size parameter $\alpha = \frac{2 \pi r_{eff}}{\lambda}$.

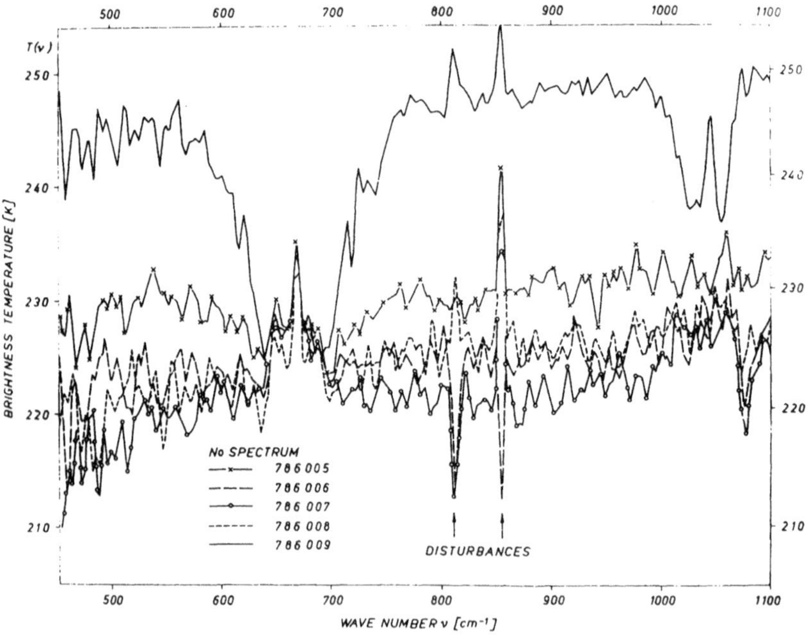

Fig. 7: Consecutive brightness temperature spectra over a mid-European low pressure system. Spectrum 786009 behind the frontal system with an estimated cloud top height of 400 hPa was taken as reference spectrum for the radiance at cirrus cloud bottom. More details in the text and in Spänkuch and Döhler (1985).

The window channels of the same measurements (Fig. 7) allow a crude estimation of r_{eff} and of $\tilde{\omega}_{ov}$ and other radiative quantities as outlined in the previous chapter. $L_{Cb}^{\uparrow}$ (Eq. 2) is taken from spectrum nr 09 behind the frontal system with an estimated cloud top pressure of 400 hPa. In using Eq. (2) we assume no change of $L_{Cb}^{\uparrow}$ across the frontal system. This is a severe assumption since one would expect $L_{Cb}^{\uparrow}$ to vary dramatically across a frontal boundary, especially if there were a broken cloud field under the cirrus shield. The derived quantities compiled in Table 1 are therefore a first guess and should demonstrate only the potential of using nearby spectral measurements also at cloudy conditions in deriving some spectral cloud characteristics. More reliable estimates are possible if corresponding information about the atmosphere beneath the cirrus shield is available.The effective beam emittance in the window is between 0.7 and 0.8 and the mass absorption coefficient ratio K_{10}/K_{12} between 1.1 and 1.2. At such values r_{eff} and $\tilde{\omega}_{ov}$ are only slightly dependent on K_{10}/K_{12}. r_{eff} is about 60 μm or larger and $\tilde{\omega}_{ov}$ is about 0.5.

IV. Summary and Conclusions

Clouds are scattering and absorbing objects of less pronounced spectral properties than absorbing gases. Medium-resolution spectral measurements are therefore often sufficient to characterize the clouds. More important than an improvement in spectral resolving power are spectrally continous measurements.

Spectral measurements in the atmospheric window are quite amenable to the brightness temperature difference technique for the **detection** of cirrus clouds.

Estimates of **cloud top height** and temperature of high clouds are, however, better done using opaque intervals within water vapour absorption bands. In these channels the contribution of the intervening atmosphere is stronger than in the window, but the radiation emitted from the atmospheric layers above and below the cloud are of opposite sign in comparison to the radiation emitted from the cloud itself. Another promising approach is the CO_2 slicing technique using spectral channels in the wing of the 15 μm CO_2 band. Both approaches are benefited by higher spectral resolution.

The only **radiative characteristic** which can directly be estimated from infrared radiancc measurements is the effective beam emittance. However, it is the broadband flux absorptance emittance, the emissivity, which is required for energetic considerations. Platt and Stephens (1980) relating these emittances with each other could show that the narrowband effective flux emittance well represents the broadband effective flux emittance independent of the cloud particle size. A strong relation exists between the effective flux emittance and the beam absorptance emittance, the parameter which is directly related to the absorption properties of the cloud. Relating the effective beam emittance to the beam absorptance emittance or the effective flux emittance requires further research[1].

Obviously, only gross features of microphysical parameters can be derived from spectral radiance measurements in the middle ir with the window region as the most relevant informative spectral portion. A coordinated effort from both, the radiative and the microphysical disciplines, is needed to achieve more progress in this field.

References

Ackerman SA, Smith WL, Spirnhirne JD, Revercomb HE (1990) The 27-28 October 1986 FIRE IFO Cirrus Case Study: Spectral properties of cirrus clouds in the 8-12 μm window. Mon Wea Rev 118: 2377-2388

Dowling DR, Radke LF (1990) A summary of the physical properties of cirrus clouds. J Appl Meteorol 29: 970-978

Güldner J (1980) Über eine Möglichkeit der Einteilung indirekt bestimmter Temperaturprofile in verschiedene Genauigkeitsklassen auf der Grundlage von Satellitenstrahlungsmessungen. Z Meteorol 30: 226-231

Inoue T (1985) On the temperature and effective emissivity determination of semi-transparent cirrus clouds by bi-spectral measurements in the 10 μm window region. J Meteorol Soc Japan 63: 88-99

Inoue T (1987) A cloud type classification with NOAA-7 split-window measurements. J Geophys Res 92: 3991-4000

Kriebel KT, Moerl P, Reinhardt ME, Schellhase R, Koenig T, Rattei W (1983) Comparison of cloud top heights measured by airborne lidar and TIROS-N image date. Adv Space Res 2 (6): 11-13

Lee TF (1989) Jet contrail identification using the AVHRR infrared split window. J Appl Meteorol 28: 993-995

Liou KN (1986) Influence of cirrus clouds on weather and climate processes: A global perspective. Mon Wea Rev 114: 1167-1199

Liou KN, Ou SC, Takano Y, Valero FPJ, Ackerman TP (1990) Remote sounding of the tropical cirrus cloud temperature and optical depth using 6.5 and 10.5 μm radiometers during STEP. J Appl Meteorol 29: 716-726

Low RD (1978) The internal cloud radiation field and a technique for determining cloud blackness. Tellus 30: 418-428

Masuda K, Takashima T (1990) Sensitivity of shortwave radiation absorbed in the ocean to cirrus parameters. Remote Sens Environ 33: 75-86

Menzel WP, Smith WL, Stewart TR (1983) Improved cloud motion wind vector and altitude assignment using VAS. J Appl Meteorol 22: 397-384

Platt CMR, Dilley AC (1981) Remote sounding of high clouds. IV: Observed temperature variations in cirrus optical properties. J Atmos Sci 38: 1069-1082

Parol F, Buriez JC, Brogniez G, Fouquart Y (1991) Information content of AVHRR channels 4 and 5 with respect to the effective radius of cirrus cloud particles. J Appl Meteorol 30: 973-984

Platt CMR, Scott JC, Dilley AC (1987) Remote sounding of high clouds. IV: Optical properties of midlatitude and tropical cirrus. J Atmos Sci 44: 729-747

Platt CMR, Stephens GL (1980) The interpretation of remotely sensed high cloudiness emittances. J Atmos Sci 37: 2314-1322

Pollinger W, Wendling P (1984) A bispectral method for the height determination of optically thin ice clouds. Beitr Phys Atmos 57: 269-281

Prabhakara C, Dalu G (1976) Remote sensing of the surface emissivity at 9 μm over the globe. J Geophys Res 81 (21): 3719-3724

[1]For the definition of these quantities see Platt and Stephens (1980)

Prabhakara C, Fraser RS, Dalu G, Wu M-LC, Curran RJ (1988) Thin cirrus clouds: Seasonal distributions over oceans deduced from Nimbus-4 IRIS. J Appl Meteorol 27: 389-399

Saunders RW, Hunt GE (1983) Some radiation budget and cloud measurements derived from Meteosat 1 data. Tellus 35B: 177-188

Smith WL, Frey R (1990) On the cloud altitude determinations from High Interferometer Sounder (HIS) observations. J Appl Meteorol 29: 658-662

Smith WL, Platt CMR (1978) Comparison of satellite-deduced cloud heights with indications from radiosonde and ground-based laser measurements. J Appl Meteorol 17: 1796-1802

Smith WL, Revercomb HE, Howell HB, Woolf HM, La Porte DD (1986) The High resolution Interferometer Sounder (HIS). CIMSS View II, No. 3

Spänkuch D (1985) The problem of indirect sounding of high clouds. Adv Space Res 5, No. 6: 177-181

Spänkuch D, Döhler W (1985) Radiative properties of cirrus clouds in the middle ir derived from Fourier spectrometer measurements from space. Z Meteorol 35: 314-324

Stephens GL, Tsay S-C, Stackhouse Jr. PW, Flatau PJ (1990) The relevance of the microphysical and radiative properties of cirrus clouds to climate and climatic feedback. J Atmos Sci 47: 1742-1753

Stone RS, Stephens GL, Platt CMR, Banks S (1990) The remote sensing of thin cirrus cloud using satellites, lidar and radiative transfer theory. J Appl Meteorol 29: 353-366

Szejwach G (1982) Determination of semi-transparent cirrus cloud temperature from infrared radiances: Application to METEOSAT. J Appl Meteorol 21: 384-393

Whitney Jr. LF (1983) International comparison of satellite winds - an update. Adv Space Res 2 (6): 73-77

Wu M-LC (1987) A method for remote sensing the emissivity, fractional cloud cover and cloud top temperature of high-level, thin clouds. J Climate Appl Meteorol 26: 225-223

MODELLING REFLECTING AND SEMI-TRANSPARENT CLOUD FOR INFRARED SOUNDING

P.D.Watts and A.J.Baran
U.K.Meteorological Office
London Road,
Bracknell, Berkshire
RG12 2SZ

Abstract

This paper examines the effect of spectrally varying cloud reflectivity and emissivity on infrared radiances. In the first part of the paper the results of two-stream calculations for cloud bulk radiative properties are given. These are directly relevant to the use of high resolution infrared measurements. In the second part, these results are used in a simple way in a non-linear retrieval scheme using TIROS Operational Vertical Sounder (TOVS) data. This scheme is at a research and development stage in the UK Meteorological Office.

It is shown that clouds cannot be assumed to be black-body emitters even when they are completely opaque. Moreover, they cannot be assumed to be grey, especially in the shorter wavelength region 3–6 μm. The non-zero reflectivity of opaque clouds is shown to give rise to significant amounts of reflected solar radiation, again in the short wavelength region.

A parameter to represent the reflectivity of the cloud is included in the non-linear retrieval scheme and values of this parameter estimated from a NOAA-11 case study appear to be in broad agreement with the predictions of the two-stream approximation.

Also presented are results from this case study which show strong directional properties of the reflected radiation – information which is lost in the simplified two-stream calculations. Work is under way utilising more sophisticated discrete ordinates and Monte Carlo scattering calculations to predict the angular reflection properties. It may then become possible to obtain cloud microphysical properties, i.e. effective drop size, from TOVS measurements.

Introduction

Retrieval of useful atmospheric parameters from infrared sounding data is often made possible by elaborate cloud detection and, sometimes, correction schemes. Although proven successful in some cases, e.g. SST from AVHRR, Southern Hemisphere soundings from TOVS, this

NATO ASI Series, Vol. I 9
High Spectral Resolution Infrared Remote Sensing
for Earth's Weather and Climate Studies
Edited by A. Chedin, M.T. Chahine and N.A. Scott

approach necessarily rejects a large proportion of the sounding data as unusable. An alternative approach to the problem of cloud contamination of infrared radiances is to explicitly model the cloud radiative effect. This has been done with some success in the scheme developed by *Eyre* 1989a and b which iteratively maximizes the probability of an atmospheric and cloud state vector given TOVS measurements and a numerical forecast background state. The cloud state vector in the scheme consists of an effective fractional cover n and a top pressure P_c. The 'effective' cover allows for cloud which is semi–transparent but with a transparency which is constant with wavelength. There is no explicit representation of reflectivity and, indeed, the basic radiative transfer model used, RTTOV (*Eyre*, 1990), does not include reflection effects. The work in the present paper suggests that the cloud model should allow for spectrally varying cloud properties if the accuracy of the calculated radiances is to be comparable to that for clear column atmospheres. This applies both to the reflection of solar radiation and to the transmissivity of thin clouds. Thin clouds are considered only briefly in here.

The structure of this paper is as follows. The first section describes briefly the work of *Yamamoto et al.* 1970 who made detailed calculations of cloud radiative properties from 5–15 μm. This is followed by the work of the present authors on Mie and two-stream calculations over the range 3.4–15 μm.

Section 2 describes the fast radiative transfer model used to apply the results of section 1 to HIRS radiances and gives results of the fast model calculations with idealised cloud conditions. Section 3 introduces the retrieval scheme and it's enhanced version and, finally, in section 4 we describe the results of applying the scheme to a day-time overpass of NOAA-11 with cloud reflection explicitly, albeit simply, modelled and retrieved.

1. Cloud radiative properties: 3.4–15 μm

That reflection properties of clouds might be important, particularly for the 4.3 μm CO_2 band sounding region, is apparant from the literature, in particular the work of *Yamamoto et al.* who performed radiative transfer calculations for water clouds of varying thickness. They assumed an altostratus drop size distribution given by *Deirmendjian D.*, (1969), which has minimum, maximum and mode radii of 2, 26 and 9 μm respectively, a number density of 450 cm^{-3} and a water content of 0.28 gm^{-3}. They accounted for the radiative properties of the vapour present assuming saturation pressure at the cloud temperature of 263 K. Their results show reflectances for opaque clouds of up to 10%. Unfortunately the calculations extend far enough only to give an indication of what to expect in the shortwave region around 3–5 μm. Highest reflectivities occur at the shortwave end which is also where the solar flux is strongest. Also of interest is the behaviour of thin clouds (i.e. those where the transmission becomes non-zero). The emissivity naturally falls but also the variation with frequency becomes more marked. From the *Yamamoto* results it appears that clouds have to be geometrically *very* thin (e.g. 50 m) before emissivities are significantly reduced. However, this does depend on the water content assumed; the 0.28 gm^{-3} used by *Yamamoto et al.* is moderately high.

Less information is available on ice clouds because their physical properties are less well documented and the non-spherical shape and non-random orientation of the crystals makes theoretical calculations difficult. However, under the assumption of spherical crystals and realistic particle number densities, *Hunt 1973* calculated radiative properties at 2.3, 3.5, 3.8, 8.5 and 11 µm. Reflectivities of ice clouds are lower than water clouds – typically 2-5% – owing to the much reduced water content.

In order to have a complete picture of cloud radiative properties from 3.4–15 µm and to develop a suitable fast model, we have obtained Mie parameters for various representative cloud types and used a two-stream radiative transfer model to estimate the cloud radiative properties. We assume mono-dispersive scattering, that is to say we assume the scattering can be characterised by Mie calculations for a drop with a radius r_e; the effective radius of the drop size distribution n(r) of the cloud. This is defined by

$$r_e = \frac{\int_0^\infty \pi n(r) r^3 dr}{\int_0^\infty \pi n(r) r^2 dr}.$$

We recognise that significant errors could arise from the mono-dispersive approximation and intend in future to use Mie paramters calculated for a distribution of drop sizes. The two-stream method assumes approximate forms both for the scattering phase function and the upward and downward intensities in order that the general radiative transfer equation can be solved analytically. Here a Hybrid Delta-Eddington approximation has been used (*Meador and Weaver, 1980*). For a full description of this work see *Baran and Watts 1992*. Suffice to say here that results were checked where possible against the work of *Yamamoto, Hunt* and others and satisfactory agreement was found.

Calculations of this form are characterised by a drop density, an effective radius r_e, a geometrical cloud depth and a solar illumination angle. The results presented in Figures 1–4 are for some typical clouds with a solar zenith angle of 0°.

Figure 1 is a plot of the reflection r, transmission t and emissivity ε, with wavelength for a typical maritime cloud characterised by a large r_e of 14 μm, a low drop density of 50 cm^{-3}, and a depth of 50 m.

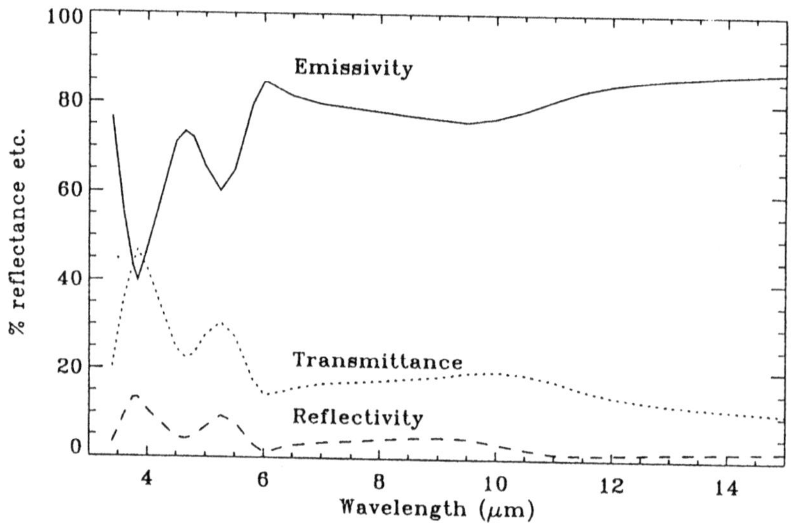

Figure 1. Radiative properties of a maritime stratus cloud with effective drop radius 14 μm, number density 50 cm^{-3} and depth 50 m.

Reflectivity is highest at around 3.8 μm and peaks again at 5.3 μm. For wavelengths greater than 6 μm reflectivity remains low, at or below 8%, and is particularly low at the 11 μm window. This behaviour is principally e result of the complex refractive index of water (see, for example, Irving and Pollack 1968). The transmittance of this cloud is also significant with a spectral variation which closely follows that of the reflectivity. Figure 2 is for the same type of cloud but with a depth of 200 m. It shows the rapidly decreasing transmission of water cloud with cloud thickness. With clouds of this or greater thickness it is a good approximation to say that the transmission is zero and the cloud has $\varepsilon = 1 - r$.

Cloud over land is often characterised by larger drop densities and smaller drops than the marine case because of the greater concentrations of cloud condensation nuclei and lower water contents present.

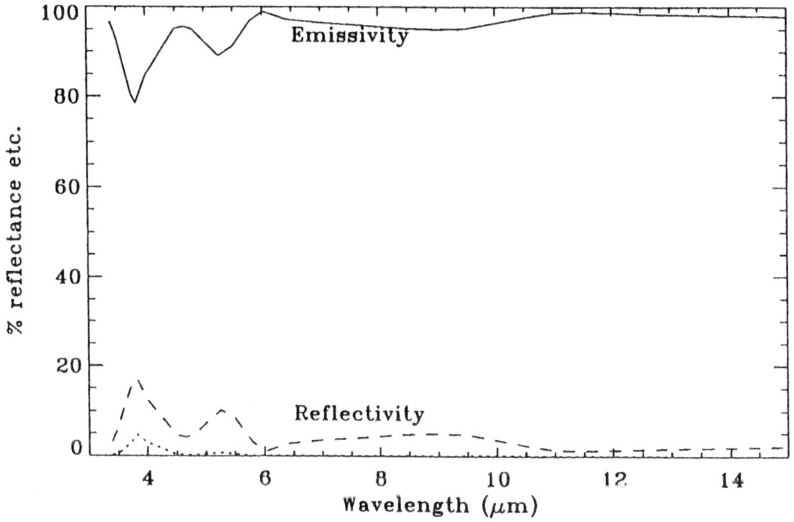

Figure 2. As figure 1 but for a cloud depth of 200 m.

Figure 3 shows results for a typical (R.W.Saunders, personal communication) land cloud with $r_e = 5.8$ μm, a drop density of 350 cm^{-3} and a depth of 200 m. Reflectances in this case are much higher than for the marine cloud — this is principally a result of the smaller drop size. Again the cloud is essentially opaque.

Finally, Figure 4 gives the radiative properties for an ice cloud with $r_e = 37$ μm, a drop density of 0.05 cm^{-3} and a depth of 1.5 km. Here we can see that the reflectivity is generally low, around 3–5%, the transmittance is high, around 50%, and the strong spectral variation found around 3-6 μm in the case of water cloud is not present.

From this albeit small selection of clouds we can conclude the following:

- For optically thick, opaque, i.e. most water clouds, we can assume negligible transmission of radiation and significant reflection especially at short wavelengths. This is principally a result of the large effective radius.
- The characteristic radiative properties of ice clouds are significantly different to those of water clouds with lower reflection and higher transmittance.

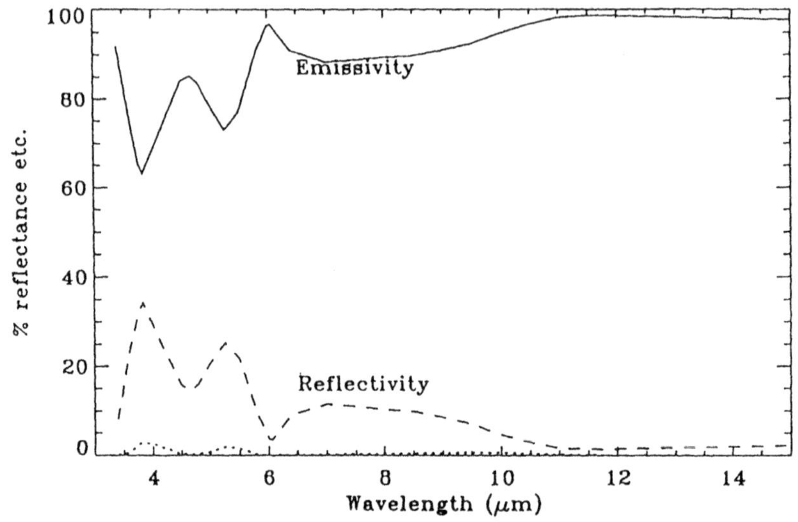

Figure 3. Radiative properties of a continental stratus cloud with effective drop radius 5.8 μm, number density 350 cm^{-3} and depth 200 m.

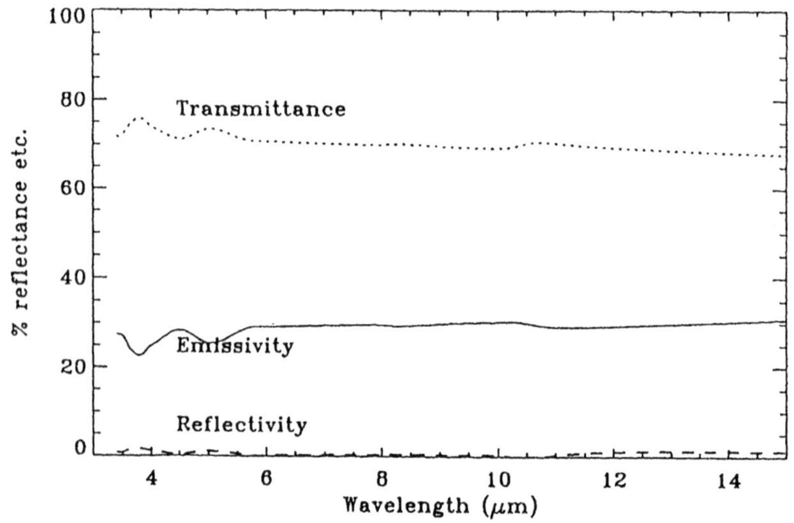

Figure 4. Radiative properties of a cirrus cloud with effective drop radius 37 μm, number density 0.05 cm^{-3} and depth 1500m.

- The peaks and troughs of the spectral reflectivity are fixed in position regardless of the cloud observed; the pattern is a result of the variation of the complex refractive index of water (or ice).

The drop size dependence of reflectivity, particularly at 3.7 μm, is well seen in Figure 5, suggesting the possibility of measuring the effective drop size of a cloud through measurements of its reflectivity. This provides motivation to include a cloud reflection/ emission model in an operational TOVS processing scheme over and above that of improving the accuracy of calculated radiances for the shortwave sounding channels.

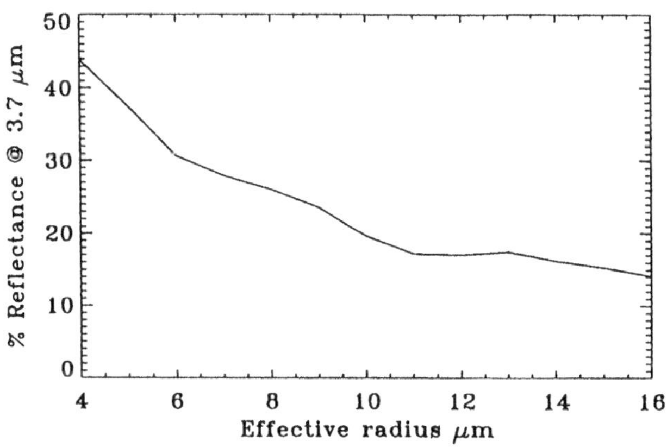

Figure 5. Variation of the 3.7 μm reflectivity of an opaque cloud with effective radius.

2. Fast radiative transfer model

The radiative transfer model used in this study was an extended version of a fast model developed at ECMWF (*Eyre 1991*) which obtains channel transmittances from levels in the atmosphere to space by a regression method. The ECMWF model was written to vectorize efficiently and was simple to extend for reflecting surfaces: downwelling atmospheric radiances for all levels are stored in a common block. The input variables to the model are temperatures at 40 standard pressure levels, humidity at 15 of the lower levels, ozone column amount, microwave surface emissivity, cloud fractional cover and cloud top pressure. The cloud fractional cover is an 'effective' amount allowing for cloud which is transmitting equally at all wavelengths. The output variables are radiances and equivalent brightness temperatures for any or all of the TOVS channels.

2.1 The enhanced fast model

The first change made to the model was to explicitly use cloud absolute fractional cover n and emissivity ε rather than an effective fractional cover (which is adequate for non-reflecting grey clouds). This is achieved in a simple way by the use of an additional input variable – a reflectivity parameter ρ – which in the model is multiplied by a 'standard' reflection profile R_i to give the cloud reflectivity at each channel *i;* $r_i = \rho R_i$. Only opaque clouds have been considered so far, so a similar extension to explicit modelling of cloud transmittance is not required. The channel emissivity of the cloud is therefore given by $\varepsilon_i = 1 - r_i$.

Any number of reflection profiles can be supplied but the one used in this study is that of the opaque marine stratocumulus given previously. This simple approach therefore assumes that the spectral reflectivity within certain cloud types, e.g. water, ice, high, low etc., can be adequately modelled by scaling a reference reflectivity for that cloud type.

A second change in the model arises from the necessity to calculate atmospheric transmittances at two angles other than the satellite view zenith angle φ. Firstly, transmittances at the solar zenith angle θ_s are required to account for attenuation of the incoming solar radiation before it is reflected off the cloud or physical surface.

Secondly, calculating reflection of downwelling infrared radiation requires, in principle, an integration over the hemisphere. An adequate approximation for a stratified atmosphere (*Goody and Yung 1989*) is to assume the radiance is arriving from a zenith angle of $\theta_d = 52.96^o$.

Since a large part of the fast model computation arises from the transmittance calculation it is very desirable not to have to perform it more than once, especially for what are second order effects on the calculated radiances. In this work transmittances at the solar and diffuse approximation angles were estimated from those calculated at the satellite view angle by assuming they were simply related by the relative path lengths. This assumption leads to the estimate for the transmittance to space from level p:

$$\tau_\theta(p) = \tau_\varphi(p)^{\sec\varphi/\sec\theta}.$$

This approximation was improved by an additional factor α in the exponent obtained empirically† by minimising the transmittance error in HIRS channel 19 for a representative atmospheric profile and for each of a discrete set of $\sec\varphi$ and $\sec\theta$:

$$\tau_\theta(p) = \tau_\varphi(p)^{\sec\varphi/\sec\theta.\alpha}.$$

Contributions to the radiance at the top of the atmosphere from reflected solar and reflected downward diffuse radiation then become

$I_s \cos\theta_s nr\tau_{\theta_s}(p_c)\tau_\varphi(p_c)$ and

$R_d(p_c)nr\tau_{\theta_d}(p_c)$ respectively.

I_s is the solar flux at the top of the atmosphere, p_c is the cloud top pressure, $R_d(p_c)$ is the downward radiance at the cloud level and the dependency of terms on wavelength (channel) is implied. The cloud emission term in the case of reflecting clouds becomes

$B(p_c)\ n(1\text{-}r)\tau_\varphi(p_c)$.

In the following section we describe results of the enhanced model and how the modified cloud emission affects the observed radiances. The calculations are performed for select HIRS channels but the conclusions are relevant to infrared sounding generally.

2.2 Model results

By comparing top of atmosphere radiances computed using the fast model with both black and reflecting clouds we can establish the degree to which we must model the cloud emissivity accurately.

Figure 6 shows the contribution of reflected solar radiation to the brightness temperature in a selection of HIRS channels under overcast conditions as a function of cloud top pressure; the cloud reflectivity used here is that for a marine stratocumulus and the solar zenith angle is 53°. The tendancy for the brightness temperature difference to increase at first slowly then rapidly

---

† α was found by minimising the difference between the approximated transmittance and the 'exact' one given by the fast model. This was done for all combinations of φ and θ hence α is a function of these angles.

with decreasing cloud pressure is a result of the decreasing atmospheric transmittance between cloud and space. HIRS 18 and 19 are strongly affected; this corresponds to the high reflectivity peak around 3.83 μm seen in Figure 2. The 4.3 μm sounding channels 13,14 and 15 are contaminated to a lesser degree but, particularly in the most transparent (13), the contamination is significant† for cloud above 800 mb. For HIRS channels with wavelengths longer than 5 μm the reflected solar contribution is negligible; this results from a combination of lower reflectivity and sharply decreasing solar flux.

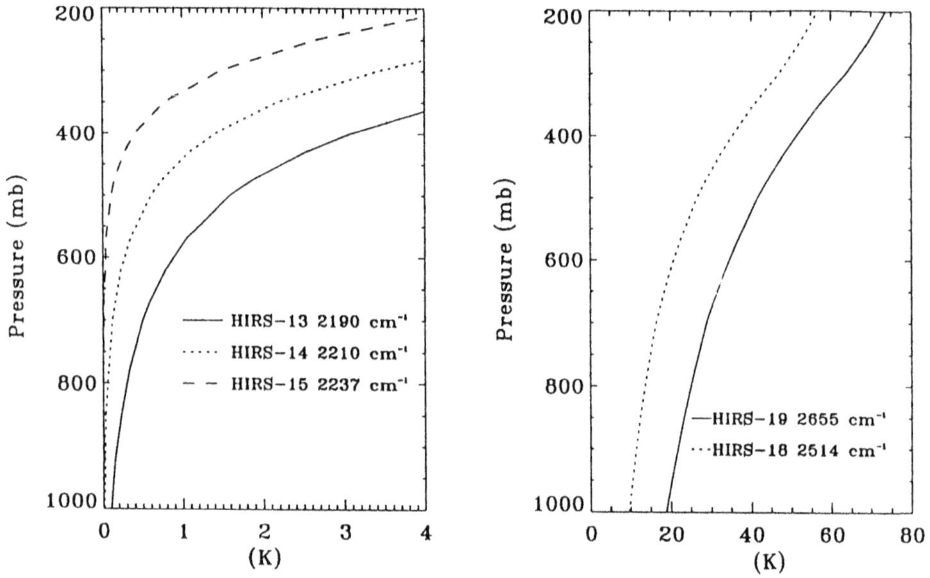

Figure 6. Contributions to HIRS channel brightness temperatures from reflected solar radiation for complete cloud cover at pressures 1000-200 mb.

The reflected diffuse downward radiation contribution is shown in Figure 7 again as a function of cloud top pressure. The effect is generally much smaller than the reflected solar radiation but increases as the cloud pressure increases and is non-negligible for the longer wavelength channels and, of course, occurs day and night. The effect is small in channels 18 and 19, despite the large reflectivities, because they are located in the 4 μm window region — a transparent atmosphere means little downward flux.

† Clear column radiances can be modelled to an accuracy of the order a few tenths of a degree – this sets the standard to which it is desirable to model cloudy radiances.

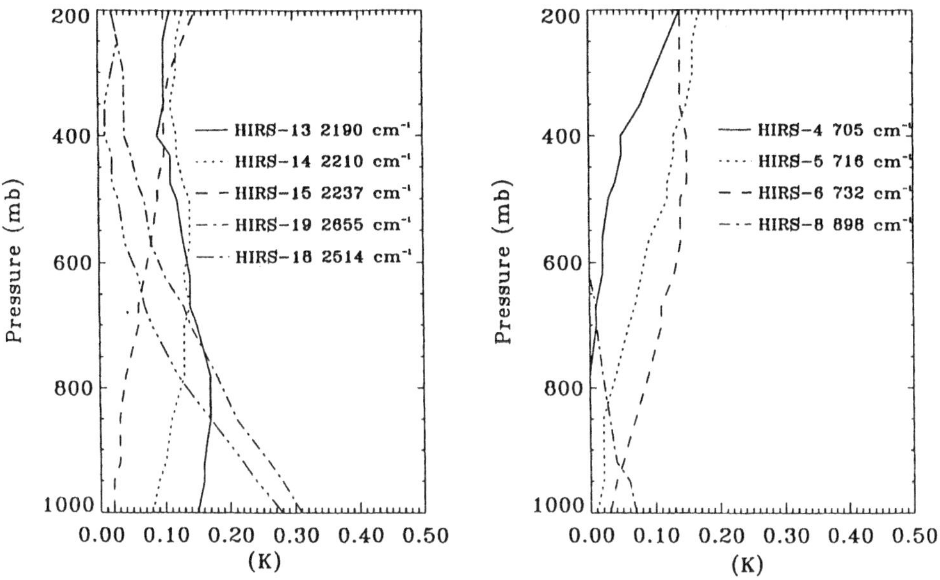

Figure 7. As figure 6 but for the reflected downwelling diffuse radiation.

Compensating the reflected radiation is a deficit in the emission because of the non-unit emissivity. This is shown in Figure 8 and the net effect — reflected and deficit — in Figure 9. The net effect in the 4 μm window channels is clearly negative by 1-2 K and that in the 11 μm window is very small. This effect has been exploited to identify low cloud and fog in Advanced Very High Resolution (AVHRR) imagery (*Eyre et al.* 1984). In the shortwave sounding channels the net effect is non negligible for cloud at most levels; for longer wavelengths only higher cloud gives a significant effect.

All the above results are subject to the following caveats. Firstly, we have presented results for water clouds up to altitudes of 200 mb where clearly the phase would normally be ice. As we saw in section 2, ice cloud usually reflects much less well than water cloud so the larger radiance effects are unlikely to occur in reality.

Secondly, two-stream predictions of cloud radiative properties refer to hemispherically averaged quantities and it is possible that, although the phenomenom we are interested in is a multiple scattering one, the reflectance retains strong directional properties. In fact, results presented in section 5 of this paper indicate that the dependence on scattering angle may be as large as that on cloud type, drop size, etc. Consequently, with suitable satellite — sun

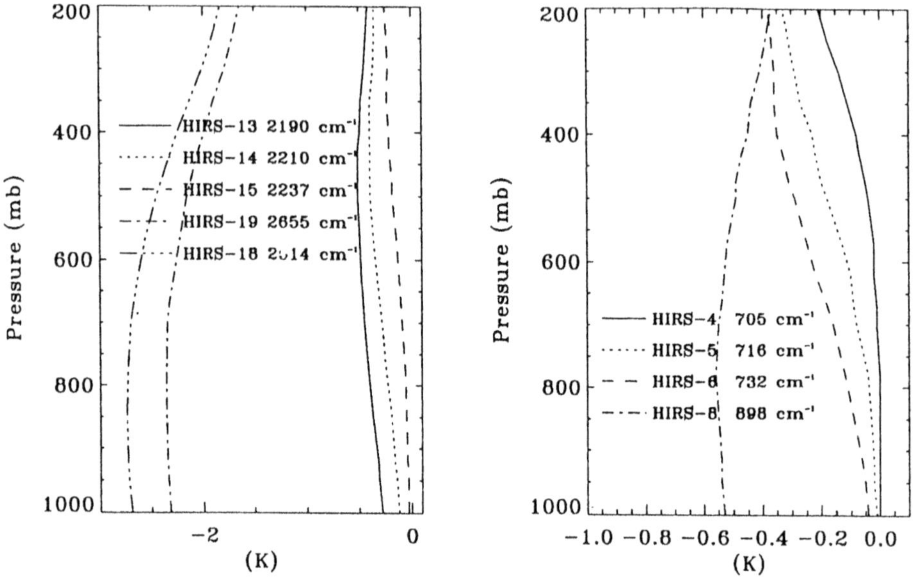

Figure 8. As Figure 6 but for the deficit in radiation emitted from the cloud.

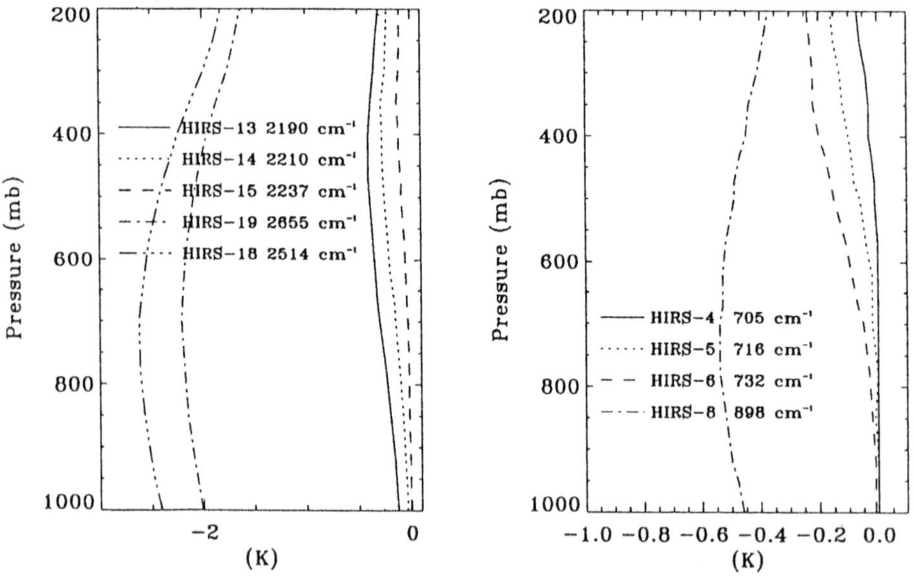

Figure 9. As Figure 6 but for the net (diffuse reflected + deficit) radiation.

geometry, i.e. back or forward scattering, the reflectivities could be well in excess of those used in the above calculations and hence the 'contamination' of the radiances even larger.

The above results then, should be seen simply as a guide as to what to expect in 'normal' conditions.

3 Enhancement of a non-linear retrieval scheme

An existing non-linear TOVS retrieval scheme TOVCFG (*Eyre* 1989a and b) which simultaneously solves the inverse radiative transfer equation for temperature and humidity profiles, cloud fractional cover† and cloud top pressure at a single sounding field of view (fov) was extended to include the enhanced fast forward model (and it's gradient version) and hence to retrieve, also, estimates of the reflectivity parameter ρ. As in the previous section, only the marine Sc R_i profile has been used so far. Radiance data from the Microwave Sounding Unit (MSU) and HIRS are used in the inversion which finds the maximum likeliehood state by Newtonian minimisation of a cost function. The cost function contains terms involving the distance of the atmospheric state from the background state and the distance of the measurements from values calculated from the state by the fast forward model. The atmospheric background state used is derived from a UK Met Office 3-6 hour regional forecast model. Background values for the cloud parameters are essentially climatological values with high error values so they effectively do not constrain the solution.

The result of the basic scheme are fields of retrieved atmospheric temperatures and humidities and, more relevant to this work, fields of retrieved cloud amount and pressure. The result of the enhanced scheme is to add cloud reflectivity as a parameter of the atmospheric state.

4 Results from a case study

The results from TOVCFG given in this section are for a NOAA-11 overpass on 17 October 1991 at 13:37 UT for which the swath of data runs from Iceland and northern Norway to the northern Mediterranean. Figure 10 is an AVHRR channel 2 (visible) image of the southern end

---

† Since we are not at this stage treating cloud transmission explicitly, the cloud fractional cover is an 'effective' fraction εn in cases of semi-transparant cloud.

of the data region and shows (although this is best seen in conjunction with infared imagery) several features we will refer to later.

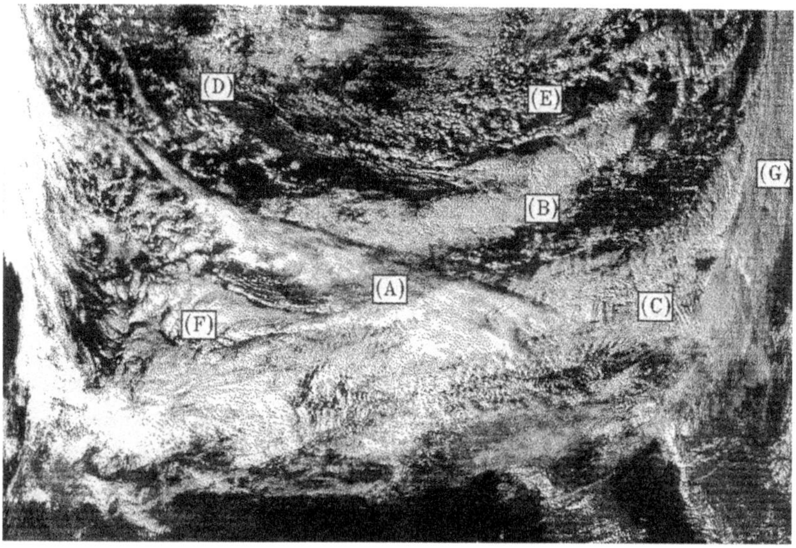

Figure 10. AVHRR channel 2 (visible) image for 17 October 1991. Annotated areas are described in the text.

Across the centre of the image there is a widening band of thin cirrus (A) which is causing a shadow on two bands of underlying stratiform cloud (B) and (C). Stratus area (B) is particularly flat and uniform but becomes convective as it curls up to end over Denmark. Very small cumulus are seen between areas (B) and (C). Streets of cumulus of increasing size can be seen from west Wales across the United Kingdom and into northern France and Belgium (D-E). Over south-west France (F) there is a solid bank of stratocumulus and on the east swath edge there is a large north-south band of thick high cloud (G).

Figures 11 and 12 are fields of the retrieved cloud top pressure P_c and cloud effective fractional cover n, for approximately the same area as the AVHRR, at the full HIRS resolution (~40 km). The ordinate and abscissa are the HIRS scan line and across-track position respectively. The pressure field clearly shows the cirrus band (A) and the n field confirms it is not opaque with values around 0.3-0.7. The stratiform areas (B) and (C) show well on the n field and the cloud top here appears to be around 600 mb. Radiosonde ascents through these areas validate these pressure estimates reasonably well. The small cumulus between (B) and (C)

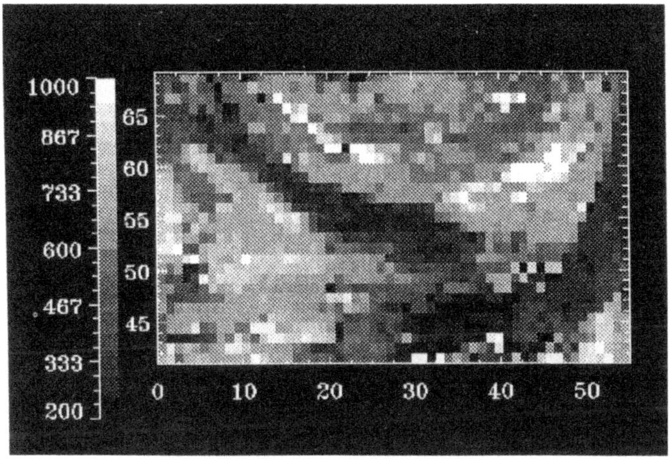

Figure 11. Cloud top pressure field (mb) retrieved from TOVS data for area similar to that of figure 10.

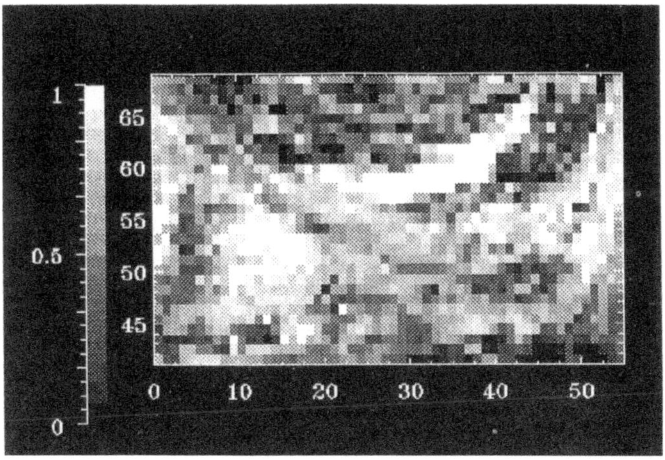

Figure 12. As Figure 11 but for retrieved effective cloud fractional cover.

and the cumulus streets (D-E) are less easily identified because they are sub-HIRS pixel size and give a very noisy picture. Both fields are consistant with the swath of high cloud (G), and with the stratocumulus area (F).

The retrieved reflectivity parameter is shown in figure 13 scaled to give the equivalent 3.7 μm percentage reflectivity.

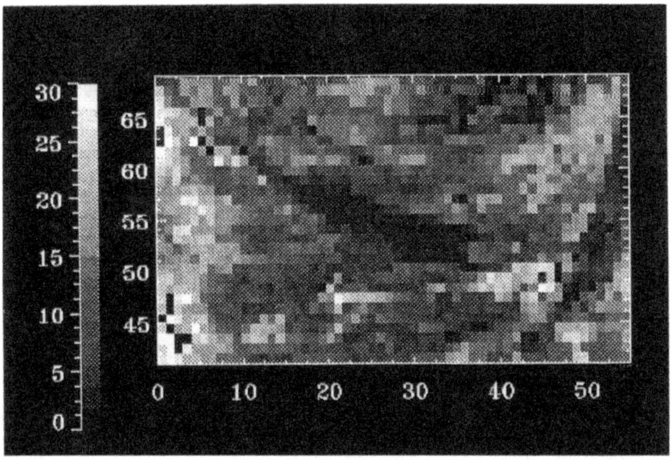

Figure 13. As figure 11 but for retrieved 3.7 μm reflectivity (%).

Again we can make some qualitative remarks about the retrieved field. Firstly, both major areas of cirrus (A and G) give estimated reflectivities of between 3 and 6% — consistant with the two-stream predictions. The stratus areas (B),(C) and (F) all have reflectivities around 15% but at the North-East end where area (B) becomes convective this drops to 6-9%. Of the cumulus streets careful scruntiny shows variation from 18-3% going from (D) to (E); from small to large cumulus. Values of 25% are found for the small cumulus between areas (B) and (C).

Other patches of high reflectivity, >20%, can be seen and seem, on inspection of the n field, to coincide with areas of low cloud cover — or small clouds. In the latter case it is possible that the assumption that the underlying surface is non-reflecting could lead to overestimates of the cloud reflection. However, rerunning the processing with an underlying surface emissivity of 0.95 does not significantly change the resulting cloud reflectivity estimates. It is unlikely that (especially sea) surface reflectivity at these wavelengths rises above 5%†. Inspection AVHRR 3.7 and 11 μm imagery confirms the strong reflection of many of the very small clouds.

A strong feature of the retrieved reflectivity field is the area of high values, >25%, which is found on the west — sun side — of the pass. A large part of the area is covered substantially by low or medium level cloud, ruling out the possibility that these values are an artifact of sea glint. In fact, it appears that the high values are a result of strong forward scattering by the cloud — a feature the two-stream solutions used are unable to predict. The directional characteristics of the reflection field are best seen in Figure 14 where reflectivity values are plotted against the scattering angle ψ†† (these data are for the whole pass — about twice the area shown in earlier figures).

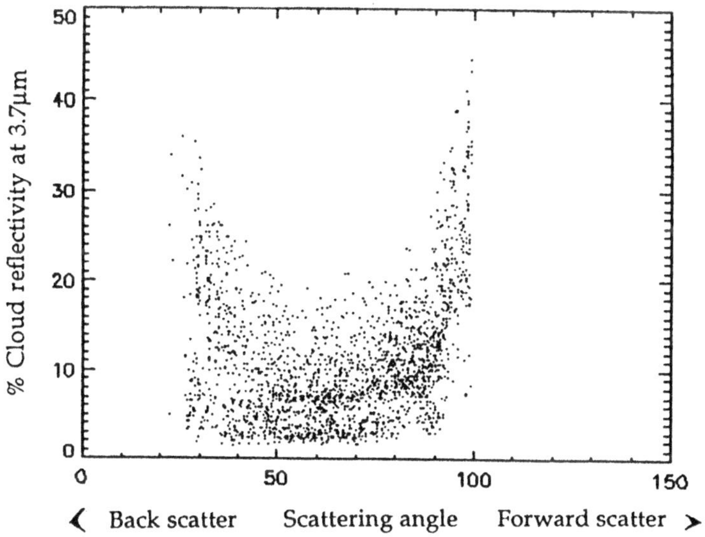

Figure 14. Cloud 3.7 μm reflectivity as a function of scattering angle. Data only shown where the retrieved cloud fraction was greater than 60%.

The solar zenith angle for this pass was typically 40-50° and the HIRS instrument geometry therefore allows minimum scattering angles of about 80°. The minimum possible is around 30°

† This probably does not hold in areas of sunglint off the ocean surface where values of 5-10% can occur. The scheme is being modified to handle these situations.

†† With the previous notation and ϕ, ϕ_S as satellite and solar azimuth angles, $\cos\psi = \cos\theta_S\cos\varphi + \sin\theta_S\sin\varphi\cos(\phi_S - \phi)$

for a setting sun. Direct backscattering could be observed with the HIRS, but on this occasion angles of up to 150° are found.

The scatter of points in Figure 14 can be compared qualitatively with the computed phase function for a 10 μm drop in Figure 15. The broad similarity of the dependence on angle strongly suggests that the scattering observed is of a low order. It is also clear that in order to interpret a reflectance measurement in terms of the effective radius of the drop distribution in the cloud, allowance for the scattering angle must be made. Whether individually the zenith and azimuth angles are important remains to be seen. More detailed scattering calculations using discrete ordinates code are under way to this end.

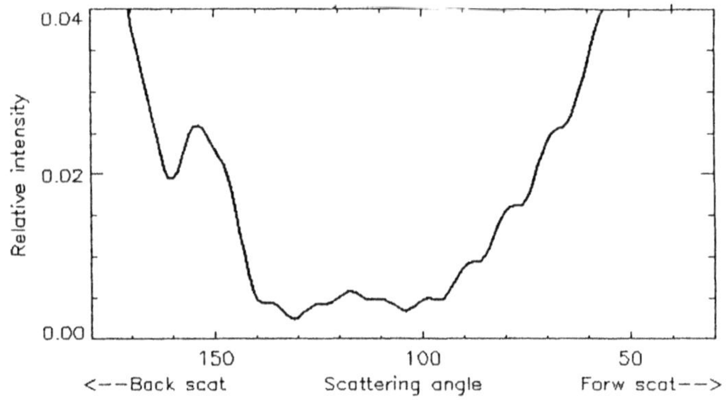

Figure 15. Phase function for scattering of 3.7 μm radiation by a 10 μm radius water drop.

Figure 14 featured FOV's where there was substantial cloud cover; the retrieved n was greater than 0.6. In these cases the cloud perhaps approximates to the plane parallel limit which exhibits predictable scattering behaviour. Reflectances for partial cloud cover conditions are less well organised and take on high values even at the scattering angle where a minimum is seen in Figures 14 and 15. The reasons for this probably lie in the broken nature of the cloud and we are currently persuing Monte Carlo methods to predict the behaviour of sub-pixel clouds.

5 Summary

Mie and two stream calculations have been performed to estimate the effect of spectrally variable radiative properties on infrared radiances. Particular attention has been drawn in this paper to the reflection of solar and downward diffuse radiation terms although a topic of equal importance is that of variable cloud transmissivity. Reflectances are found to depend strongly on both wavelength and the effective drop size of the cloud with particularly strong features at 3.83 and 5.3 μm. The position of these features is fixed because they result from the behaviour of the refractive index of the water. Ice clouds generally have lower reflectivity and higher transmission but again, any features are fixed by the refractive index.

The stable nature of the reflectance pattern has allowed a simple model of reflectance to be used in a fast radiative transfer model for the HIRS instrument. The effect of a typical water cloud reflectivity was shown for several HIRS channels under certain conditions. The largest effect is reflection of solar radiation by channels near the 3.83 μm feature. Smaller, though non-negligible, effects are found for even the longer wavelength channels through the lower emission and reflection of downward diffuse radiation by a non-black cloud.

The enhanced fast model for HIRS was incorporated into a non-linear TOVS retrieval system which explicitly models and retrieves the cloud present in the field of view. Applied to a case study, the resulting reflectivity field was compared, in conjunction with fields of retrieved cloud pressure and amount, to qualitative judgments made using AVHRR imagery. Retrieved reflectivities were found to be consistant with the two stream predictions; cirrus and other ice cloud can be identified with very low values, stratiform water cloud with medium and small cumulus with very high values. This result lends encouragment to the possibility of estimating effective drop size from HIRS measurements.

However, there are complicating factors which we are currently tackling; strong forward and back scattering and anomalous scattering by small clouds are suggested by the results. More detailed calculations using the discrete ordinates and Monte Carlo methods should provide a clearer connection between retrieved reflectivity and cloud effective drop size.

REFERENCES

Baran AJ and Watts PD, (1992) Radiative properties of water and ice clouds at wavelengths appropriate to the HIRS instrument. Met. Office Short Range Forecasting Division Internal report in preparation.

Deirmendjian D, (1969) Electromagnetic scattering on spherical polydispersions. Elsevier. New York.

Eyre JR, Brownscombe JL and Allam RJ (1984) Detection of fog at night using Advanced Very High Resolution Radiometer imagery. Met Mag (UK) **113** no. 1346 p. 264-271

Eyre JR (1989a and b) Inversion of cloudy TOVS radiances by non-linear optimal estimation. I: Theory and simulation for TOVS. II: Application to TOVS data. Q.J.R.Meteorol.Soc. **115**, pp.1001-1037

Eyre JR (1991) A fast Radiative transfer model for satellite sounding systems.Technical Memorandum no.176, European Centre for Medium range Weather Forecasts, Shinfield Park, Reading, U.K.

Hunt GE, (1973) Radiative properties of terrestrial clouds at visible and infrared thermal window wavelengths. Q.J.R.Meteorol.Soc. **19**, pp.346-369

Irving WM and Pollack JB, (1968) Infrared optical properties of water and ice spheres. Icarus, **8** pp 324-360

Meador WE and Weaver WR, (1980) Two-stream approximations to radiative transfer in planetary atmospheres: A unified description of existing methods and a new improvement. J.Atmos.Sci. **37** p 630-643

Yamamoto G, Masayuki T and Shoji A (1970) Radiative transfer in water clouds in the infrared region. J.Atmos.Sci. **27**, pp.282-292

Chapter 4

Next generation IR Sounder and surface characteristics analysis

"Infrared remote sensing of surface temperature and surface spectral emissivities"

F. BECKER, Z. LI - ENSPS, France

"Effects of spectral resolution on satellite surface temperature retrieval"

A.E. LIPTON - Phillips Lab./USAF, USA

"Investigation of infrared emissivities of Sahara dust powders and quartz from space"

T. TAKASHIMA, K. MASUDA - Meteor. Research Inst., Japan

INFRARED REMOTE SENSING OF SURFACE TEMPERATURE AND SURFACE SPECTRAL EMISSIVITIES

François BECKER and Zhao-Liang LI
Groupement Scientifique de Télédétection Spatiale (GSTS)
LSIT[1] / ENSPS
7 rue de l'Université
67000 Strasbourg, France

ABSTRACT

This paper presents a method based on the concept of temperature-independent spectral indices (TISI) to retrieve both the spectral emissivities and land surface temperature from AVHRR data. First, the bi-directional reflectivity (BDR) in channel 3 of AVHRR (3.7μm) is determined by extracting the solar reflection occuring in this channel during the day thanks to an elimination of the emitted radiation by a particular intercomparision between day and night TISI. Then, assuming that the angular dependence of this BDR is known, the emissivity in channel 3 can be derived, leading to a straightforward determination of the emissivities in channels 4 and 5 using the corresponding TISI. Kowning the emissivities in channels 4 and 5, the land surface temperature (LST) is deduced by the local split-window method. The sensibility of this method to atmospheric uncertainties is discussed. First results obtained from AVHRR data over Oran area in Northwestern Algeria are presented.

INTRODUCTION

One of the key parameters controlling biospheric processes and heat exchange between the Earth's surface and its atmosphere is the surface temperature. However, the measurement and monitoring of this parameter from space radiometry is a very difficult task because there are two types of corrections to be made, besides the radiometric calibration and the cloud screening procedures, which are outside the scope of this work. One type is called emissivity corrections and the other type is called atmospheric corrections. For sea surfaces, those problems are relatively easy to deal with because the emissivities are close to unity and are almost constant in

---

[1] URA 1207 associated to CNRS

NATO ASI Series, Vol. I 9
High Spectral Resolution Infrared Remote Sensing
for Earth's Weather and Climate Studies
Edited by A. Chedin, M.T. Chahine and N.A. Scott

the 10 - 12 μm spectral domain. In fact, a large amount of work has been done for sea surface temperature (SST) determination from space (Prabakhara et al., 1974; McClain et al., 1985; Walton, 1988). As for the land surface temperature (LST), much less effort has been undertaken because of specific difficulties encountered in LST retrievals which do not appear for SST determination:

i) The emissivity of land surfaces may be substantially lower than one, may vary strongly from one point to another one and, furthermore, may have large spectral variations (Lyon, 1968; Nerry et al., 1990; Labed, 1990; Schmugge et al., 1991), inducing errors in retrieved land surface temperature that may be large (Wolfe, 1975; Becker, 1987).

ii) The surface temperature itself may be strongly variable over small distances: namely within a pixel, leading to a difficult averaging procedure and to some problems in the definition of the land surface temperature itself.

iii) The difference between near-surface air temperature and LST may be much larger than for oceans, reducing the accuracy of some approximations used for SST determination from space.

Due to these difficulties, it is not possible to determine the land surface temperature only from passive radiometry without making assumptions regarding emissivity, because the number of unknowns to determine is always larger than the number of independant measurements (the number of channels used). This is the so-called "missing equation" concept. Therefore, each time one wants to obtain the surface temperature using only passive radiometry, an assumption had to be made in order to obtain the missing equation. For instance, one assumes that the emissivity in one channel is unity or constant (Kahle et al., 1980); or one assumes that the emissivities are known from other measurements (Wan and Dozier, 1989; Becker and Li, 1990a).

Once the emissivities are known, there are two ways to determine land surface temperature from space. First, if the atmospheric profile is available from either conventional radiosoundings or satellite soundings (Susskind et al.,1984; Chédin et al., 1985), one may invert the radiative transfer equation using one-channel models; for exemple, 4A (Scott and Chédin, 1981) or LOWTRAN-7 (Kneizys et al., 1987). Second, if such an atmospheric sounding is not available, one must use multispectral methods, for example, the modified split window, while taking into account the emissivity effects (Wan and Dozier, 1989; Becker and Li, 1990a). However, in most cases, the emissivities are not known; therfore, it is necessary to determine both the emissivities and LST from space.

In this paper, we present a method to retrieve both the spectral emissivities and land surface temperature from AVHRR data. In order to evaluate the utility of this method when a sounding of the atmosphere is not available, the sensibility of this method to uncertainties in atmospheric characteristics (Li and Becker, 1992) will be briefly presented. The preliminary results obtained using the proposed method from AVHRR data over Oran area in Northwestern Algeria will be presented.

PRINCIPLE OF THE METHOD

It is well known that in channel 3 of AVHRR, i.e. around 3.7μm, the radiance emitted by the land surface itself and the reflected radiance due to solar radiation during the day are of the same order of magnitude if the surface albedo is about 0.1. Using this very interesting property, Becker and Li (1990b) proposed a method for determining the surface temperature and surface emissivities from AVHRR using only the sun as an active source.The general idea of this method is to use the emitted radiation during the night to evaluate the emitted radiation during the day thanks to the independence of Temperature-Independent Spectral Indices (TISI) on surface temperature. From this intercomparison between day and night TISI, the bidirectional reflectivity $\rho_{b3}(\theta_s,\theta)$ in channel 3 is obtained and then the emissivities in channels 4 and 5 are derived from the TISI's , enabling us to use the local split-window method (Becker and Li, 1990a) for determining the land surface temperature.

DEFINITION OF TISI

We have shown (Becker and Li, 1990b) that it is possible to construct particular ratios of radiance that are almost independent of surface temperature (TISI). One particular TISI of interest for determining emissivities is constructed with three channels (3, 4 and 5) of AVHRR and is given by

$$\mathrm{TISI} = M \frac{B_3(T_{g3})}{B_4(T_{g4})^{a_4} B_5(T_{g5})} \qquad (1)$$

where a_4 and M are constant (for NOAA-11 a_4 = 1.916 and M=0.68257 x 10^{-7}), and $B_i(T_{gi})$ is the radiance at ground level in channel i given by

$$B_i(T_{gi}) = \varepsilon_i B_i(T_s) + (1-\varepsilon_i) R_{ati\downarrow} \qquad (2)$$

In this expression, T_{gi} is the surface brightness temperature in channel i at ground level, $R_{ati\downarrow}$ is the downwelling atmospheric radiance in channel i, and $B_i(T_s)$ is the radiance emitted by a blackbody surface with temperature T_s. The power a_4 is determined in the following manner such as to eliminate the largest part of the effect of T_s on TISI.

The radiance $B_i(T_s)$ can be written to a good approximation, around a reference temperature T_0, (Slater,1980) as

$$B_i(T_s) = \alpha_i T_s^{\,n_i} \tag{3}$$

where α_i and n_i are channel-specific constants depending on T_0.

Using approximation (3), it is possible to rewrite (2) as

$$B_i(T_{gi}) = \varepsilon_i\, \alpha_i T_s^{\,n_i}\, C_i \tag{4}$$

where C_i is a corrective term that takes into account the atmospheric reflected radiance: i.e.

$$C_i = 1+\beta_i \frac{1-\varepsilon_i}{\varepsilon_i}, \text{ with } \beta_i = \frac{R_{ati\downarrow}}{B_i(T_s)} .$$

Inserting (4) into (1), we get:

$$\mathrm{TISI} = M \frac{\alpha_3}{\alpha_4^{a4}\, \alpha_5} \cdot \frac{\varepsilon_3}{\varepsilon_4^{a4}\varepsilon_5} \cdot \frac{T_s^{n3}}{T_s^{\,a_4 n_4}\, T_s^{n5}} \cdot \frac{C_3}{C_4^{a4} C_5} \tag{5}$$

It is easy to see that if we choose the a_4 such that

$$n_3 - a_4 n_4 - n_5 = 0 , \tag{6}$$

the effect of surface temperature T_S is strongly reduced in (5). If we further choose M such that

$$M = \frac{\alpha_4^{a4}\, \alpha_5}{\alpha_3} ,$$

then

$$\mathrm{TISI} = C_{345} \cdot \mathrm{TISIE}, \tag{7}$$

with

$$C_{345} = \frac{C_3}{C_4^{a4} C_5}, \text{ and } \mathrm{TISIE} = \frac{\varepsilon_3}{\varepsilon_4^{a4} \varepsilon_5} .$$

It should be noted that other TISI's constructed from channel 3 and 4 or 3 and 5 may be used. This point is discussed elsewhere.

DETERMINATION OF SURFACE EMISSIVITIES

As there is no solar reflection during the night, the night TISI referred to as $TISI_n$ is given according to (1) and (7) as:

$$TISI_n = M \frac{B_3(T_{g3n})}{B_4(T_{g4n})^{a4}\, B_5(T_{g5n})} = C345n \cdot TISIE_n = C345n \frac{\varepsilon_3}{\varepsilon_4^{a4}\varepsilon_5} \quad (8)$$

where the emissivities take on their night values.

During the day, since the solar reflection is important in channel 3 of AVHRR, the radiance $B_3(T_{g3d})$ at ground level in this channel is no longer given by (2) but by

$$B_3(T_{g3d}) = D_3(T_{3d}) + \rho_{b3}(\theta_s,\theta)\, R^s_{g3}(\theta_s) \cos\theta_s \, , \quad (9)$$

where $D_3(T_{3d})$ is the channel radiance that would have been measured in channel 3 if there were no solar reflection, $\rho_{b3}(\theta_s,\theta)$ is the bidirectional reflectivity of the surface in channel 3 in the direction of θ_s, θ, $R^s_{g3}(\theta_s)$ is the solar radiance at the surface in channel 3, and θ_s, θ are respectively the zenith angle of solar irradiation and the angle of observation.

From the previous discussion, we can write

$$M \frac{D_3(T_{3d})}{B_4(T_{g4d})^{a4}\, B_5(T_{g5d})} = C345d \cdot TISIE_d = C345d \frac{\varepsilon_3}{\varepsilon_4^{a4}\, \varepsilon_5} \, , \quad (10)$$

where the emissivities take on their day values.

Therefore, the day value of TISI (referred to as $TISI_d$) is obtained by inserting expressions (9) and (10) into the definition of TISI (Eqn. 1): namely,

$$TISI_d = C345d \cdot TISIE_d + M \frac{\rho_{b3}(\theta_s,\theta)\, R^s_{g3}(\theta_s) \cos\theta_s}{B_4(T_{g4d})^{a4}\, B_5(T_{g5d})} \, . \quad (11)$$

From now on, we assume that the emissivity ratios do not vary substantially from day to night, so that

$$TISIE_d = TISIE_n \, .$$

As emissivity data set are not yet available to check accurately this assumption, we just checked that it was valid for a particular data set measured in situ (Labed 1990) which was available to the authors, i.e.for $TISIE_{54}$ ($TISIE_{54}=\varepsilon_5/\varepsilon_4^{0.913}$). Day and night surface emissivities measured in-situ with their corresponding $TISIE_{54}$ are displayed in Table 1. It shoud be noted that this result is not sufficient to validate the assumption used in this paper, it has to be checked more carefully when larger data set are available.

Parameters	In situ measurements	In situ measurements
	Day	Night
ε_4	0.978	0.957
ε_5	0.997	0.976
$TISIE_{54}$	1.014	1.016

Table 1 Comparison of $TISIE_{54}$ between day and night using the emissivities measured by Labed (1990)

Using (8) and (11), we can obtain the bidirectinal reflectivity $\rho_{b3}(\theta_s,\theta)$ in channel 3:

$$\rho_{b3}(\theta_s,\theta) = \frac{(TISI_d - P \cdot TISI_n)\, B_4(T_{g4d})^{a_4}\, B_5(T_{g5d})}{M\, R^s_{g3}(\theta_s)\cos\theta_s}, \quad \text{with } P = \frac{C_{345d}}{C_{345n}}.$$

From Kirchoff's theorem, it is possible to write $\rho_{b3}(\theta_s,\theta)$ for opaque media as

$$\rho_{b3}(\theta_s,\theta) = \frac{1-\varepsilon_3}{\pi} f_3(\theta_s,\theta) = \frac{\rho_{h3}}{\pi} f_3(\theta_s,\theta) \qquad (12)$$

where $f_3(\theta_s,\theta)$ is the angular form factor of the surface which is equal to unity if the surface is a Lambertian reflector and ρ_{h3} is the hemispherical reflectivity of channel 3.

We finally obtain:

$$\varepsilon_3(\theta) = 1 - \frac{\pi\,(TISI_d - P \cdot TISI_n) B_4(T_{g4d})^{a_4}\, B_5(T_{g5d})}{M\, R^s_{g3}(\theta_s)\cos\theta_s\, f_3(\theta_s,\theta)} \qquad (13)$$

Then, from the definition and properties of TISI (Becker and Li,1990b), it is very easy to obtain the emissivities in channels 4 and 5 using the following formulae:

$$\varepsilon_4 = \{\frac{\varepsilon_3}{TISI_{54n}\cdot TISI_n} C_{54n}C_{345n}\}^{(n_{54}+a_4)^{-1}} \tag{14}$$

$$\varepsilon_5 = \varepsilon_4^{\,n_{54}} TISI_{54n}\frac{1}{C_{54n}} \tag{15}$$

with $n_{54} = 0.913$ and $TISI_{54} = 2.383\,\frac{B_5(T_{g5n})}{B_4(T_{g4n})^{n_{54}}}$

It is worth noting that, although C_{345} and C_{54} vary with surface emissivities, surface temperature and atmospheric conditions as shown respectively in figure 1 and 2, taking $P = 1$, $C_{345n} = 1$ and $C_{54n} = 1$ in (13), (14) and (15) is a good approximation because the relative error introduced on the retrieved emissivities by this approximation is less than 1%. This error is estimated by the following formulea derived from (13), (14) and (15):

$$\frac{\Delta\varepsilon_3}{\varepsilon_3} = \frac{1-\varepsilon_3}{\varepsilon_3}\,\frac{1}{\frac{TISI_d}{TISI_n}-1}\,\frac{\Delta P}{P}$$

$$\frac{\Delta\varepsilon_4}{\varepsilon_4} = \frac{1}{a_4+n_{54}}\left(\frac{\Delta\varepsilon_3}{\varepsilon_3}+\frac{\Delta C_{54n}}{C_{54n}}+\frac{\Delta C_{345n}}{C_{345n}}\right),$$

$$\frac{\Delta\varepsilon_5}{\varepsilon_5} = \frac{n_{54}}{a_4+n_{54}}\left(\frac{\Delta\varepsilon_3}{\varepsilon_3}+\frac{\Delta C_{345n}}{C_{345n}}\right) - \frac{a_4}{a_4+n_{54}}\,\frac{\Delta C_{54n}}{C_{54n}}\,.$$

For the worst situation in figures 1 and 2 ($T_{sd} = 320$ K, $T_{sn} = 290$ K for model 2 and $\varepsilon_3 = 0.80$), $\frac{TISI_d}{TISI_n} \cong 2$, $\frac{\Delta P}{P} \cong -3\%$, $\frac{\Delta C_{54n}}{C_{54n}} \cong 0.6\%$, and $\frac{\Delta C_{345n}}{C_{345n}} \cong -1.3\%$, which means $\frac{\Delta\varepsilon_3}{\varepsilon_3} \cong -0.8\%$, $\frac{\Delta\varepsilon_4}{\varepsilon_4} \cong -0.6\%$ and $\frac{\Delta\varepsilon_5}{\varepsilon_5} \cong -1.0\%$.

Therefore, in practice, we can take $P = 1$, $C_{345n} = 1$ and $C_{54n} = 1$ in (13), (14) and (15) to derive surface emissivities from space: namely,

$$\varepsilon_3(\theta) = 1 - \frac{\pi\,(TISI_d - TISI_n)B_4(T_{g4d})^{a_4}\,B_5(T_{g5d})}{M\,R^s_{g3}(\theta_s)\cos\theta_s\,f_3(\theta_s,\theta)}\,, \tag{13a}$$

$$\varepsilon_4 = \{\frac{\varepsilon_3}{TISI_{54n}\cdot TISI_n}\}^{(n_{54}+a_4)^{-1}}, \tag{14a}$$

$$\varepsilon_5 = \varepsilon_4^{\,n_{54}} TISI_{54n}\,. \tag{15a}$$

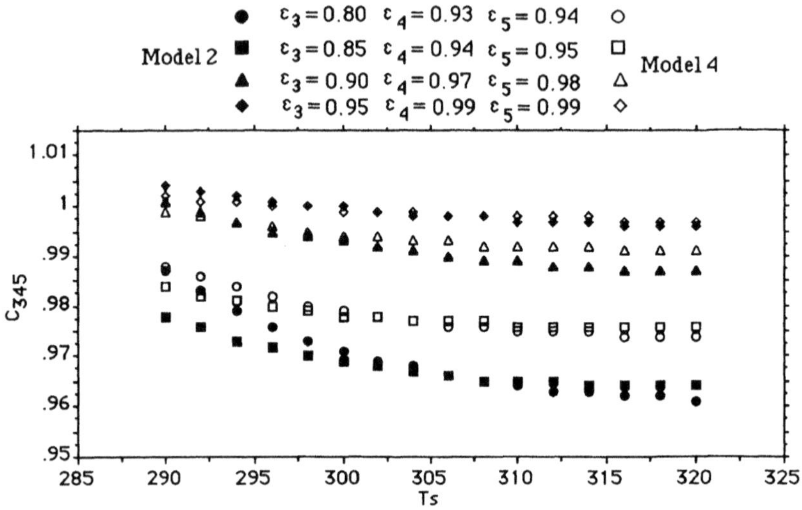

Figure 1 Variation of C_{345} with surface temperature for different surface emissivities and atmospheric conditions (Model 2: midlatitude summer model; Model 4: subarctic summer model)

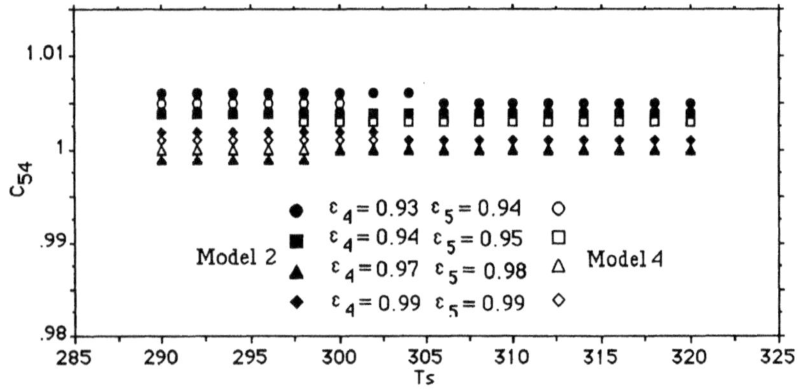

Figure 2 As in figure 1, but for C_{54}

Expressions (1), (13a), (14a) and (15a) show that the emissivities can be retrieved directly from space, provided that $f_3(\theta_s,\theta)$ and $R^{s}_{g3}(\theta_s)$ are known. However, in order to obtain the ground brightness temperature T_{gi} appearing in these formulae, the atmospheric corrections have to be performed. As it has been shown in Li and Becker (1992), it is not necessary to have a very accurate description of the atmosphere to perform atmospheric corrections for emissivity determination from AVHRR data with the proposed method. Corrections can therefore be made using an approximate knowledge of atmosphere (using a climatological atmosphere or, better, using atmospheric parameters retrieved from satellite sounding) if radiosoundings are not available, as it will be discussed below.

DETERMINATION OF LAND SURFACE TEMPERATURE

As mentioned in the introduction and as shown also in Li and Becker (1992), the determination of land surface temperature is a very complicated task because it is difficult to separate the effects of the atmosphere and those of the emissivity from the effects of the surface temperature in the measured radiance. However, Wan and Dozier, (1989) and Becker and Li, (1990a) have shown that LST can be retrieved as it was done for the retrieval of sea surface semperature using a modified split-window method with coefficients depending only on spectral emissivities but not on atmospheric conditions. Therefore, once the emissivities in channels 4 and 5 are known, the surface temperatures are then easily derived from the brightness temperatures T_4 and T_5 measured by AVHRR using, for example, the local split-window method (Becker and Li 1990a), which reads

$$T_s = A_0 + P\,\frac{T_4+T_5}{2} + M\,\frac{T_4-T_5}{2} \tag{16}$$

with, for NOAA 11,

$$A_0 = 1.274$$

$$P = 1 + 0.15616\,\frac{1-\varepsilon}{\varepsilon} - 0.482\frac{\varepsilon_4 - \varepsilon_5}{\varepsilon^2}$$

$$M = 6.26 + 3.98\,\frac{1-\varepsilon}{\varepsilon} + 38.33\frac{\varepsilon_4 - \varepsilon_5}{\varepsilon^2}$$

$$\varepsilon = \frac{\varepsilon_4 + \varepsilon_5}{2}\,.$$

It should be noted that this method to account for atmospheric effects is different than that used in (13a), (14a) and (15a). As shown in Li and Becker (1992), the retrieval of emissivity using (13a), (14a) and (15a) is less sensitive to atmospheric uncertainties than the temperature itself. If a radiosounding is not available, an approximate knowledge of the atmosphere may be used to obtain the emissivity with enough accuracy, but in that case the use of the split-window improves the accuracy of LST retrievals

DISCUSSION OF METHOD AND ERROR BUDGET

The method that we propose here has some advantages. First, it allows retrieval of both the spectral emissivities and surface temperature from space using only passive radiometry. In fact, once $R_{g3}^{s}(\theta_s)$ and $f_3(\theta_s,\theta)$ are known, all the quantities appearing in formulae (13a), (14a), (15a) and (16) are measurable from space.

Second, this method reduces strongly the impact of errors of measurement on the retrieval of emissivity.

(i) The relative accuracy of the emissivity in channel 3 is related to the relative accuracy of the reflectivity by

$$\frac{\Delta\varepsilon_3}{\varepsilon_3} = -\frac{1-\varepsilon_3}{\varepsilon_3}\left(\frac{\Delta\rho_{b3}}{\rho_{b3}} + \frac{\Delta f_3}{f_3}\right) = -\frac{1-\varepsilon_3}{\varepsilon_3}\frac{\Delta\rho_{h3}}{\rho_{h3}},$$

as follows from a perturbation analysis of (12). Thus the relative error of ε_3 may be much smaller than the relative error of the reflectivity ρ_{h3} by a factor 5 for $\varepsilon_3 = 0.83$ to 50 for $\varepsilon_3 = 0.98$.

(ii) There is a strong reduction of the atmospheric effects on the retrievals of emissivities and LST. In fact, besides the errors which appear in all measurements from space (calibration, etc.), there are four main sources of error in the determination of emissivities from space by the proposed method: (A) errors due to the approximations ($P=1$, $C_{345n} = 1$, $C_{54n} = 1$) made in formulea (13a), (14a) and (15a), (B) errors due to the uncertainties of atmosphere in atmospheric corrections (using an approximate atmosphere instead of an actual atmosphere), (C) errors due to the unknowledge of surface description (f_3) and solar incoming flux R_{g3} (a rough estimation showed $\Delta\varepsilon_3/\varepsilon_3 < 1.5\%$ if $\Delta R_{g3}/R_{g3} < 10\%$ (see Schmugge et al., 1991; Li and Becker, 1992)), (D) errors due to the choice of TISI in determination of emissivities (a rough estimation to be improved led to $\Delta\varepsilon_3/\varepsilon_3 < 1\%$). To demonstrate this reduction of atmospheric effects and the impacts of approximations on the retrieval of emissivities and temperature, we simulated the radiances L_i measured by AVHRR using LOWTRAN-6 (Kneizys et al. 1983) and the radiosonde data of 16 June 1986 over Hapex (André et al., 1986) with day surface temperature T_{sd}=320, 310, 300K, night surface temperature T_{sn}=295, 285K and a set of values of emissivities. Then we corrected those radiances for atmospheric effects to get the radiances at ground level in channel i, $B_i(T_{gi})$ by

$$B_i(T_{gi}) = \frac{L_i - R_{ati\uparrow}}{\tau_i} \qquad (17)$$

where the transmittance of the atmosphere τ_i and the upwelling atmospheric radiance $R_{ati\uparrow}$ in channel i were computed by LOWTRAN-6 with the different atmospheric models listed in Table 2. Using the formulae (1), (13a), (14a) and (15a) with the different atmospheric models given in Table 2, we computed the retrieved surface emissivities. The relative errors $(\frac{\text{retrieved value - true value}}{\text{true value}})$ of ρ_{b3} and of surface emissivities in three channels of AVHRR are shown in Table 3 for different emissivities in channel 3 and for different surface temperatures. It should be noted that those relative errors in table 3 include both the errors due to uncertainties of atmosphere (error (B)) and the errors due to the approximations made in formulae (13a), (14a) and (15a) (error (A)). In order to estimate the overall error, one should add the errors (C) and (D) mentioned above.

From the table 3, one can make some remarks:

i) The impacts of both the uncertainties of the atmosphere and the approximations made in (13a), (14a) and (15a) on the retrieved emissivities in channels 4 and 5 are less than 3%, and this may be reduced to 1% if the atmospheric model near to the actual atmosphere were used (for example, model 2 in this table).

ii) Although the impact on ρ_{b3} is important, the relative error of ε_3 is less sensitive to atmospheric states; it is less than 5% .

iii) The relative emissivity errors generated by the proposed method are essentially independent of the absolute emissivity values.

iv) As was shown in Li and Becker (1992), the error on LST, ΔT_s due to the errors of emissivities could be estimated by:

$$\Delta T_s \cong -52\ \Delta\varepsilon - 110\ \Delta(\varepsilon_4 - \varepsilon_5)$$

with

$$\Delta\varepsilon = \Delta(\frac{\varepsilon_4 + \varepsilon_5}{2}),$$

when using the local split window (formula 16).

Inserting the values of $\Delta\varepsilon$ and $\Delta(\varepsilon_4 - \varepsilon_5)$ given by Table 3 into this formula, one gets ΔT_s which is of order of 0.5K. It should be noted that this relatively low value of ΔT_s is due to the behaviors of the errors generated on the emissivities and that the error of 0.5K did not take into account the error generated by the local split window discussed in Becker and Li (1990a).

Model	Atmospheric model	Precipitable water W (g cm^{-2})	$\Delta W = W_7 - W_i$ (g cm^{-2})
2	Midlatitude summer	2.38	- 0.38
6	Midlatitude summer + U.S. Standard 1962water vapor profile	1.14	0.86
7	Radiosonde data of 16 June 1986	2.00	0

Table 2 Atmospheric models used to simulate the effects of the atmosphere on the retrieval of surface emissvity.

ε_3	Day T_s	Night T_s	Model 2				Model 6			
			$\frac{\Delta\rho_{b3}}{\rho_{b3}}$	$\frac{\Delta\varepsilon_3}{\varepsilon_3}$	$\frac{\Delta\varepsilon_4}{\varepsilon_4}$	$\frac{\Delta\varepsilon_5}{\varepsilon_5}$	$\frac{\Delta\rho_{b3}}{\rho_{b3}}$	$\frac{\Delta\varepsilon_3}{\varepsilon_3}$	$\frac{\Delta\varepsilon_4}{\varepsilon_4}$	$\frac{\Delta\varepsilon_5}{\varepsilon_5}$
0.85	320	295	-13.6	2.4	1.1	1.1	9.6	-1.7	-1.6	-2.6
		285	-25.7	4.5	0.4	-0.3	20.4	-3.6	-0.4	-0.7
	310	295	-3.2	0.6	0.5	0.6	-1.5	0.3	-0.9	-2.0
		285	-10.5	1.8	-0.4	-1.0	5.5	-1.0	0.4	0.1
	300	285	-1.7	0.3	-0.9	-1.5	-3.5	0.6	1.0	0.6
0.94	320	295	-44.7	2.8	1.2	1.2	43.3	-2.8	-1.9	-2.9
		285	-75.6	4.8	0.4	-0.3	70.8	-4.5	-0.7	-0.9
	310	295	-16.5	1.0	0.6	0.7	13.3	-0.8	-1.2	-2.3
		285	-35.6	2.3	-0.4	-1.0	31.5	-2.0	0.2	-0.2
	300	285	-12.2	0.8	-0.9	-1.4	7.7	-0.5	0.7	0.3

Table 3 Relative errors in percent of emissivities and reflectivitiy in channel 3 generated by the proposed method due to both the uncertainties in atmospheric corrections and the approximations made in formulea (13a), (14a) and (15a)

As a conclusion, if an approximate atmosphere is not far from the actual atmosphere, the use of an approximate atmosphere for emissivity determination instead of an actual atmosphere may lead to 3% or less of errors in emissivities; as for LST determination, the use of the split-window method with the emissivities derived by using an approximate atmosphere lead to 1.0Kerrors in the retrieved LST (0.5K due to the errors of retrieved emissiviries + 0.5K due to the accuracy of local split window itself). Consequently, if radiosoundings are not available, the atmospheric corrections required in the proposed method for determining the emissivity can be performed as a first approximation using climatological atmospheres (or better using atmosperic parameters retrieved from satellite soundings). In that case, the land surface temperature must be determined using the local split window method with the retrieved emissivity values.

This method has two drawbacks (which appears also in the determination of the spectral albedo). It needs a knowledge of both the angular form factor, $f_3(\theta_s,\theta)$ and the solar radiance at the surface in channel 3, $R^s_{g3}(\theta_s)$.

PRELIMINARY RESULTS

Two successive day and night AVHRR images from NOAA-11 acquired 1 January 1989 were chosen to apply the proposed method. These two AVHRR images have been provided by the Centre de Météorologie Spatiale (CMS) in Lannion (France), where they were calibrated, navigated and mapped in a polar stereographic projection. As an example, we considered 250 x 250 cloud-free pixel extracts over the Oran area in Northwestern Algeria.

Since the first registration method performed in CMS Lannion has an accuracy of about 5 pixels r.m.s. (Brunel and Marsouin, 1987), which is not sufficient for our purposes, we superimposed the night image on the day image by means of the software developed in LERTS (Toulouse, France). The final error of registration over the Oran area is of the order of one pixel. In order to compensate for this error, we smoothed the images with a mask of 2 x 2 pixels.

The atmospheric corrections required in Eqs. (13a), (14a) and (15a) for determining surface emissivities were performed using LOWTRAN-6 with the standard model atmosphere (Midlatitude winter).

As for the solar radiance in channel 3 at the ground level, $R^s_{g3}(\theta_s)$, we estimated it using a first order of approximation

$$R^{s}_{g3}(\theta_s) = (\frac{\alpha}{2})^2 \pi \int_0^{\infty} N_3(\lambda) B_\lambda(5784) \tau_\lambda(\theta_s) d\lambda$$

where $N_3(\lambda)$ is the spectral response of the radiometer of AVHRR in channel 3, α is the solar apparent diameter ($\alpha = 9.3 \times 10^{-3}$), $\tau_\lambda(\theta_s)$ is the total spectral atmospheric transmission estimated here by Lowtran-6 with the standard model atmosphere, and the temperature 5784K is the one which best fits the solar spectrum near 3.7μm (Kondratyev, 1969).

PRELIMINARY RESULTS CONCERNING SURFACE EMISSIVITIES

Applying the proposed method to the AVHRR data described above and assuming that the surface was Lambertian (i.e. we use $f_3 = 1$), we obtained from (13a), (14a) and (15a) three images of emissivity for channels 3, 4 and 5, which are coded in color and displayed, respectively, in Figures 3, 4, and 5. The color scales are attached to each image. It should be noted from these figures that in the region of 250x250 km^2 over the image of Oran region the emissivity in channel 3 varies mainly from 0.60 to 0.98, and the emissivities in channels 4 and 5 vary mainly from 0.86 to 1.00, as is shown in the histogram displayed in Figure 6. This confirms that the variation of the emissivities of various media is larger in channel 3 than in channels 4 and 5, as was observed by Nerry et al. (1990) from laboratory measurements and by Li and Becker (1992) from satellite measurements. If we look at the values of emissivities in Figures 3, 4 and 5, we find that the emissivities in the south-eastern part of this image are lower than the values we expected. For lack of in-situ measurements and soil composition information concerning this area, we do not know whether these values correspond to actual values or are false values; for instance due to angular effects on emissivity determination ($f_3 \neq 1$) since this location seems to be a montain. This aspect must be studied in the future.

IMPACT OF EMISSIVITIES ON SURFACE TEMPERATURE

As it has been pointed out by many authors (for example, Becker, 1987), The impact of emissivity uncertainties on land surface temperature retrievals is significant and, assuming that the emissivities in channels 4 and 5 are equal to 1, the error generated on LST while correcting atmospheric effects using the split window method may be of the same order of magnitude as those attributed to the atmosphere itself.

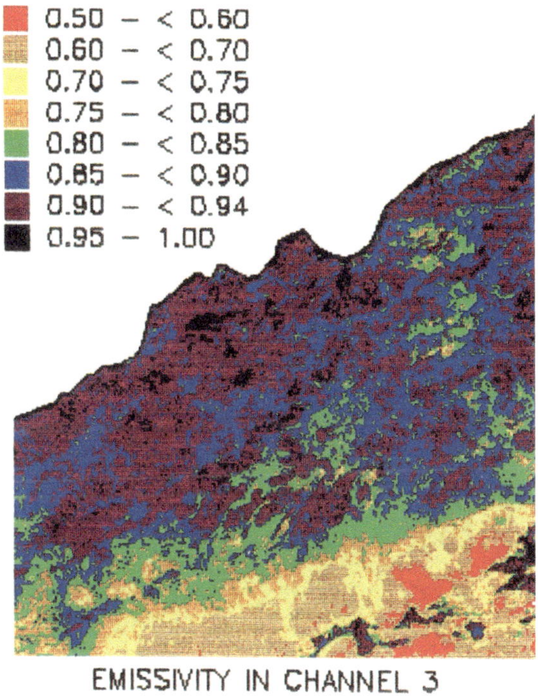

Figure 3 Map of day surface emissivity in channel 3 over the Oran area. The color scales of this map show the correspondance between colors in the map and the values of emissivity in channel 3.

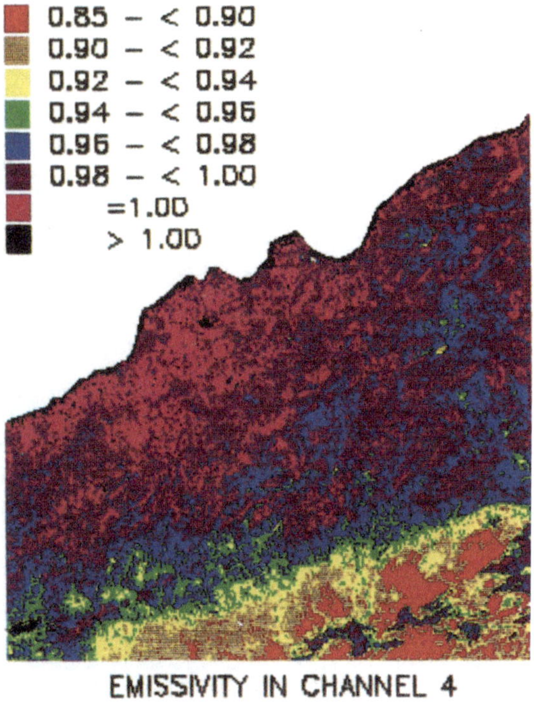

Figure 4 As in figure 3, but for surface emissivity in channel 4

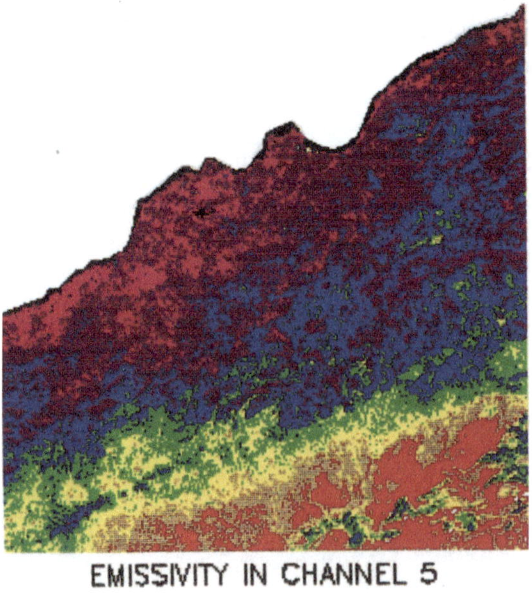

Figure 5 As in figure 3, but for surface emissivity in channel 5

To demonstrate this, we display in Figure 7 the map of day land surface temperature (LST) derived using the local split window method (Eqn. 16) with the retrieved day surface emissivities ε_4 and ε_5 (Figures 4 and 5), and in Figure 8 the temperature differences between LST and ST (the surface temperature obtained using the local split window method but with the assumption that $\varepsilon_4 = \varepsilon_5 = 1$). From this figure, we noticed that the temperature differences due to the emissivity effects are from 0 to 6 K.

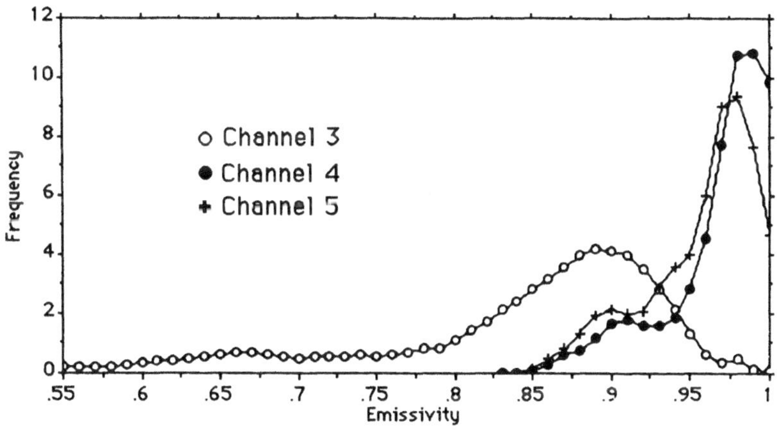

Figure 6 Histograms of surface emissivities in channels 3, 4 and 5 of AVHRR for the data of figues 3, 4 and 5.

CONCLUSIONS

This paper presents a first trial to determine both land surface temperatures and surface emissivities directly from AVHRR data.

This determination has been shown to be possible thanks to the comparison of day and night radiances in a channel around 3.7mm, and thanks also to the concept of TISI's that we introduced in Becker and Li (1990b).

We have shown that the atmospheric corrections required for the determination of surface emissivities from AVHRR data with the proposed method do not need a description of the atmosphere as accurate as for the direct retrieval of LST. They can therefore be performed using climatological data or better using atmospheric parameters retrieved from satellite vertical

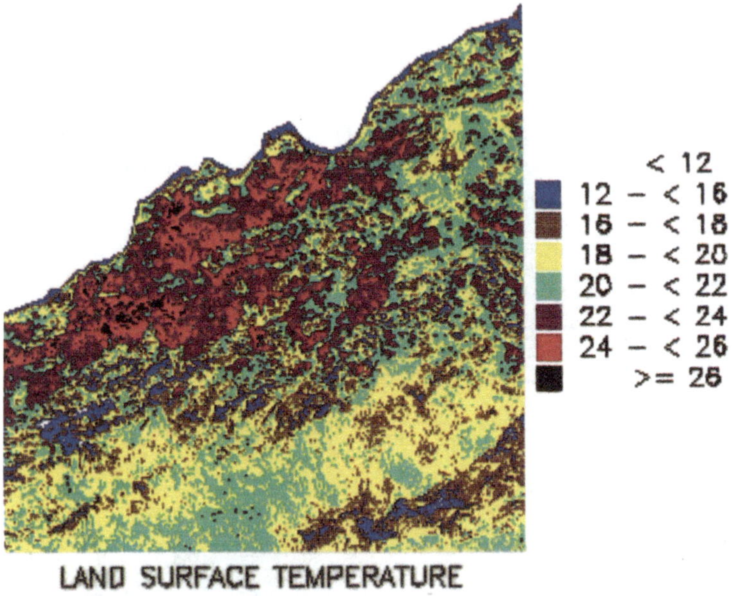

Figure 7 Map of day surface temperature (LST) over the Oran area obtained using the local split window method with the retrieved surface emissivities ε_4 and ε_5 presented in Figures 4 and 5

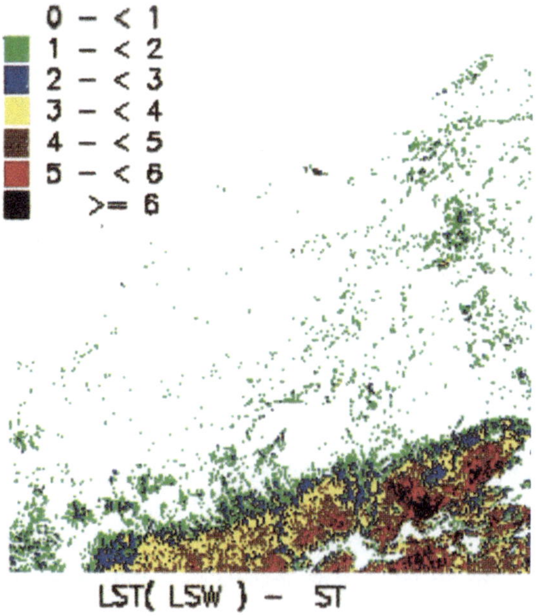

Figure 8 Map of temperature differences between LST and ST due to the effects of emissivity on retrieval of surface temperature by the split window method (see text)

sounders if atmospheric radiosoundings are not available. Once the emissivities have been derived, the LST can be obtained either by using the formulae (17) and (2) with the retrieved emissivities if there is a coincident radiosounding data, or by using the local split-window method given by formula (16) with the retrieved surface emissivities and measured brightness temperatures T_4 and T_5 if atmospheric radiosoundings are not available (which is the most common situation), since the local split window takes care of the atmospheric effects.

We have shown also that the emissivities can be determined within 3% from space using the proposed method with an approximate atmosphere, and that LST can be obtained with an accuracy of 1.0K using the split-window method and the retrieved emissivities.

We have applied this method to actual AVHRR data acquired over the Oran area assuming that the surface is a Lambertian reflector. The results obtained are preliminary, need to be studied in the future.

From the studies accomplished in this paper, we recommend strongly that the channel at 3.7μm continue to be included in the forthcoming radiometers flown on future satellites, and that day and night measurements at this channel must be made.

Acknowledgements

We are indebted to Professor M. P. Stoll for constructive discussions and to Ms S. de Brisis at LERTS for preparing the NOAA registered images in this work. We wish also to express our appreciation to Centre de Météorologie Spatiale, Lannion for providing us with AVHRR data.

REFERENCES

André JC, Goutorbe JP, Perrier A (1986) HAPEX-MOBILHY: A hydrologic atmospheric experiment for the study of water budget and evaporation flux at climatic state. Bull Amer Meteorol Soc 67(2):134-144

Becker F (1987) The impact of spectral emissivity on the measurement of land surface temperature from a satellite. Int J Remote Sens 10:1509-1522

Becker F, Li Z-L (1990a) Towards a local split window method over land surface. Int J Remote Sens 3:369-393

Becker F, Li Z-L (1990b) Temperature-independent spectral indices in thermal infrared bands. Remote Sens Environ 32:17-33

Brunel P, Marsouin A (1987) An operational method using ARGOS orbital element for navigation of AVHRR imagery. Int J Remote Sens 8:569-578

Chédin A, Scott NA, Wahiche C, Moulinier P (1985) The improved initialization inversion method: a high resolution physical method for temperature retrievals from the TIROS-N series. J Clim and Appl Meteorol 24:124-143

Kahle AB, Madura DP, Sola JM (1980) Middle infrared multispectral aircraft scanner data: analysis for geological applications. Appl Opt 19(14):2279-2290

Kneizys FX, Shettle EP, Gallery WO, Chetwynd JR, Abreu LW, Selby JEA, Clough SA, Fenn RW (1983) Atmospheric Transmittance/Radiance: Computer Code LOWTRAN 6. Optical Physics Division, U.S. Air Force Geophysics Laboratory, Hanscom, AFB, MA, AFGL-TR-83-0187

Kondratyev K.YA (1969) Radiation in the atmosphere, Academic, New York

Labed J, (1990) Approche expérimentale des problèmes liés à l'émissivité et à la température de surface, ainsi qu'à leur variabilité spatio-temporelle, dans le cadre de la télédétection spatiale dans l'infrarouge thermique. Thèse Université de Strasbourg, Septembre.

Li Z-L, Becker F (1992) Feasibility of land surface temperature and emissivity determination from AVHRR data. Remote Sens Environ (to be published)

Lyon RJP (1965) Analysis of rocks by spectral infrared emission (8-25μm). Econ Geol 60:715-736

McClain EP, Pichel WG, Walton CC (1985) Comparative performance of AVHRR-based multichannel sea surface temperatures. J Geophys Res 20:11587-11601

Nerry F, Labed J, Stoll MP (1990) Spectral properties of land surfaces in the thermal infrared band. Part I : Laboratory measurements of absolute spectral emissivity and reflectivity signatures. J Geophys Res 95:7027-7044

Prabhakara C, Dalu G, Kunde VG (1974) Estimation of sea surface temperature from remote sensing in the 11 and 13μm window region. J Geophys Res 79:5039-5044

Schmugge TJ, Becker F, Li Z-L (1991) Spectral emissivity variations observed in airborne surface temperature measurements. Remote Sens Environ 35:95-104

Scott NA, Chédin A (1981) A fast line by line method for atmospheric absorption computations: The Automatized Atmospheric Absorption Atlas. J Clim and Appl Meteorol 20:801-812

Slater PN (1980) Remote sensing optics and optical system, Addison-Wesley, Reading, MA.

Susskind J, Rosenfield J, Renter D, Chahine MT (1984) Remote sensing of weather and climate parameters from HIRS2/MSU on TIROS-N. J Geophys Res 89:4677-4697

Walton CC (1988) Nonlinear multichannel algorithms for estimating sea surface temperature with AVHRR satellite data. J Appl Meteorol 27:115-124

Wan Z, Dozier J (1989) Land surface temperature measurement from space: physical principles and inverse modeling. IEEE Trans on Geosci and Remote Sens 27:268-278

Wolfe, W.L. (1975), Differences in radiance: relative effects of temperature changes and emissivity changes. Appl Opt 14: 1937-1939.

EFFECTS OF SPECTRAL RESOLUTION ON SATELLITE GROUND SURFACE TEMPERATURE RETRIEVAL

Alan E. Lipton
Atmospheric Sciences Division
Geophysics Directorate
Phillips Laboratory, PL/GPAS
Hanscom Air Force Base, MA 01731
USA

1. Introduction

Ground surface temperatures play a large role in determining surface energy fluxes by way of their effects on sensible, latent, and radiative energy transfer. It is therefore necessary to estimate surface temperatures when studying atmospheric processes, such as mesoscale circulations, that depend heavily on surface fluxes. Short-range forecasts and climatological analyses can be particularly reliant on surface temperature estimation.

Widespread and frequent measurements of ground surface temperature can be made using satellite-based infrared sounders. The primary drawback of this method is that the accuracy of surface temperature retrievals is sensitive to cloud cover, and they cannot be made at all under overcast conditions. Errors in the assumed surface emittance can be another significant source of retrieval error (Kornfield and Susskind, 1977; Sutherland, 1986; Becker, 1987). Nevertheless, satellite-retrieved surface temperatures can be very useful for mesoscale analysis when cloud cover is incomplete.

Surface temperature information from the VISSR Atmospheric Sounder (VAS) has been used by Lipton and Vonder Haar (1990) in four-dimensional numerical analysis of circulations in the mountains and plains of the northeastern Colorado region. Their analysis system consists of retrieval algorithms coupled with a mesoscale numerical model. Some parameters of the gridded model data, including ground surface temperatures, are adjusted each time satellite observations are available, such that radiances computed from the gridded data come to agree more closely with the observed radiances as a result of the adjustment. In the Colorado case study, adjustments were made every 3 hours, and the resulting analysis accurately indicated where convective cloud development was most likely to occur later in the day.

NATO ASI Series, Vol. I 9
High Spectral Resolution Infrared Remote Sensing
for Earth's Weather and Climate Studies
Edited by A. Chedin, M.T. Chahine and N.A. Scott

One of the challenges in remote sensing of surface temperatures for mesoscale applications is to distinguish clearly the influence of the ground surface on the measured radiances from the influence of the overlaying atmosphere. It is common, particularly in cloud-free situations, to have large differences in temperature over very short vertical distances from the ground up through the lowest kilometer of the atmosphere. The research reported here focused on the vertical resolution capabilities of satellite sounders in this lowest part of the temperature profile.

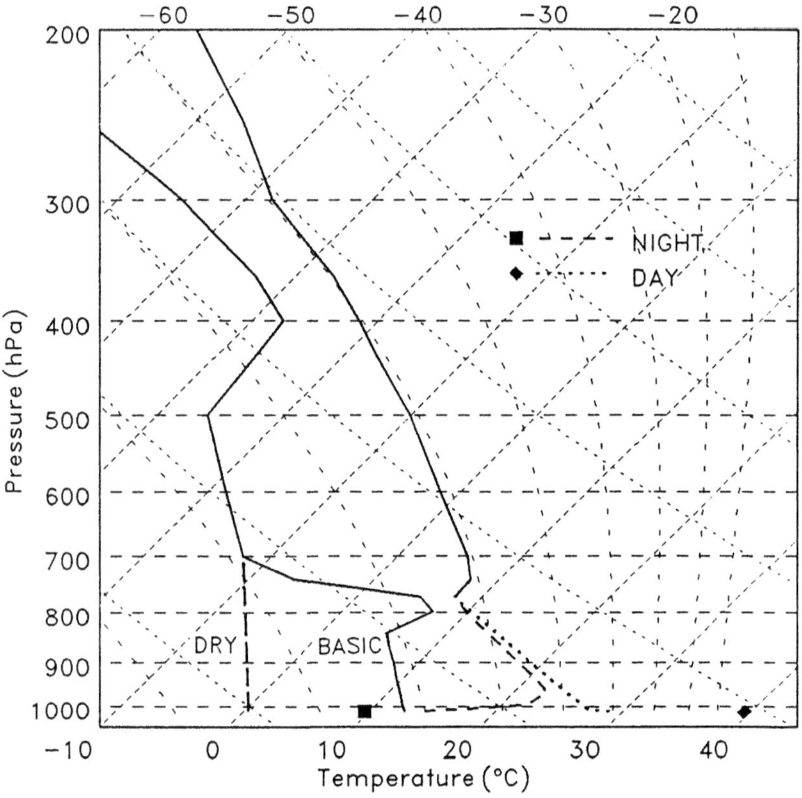

Figure 1. The reference temperature (right curves) and dewpoint temperature (left curves) profiles, with marker symbols indicating ground surface temperatures. The two versions of the bottom portion of the temperature curve correspond the the night and day reference conditions. The solid and dashed versions of the bottom of the dewpoint temperature curve correspond to the basic and dry profiles, respectively.

The meteorological conditions we considered were representative of a midlatitude, summertime, potentially convective environment (Fig. 1). The two temperature profiles that were used differed only in their low-level air and ground surface temperatures. One was characteristic of nighttime conditions, just before dawn, with a strong surface-based temperature inversion. The other was characteristic of midday conditions in the presence of strong solar heating of the ground. For radiative transfer calculations it was assumed that there was a temperature discontinuity between the bottom of the atmosphere and the ground surface, instead of trying to model the actual gradient of temperature between the ground and the ~2-m height at which "surface" air temperature observations are typically made. The specified temperature differences between the ground surface and the bottom of the atmosphere were appropriate for these meteorological conditions (Geiger, 1965; McNider and Pielke, 1981), although plotting ground surface temperatures on a skew-T/log-p diagram (as in Fig. 1) makes them appear to be extreme.

Midlatitude summer situations typically have water vapor concentrations somewhat less than in the tropics, but considerably greater than in cooler seasons and latitudes. Given the large effect of water vapor on infrared radiative transfer in the lower troposphere, an alternate low-level water vapor profile was considered. This "dry" water vapor profile (dashed in Fig. 1) was specified to be equal to the "basic" midlatitude summer profile (solid in Fig. 1) except that its dew-point temperature was linear (in log-pressure) from the value at 700 mb to a value of 0°C at the surface. The dry profile had about half as much integrated precipitable water as the basic profile — 14 mm versus 27 mm.

2. Satellite instruments

Only geostationary satellite sounders were considered in these experiments because they can provide the frequent observations needed for mesoscale analysis. The capabilities of the operational VAS filter radiometer, which has relatively low spectral resolution (Menzel *et al.*, 1983), were compared with a proposed instrument, the GOES High-resolution Interferometer Sounder (GHIS; Smith *et al.*, 1990). While there is no assurance that the particular design of the GHIS will ever be implemented on GOES, the performance specifications Smith *et al.* provided are taken as representative of high spectral resolution sounder technology.

The satellite data for both the operational and proposed instruments were simulated, so that the instruments could be compared on an equivalent basis. Use of simulations also made it possible to treat the vertical resolution issue in isolation from the other potential sources of retrieval error. Simulated radiances for VAS were computed with MODTRAN software (Berk *et al.*, 1989), while for GHIS simulations it was necessary to use the higher-resolution, but computationally slower, FASCODE-3 (Clough *et al.*, 1989).

Both VAS and GHIS have surface-sensitive channels in both the 10-14-μm band and the 3-5-μm band, but only the former band was considered here because it is free of the effects of solar radiation. Another advantage of the 10-14-μm band is that variations of surface emittance from unity tend to be smaller than in the 3-5-μm band (Sutherland, 1986; Nerry, *et al.*, 1990); nevertheless, such variations were not considered in these experiments. The instrument noise standard deviations for the channels in the 10-14-μm band are listed in Table 1 in terms of radiance. Noise values for VAS were based on the typical spin budgets used in dwell sounding mode (Table 1), where the spin budget represents the number of observations taken for a given field of view and the noise values have been reduced from single-observation values by a factor of the square root of the spin budget.

Table 1. Noise levels of sounding instruments.

VAS *			GHIS †		
Channel	Spin budget	Noise standard deviation $\left(\frac{\times 10^{-7}\ \mathrm{W}}{\mathrm{cm}^2\ \mathrm{sr}\ \mathrm{cm}^{-1}}\right)$	Mode	Spectral resolution (cm^{-1})	Noise standard deviation $\left(\frac{\times 10^{-7}\ \mathrm{W}}{\mathrm{cm}^2\ \mathrm{sr}\ \mathrm{cm}^{-1}}\right)$
5	2	0.5	low	3.33	0.14
7	2	0.6	high	0.32	0.46
8	1	0.3			

* (Chesters and Robinson, 1983) † (Smith *et al.*, 1990)

The proposed GHIS has three spectral resolution modes — low, medium and high. Only the low and high-resolution modes were considered here, and will hereafter be referred to as GHIS(low) and GHIS(high), respectively. Even the low-resolution mode has much finer spectral resolution than VAS. The spectral resolving powers of VAS and GHIS are illustrated in Fig. 2. VAS spans nearly all of the wavenumber range 720-1010 cm^{-1} with just three channels (Fig. 2a). The plot of surface transmittance (Fig. 2b) illustrates the spectral detail available with GHIS(high). The ability to observe between individual absorption lines resulted in a maximum computed surface transmittance for the basic water vapor profile of 0.78 (at 964 cm^{-1}) for GHIS, whereas the maximum transmittance computed for VAS (in channel 8) was 0.63. Given that

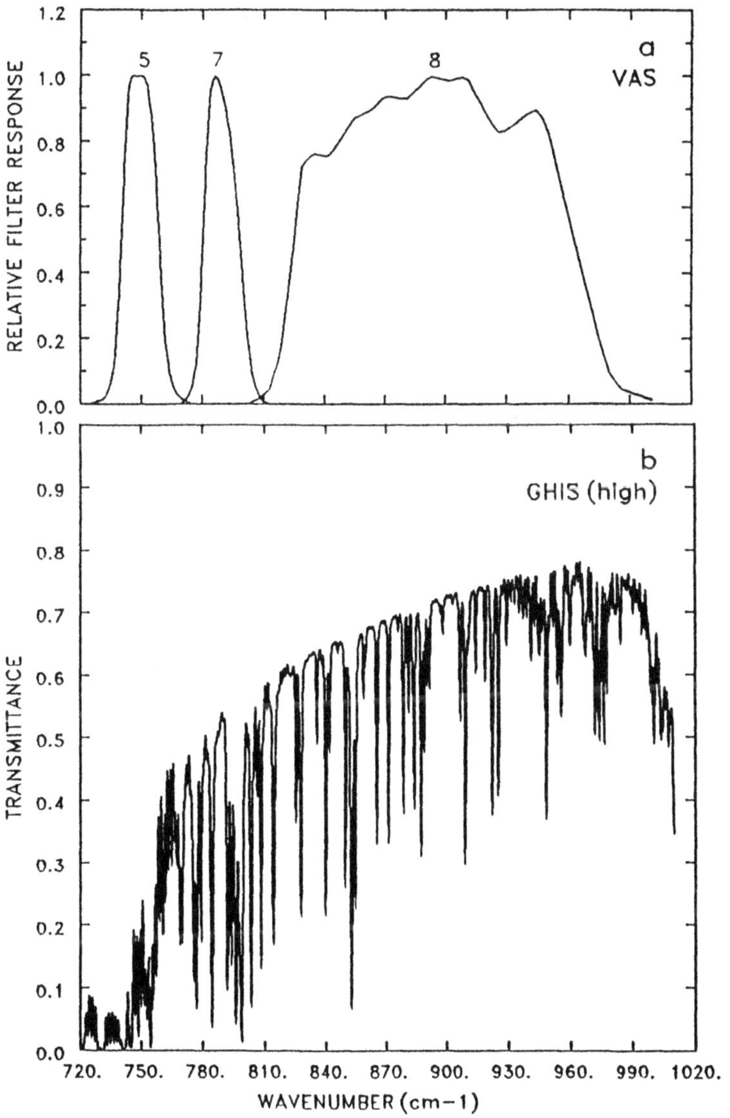

Figure 2. Relative filter response for VAS (a), and atmospheric transmittance from the ground to space for the night reference condition, with the basic water vapor profile, for GHIS(high) (b). The curves in (a) are labeled according to VAS channel number.

the atmospheric contribution to a radiance is indicated by one minus the transmittance, these transmittance values imply about 40% reduction in atmospheric contribution, going from VAS to GHIS. Such a reduction can be expected to facilitate distinguishing the thermal signature of the ground surface from any atmospheric effects.

In comparison with the situation of the basic water vapor profile, GHIS and VAS both had a much clearer view of the surface for the dry conditions, with maximum transmittance values of 0.94 and 0.85, respectively. This result implies that both instruments would give more-accurate surface temperature retrievals under dry conditions than under moist ones.

3. Sensitivity experiments

The question of how well VAS and GHIS can resolve the ground surface versus the air just above was addressed by considering the changes in computed radiances caused by perturbing the reference temperature profiles of Fig. 1 in two specific ways. The first type of perturbation consisted of a change in the ground surface temperature ΔT_{sfc}. The second was a systematic change in the low-level air temperature, represented by the ratio f, such that

$$f = \frac{T_p(p) - T_o(p)}{T_r(p) - T_o(p)} . \tag{1}$$

Temperatures are given as functions of pressure, subscripts p and r indicate the perturbed and reference profiles, respectively, and T_o is a profile extending linearly (in $\log p$ coordinates) from the perturbed ground surface temperature to the base of the isothermal layer at 770 mb. These perturbations are illustrated for the night situation in Fig. 3. As examples, if $f = 0$ the warm layer just above the surface is completely destroyed ($T_p = T_o$), if $f = 1$ the air has the same temperature as with the reference profile, and if $f \geq 1$ the warm layer is enhanced.

The perturbations were defined in terms of these two specific free parameters (ΔT_{sfc} and f) to allow for representing the ambiguity that can occur when interpreting radiances from channels sensitive to the ground. Any increase in radiance caused by warming of the ground can be partly compensated for by a cooling of the air just above the ground, and vice versa. This particular definition of the perturbations also limits the results to profiles that are meteorologically realistic.

Experiments were conducted to determine how much these two free parameters could be perturbed without causing any computed radiances to differ from reference-profile radiances by more than one standard deviation of the instrument noise. The results can be viewed as indicating how much retrievals could differ from the true (reference) profile, in terms of compensating ground/air temperature errors, without exceeding the noise limit. The signs of the perturbations were arbitrarily chosen such that the ground surface would warm and the lower atmosphere would cool for the night situation, and the opposite would occur for the day situation.

Figure 4 is an illustration of how the sensitivity experiments were done for the night situation, basic water vapor profile, with the high-resolution mode of GHIS. The radiance differences plotted correspond to perturbation parameter values $\Delta T_{sfc}=0.60$ and $f=0.86$. Differences are positive in spectral regions where the atmospheric transmittance is largest and are negative in regions where sensitivity is concentrated in the 800 to 950-mb layer. This is the noise-limited condition in the sense that the noise level (taken from Table 1) would be exceeded near 960 cm^{-1} if ΔT_{sfc} were increased or near 800 cm^{-1} if f were decreased any further.

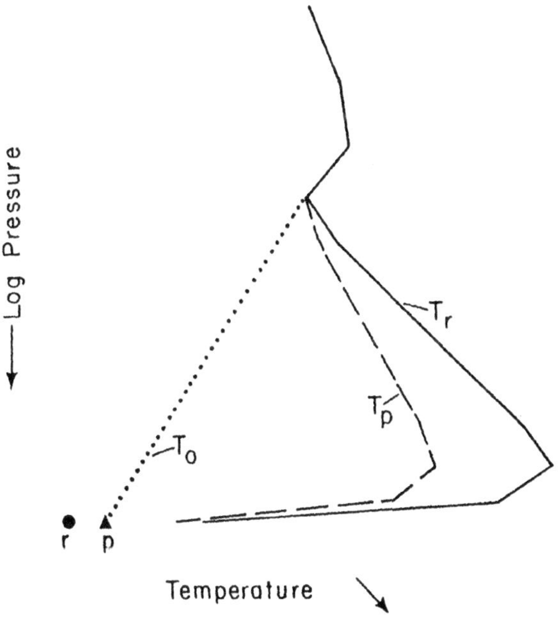

Figure 3. An illustration of the profile perturbation process, for the night situation. The dot and triangle indicate the reference and perturbed ground surface temperatures, respectively. The solid and dashed curves indicate the reference and perturbed temperature profiles, respectively.

The noise-limited perturbations for VAS and GHIS, with the basic water vapor profile, are plotted in Fig. 5 and summarized in Table 2. The vertical resolution and noise characteristics of VAS are such that it allowed perturbations about twice as large as for GHIS(high) and about five times as large as for GHIS(low). Table 2 also includes results for the dry water vapor profile for the night situation. As expected, the unresolvable perturbations of surface temperature were smaller for the dry atmosphere than for the moister one. Furthermore, the relative advantage of GHIS over VAS with regard to surface temperature perturbations is smaller for a drier atmosphere, particularly for GHIS(high). This result is consistent with the smaller GHIS(high)−VAS difference in maximum transmittance (0.94−0.85=0.08 versus 0.78−0.63=0.15) for the dry case as compared to the basic case. The unresolvable air temperature perturbations (rightmost column) became larger for the drier atmosphere, as would be expected when there is a weaker signal from the lower atmosphere in the radiances.

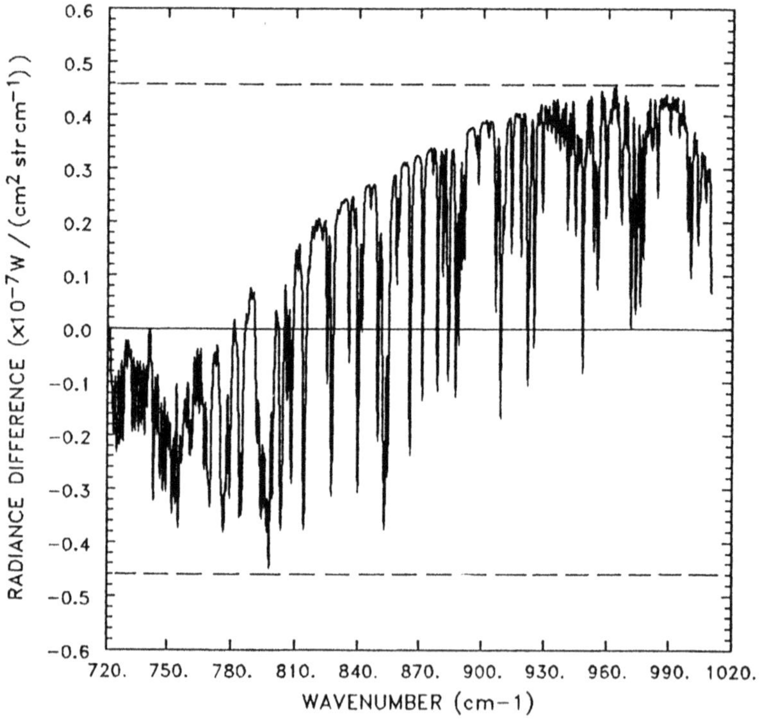

Figure 4. Difference in spectral radiance for GHIS(high) and the night basic situation — perturbed profile (ΔT_{sfc} =0.60, f =0.86) minus reference profile. The dashed lines denote the noise level for GHIS(high).

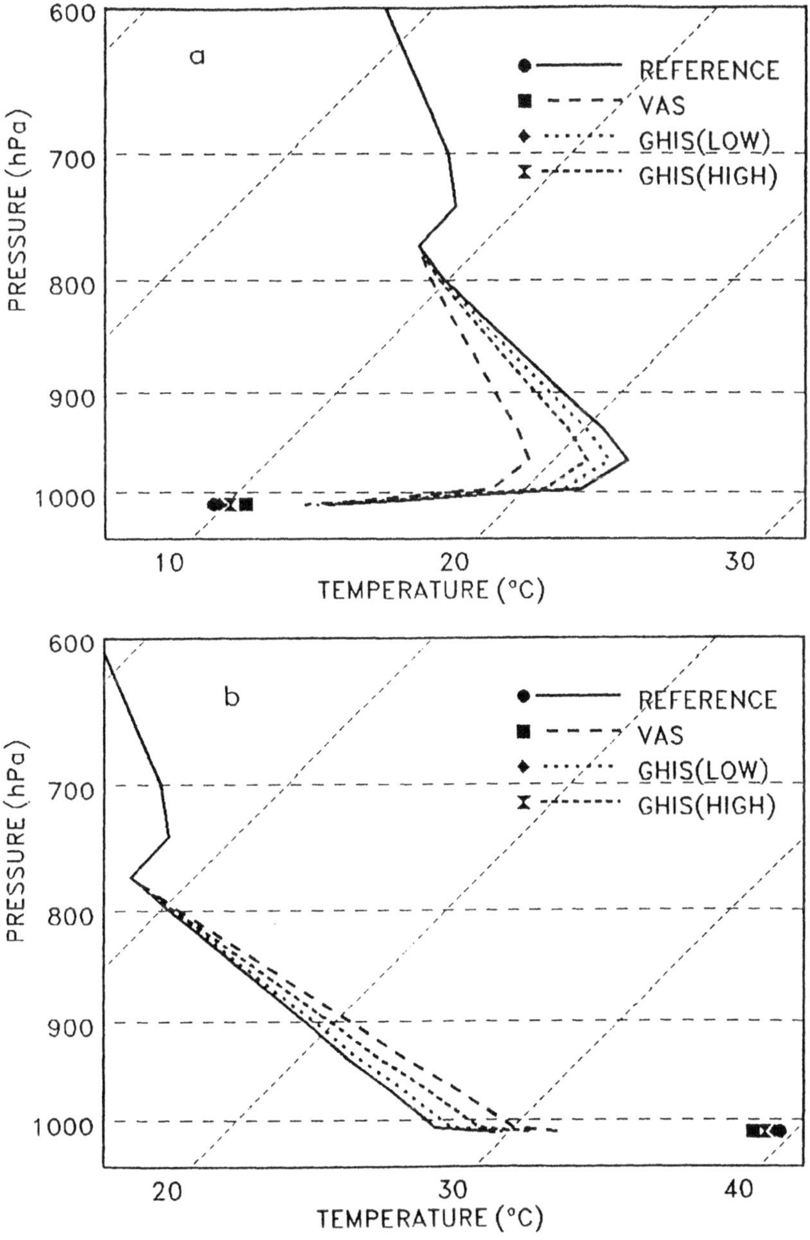

Figure 5. Noise-limited perturbed temperature profiles for the (a) night and (b) day situations with the basic water vapor profile, labeled according to satellite instrument. Marker symbols indicate ground surface temperatures, where the GHIS(low) and reference markers overlap.

Table 2. Absolute differences of temperature between noise-limited perturbed and reference profiles.

Instrument	Ground surface temperature difference (°C)	Largest air temperature difference (°C)
Night (basic):		
VAS	1.2	3.5
GHIS(low)	0.2	0.7
GHIS(high)	0.6	1.4
Day (basic):		
VAS	1.0	2.8
GHIS(low)	0.2	0.6
GHIS(high)	0.5	1.6
Night (dry):		
VAS	0.5	4.0
GHIS(low)	0.1	0.9
GHIS(high)	0.4	1.9

GHIS(low) outperformed GHIS(high) in these experiments because its lower noise level more than compensated for its lower vertical resolving power. A limitation of these experiments, however, is that they do not account for the vast differences in the number of channels available with each of these instrument configurations. GHIS(low) has about 90 channels in the spectral band considered here, while GHIS(high) has about 900. Given a radiance error limit of one noise standard deviation, retrievals with errors as large as these perturbation–reference differences would occur only if observed radiances had the property that all channels most sensitive to the ground had errors of one sign and all channels most sensitive to the lower atmosphere had errors of the opposite sign. This represents a particular pattern of channel-to-channel error correlation that is very unlikely to ever be realized with 90 channels, and is even less likely with 900 channels. In contrast, such a correlation pattern is likely to be realized occasionally when there are just 3 channels, as VAS has in this band. Retrieval experiments were conducted to address this issue.

4. Retrieval experiments

Retrievals were performed only for the night reference situation with the basic water vapor profile, under the assumption that retrieval experiments for the day situation and for the dry profile would offer little information beyond what can be inferred from these results and those of the sensitivity experiments. The retrievals employed simulated radiances with pseudorandom noise added according to the standard deviations listed in Table 1. Ten retrievals were performed for each instrument configuration, each with a different set of radiances based on different noise realizations.

The retrieval formulation was chosen appropriately for this study's focus on resolvability of the ground. A non-linear, iterative, least-squares solution for the perturbation parameters was used, where

$$\mathbf{x}_{n+1} = \mathbf{x}_n + \left(\mathbf{K}_n^{\mathrm{T}} \cdot \mathbf{E}^{-1} \cdot \mathbf{K}_n\right)^{-1} \cdot \mathbf{K}_n^{\mathrm{T}} \cdot \mathbf{E}^{-1} \cdot \left\{\mathbf{y}^m - \mathbf{y}(\mathbf{x}_n)\right\} \tag{2}$$

is the retrieval equation, the solution vector is

$$\mathbf{x} = \begin{bmatrix} \Delta T_{sfc} \\ f \end{bmatrix}, \tag{3}$$

n is the iteration number, $\mathbf{y}^m$ is the vector of simulated measurement radiances, $\mathbf{y}(\mathbf{x})$ is the vector of computed radiances, $\mathbf{E}$ is the measurement error covariance matrix, and $\mathbf{K}$ is the matrix of partial derivatives of the elements of $\mathbf{y}$ with respect to the elements of $\mathbf{x}$ (Deutch, 1965, pp 61-75; with notation as in Eyre, 1989). The initial guess $\mathbf{x}_0$ was the night reference profile. The elements of $\mathbf{K}$ were computed by the brute force method, which consists of computing radiances for a given $\mathbf{x}$, then with each of the two elements of $\mathbf{x}$ altered by a small amount, and finally computing derivatives in terms of finite differences.

Note that this formulation does not include any constraint tying the solution to a background profile and, thus, the solution to which the profile parameters converge is dictated fully by the measured (simulated) radiances. In addition, the measurement errors were assumed to be completely uncorrelated, so $\mathbf{E}$ is diagonal. Error variances for VAS were taken from the instrument noise statistics (Table 1). For GHIS the specified noise is the same for all channels in the relevant band and, therefore, $\mathbf{E}^{-1}$ is just the product of a scalar and an identity matrix and can be eliminated from Eq. 2.

The worst-case results from among the ten noise realizations are plotted in Fig. 6 and summarized in Table 3. The retrieval–profile differences for VAS and GHIS(high) happened to have signs opposite from those of the sensitivity experiment differences (Fig. 5) but, for VAS, the magnitudes of the differences were very similar in the two sets of experiments. In contrast, the GHIS retrieval results were much closer to the reference profiles than were the sensitivity experiment results.

Another significant feature of the retrieval results is that GHIS(high) outperformed GHIS(low) by about a factor of two, demonstrating how the greater number of channels can offset the greater noise level in the high resolution mode.

Table 3. Absolute differences of temperature between the worst of ten retrievals and the reference profile.

	Instrument	Ground surface temperature difference (°C)	Largest air temperature difference (°C)
Night:			
	VAS	1.01	3.68
	GHIS(low)	0.11	0.40
	GHIS(high)	0.05	0.23

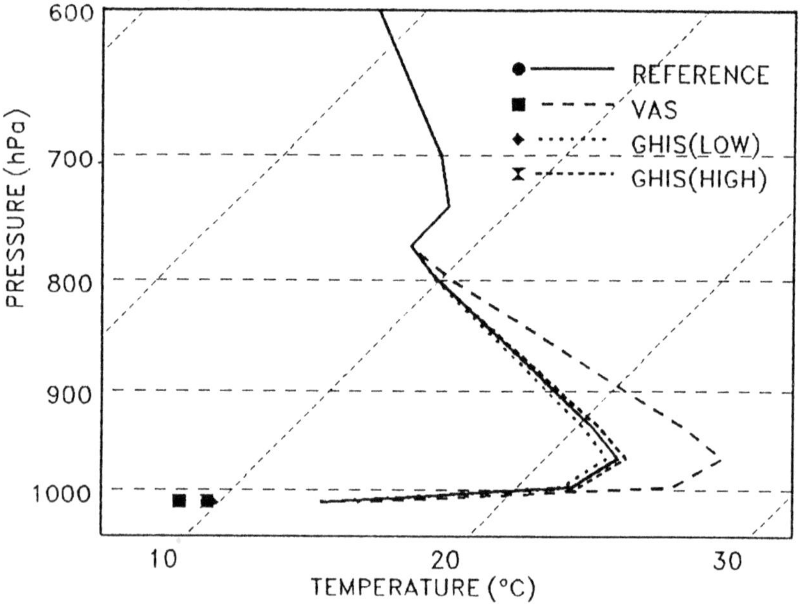

Figure 6. Retrieved temperature profiles (the worst of ten simulations) for the night situation with the basic water vapor profile, labeled according to satellite instrument. Marker symbols indicate ground surface temperatures, where the GHIS markers almost directly overlay the reference marker.

5. Conclusion

The results of simulated retrievals and experiments on the sensitivity of radiances to profile perturbations gave an indication of how well VAS and a proposed high spectral resolution sounder, GHIS, are able to resolve the ground surface from the air just above ground. Interpretation of the results must be constrained by the fact that only two water vapor profiles and two, midlatitude summertime, temperature profiles were considered.

The two sets of experimental results agreed that, for a moderately moist atmosphere, ground surface temperature errors of about 1°C and low-level air temperature errors of about 3°C can occur in VAS retrievals due specifically to vertical resolution limitations and instrument noise. For a drier atmosphere the surface temperature errors tend to be smaller and the low-level air temperature errors tend to be larger. These values do not account for other sources of retrieval error, such as interference by clouds, uncertainty of the ground surface emittance, or deficiencies in the radiative transfer computation method.

The results of the retrieval and sensitivity experiments were more varied with regard to GHIS, stemming from the experiments' different assumptions about channel-to-channel correlations of data noise. Nevertheless, it appears that these vertical-resolution-related retrieval errors can be reduced by a factor of about 70-90% by going from VAS to an instrument with performance specifications such as those of GHIS.

Acknowledgments. The author thanks J. Chetwynd, M. Griffin, and T. Kleespies for instruction on the use of FASCODE and MODTRAN, J. W. Snow for reviewing the manuscript, and B. Main for help in drafting figures.

References

Becker F (1987) The impact of spectral emissivity on the measurement of land surface temperature from a satellite. *Int J Rem Sens* 8:1509–1522

Berk A, Bernstein LS, Robertson DC (1989) MODTRAN: A moderate resolution model for LOWTRAN 7. GL-TR-89-0122, Geophysics Laboratory, Air Force Systems Command, Hanscom AFB, MA, USA

Chesters D, Robinson WD (1983) Performance appraisal of VAS radiometry for GOES-4, -5 and -6. NASA Tech Memo 85125, Goddard Space Flight Center, Greenbelt, MD, USA

Clough SA, Kneizys FX, Anderson GP, Shettle EP, Chetwynd JH, Abreu LW, Hall LA, Worsham RD (1989) FASCOD3: Spectral simulation. In: Lenoble J, Geleyn JF (eds) *IRS '88: Current problems in atmospheric radiation.* A. Deepak, Hampton, VA, USA, p 372

Deutch R (1965) *Estimation theory.* Prentice-Hall, Englewood Cliffs, NJ, USA

Eyre JR (1989) Inversion of cloudy satellite sounding radiances by nonlinear optimal estimation. I: Theory and simulations for TOVS. *Quart J Roy Meteor Soc* 115:1001–1026

Geiger R (1965) *The climate near the ground.* Harvard Univ Press, Cambridge, MA, USA

Kornfield J, Susskind J (1977) On the effect of surface emissivity on temperature retrievals. *Mon Wea Rev* 105:1605–1608

Lipton AE, Vonder Haar TH (1990) Pre-convective mesoscale analysis over irregular terrain with a satellite/model coupled system. *Mon Wea Rev* 118:1330–1358

McNider RT, Pielke RA (1981) Diurnal boundary layer development over sloping terrain. *J Atmos Sci* 38:2198–2212

Menzel WP, Smith WL, Wade GS, Herman LD, Hayden CM (1983) Atmospheric sounding from a geostationary satellite. *Appl Opt* 22:2686–2689

Nerry F, Labed J, Stoll MP (1990) Spectral properties of land surfaces in the thermal infrared. 1. Laboratory measurements of absolute spectral emissivity signatures. *J Geophys Res* 95:7027–7044

Smith WL, Revercomb HE, Howell HB, Huang H-L, Knuteson RO, Koenig EW, Laporte DD, Silverman S, Sromovsky LA, Woolf HM (1990) GHIS – The GOES High-Resolution Interferometer Sounder. *J Appl Meteor* 29:1189–1204

Sutherland RA (1986) Broadband and spectral emissivities (2-18 μm) of some natural soils and vegetation. *J Atmos Ocean Tech* 3:199–202

INVESTIGATION OF INFRARED EMISSIVITIES OF SAHARA DUST POWDERS AND QUARTZ FROM SPACE

Tsutomu Takashima and Kazuhiko Masuda
Meteorological Research Institute,
Tsukuba, Ibaraki 305, JAPAN

KEYWORDS/ algorithm/ satellite observation / radiative transfer/ ASTER-EOS/ / NOAA-AVHRR / emissivity / quartz / Sahara dust / / remote sensing / land surface temperature/

Abstract -- A method to measure emissivities of quartz and Sahara dust powders from space is discussed based on the simulation in the infrared wavelength region(7-17μm). The doubling method is adopted for the emissivity computation under the assumption of cloudy atmosphere of substances. Furthermore, the discrimination of the surface characteristics is forcussed in the wavelenths to be provided by the ASTER-EOS and NOAA-AVHRR channels.

1. Introduction

A precise knowledge of emissivities over the land surface is essential for the investigation of the earth radiation budget. This knowledge is needed for a global scale investigation using satellite observations. Emissivities of quartz and Sahara dust powders can be investigated as representative examples of the surface composition. Thus as a first step, a simulation of radiative characteristics of these substances is carried out in the infrared region(7-17μm).

2. EOS program

NASA plan to launch the EOS-AM in the morning orbit at 10:30$\pm$15M. This will carry 5 radiometers such as ASTER, MODIS-N, CERES, MISR, and MOPITT. Among them, the ASTER(Advanced Spaceborne Thermal Emission and Reflection radiometer) is composed of three radiometers covering the visible, near IR and infrared region of spectrum. One of them, the TIR(Thermal Infrared Radiometer) provides 5 channels in the wavelength region ranging from 8.125 to 11.65μm with the spacial resolution of 90m. The total spectral response(10-14 channels) of the ASTER-TIR is shown in Fig.1. Furthermore, the performance of the TIR is shown at Table 1. Note that the 17 channels of the thermal bands of the MODIS-N ranges from 3.75 to 14.24μm for atmospheric and surface temperature with the spacial resolution of 1km.

NATO ASI Series, Vol. I 9
High Spectral Resolution Infrared Remote Sensing for Earth's Weather and Climate Studies
Edited by A. Chedin, M.T. Chahine and N.A. Scott

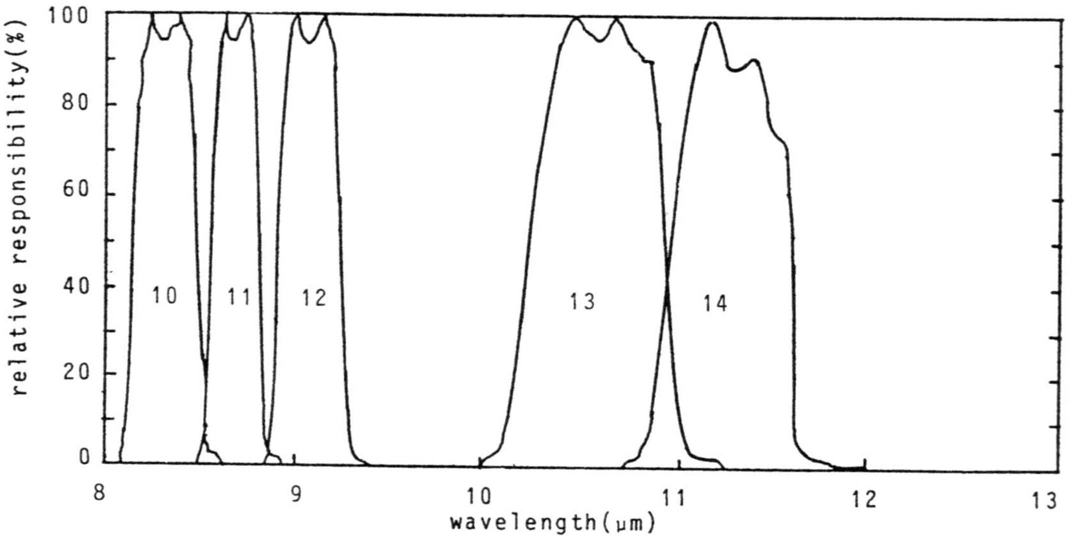

Fig. 1. Total Spectral Response of Bandpass Filter of ASTER-TIR on EOS-AM. It is normalized by peak value of each band(Fujitsu,1992).

Item		Required specification	Performance at present
Spectral coverage	Band 10	8.125 - 8.475 um	Same as at left
	Band 11	8.475 - 8.825 um	
	Band 12	8.925 - 9.275 um	
	Band 13	10.25 - 10.95 um	
	Band 14	10.95 - 11.65 um	
Swath width		60 km	60 km (FOV:4.9deg)
Ground resolution		90 m	90 m (IFOV:127.6urad)
MTF (at Nyqust)		> 0.20 : along track > 0.25 : closs track	0.31 0.25
Scan method		————	Mechnical scanning
Quantization level		12 (bits)	12 (bits)
Optical system		————	Catadioptric system
Cooling method		Cryocooler or Alternative	Liner drive, Stirling cycle cryocooler
Pointing		± 8.55 deg	± 8.55 deg
Calibration		Accuracy 3 K at 200 - 240 K 2 K at 240 - 270 K 1 K at 270 - 340 K 2 K at 340 - 370 K	Accuracy 1.40K at 200 - 240 K 0.97K at 240 - 270 K 0.65K at 270 - 340 K 0.81K at 340 - 370 K
NEΔT	BAND 10	less than 0.3 K (AT 300K)	0.13K
	BAND 11		0.13K
	BAND 12		0.14K
	BAND 13		0.13K
	BAND 14		0.15K
Lifetime		5year (Cryocooler > 50000h)	5 year (Cryocooler > 50000h)

Table 1. Performance of ASTER-TIR on EOS-AM(Fujitsu,1992).

3. Investigation of surface characteristics in IR region

The thermal emission from the surface is attenuated through the atmosphere. Atmospheric temperature profile should firstly be determined to derive the radiative property of the atmosphere. In the atmosphere, the influence of absorbend gases such as water vapor and ozone, cloud, aerosol, volcanic ashes should simultaneously be investigated. As for the surface, a separation of the emissivity property from the surface temperature field is essential to identify the minerals. In addition partial vegetation coverage, moisture content, variety of minerals or non-uniform surface in the instantaneous field of view should properly be taken into account.

4. Atmospheric transmittance

Fig.2 shows the atmospheric transmittance in the mid-latitude summer model (Kneizys,F.X., et al.,1983) There is strong absorption band in the region(9.3-10 μm) centered at near 9.5μm. The regions, 8-9.3μm and 10-13μm are considered as atmospheric windows. But there is strong contamination due to water vapor absorption, in particular, in the mid-latitude in summer and tropics. The channels of NOAA-AVHRR and ASTER-EOS are located within the atmospheric windows, avoiding the strong ozone absorption band. An efficient and practical algorithm for atmospheric correction is needed for a quantitative investigation of surface characteristics.

5. Emissivities of quartz and Sahara dust powders

The characteristics of thermal emission across the spectrum is due to composition and surface roughness or particle size of substances(Conel,1969). Barton and Takashima(1986), for example, have measured the sand surface emissivity as a function of zenith angle. It shows the emissivity between 0.97 and 0.98 at the zenith and decreases monotonically with the increase of zenith angle. This trend is agreeable well to numerical simulation. More precisely if the refractive index and size distribution of substances are known, their emissivity can be calculated theoretically. As for a practical aspect, a cloudy atmospheric model involving independent scatters was adopted under the assumption of spherical particles. Thus emissivities of quartz(ε_q) and Sahara dust powders(ε_s) have been computed numerically under the assumption of a cloudy atmosphere with respect to substances using the doubling method (Takashima and Masuda,1987). Note that the optical thickness of the initial sublayer was set to 0.00025 and the doubling procedure was undertaken 19 times before the emissivity converged numerically. The corresponding optical thickness is around 130. Where the refractive index of quartz and Sahara dust powders are given by Spitzer and Kleinman(1961) and Volz (1973), respectively(Fig.3). Furthermore, the size distribution is given by Conel (1969). Emissivities of quartz and Sahara dust powders are shown in Fig.4. Their angular dependencies are also described in Fig.5. Main results of numerical simulation is following: (1) With the increase of particle size, (a) major minimums of ε_q in the regions 8.3-9.3μm with peak value at the center near 8.6μm,

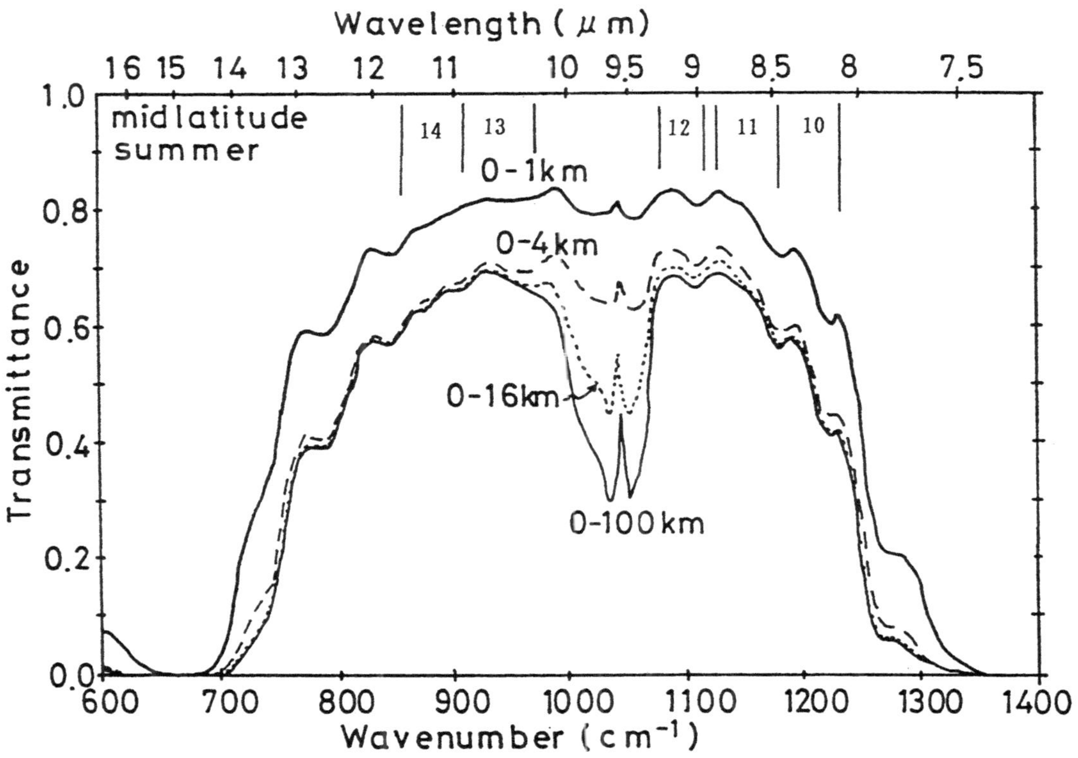

Fig. 2. Total Atmospheric Transmittance by the LOWTRAN-6. Mid-latitude summer model(Kneizys, F. X. et al, 1983). Numerals 10-14 correspond to the channels of the TIR-ASTER.

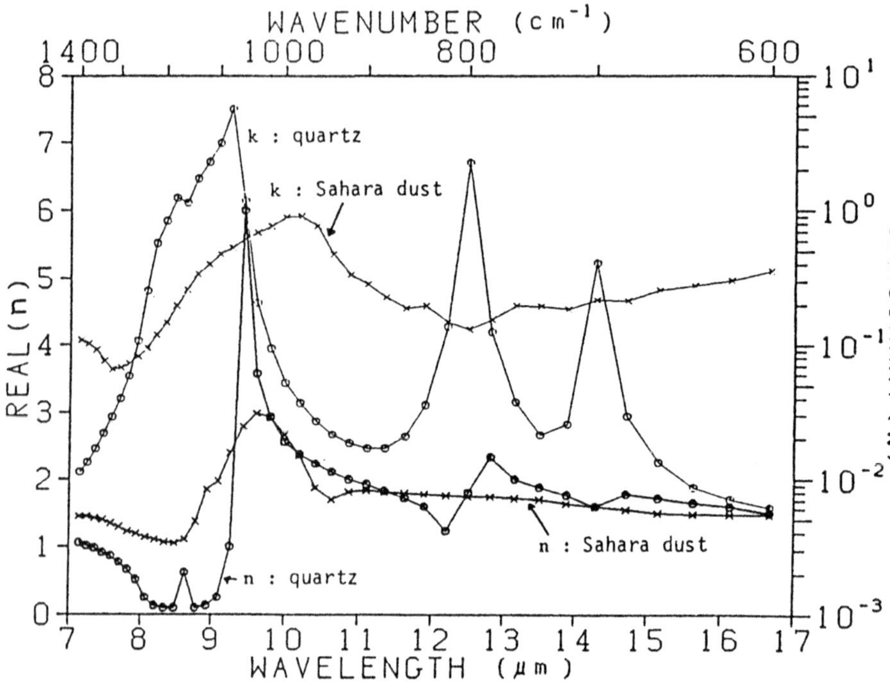

Fig.3. Refractive indices of quartz(o) based on data by Spitzer and Kleinman (1961) and those for Sahara dust(x) by Volz(1973) in the infrared wavelength region(7-17 μm). Symbols n and k represent real(heavy lines) and imaginary part(fine lines) of refractive indices(m=n-ik), respectively.

and 12.5 μm decreases. (b) ε_q increases in the regions, 10-12 μm, 13-14 μm, and 15-17 μm. (2) The emissivity of plane surface is smaller than ε_q. (3) At wavelengths, 7-8 μm, near 8.6 μm, 12.2 μm, and 14. μm, ε_q shows little change with particle size. (4) ε_S increases with the increase of particle size, with the exception in the region 9.0-10.5 μm. ε_S is over 0.94 in the entire spectrums. (5) ε_q and ε_S decrease monotonically with nadir angle(θ) in all spectrums. (6) At NOAA AVHRR channels, (a) ε_q tends to increase with the increase of particle size. (b) ε_{12} is generally higher than ε_{11}, where ε_{11} and ε_{12} denote average emissivities of 10.5-11.5 μm and 11.5-12.5 μm, respectively. (c) The apparent brightness temperature difference derived in these channels, ΔT_q, changes slowly with θ. (d) ΔT_q decreases with the increase of particle size. (e) ΔT_q depends more on particle size than on surface temperature. (f) Sahara dust powders show rather small change of ε and ΔT_S.

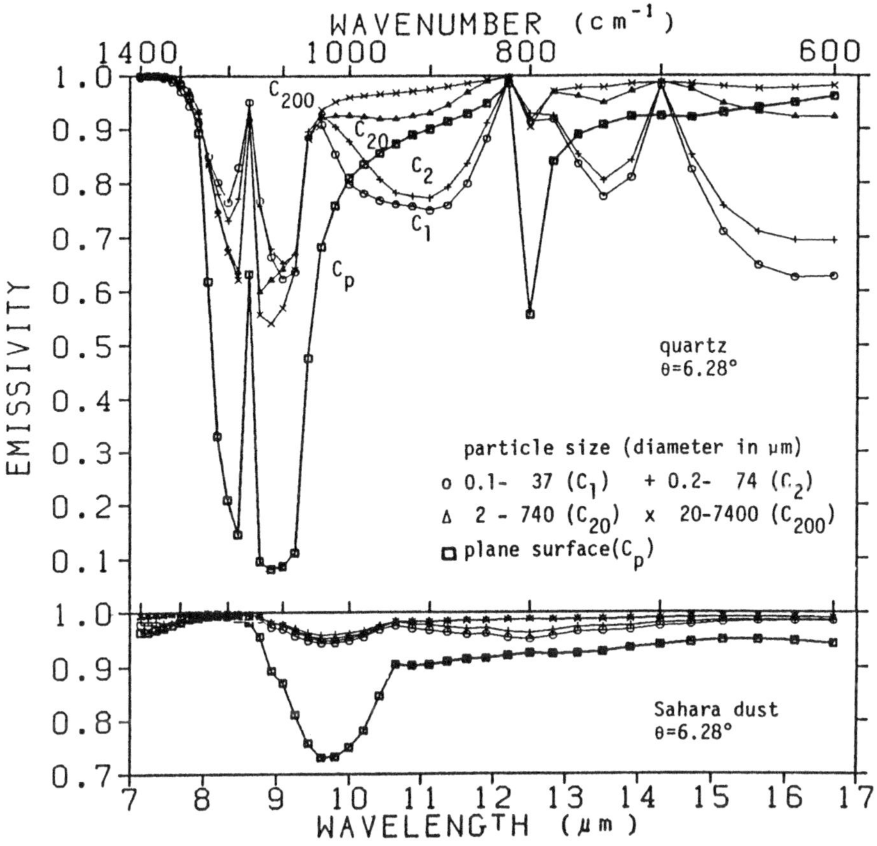

Fig.4. Emissivity of quartz and Sahara dust powders as a function of particle size at the zenith angle θ=6.28° . Emissivity for a plane surface is also shown(C_P). After Takashima and Masuda, 1987.

6. conclusion

To identify quartz, high spectral resolution is required in the region ranging from 8.3 to 9.3μm. Emissivity depends on the size distribution, but ε_q shows little change with the size distribution in the regions, 7-8μm, near 8.6 μm, 12.2μm, and 14.3μm. In the split window channel of the NOAA-AVHRR, the apparent temperature difference between these channels may be large enough to measure from space, if the particle size is small. ε_s is over 0.94, showing little change with particle size and wavelengths. A separation of emissivity of substances from its temperature field seems to be tough for satellite measurements. Thus a combined use of visible channels with infrared ones would be better for the investigation of surface characteristics. The reflectance of the

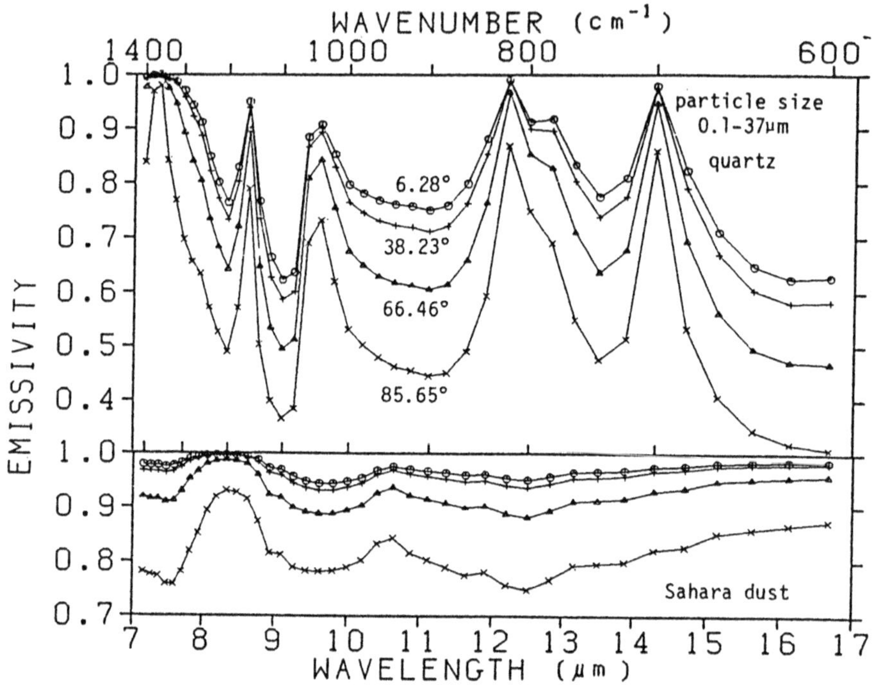

Fig.5. Emissivity of quartz and Sahara dust powders as a function of zenith angle. The Conel size distribution(0.1-37 μm) is assumed. After Takashima and Masuda, 1987.

substances would provide an additional information with respect to the surface characteristics.

References

Barton IJ and Takashima T (1986) An AVHRR investigation of surface emissivity near lake Eyre, Australia, Remote Sen. Environ. 20:153-163.

Conel JE (1969) Infrared emissivities of silicates: Experimental results and a cloudy atmosphere model of spectral emission from condensed particulate mediums, J.Geophys.Res. 74:1614-1634.

Fujitsu,Co (1992) ASTER-US-JAPAN Meeting, Pasadena.

Kneizys,F.X., et al.(1983) Atmospheric transmittance/ radiance: computer code LOWTRAN 6, Air Force Geopysics Laboratory, Massachusetts, AFGL-TR-83-0187.

Spitzer WG, and Kleinman DA (1961) Infrared lattice bands of quartz, Phys.Rev. 121:1324-1335.

Takashima T, and Masuda K (1987) Emissivities of quartz and Sahara dust powders in the infrared region(7-17 μm), Remote Sen. Environ. 23:51-63.

Volz FE (1973) Infrared optical constants of ammonium sulfate, Sahara dust, volcanic pumica, and flyash, Appl.Opt. 12:564-568.

Chapter 5

Next Generation IR sounders and trace gas analysis

"Trace gas remote sounding from near IR sun glint observation with tunable etalons"

Tad. AOKI, M. FUKABORI, Ter. AOKI - Meteor. Research Inst., Japan

"High resolution observation of the Earth's atmosphere in the far infrared"

B. CARLI - IROE, Italy

"Remote sensing of atmospheric trace constituents using MID-IR Fourier transform spectrometry"

H. FISCHER - IMK/KfK, Germany

"The measurement of global carbon monoxide using the Atmospheric Infrared Radiation Sounder (AIRS) "

L. LARRABEE STROW - Univ. of Maryland, USA

Trace gas remote sounding from near IR sun glint observation with tunable etalons

Tadao Aoki, Masashi Fukabori and Teruo Aoki
Meteorological Research Institute
Tsukuba, Ibaraki 305, Japan

Abstract

A method of remotely sounding the tropospheric trace gases by measuring the surface reflected near infrared solar radiation is described. The absorption lines of trace gases are selected by a very narrow band pass filter and scanned with a tunable etalon. The interferogram is directly put into an inversion program to obtain the vertical profiles of trace gases. According to a simulation study the error of retrieved vertical profile of trace gases in the lower atmosphere is a few percent.

I. Introduction

The troposphere together with the earth surface constitutes a medium that is vital for the completion of major biogeochemical cycles essential to life. The composition of the tropospheres is also important to determine the climate system of the earth, which in turn affects the biosphere. Activities of man now constitute a major perturbation to the chemistry of the troposphere. The assessment of the impact of human activities is hampered by a lack of understanding of underlying physical, chemical and biological processes, which interact with each other on local and global scale.

To understand the fundamental processes which control these global system it is necessary to obtain global and long-term information on atmospheric constituents. If the accuracy requirements are met, space borne observations are ideal in the sense that they provide homogeneous data in the three dimensional global distribution of trace species. So far the remote sounding of the tropospheric gases from satellite have been made by measuring the emission from the atmosphere and surface. For more than a decade the TIROS-N satellite series has been operationally obtaining the water vapor vertical profile by measuring the infrared emission from the atmosphere with a filter type radiometer, HIRS(High Resolution Infrared Radiometer Sounder). The error is more than 20 percent near the ground and worse in the higher altitude (Timchak, 1986). The remote measurement of CO has been made by MAPS(Measurement of Air Pollution from Satellites) on board the space shuttle, also by measuring the infrared emission through gas filter. The accuracy is expected to be an order of 20 percent (Reiche et al., 1982).

In the future satellite program some new type of instruments such as a Fourier transform spectrometer and a grating spectrometer are proposed for remote sounding the tropospheric trace gases(see other sections of this book). These are again based on the measurement of thermal emission from the atmosphere. The spectral resolution of these instruments are the order of 0.1 cm^{-1} or larger.

NATO ASI Series, Vol. I 9
High Spectral Resolution Infrared Remote Sensing
for Earth's Weather and Climate Studies
Edited by A. Chedin, M.T. Chahine and N.A. Scott

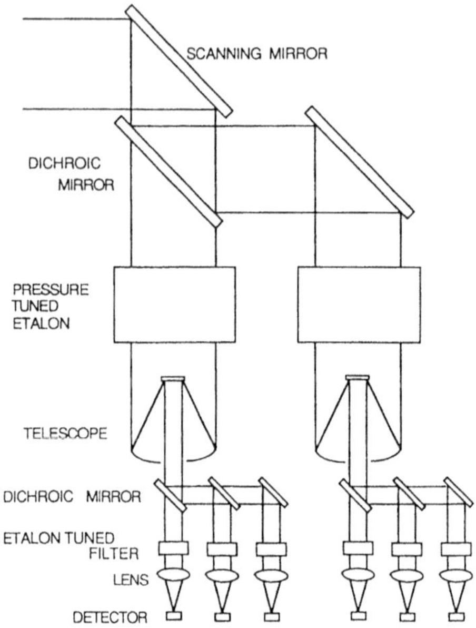

Fig.1 Conceptual design of the TERSE.

In this paper we describe a new vertical sounder for tropospheric trace gases, TERSE(Tunable Etalon Remote Sounder for the Earth) proposed by Aoki(1989a,b). The essential part of this instrument is composed with a narrow band pass filter and tunable etalon. TERSE measures the near infrared solar radiation reflected at the earth surface or cloud. An algorithm for the retrieval of vertical density profile and a result from a simulation study are presented.

II. Conceptual design of the instrument

A conceptual design of the TERSE is shown in Fig.1. The near infrared radiation between $1.25 \sim 2.5\mu m$ is divided into two parts, $1.2 \sim 1.7\mu m$ and $2.0 \sim 2.5\mu m$. These radiations are modified by the tunable etalon which is enclosed in a box pressurized with a gas. The refractive index of the gas between two plates of the etalon is changed by periodically changing the pressure of the gas. (In the original proposal of Aoki(1989a,b), the spacing of the two plates was changed with a piezo electric material or the incident angle of the light was changed by turning the etalon whose rotational axis was slightly tilted with respect to the incident light and the normal of the etalon plate is also tilted with respect to the axis of the rotation). An example of the transmittance spectra of the tunable etalon for some different values of the refractive index is shown in the top three panels of Fig.2.

By changing the refractive index one of lines of the tunable etalon scans the absorption line concerned(in the case of Fig.2 it is the one at $6547.26\ cm^{-1}$).

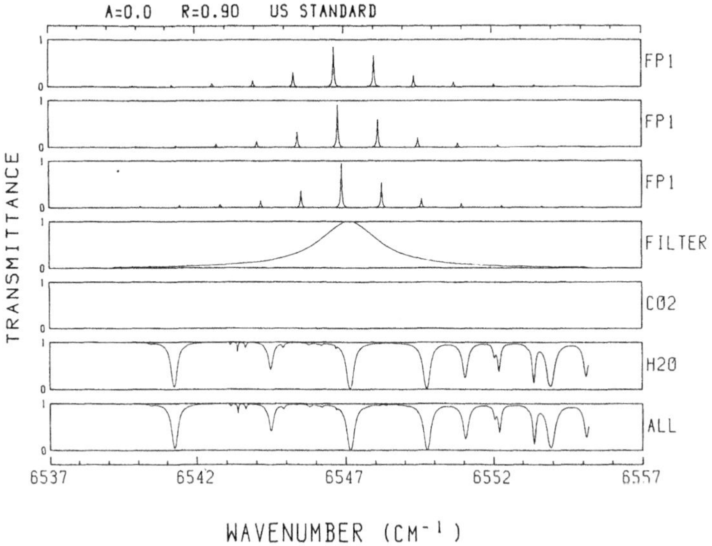

Fig.2 Spectra of the wavenumber region for the H_2O sounding. The lowest panel is the transmittance of the column atmosphere, the next two are those of the H_2O and CO_2, respectively. The panel signed as "FILTER" is the transmittance of the band pass filter and the upper three panels are those of the tunable etalon for different refractive indices.

The scan is made over about one free spectral range, which is about 1.3 cm^{-1}. The finess is about 30 and the sampling is made at 90 points in constant interval. An etalon and an interference filter together constitute a very narrow band pass filter as shown in the panel signed "FILTER" in the Fig.2. This band pass filter only covers a small wavenumber region which contains a few absorption lines of a trace gas concerned.

III. Radiative transfer

The signal output of the sensor at the i-th sample during a scan of pressure in the etalon box is written in the form:

$$V_i = C \int_{\nu_1}^{\nu_2} d\nu \phi_{FP}(\nu, t_i) \phi_B(\nu)[I_{refl}(\nu) + I_{path}(\nu)], \tag{1}$$

where C is the constant, ϕ_{FP} is the transmission function of TE, ϕ_B is the transmission function of the narrow band pass filter, ν_1 and ν_2 are the lower and upper limits of the function ϕ_B, respectively, I_{refl} is the radiance reflected at the surface and I_{path} is that scattered by the atmosphere. For an ideal etalon ϕ_{FP} is given by the Airy function

$$\phi_{FP}(\nu, t) = \frac{(1 - r_{FP} - A_{FP})^2}{1 - 2r_{FP}\cos(2\pi\nu t) + r_{FP}^2}, \tag{2}$$

$$t = 2n_{FP}\delta \cos\gamma, \tag{3}$$

where A_{FP} is the absorptance of the etalon, r_{FP} is the reflectance of the surface of the etalon plate and n_{FP} is the refractive index of the medium between the etalon plates. δ is the spacing between the two etalon plates and γ is the incident angle of the light. In this study n_{FP} is changed by changing the pressure of the gas between the etalon plates. The reflected component of solar radiation at the surface is written as

$$I_{refl}(\nu) = I^{\odot}\phi^{\odot}(\nu)r_s\Omega{\tau_e}^A(TG+GO)T_{refl}(\nu), \tag{4}$$

$$T_{refl}(\nu) = {\tau_e}^M(TG+GO), \tag{5}$$

where $I^{\odot}\phi^{\odot}(\nu)$ is the solar radiation outside the atmosphere, which may have a strong wavenumber dependence in the range between ν_1 and ν_2 due to absorption by the solar atmosphere. We assume that the absorption due to the solar atmosphere is well known so that we can give the solar radiation outside the atmosphere with a constant $I^{\odot}$ and a known function $\phi^{\odot}(\nu)$. $r_s\Omega$ is the fraction of the incident solar radiation that is reflected at the surface into the view angle of the sensor, Ω. r_s is the bidirectional reflection coefficient of the earth surface, which depends on the incident angle, θ_{sun}, of the solar radiation and the reflected angle, θ_{refl}. However, during the observation of one spot these values can be assumed constant.

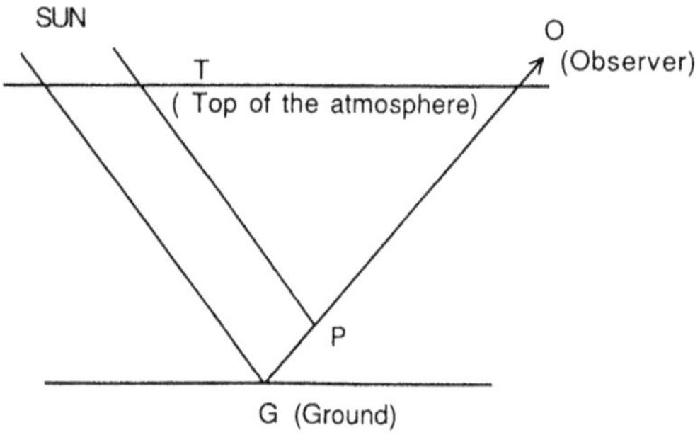

Fig.3 Schematic representation of the radiative transfer geometry.

The wavenumber dependence of r_s may be neglected since the interval between ν_1 and ν_2 is the order of 10 cm^{-1} (see Fig.2). Thus, we can make assume r_s constant with respective to the wavenumber. ${\tau_e}^{\chi}(TG+GO)$ is the extinction due to the absorption and scattering by the medium χ during the transfer of the light through the combined path TG from the top of the atmosphere to the ground and the path GO from the ground to the observation point(satellite, airplane, etc.), where χ stands for the aerosols(A) or the molecules(M). The geometry of the path of the solar radiative transfer is shown in Fig.3, where T denotes the top of the atmosphere, G the ground, O the observation point and P the scattering point, respectively. The transmission function for the absorption due to molecules in a slant path from a point $P_1(z = z_1)$ to $P_2(z = z_2)$ is given by

$$\tau_a^M(\nu, \theta; z_1, z_2) = exp\Big[-\mu \int_{z_1}^{z_2} \sum_g k_{\nu,g}(z)\rho_g(z)dz\Big], \tag{6}$$

$$\mu = \frac{1}{\cos\theta}, \tag{7}$$

where θ is the zenith angle of the slant path, g stands for the species of the molecule, k is the absorption coefficient and ρ is the density. The calculation algorithm for τ_a^M is a slightly modified version of Aoki(1988).

The effect of multiple scattering can be neglected in the near infrared region concerned in this paper, and the path radiance is well calculated with the single scattering approximation(Kneizys et al., 1983):

$$I_{path}(\nu) = I^{\odot}\phi^{\odot}(\nu)\Omega T_{path}(\nu), \tag{8}$$

$$T_{path}(\nu) = \int_0^{z_{obs}} dz \tau_e{}^{A+M}(TP + PO) \sum_{\chi=A,M} P^{\chi}(z_P)\frac{1}{\tau_s{}^{\chi}(PO)}\frac{\partial \tau_s{}^{\chi}(PO)}{\partial z}, \tag{9}$$

where z_{obs} is the height of the observation point, $P^{\chi}(z_P)$ is the scattering phase function at the height z_P and $\tau_s{}^{\chi}(PO)$ is the scattering transmittance of the path from P to O due to the medium χ. In this paper we use the following analytic function for the molecule and aerosol phase functions:

$$P^M = \frac{3}{16\pi}[1 + \cos^2\Theta], \tag{10}$$

$$P^A = \frac{1}{4\pi}\frac{(1-\eta^2)}{(1 - 2\eta\cos\Theta + \eta^2)^{3/2}}, \tag{11}$$

where Θ is the scattering angle and η is a constant.

In the near infrared region the magnitude of I_{path} is the the order of one tenth of I_{refl} over the Lambertian surface whose albedo is 0.05 when there is no absorption. However, in the wavenumber region of strong absorption I_{path} can be comparable to or even more than I_{refl}. Since the distribution of the aerosol and their optical properties may not be well known even if a sensor for the aerosol measurement is on board the same satellite, an accurate estimation of the term I_{path} is difficult. To minimize the uncertainty of the effect of the aerosols it is better to use the sun glint. The intensity of sun glitter radiance that is reflected into a view angle of the sun, which is about 0.5 degree, can be the order of $10^{-3} \sim 10^{-5}$ of the incident radiation(Plass et al., 1977;

Guinn, et al., 1979). This is $10^4 \sim 10^2$ times larger than that over the Lambertian surface. Then in the case of sun glint I_{path} can be neglected compared to I_{refl}, or at a better choice, it may be taken into account based on, e.g., climatic values of aerosol distribution and optical properties.

IV. Sensitivity to the density profile

The advantage of using the absorption method which measures the reflected solar radiation is described in this section by comparison with the emission method which measures the emitted radiance from the atmosphere and surface. To simplify the problem we assume a constant mixing ratio for a gas, whose absorption coefficient is independent of the temperature. Only a single absorption line of a molecule is taken into account, neglecting the wing effect of other lines, path radiance, or other absorption such as that due to aerosols or the solar atmosphere. The voigt shape was used for the line shape. The pressure broadening coefficient of this line was assumed to be 0.1 cm^{-1}/atm and the Doppler width of carbon dioxide was taken as representative. The observation is made from satellite($z_{obs} = \infty$) and we assume $\theta_{sun} = \theta_{refl}$ for simplicity.

Based on these assumption we examine the sensitivity of I_{refl} to the change of the density. From the equations(4), (5) and (6) we have:

$$I_{refl}(\nu) = I^{\odot}\phi^{\odot}(\nu) r_s \Omega \tau_\nu^2(0), \tag{12}$$

$$\tau_\nu(z) = exp\left[-\mu \int_z^\infty k_\nu \rho dz\right]. \tag{13}$$

Rewriting ρ with the initial value ρ° and a parameter x as

$$\rho(z) = \rho^\circ(z)[1 + x(z)] \tag{14}$$

the change of I_{refl} due to the change of ρ by a fraction $x(z)$ is given by

$$\frac{\partial I_{refl}}{\partial x(z)} = -2\mu k(z)\rho^\circ(z) dz I_{refl}. \tag{15}$$

Let us assume the dynamic range of the sensor spans from 0 to $I^{\odot}\phi^{\odot} r_s \Omega$ and define the sensitivities of I_{refl} to the unit fractional change of the density

$$W_{refl}(z, \nu, \tau(0)) \equiv \frac{\partial I_{refl}}{\partial x} \frac{1}{I^{\odot}\phi^{\odot}(\nu) r_s \Omega}$$

$$= -2\mu k(z)\rho^\circ(z) dz \tau^2(0). \tag{16}$$

This quantity was calculated changing the values of z, wavenumber from line center and line intensity.

On the other hand the radiative transfer of the thermal emission in the infrared region is written as

$$I_{emi}(\nu) = B_\nu(T_s)\tau_\nu(0) + \int_0^\infty B_\nu[T(z)]\mu k_\nu(z)\rho(z)\tau_\nu(z)dz, \tag{17}$$

The derivative of I_{emi} becomes:

$$\frac{\partial I_{emi}}{\partial x(z)} = -\mu k(z)\rho^\circ B(T_s)\tau(0)dz + B(T(z))\mu k(z)\rho^\circ(z)\tau(z)dz$$

$$- \mu k(z)\rho^\circ(z)dz \int_0^z B(T(z'))\mu k(z')\rho^\circ(z')\tau(z')dz', \tag{18}$$

where it should be noted that the upper limit of the integral of the third term has been changed since

$$\frac{\partial \tau(z')}{\partial x(z)} = 0 \quad for \quad z' > z. \tag{19}$$

Approximating the integral in (18) by

$$B(\bar{T}) \int_0^z \frac{\partial \tau}{\partial z} dz = B(\bar{T})[\tau(z) - \tau(0)], \tag{20}$$

we have following the equation for the sensitivity of the emission method:

$$W_{emi}(z, \nu, \tau(0)) \equiv \frac{\partial I_{emi}}{\partial x} \frac{1}{B(Ts)}$$

$$= -\mu k\rho^\circ dz \left[\tau(0)\frac{B(T_s) - B(\bar{T})}{B(T_s)} + \tau(z)\frac{B(\bar{T}) - B(T(z))}{B(T_s)}\right], \tag{21}$$

where $\bar{T}$ is the mean temperature of the layer from the surface to the height z. We assumed the dynamic range of the sensor to be from 0 to $B(T_s)$ and normalized the derivative of I_{emi} by $B(T_s)$ as before.

Now consider the remote measurement of the molecular density at $z = 1km$. The mean temperature $\bar{T}$ may be smaller than T_s by about 3 K and larger than $T(z)$ by the same amount, so that the order of factors in (21) are:

$$\frac{B(T_s) - B(\bar{T})}{B(T_s)} \simeq \frac{B(\bar{T}) - B(T(z))}{B(T_s)} \simeq 4 \times 10^{-2},$$

$$for \quad z \simeq 1\,km. \tag{22}$$

For sounding the lower troposphere the transmittance must be sufficiently high that the factors $2\mu k\rho^\circ\tau^2(0)$ in (16) and $\mu k\rho^\circ\tau(0)$ or $\mu k\rho^\circ\tau(z)$ in (22) are of the same order. Thus the sensitivity of the absorption method is higher than that of the emission method by more than one order of magnitude. For the higher altitudes W_{refl} becomes small because the density ρ decreases. If a larger value of k is used to compensate this decrease in ρ the value of $\tau(0)$ decreases and W_{refl} again becomes small. For the emission method, however, W_{emi} increases due to the second term in the bracket of (21). More detailed behavior of the dependence of the sensitivities on the distance from the line center and line strength are shown in Figs.4-5.

In these figures the abscissa is the distance from the line center and the ordinate is the optical depth of the atmospheric column at the line center rather than the line strength. Figs.4a and b show the sensitivity of the radiances to the unit change of the density within a 1km layer thickness(dz=1km) for the absorption method and emission method, respectively, at z= 0.4km. The sensitivity of the absorption method is about two order higher than that of the emission method. For the higher altitudes the former gradually decreases whereas the latter increases as shown in Figs.5a and b. It should be noted that in addition to the low sensitivity of the emission method at lower altitudes the accuracy of this method strongly relies on that of the vertical temperature profile which may also be measured from satellite. The equation(21) suggests that the error of 1K in estimated temperature profile will result in about 60 percent of error in sensitivity function when $\bar{T}$ - $T(z)$ and T_s - $\bar{T}$ are 3K.

V. Retrieval algorithm for the vertical density profiles

We define a nondimensional quantity for the i-th sampling by

$$F_i = \frac{V_i}{\bar{V}}. \tag{23}$$

For the normalization factor, $\bar{V}$, we take an average of some measurements:

$$\bar{V} = \frac{1}{M'}\sum_{i=1}^{M'} V_i, \tag{24}$$

$$1 \leq M' \leq M, \tag{25}$$

where M is the total number of the samples. (In the present work $M' = M$.) Using (1) this can be rewritten as:

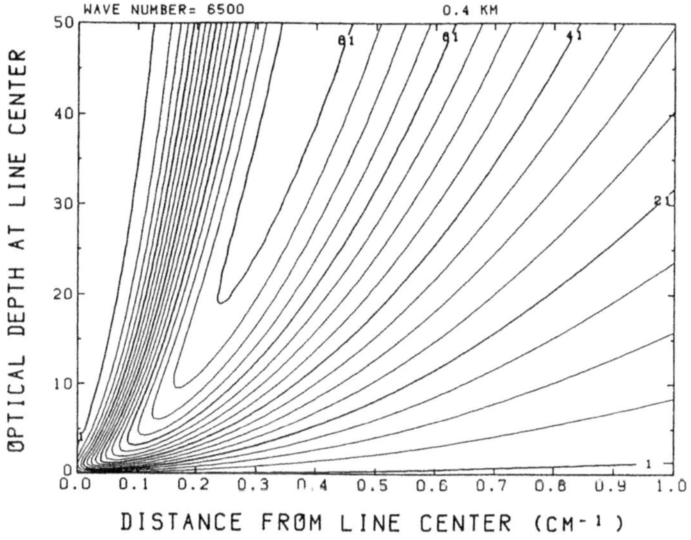

Fig.4a The sensitivity of the absorption method to the density at height 0.4km.

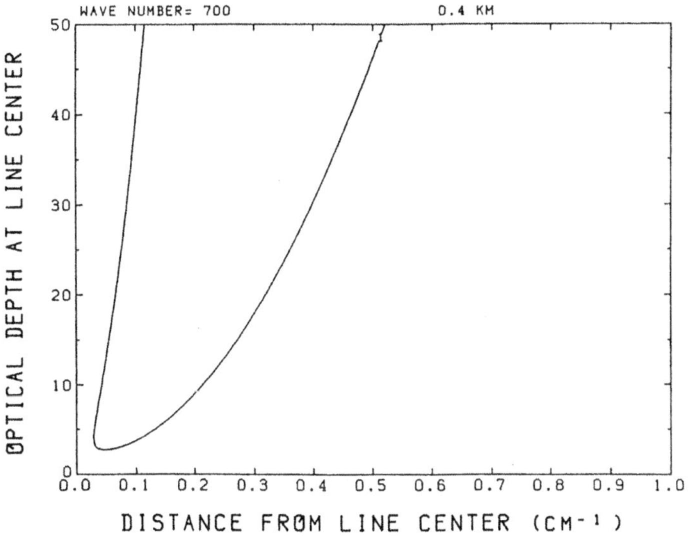

Fig.4b As in Fig.4a except for the emission method.

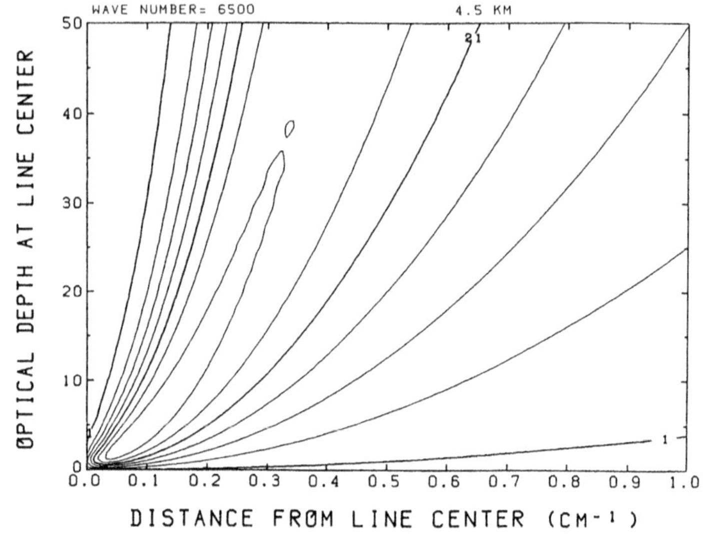

Fig.5a As in Fig.4a except for the height 4.5km.

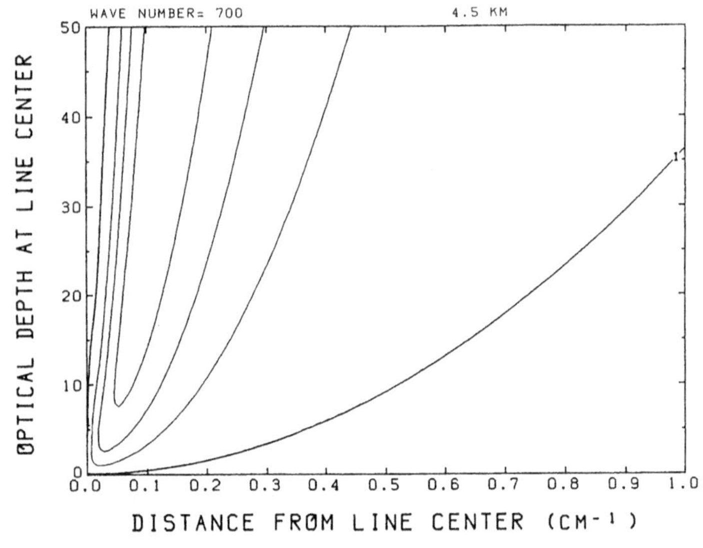

Fig.5b As in Fig.4b except for the height 4.5km.

$$F_i = \frac{R_i}{\bar{R}}, \tag{26}$$

$$\bar{R} = \frac{1}{M'}\sum_{i=1}^{M'} R_i, \tag{27}$$

where it has been defined

$$R_i = \int_{\nu_1}^{\nu_2} d\nu\phi_{FP}(\nu, t_i)\phi'_B(\nu)\left[T_{refl}(\nu) + \frac{1}{r_s\tau_e^A(TG+GO)}T_{path}(\nu)\right], \tag{28}$$

$$\phi'_B(\nu) = \phi_B(\nu)\phi^{\odot}(\nu). \tag{29}$$

The density profiles are iteratively determined by approaching the calculated quantity of (26), to the observed one in (23). It should be noted that F_i in (23) does not contain the constant C so that the absolute measurement of the radiance is not required in this algorithm. When we use the sun glitter it may be possible to neglect the second term in the bracket of (28). This means that we can perform the retrieval without the knowledge of the values of the surface reflectance r_s, absolute solar radiation $I^{\odot}$ and transmission function of the aerosol $\tau_e^A(TG+GO)$. Thus the advantage of the sun glitter measurements is not only to retain the high signal-to-noise-ratio but also to avoid the difficulty of the absolute measurement of the radiance, surface reflectance and aerosol.

By expanding F_i around an initial guess we obtain

$$F_i(\delta_i) = F_i^{\circ}(\delta_i) + \sum_{j=1}^{N_0}\frac{\partial F_i}{\partial x_j}x_j, \tag{30}$$

$$y_i = F_{i,obs} - F_i^{\circ} = \sum_m K_{im}x_m, \tag{31}$$

$$K_{im} = \frac{\partial F_i}{\partial x_m}, \tag{32}$$

where F° is the F theoretically calculated under the condition of the initial guess. x is the unknown parameter to be determined such as the deviation of the density, temperature or surface pressure from their initial guess values. The unknown parameters are iteratively determined with the use of the maximum likelihood solution(Rodgers, 1976):

$$X_{(n+1)} = [S_x{}^{-1} + K_{(n)}{}^t S_\epsilon{}^{-1}K_{(n)}]^{-1}K_{(n)}{}^t S_\epsilon{}^{-1}[Y - Y_{(n)} + K_{(n)}X_{(n)}], \tag{33}$$

where $X_{(n)}$ is the vector of x at the n-th step of the iteration, S_x is the covariance matrix of X, $K_{(n)}$ is the K appeared in (32) at the n-th step and S_ϵ is the covariance matrix of the measurement error.

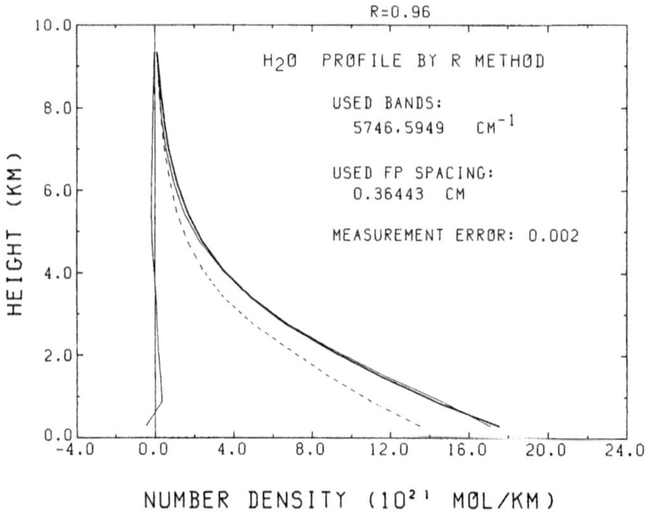

Fig.6a Comparison between the retrieved density profile of H_2O(thin solid line) and the true one(thick solid line). The dashed line is the initial profile. The profile of the lefthand side shows the error.

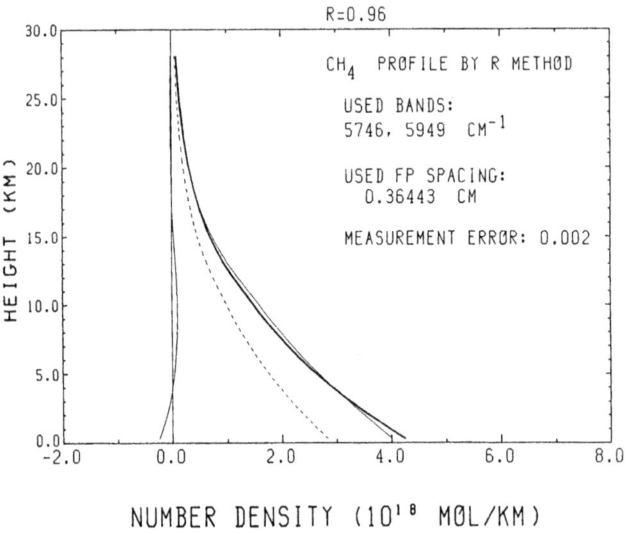

Fig.6b As in Fig.6a except for CH_4.

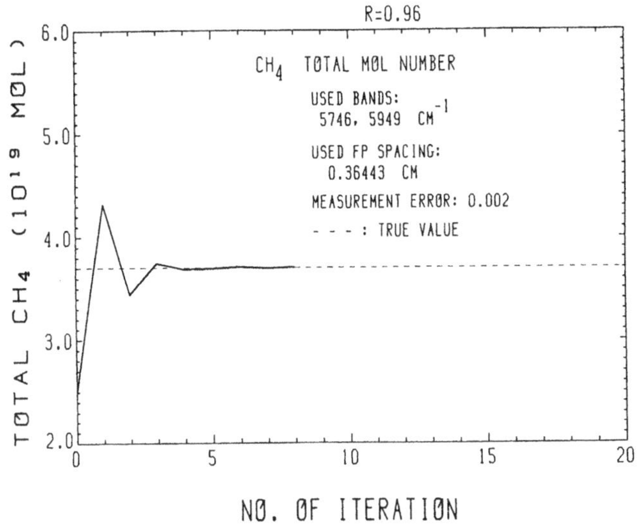

Fig.7 Convergence of the total methane amount with the iteration.

VI. Simulation of the retrieval

The outputs represented in (1) were calculated for M samples(i.e., M of refractive indices) and random errors, whose RMS value is 0.002 of the average of these values, are added to them. Both the matrixes of S_x and S_ϵ in (33) were assumed diagonal. The iteration was continued until the fractional changes of the density from the value of the last iteration become less than 0.01 at all levels of altitude. Examples of the result of simulation for H_2O and CH_4 are shown in Fig.6a and b, respectively. It can be seen that the errors in the lower troposphere are about a few percent. The total column amount is generally retrieved in better accuracy as shown in Fig.7.

VII. Summary

The surface reflected solar radiation in the near infrared region has two times higher sensitivity to trace gas density in the lower atmosphere than the far infrared thermal emission. The former is far less sensitive to the atmospheric temperature profile than the latter. These two advantages of the reflected solar radiation provide us high accuracy measurements of tropospheric chemical and physical parameters. TERSE is a radiometer to measure the surface reflected solar radiation in the near infrared region with high spectral resolution.

The scientific challenges of the TERSE are:

1. Measurements of H_2O, CO_2, N_2O, CH_4, CO, HDO, O_2.
2. Spacial distribution in 40 km mesh.

3. Vertical distribution with an accuracy of a few percent in the lower troposphere and 10 percent at the higher levels.
4. Total absorber amount with an accuracy of a few percent.
5. Measurement of water vapor and methane above the cloud top level.
6. Surface and cloud top pressure with an error less than 5hPa.
7. HDO/H_2O ratio for the global water cycle study.

The characteristics of the TERSE is briefly summarized as follows:

1. Sun glint measurements in the 1.2 - 2.5 μm region
2. Pressure tuned etalon + etalon tuned narrow band pass filter
3. Swath: 500 km
4. Instantaneous field of view: 0.2 degree
5. Time duration: 0.4 sec/point
6. No calibration (no absolute measurement of radiance)
7. Weight: 100 kg
8. Power consumption: TBD

References

Aoki T (1988) Development of a line-by-line model for the infrared radiative transfer in the earth's atmosphere. Papers in Meteorology and Geophysics 39:53-58.

Aoki T (1989a) A remote sounder for the vertical profiles of the atmospheric gases: Simulation. Proc. of Spring Meeting of Japan Meteorological Society, May 22-24, Tokyo. (in Japanese)

Aoki T (1989b) Satellite remote sensing for the vertical profiles of the atmospheric trace gases with using near infrared spectrometer. Proc. of Remote Sensing Symposium of the Society of Instrument and Control Engineers, October 13-14, Tokyo. (in Japanese)

Kneizys FX, Shettle FP, Gallery WO, Chetwynd Jr JH, Abreu LW, Selby JEA, Clough SA, Fenn RW (1983) Atmospheric transmittance/radiance: Computer code LOWTRAN 5, AFGL-TR-83-0187, Air Force Geophysics Laboratory, 1 August.

Guinn JA Jr. Plass GN, Kattawar GW (1979) Sunglint glitter on a wind-ruffled sea: further studies. Applied Optics 18:842-849.

Plass GN, Kattawar GW, Guinn JA Jr (1977) Isophotes of sunglint glitter on a wind-ruffled sea. Applied Optics 16:643-653.

Reichle HG Jr, Beck SM, Haynes RE, Hesketh WD, Holland JA, Hypes WD, Orr III HD, Sherrill RT, Wallio HA, Casas JC, Saylor MS, Gormsen BB (1982) Carbon monoxide measurements in the troposphere. Science 218:1024-1026.

Rodgers CD (1976) Retrieval of atmospheric temperature and composition from remote measurements of thermal radiation. Review of Geophysics and Space Physics 14:609-624.

Timchak A(1986) Satellite-derived moisture profiles. NOAA Tech. Rep. NESDIS 24.

HIGH-RESOLUTION OBSERVATION OF THE EARTH'S ATMOSPHERE IN THE FAR INFRARED

Bruno Carli
Istituto di Ricerca sulle Onde Elettromagnetiche
via Panciatichi, 64
50127 Firenze
Italy

Abstract
Limb-sounding high-resolution emission spectroscopy in the far-infrared can be used for measuring the composition of the Earth's stratosphere. The results obtained with balloon-borne instruments are reviewed; they give proof that this relatively unexploited spectral region provides valuable new measurements with reliable accuracy. The measurement capabilities that could possibly be provided in the future by space-borne instruments are briefly evaluated.

1 - The characteristics of far-infrared atmospheric sounding

Far-infrared atmospheric observations are typically made in the 30 - 300 cm^{-1} spectral region. At these frequencies, heterodyne measurement may be possible in the future; but only photometric measurements have been made up to now and this paper will consider only this case.

Remote sensing of the Earth's atmosphere in the far-infrared spectral region has interesting characteristics which make this type of observation scientifically very informative, although it may pose some technical challenges.

NATO ASI Series, Vol. I 9
High Spectral Resolution Infrared Remote Sensing
for Earth's Weather and Climate Studies
Edited by A. Chedin, M.T. Chahine and N.A. Scott

A first general characteristic is the use of long wavelength radiation which, correspondingly, requires large optical components and instruments. This difficulty is counterbalanced by the fact that the mechanical and optical tolerances allowed in the construction of the instrument are also proportional to the wavelength. Furthermore, long wavelengths are less affected by the scattering of stratospheric aerosols and can be used also in presence of dense clouds such as the one recently released by the volcanic eruption of Pinatubo.

A second important characteristic is the fact that, in the far infrared, the Earth's atmosphere emits as a black-body and its emission is in the Rayleigh-Jeans region of the Plank distribution. Observation of the atmosphere in emission provides unconstrained measurements in both direction and time so that the best geographical and diurnal coverage may be obtained. Furthermore, the Rayleigh-Jeans emission is linearly dependent on temperature and the spectral intensity is not critically at variance with this parameter. The available signal is less than that at higher frequencies and for solar absorption measurements, but the use of cryogenically-cooled detectors ensures adequate sensitivity.

Finally, a third characteristic is the presence of the pure rotational spectra of most molecules. Differently from vibrational spectra, which tend to display well-separated bands and require a broad spectral coverage for the simultaneous observation of different species, rotational spectra are on the average more evenly distributed and occur in a spectral range which is smaller than that of vibrational spectra, even when measured on a logarithmic scale. The density and ubiquitousness of the rotational features make it possible to simultaneously observe many atmospheric species with a single instrument operating in a relatively small spectral interval; at the same time, however, high spectral resolution is required in order to limit the spectral blending that may be caused by the instrumental function. Furthermore, the pure rotational features are often due to a single transition within the ground state and can be described with a very simple spectroscopic model.

2 - The atmospheric spectrum

In the far infrared, sounding of the atmosphere is possible only at high altitudes, preferably above the tropopause where, because of the reduced water concentration, many transparent spectral windows can be observed among strong but isolated water-absorption features. In these windows, measurements can also be made with an optical path as long as the one used for limb sounding observations, so that it is possible to detect features of minor atmospheric constituents.

Measured spectra show that the most prominent features, irregularly scattered over the spectrum, are due to strongly-saturated rotational water lines, as well as to Q-type branches belonging to the rotational spectrum of ozone in its ground state, and to molecular oxygen lines.

Because of their dominant presence in the spectrum, the three molecular species H_2O, O_3, O_2 can be considered to be the "main spectroscopic constituents" of the far-infrared stratospheric spectrum, in contrast to the other molecular species which, because of either their isolated or their weak features, can be considered as "minor spectroscopic constituents". This classification is based only on the consistency of the spectral features, and leaves out of consideration the actual quantity of each constituent in the atmosphere.

An example is given in Fig. 1, where the spectrum of the atmospheric emission in the 61-63 cm^{-1} interval recorded with a balloon-borne spectrometer at 38 km altitude is shown. The broad water-line, centered at about 62.3 cm^{-1}, is well visible; the deep spike penetrating the feature in the center is an

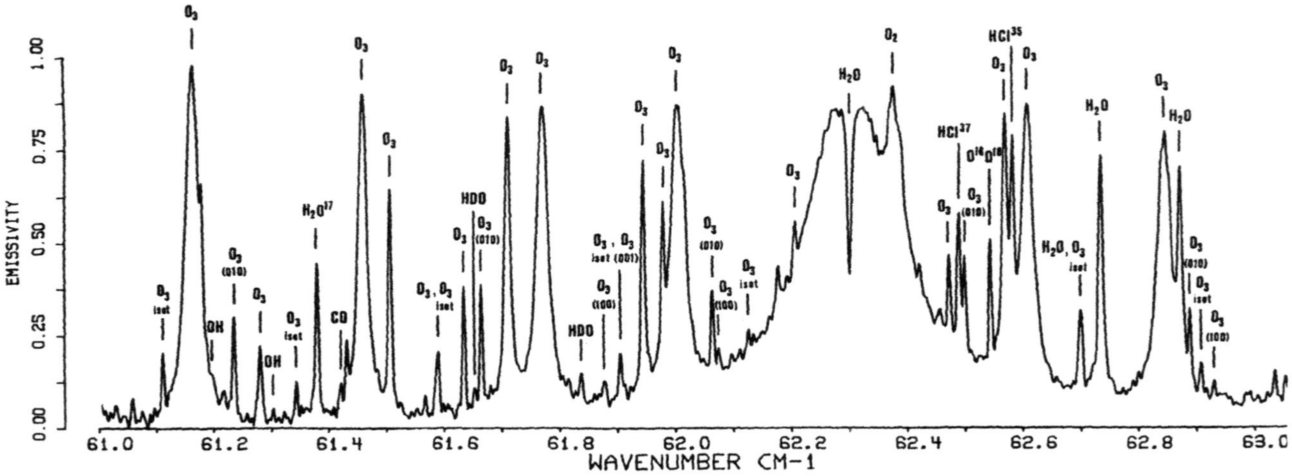

Figure 1. The atmospheric emissivity in the 61 - 63 cm^{-1} spectral interval measured from an altitude of 38 km, at a zenith angle of $93,2^{o}$, with an unapodized spectral resolution of 0.0033 cm^{-1}

instrumental effect caused by absorption of the water vapor present inside the instrument. One of the molecular-oxygen triplet components emerges on the high-frequency slope of the feature. Several resolved ground-state ozone transitions are also clearly visible.

If the spectral resolution of the measurement is sufficiently high, as in the case of the example shown in Fig. 1, it is possible to observe several signals due to the minor spectroscopic constituents among the features of the main spectroscopic constituents.

The isotopic abundance of the atoms involved in the three main spectroscopic constituents are sufficient to provide several molecular combinations that are detectable in the far-infrared spectrum. These are: $^{16}O^{16}O^{18}O$, $^{16}O^{18}O^{16}O$, $^{16}O^{16}O^{17}O$, HDO, $H_2{}^{18}O$, $H_2{}^{17}O$, $^{16}O^{18}O$.

Furthermore, the population of the low-lying, vibrationally-excited states of the main spectroscopic constituents is sufficient in some cases to produce detectable pure rotational transitions. Rotational transition of isotopic species of water vapor and ozone, as well as pure rotational lines of vibrationally-excited ozone, can be seen in Fig. 1.

Other molecular species that have been positively observed in the far infrared are HF, HCl, HCN, HNO_3, N_2O, CO, OH, HOCl, H_2O_2 and HO_2 (see, for instance, Carli [1984]). Measurements which are possible in the future, but which require an improvement in the signal-to-noise ratio, are ClO and HBr.

Fig. 1 provides an example of features due to OH and HCl. The two weak features of OH that are observed at about 61.2 and 61.3 cm^{-1} are the first of a sequence of doublets that this molecule displays. These weak features can be compared to the

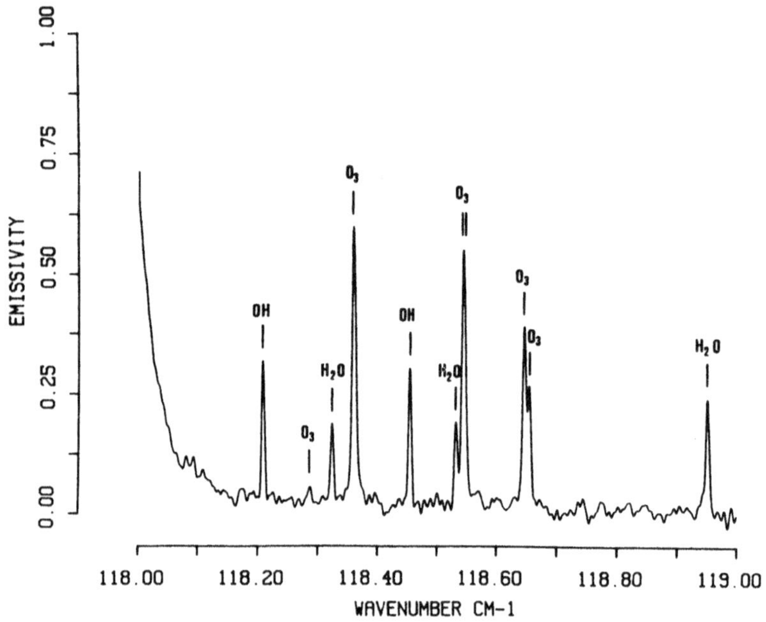

Figure 2. The atmospheric emissivity around 118 cm^{-1}, where two strong and well-resolved OH transitions are observed. All other observational conditions are identical to tose of Fig. 1

strong features that are shown in Fig.2 at about 118.2 and 118.4 cm^{-1}. In Fig. 1 HCl is present with two features due to the two chlorine isotopes. An example of some of the other HCl transitions that can be observed in the far-infrared are shown in spectra at different limb angles in Fig. 3.

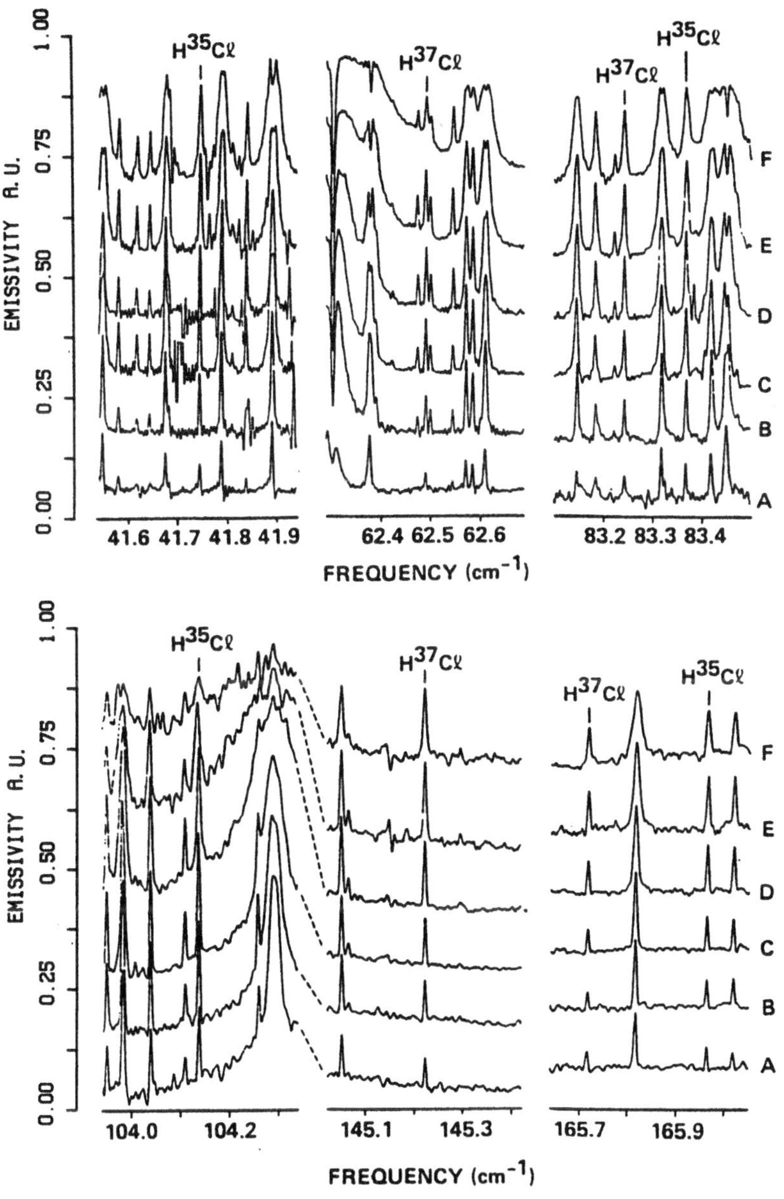

Figure 3. The far-infrared emission spectrum at different zenith angles in those spectral intervals that contain HCl transitions. For the sake of clarity,an offset is introduced into the emissivity scale

3 - The spectrometer

High-resolution far-infrared observations of the Earth's atmosphere have been made with two balloon-borne Fourier transform spectrometers developed, respectively, at the "Smithsonian Astrophysical Observatory" (SAO) and at the "Istituto di Ricerca sulle Onde Elettromagnetiche" (IROE). The IROE instrument [Carli et al., 1984] is a Michelson interferometer equipped with polarizing optics (Martin-Puplett interferometer [Martin and Puplett, 1970]), which operates in the 7 - 180 cm^{-1} spectral region with an unapodized resolution of 0.0033 cm^{-1}. Two detectors are used simultaneously at the two outputs of the spectrometer in order to observe complementary spectral regions. Different combinations of detectors were used on different flights. Photoconductors cooled by a liquid helium bath at 4.2 K were used at frequencies higher that 90 cm^{-1}. Bolometers cooled either at 1.6 K with pumped liquid helium, or at 0.4 K with the isotope 3 of helium, were used at lower frequencies.

The spectrometer observes the atmosphere from an altitude of about 40 km at a sequence of different zenith angles in the limb direction (see, for instance, Fig. 3). Each observed spectrum is therefore characterized by the direction of observation (measured with either the zenith angle or the limb angle) or, equivalently, by the tangent altitude, that is, the minimum altitude from the sea level reached by the optical path. Limb scanning is used to retrieve the vertical distribution of the concentration of the observed species (concentration profile). A field of view of 1.5^{o} was used in some early observations that gave priority to the spectroscopic survey; a narrower field of view of 0.35^{o} is used in more recent observations for an improved vertical resolution of the retrieved profiles.

The pointing direction is controlled with an attitude system that combines the signals of an accelerometer and a gyroscope. Accuracy of the order of a few minutes of arc is typically required for the zenith angle, in order to determine with sufficient accuracy the tangent altitude of the limb-sounding observation.

Three balloon flights of this instrument, respectively in 1979, 1982 and 1983, provided the results that are reviewed in this paper.

4 - Results

4.1 - Understanding the observed spectra

The possibility of exploiting spectroscopic measurement as a remote-sensing technique depends on our ability to describe the observation with a correct model. The model necessary for describing far-infrared emission spectra is relatively simple. As we have already noted in Sect. 1, the radiative emission of the stratosphere occurs in the "well-behaved" Rayleigh-Jeans region and pure rotational spectroscopy is characterized by single transitions within the ground state. Furthermore, the relatively long mean lifetime of rotational levels makes the hypothesis of local thermodynamic equilibrium (LTE) valid at most altitudes. Scattering and diffraction corrections are negligible.

The simplicity of the physics involved makes the development of a correct model easier and reduces the possible sources of systematic errors.

Synthetic spectra calculations based on a line-by-line model of radiative transfer along the limb path reproduce the observed spectra with good accuracy. Fig. 4 provides a comparison between the observed spectrum (dots) at a tangent altitude of 35 km and a fitted synthetic spectrum (continuous line) that includes a simulation of the instrumental function.

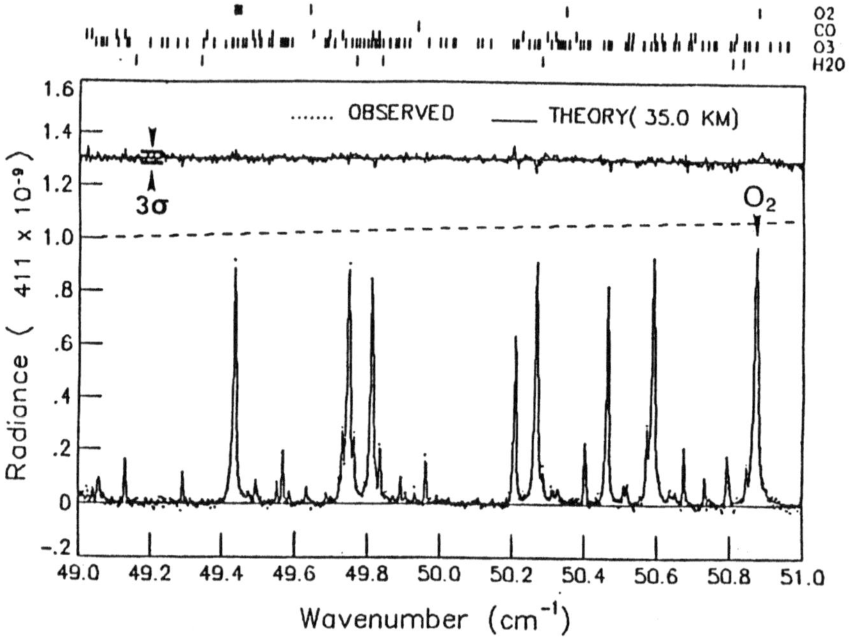

Figure 4. Comparison between observed spectrum (dots) and calculated synthetic spectrum (continuous line)

The difference between the two spectra is plotted at the top of the figure and is compared with the standard deviation of measurement noise. The agreement between observation and theory is good and indicates the correctness of the model.

The fit of spectra at tangent altitudes lower than 35 km tends to present larger residuals. This is caused by the insufficient accuracy with which some pressure-broadening parameters are known. Laboratory measurements of these parameters are highly desirable. As the accuracy of the field measurements improves the accuracy of some other spectroscopic parameters, such as line strength, will also require some improvement (see also Sect. 4.3).

4.2 - Retrieved profiles

The concentration profile of the atmospheric species is retrieved from the observed limb-scanning sequence using a non-linear fit of the observations with synthetic spectra. More than one procedure has been developed for this analysis [e.g. Abbas et al, 1984; Park and Carli, 1986]. The most recent one is the global-fit method [Carlotti, 1988], which simultaneously analyzes all the available observations to retrieve the whole profile, as well as some instrumental and observational parameters. In this method, the fact that an observation depends not only on the concentration at the tangent altitude, but also on the concentration at higher altitudes, is taken into account for the retrieval at all relevant altitudes, and is not the cause of the "feedback" errors typical of the more classic "onion-peeling" method (for a fuller discussion, see Carlotti [1988]).

Figure 5 provides an example of the results in the case of

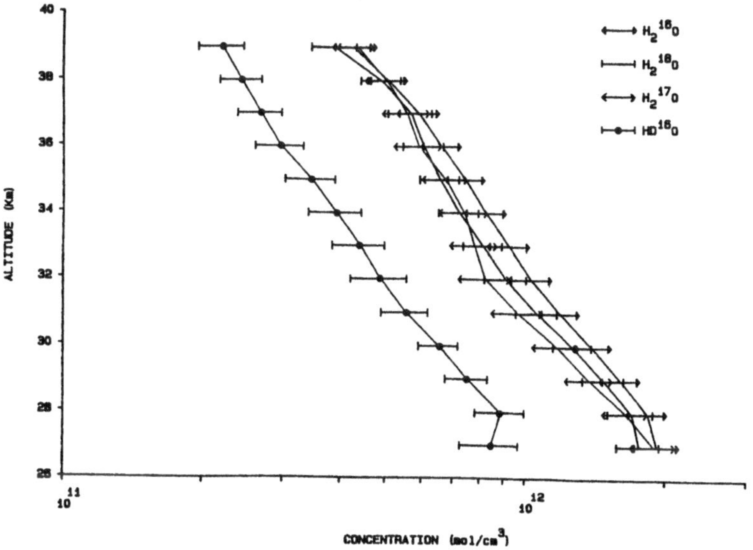

Figure 5. Concentration profiles of water vapor and its isotopes retrieved with the global-fit method

water vapor and its isotopes [Dinelli et al. 1991]. Similar profiles have been retrieved in the case of all the species that have been spectroscopically identified in the atmospheric spectra (see Sect 2), exceptions being ClO and HBr, for which only an average concentration has been obtained, and HNO_3 and N_2O, which have not been analyzed because better measurements are expected to be provided by middle-infrared observations.

For a discussion of the individual profiles, see Carli and Park [1988], Carli et al. [1988 and 1989], and Carlotti et al. [1989].

4.3 - Error budget

The broad-band spectra obtained with photometric techniques make possible the simultaneous observation of several constituents, as well as of more than one feature for each constituent. Fig. 3 shows an example in the case of HCl; other typical species with ubiquitous features are water vapor and ozone. The availability of more than one feature provides an important tool for assessing the internal consistency of the measurements and for performing statistical analyses.

Fig. 6 shows the results obtained from the analysis of a set of ozone lines in a set of limb-scanning sequences [Carlotti et al, 1989]. Each line in the figure is a profile retrieved from one spectral interval and one sequence. The statistical distribution of the values makes it possible to identify the source and the entity of the errors. In the final error budget, the error of the measured spectra turns out to be a relatively small component. Systematic errors due to uncertainties in the spectroscopic parameters make a significant contribution, but the major source of error in this case lies in the uncertainties in the observation geometry (flight altitude and limb scanning angle).

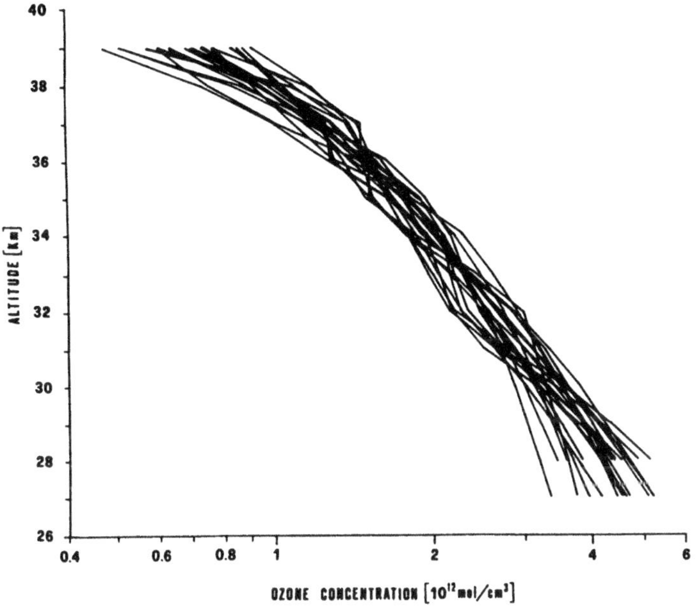

Figure 6. Set of ozone concentration profiles obtained with the same balloon flight from different limb-scanning sequences and different spectroscopic features

4.4 - Diurnal variability

An important species that has been successfully observed in the far infrared is the hydroxyl radical OH [Carli et al. 1985 and 1989]. This radical interferes with all three of the main catalytic cycles that lead to the depletion of ozone: the hydrogen cycle, the chlorine cycle and the nitrogen cycle. In the atmosphere, the hydroxyl radical shows a marked diurnal variation; hence, emission measurements which are possible with no time constraint are ideal for its detection in the atmosphere. The far infrared spectral region provides the only opportunity for passive remote sensing of this molecule in emission.

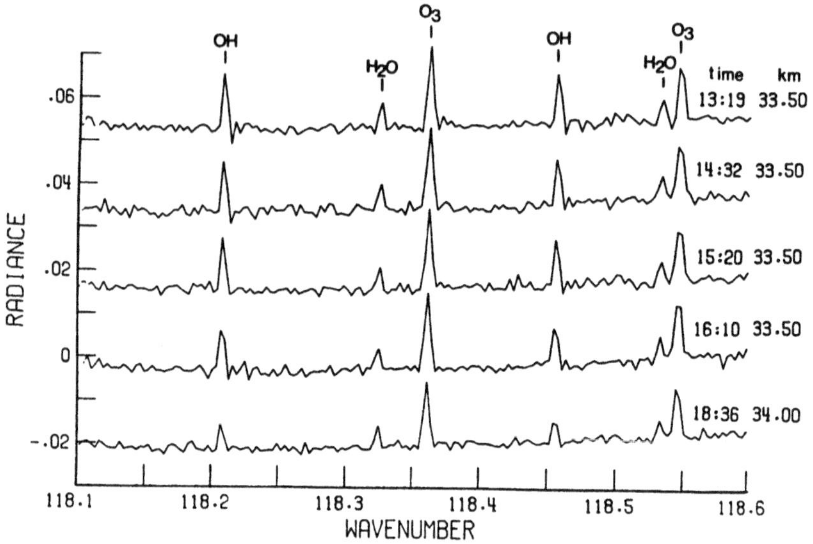

Figure 7. A sequence of spectra at almost constant tangent altitude, taken at different times of the day

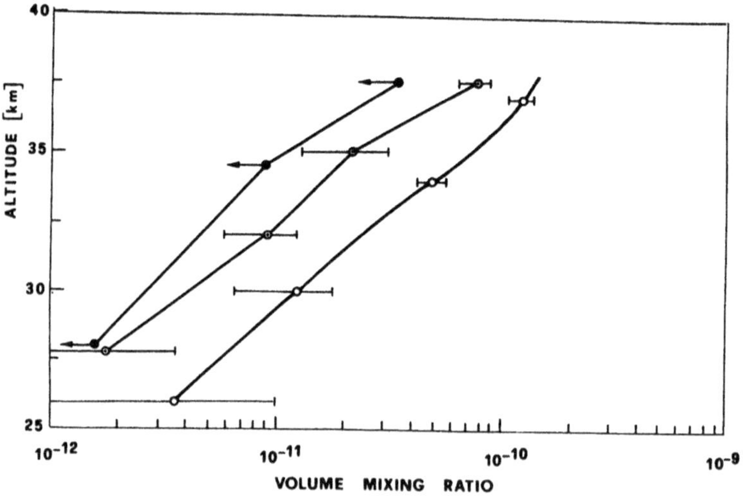

Figure 8. Mixing ratio profile of OH measured in the early afternoon (open circle), at sunset (circle with dot) and after sunset (full circle)

Fig. 7 shows the emission spectrum of the atmosphere at an almost constant tangent altitude of 33.5 km, at different times of the day. The features due to ozone and water vapor show no variability. A systematic decrease is observed in the intensity of the two OH features. Consequently, the same type of variability is observed also in the profiles that are retrieved. Figure 8 shows the two profiles that were measured in the early afternoon and at sunset, as well as an upper limit obtained after sunset.

5 - Plans for space-borne observations

The scientific importance of the measurements that are possible in the far infrared and the maturity of both instrumentation and data interpretation make it desirable to perform these observations also from space platforms.

A first opportunity to use far-infrared spectroscopy for the study of the Earth's atmosphere may be provided by the SAFIRE (Spectroscopy of the Atmosphere using Far-InfraRed Emission) experiment, which is scheduled to fly as part of the EOS (Earth Observing System) program.

SAFIRE combines two instruments that operate in emission and observe the atmosphere in two different spectral regions: the far infrared and the middle infrared.

The SAFIRE far-infrared instrument is an FTS based on the experience acquired with the balloon experiments. It implements a folding of the optical path in order to minimize the size of the instrument and to obtain a better indemnity from alignment errors [Carli, 1991]. The cooling requirements of the far-infrared detectors are met by a cryogenic system which exploits the passive cooling of liquid helium for low temperatures and the active cooling of a double-stage mechanical cooler for extended lifetime. Only photoconductors are used: these detectors have a low frequency cut-off at about 80 cm^{-1}, but

Table I - SAFIRE far-infrared channels

Channel	Spectral interval (cm-1)	Primary gases	Other gases
1	82 - 84.4	O_3, OH, HCl	H_2O, O_2, $^{50}O_3$
2	93.8 - 96	H_2O_2, HO_2	O_3, O_3, HDO
3	98.5 - 100	HOCl	H_2O, O_3, O_3, HBr
4	111 - 112.6	H_2O_2	H_2O, O_3
5	117.8 -119.6	OH	H_2O, O_3, O_2, HBr
6	157 - 159 310 - 390	H_2O, $O(^3P)$ N_2O_5	

their relatively higher operating temperature ensures a longer life time of the cryogenic system (a life of five-year is planned at present).

In order to exploit the low intrinsic noise of the detectors, photon noise is reduced by limiting the radiation flux from the warm environment. This is achieved with a cold pupil and a cold narrow-band filter in front of the detector. No cooling of the spectrometer is required. The capacity to observe simultaneously several atmospheric species is preserved by the simultaneous use of several detectors with pass-band filters centered at different selected frequencies. A list of the selected spectral intervals and of the species that can be observed in these intervals is provided in Table I.

6 - Conclusions

High resolution far infrared spectroscopy of the Earth's atmosphere has proved to be a valuable technique for the passive remote sensing of atmospheric composition. Balloon-borne Fourier transform spectrometers have provided the accurate and well-understood measurement of many species such as ozone and its isotopes, water and its isotopes, HF, HCl, HCN, CO, OH, HOCl, H_2O_2 and HO_2. The technique is mature for the development of measurements from space platforms for continuous observations and a comprehensive geographical coverage.

REFERENCES

Abbas MM, Jing Guo, I Nolt, B Carli (1984) Far infrared remote sounding of stratospheric temperature and trace gas distibution. J Atm Chem 2:145-162

Carli B (1984) The high resolution submillimetre spectrum of the stratosphere. J Quan Spactrosc Radiat Transfer 32:397-405

Carli B, F Mencaraglia, A Bonetti (1984) Submillimeter high-resolution FT spectrometer for atmospheric studies. Appl Opt 23:2594-2603

Carli B, F Mencaraglia, A Bonetti, M Carlotti, I Nolt (1985) Detection of atomic oxygen and further line assignments in the far-infrared stratospheric spectrum. Int J of Infrared and Millimeter Waves 6:149-176

Carli B, JH Park (1988) Simultaneous measurement of minor stratospheric constituents with emission far infrared spectroscopy. J Geophys Res 93 D4:3851-38

Carli B, F Mencaraglia, M Carlotti, BM Dinelli, I Nolt (1988) Submillimeter measurement of stratospheric chlorine monoxide. J Geophys Res 93 D6:7063-7068

Carli B, M Carlotti, BM Dinelli, F Mencaraglia, JH Park (1989) The mixing ratio of stratospheric hydroxyl radical from far-infrared emission measurements. J Geophys Res 94:11049-11058

Carli B (1991) High-resolution far-infrared FT spectroscopy of the stratosphere: Optimization of the optical design of the instrument. SPIE 1145:93-98

Carlotti M (1988) Global-fit approach to the analysis of limb-scanning atmospheric measurements. Appl Opt 27:3250-3254

Carlotti M, A Barbis, B Carli (1989) Statospheric ozone vertical distribution from far-infrared balloon spectra and statistical analysis of the errors. J Geophys Res 94:16,365-16,372

Dinelli BM, B Carli, M Carlotti (1991) Measurement of stratospheric distribution of $H_2{}^{16}O$, $H_2{}^{18}O$, $H_2{}^{17}O$, and HDO from far-IR spectra. J Geophys Res 96:7509-7514

Martin DH, EF Puplett (1970) Polarized interferometric spectrometry for the millimeter and submillimeter spectrum. Infrared Phys 10:105-109 (1970)

Park JH, B Carli (1986) Analysis of far-infrared emission Fourier spectra. Appl Opt 25:3490-3501

REMOTE SENSING OF ATMOSPHERIC TRACE CONSTITUENTS USING MID-IR FOURIER TRANSFORM SPECTROMETRY

H. Fischer
Institut für Meteorologie und Klimaforschung
Kernforschungszentrum/Universität Karlsruhe
P.O. Box 3640
7500 Karlsruhe 1
Germany

Introduction

The impact of industrial and agricultural human activities on the environment has now reached such proportions that even global scale effects are clearly observed. These activities lead to a steady growth in the atmospheric abundance of several radiatively and chemically active trace gases which influence the greenhouse effect as well as tropospheric and stratospheric ozone.

An understanding of ozone changes demands the global measurements of many trace species which are chemically coupled with ozone. Measurements of radiatively important trace gases with relatively short atmospheric life times, as e.g. H_2O and O_3, are also needed to study the increasing greenhouse effect.

These scientific objectives define the necessary instrument performance requirements to achieve meaningful atmospheric measurements. First, measurements have to be carried out simultaneously in both time and space and complementary , i.e. all relevant parameters are measured together. Furthermore, the observations must be global in domain and performed during day and night for several years. The derived vertical profiles of trace species should have a reasonable vertical resolution in order to be able to investigate the various atmospheric processes. In the following, appropriate instruments for trace gas measurements will be described. Also, previous experiments performed by different scientific groups (overview see Fischer 1992) will be discussed briefly. The MIPAS (Michelson Interferometer for Passive Atmospheric Sounding) experiments and results derived from corresponding measurements will be the main part of this article.

Appropriate instrumentation

Many atmospheric trace species exhibit rovibrational absorption bands in the infrared (IR). In order to measure a large number of these trace gases simultaneously with redundant information, the required instrument has to detect a wide spectral range in the infrared with high spectral resolution.

NATO ASI Series, Vol. I 9
High Spectral Resolution Infrared Remote Sensing
for Earth's Weather and Climate Studies
Edited by A. Chedin, M.T. Chahine and N.A. Scott

Obviously, Fourier Transform Spectrometers (FTS) are especially appropriate for this measurement task. Due to technical reasons the spectral range has to be limited to a certain degree, e.g. to the mid-infrared. Time-independent measurements can be performed with a cooled IR-spectrometer operating in the emission mode. Reasonable vertical resolution can be achieved by limb sounding. Last, but not least, global measurements can be performed if the instrument is flown aboard a polar orbiting satellite. Consequently, the basic requirements of trace constituent measurements can be met qualitatively by IR-Fourier Transform Spectrometers (FTIR).

FTIR instruments are only commercially available for the laboratory environments. For detecting atmospheric emission spectra cooled FTIR are needed with demanding specifications. Using these instruments aboard balloon gondolas or aircrafts increases the technical requirements substantially. Balloon experiments and more recently also aircraft experiments with cooled interferometers (Blom, Fischer et al. 1993) have already been used in field campaigns. Mostly, these interferometers are based on commercially available instruments which have been modified to some degree.

A definition study for a cooled FTIR with low power consumption, compact size and short scantime was carried out in cooperation with industry in the beginning of the 1980's in Southern Germany (Burkert et al. 1983). This investigation led to the so-called Double Pendulum Interferometer (DPI) concept which has been applied for different MIPAS (Michelson Interferometer for Passive Atmospheric Sounding) experiments (e.g. Fischer et al. 1983) as the central part of complex measurement systems.

Technical limitations with respect to size of the instrument and measurement time make it advisable to use a spectral resolution as low as possible, but sufficient to avoid blending of spectral lines. These constraints lead to a requirement that depends very much on the spectral region and on the specific objectives of the measurement. A feasibility study for the detection of minor constituents in the middle atmosphere by limb emission spectroscopy performed by Oelhaf and Fischer (1983) showed that more than twenty trace species can be measured if the spectral resolution is at least 0.1 cm^{-1} in the mid-infrared. These theoretical considerations are confirmed by the results of several instruments (see below).

<u>Previous experiments</u>

During the last ten years FTIR experiments for measuring trace species have already been performed by different scientific groups (overview see Fischer 1992). A considerable number of uncooled mid-IR experiments yielded very interesting scientific results in the past but can not meet all the requirements specified above. With one exception all investigators have applied the limb sounding geometry. Smith and his colleagues have flown the HIS-experiment aboard an aircraft using the nadir viewing geometry (Revercomb et al. 1988). Due to the relatively warm and dense troposphere, emission spectra with comparatively low spectral resolution can be detected with good signal to noise with an uncooled experiment.

The best known FTIR experiment is the ATMOS (Atmospheric Trace Molecule Spectroscopy, Farmer et al. 1987) Project. Twenty occultations were recorded during the Spacelab 3 flight which resulted in concentration profiles for about 30 trace gases in both the Northern and Southern hemisphere. A second space mission of ATMOS took place more recently in connection with the first ATLAS (Atmospheric Laboratory of Applications and Science) flight. Again, a large number of mid-IR solar spectra with high spectral resolution were recorded at different geographical locations. Data processing is underway.

Cooled Michelson-type interferometers have already been operated aboard stratospheric balloon gondolas by several groups (see Fischer 1992). The limb emission spectra are of high quality and allow the determination of many trace species. The nighttime partitioning of the main reactive and reservoir nitrogen compounds has been derived simultaneously the first time (Abbas, Kunde et al. 1991) using the 1986 SIRIS (Stratospheric Infrared Interferometer Spectrometer) data.

More recently, a cooled FTIR flew for the first time aboard an aircraft in order to determine column amounts of stratospheric trace species (see next chapter). These contributions of FTIR experiments to atmospheric research were essential even if no cooled interferometer for time-independent global emission measurements has been operated in space so far.

MIPAS (Michelson Interferometer for Passive Atmospheric Sounding) Experiments

MIPAS has been designed to meet the requirements for the measurement of atmospheric trace gases as defined above. The central part of MIPAS is the rapid scanning Double Pendulum Interferometer which operates in the 3 to 15 μm region and records one interferogram within a few seconds.

Operationally, MIPAS is not a single instrument but a measurement programme. The objective is to develop several instruments that have systematically increasing complexity. Finally, the MIPAS space experiment will meet all the scientific requirements as specified above. In the following results of the first three MIPAS-instruments will be described. An uncooled version of this interferometer was developed during the first half of the 1980's. This MIPAS-LM (Laboratory Model, unapodized spectral resolution 0.015 cm^{-1}) is being used as a ground-based device for measuring column amounts of trace gases at polar stations by detecting the attenuated solar radiation. During the CHEOPS III campaign (1990) and during EASOE (European Arctic Stratospheric Ozone Experiment), time series of zenith column amounts of several trace gases (e.g. O_3, N_2O, CH_4, HNO_3, NO_2, HCl, HF) have been derived from the recorded interferograms and analysed with respect to dynamical and chemical processes in the arctic stratosphere (Adrian et al. 1991).

For a certain time period in 1990 a detailed study of the spectral line shape of HCl at 2775,7 cm^{-1} obtained from ground-based spectral measurements was performed to determine some information on the vertical distribution of HCl (Adrian et al. 1993). Various shapes of the vertical volume mixing ratio profiles, differing in tropospheric concentration, tropopause height and concentration gradient in the stratosphere, were used to fit the measured line profile. The derived mixing ratio profiles are presented in Figure 1. The shaded areas mark the range of uncertainty, in which the line shapes are fitted equally well. The weighting functions are calculated with the enveloping profiles of February 4. Obviously, these results show that the MIPAS-LM spectra contain some information about the HCl concentration and its gradient in the stratosphere. The sensitivity to tropospheric concentration is comparatively low.

Both the derived integrated zenith column amounts of trace species as well as the low resolved vertical distributions (about two independent pieces of information) have been interpreted in terms of dynamical and chemical processes. During the late winter 1990, periods of strongly perturbed stratospheric chemistry have been identified in the Arctic. Our results indicate evident similarities to ground based Antarctic measurements reported by other groups but distinct differences regarding the behaviour of the HNO_3 column amounts. More detailed information is found in Adrian et al. (1991) and Adrian et al. (1993).

A cooled version of MIPAS has been constructed to observe the composition and structure of the stratosphere as well as their diurnal variations with good vertical resolution from balloon-borne platforms. Field experiments have been performed during balloon flights in France in May 1989 and May 1990 as well as in Kiruna/Northern Sweden in January and March 1992. Limb emission spectra were measured within the altitude range between 5 and 40 km. A considerable number of

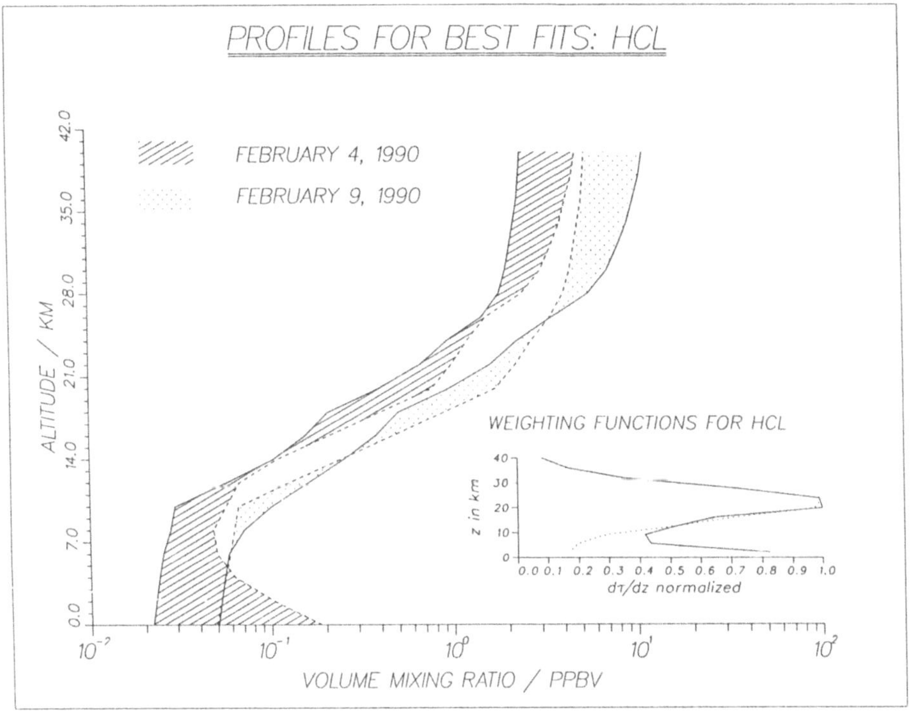

Fig. 1: Vertical concentration profiles of HCl retrieved from ground-based FTIR measurements at February 4, and 9, 1990; also weighting functions are presented.

atmospheric trace species are identified in the spectra, namely CO_2, O_3, H_2O, CH_4, N_2O, CCl_4, CF_2Cl_2, $CFCl_3$, CHF_2Cl, $ClONO_2$, HNO_3, N_2O_5 and C_2H_6.

The wealth of information contained in the MIPAS balloon spectra can be illustrated by observing complete limb sequences (Oelhaf et al. 1991). Figure 2 shows eight selected individual limb spectra covering tangent heights between the upper stratosphere and the middle troposphere. In the uppermost spectrum corresponding to a tangent height of about 37 km, only the more abundant species CO_2 and O_3 can be detected. Below about 35 km tangent height the nitric acid bands are appearing with increasing radiances towards decreasing altitude. The HNO_3 bands reach a radiance maximum near 20 km, consistent with the vertical mixing ratio profile of HNO_3 as a stratospheric reservoir and sink gas. Because of the steepness of the vertical mixing ratio profiles of the halogenated source gases in

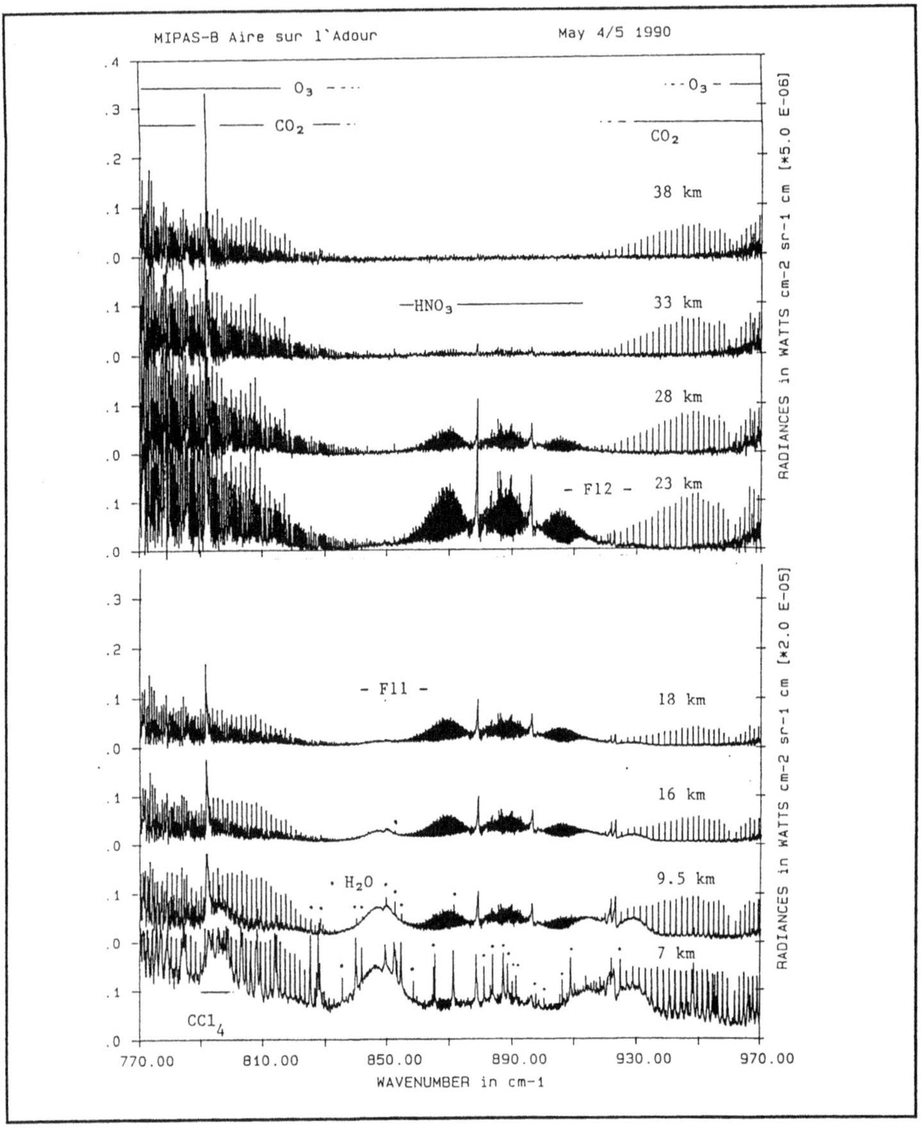

Fig. 2: Measured limb sequence from 38 km observer altitude illustrating the growth of emissions of various molecules depending on their altitude distribution. Individual spectra are shifted by $1 \cdot 10^{-6}$ cm^{-2} sr^{-1}/cm^{-1} in the upper frame and by $4 \cdot 10^{-6}$ W cm^{-2} sr^{-1}/cm^{-1} in the lower frame for clarity. Spectral resolution is about 0.07 cm^{-1} apodized. Note that the ordinate scale is changing from the upper four to the lower four traces. Tangent heights indicated are preliminary.

the stratosphere, the features of CCl_4, $CFCl_3$ and CF_2Cl_2 are disappearing in the spectra above about 15 km, 20 km and 25 km respectively. Water vapor lines are generally rather weak in this atmospheric window, therefore they cannot be separated from noise in the dry stratosphere. However, they appear just below the hygropause and increase rapidly when scanning downwards into the troposphere. Consequently, these H_2O lines should be very suitable for deriving the height of the hygropause and the water vapor mixing ratio in the upper troposphere. In addition to the species mentioned above, spectral features of $ClONO_2$, CHF_2Cl and C_2H_6 have been identified.

Within the second spectral channel of the balloon instrument between 1170 cm^{-1} and 1380 cm^{-1} the species O_3, CO_2, H_2O, CH_4, N_2O, HNO_3 and N_2O_5 have been identified so far. The most interesting information in this region is the unresolved feature of N_2O_5 (Oelhaf et al. 1991).

The capability of MIPAS-B has been proven by retrieving vertical concentration profiles of different trace species (Clarmann, et al. 1993). Figure 3 shows as example the HNO_3 profile as derived from the 1989 campaign in Aire sur l'Adour/France and a quicklook analysis of the MIPAS data recorded on January 14, 1992 over the gulf of Bothnia inside the polar vortex. Error bars for the 1989 data have been determined by extensive Monte Carlo runs for various spectral microwindows that take into account the measurement uncertainties. MIPAS results are compared with a set of measurements and model results for midlatitude summer conditions. Obviously, there is a high amount of HNO_3 in the lower stratosphere in the polar winter measurement without layers of low mixing ratio which is an important fact for the study of the denitrification process in the Arctic.

From the March 1992 flight a limb sequence with 8 different tangent heights was analysed with respect to O_3, HNO_3 and $ClONO_2$ (Oelhaf et al. 1993). Already the visual inspection of quick-look spectra during flight revealed that the $ClONO_2$ ν_4 Q branch feature was strongly enhanced in the lower stratosphere. The retrieved $ClONO_2$ profile has been compared with the few previously reported nighttime and sunrise measurements from midlatitudes. The $ClONO_2$ profile of March 1992 is in marked contrast to the other reported profiles. The value of the maximum is much higher and the corresponding altitude level is considerably lower (around 19 km). Evidently, nearly all of the available chlorine below about 20 km is converted into $ClONO_2$ in spring time in the Arctic (Oelhaf et al. 1993).

Besides the MIPAS ground based and the balloon experiment, our new cooled MIPAS aircraft experiment was flown during the EASOE campaign in winter

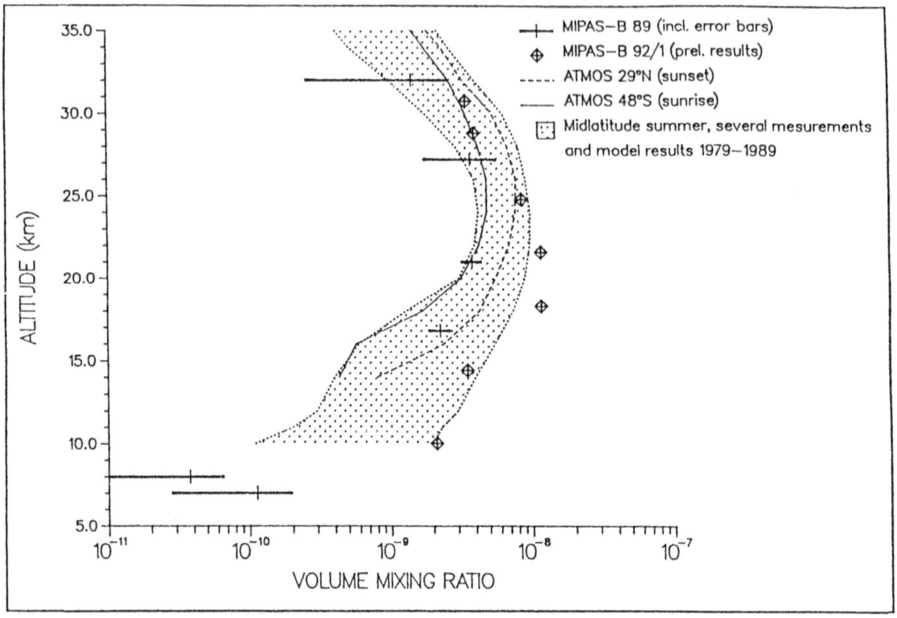

Fig. 3: HNO_3 profiles as retrieved from MIPAS-B data recorded on May 15, 1989 in midlatitudes and on January 13, 1992 in North Sweden (preliminary) inside the polar vortex; these results are compared with a set of measurements and model results for midlatitude summer conditions; error bars for 1992 data are expected to be significantly smaller than for 1989 data due to considerably improved instrument performance.

1991/92. The new atmospheric sounder MIPAS-FT (Flugzeug Transall) detects infrared radiances emitted by various atmospheric trace constituents in the spectral range between 750 cm^{-1} and 1250 cm^{-1}. Due to the low flight level of the Transall aircraft only column amounts of stratospheric trace gases could be determined. As first results zenith column amounts of O_3 and HNO_3 are presented for the time period between January 30 and February 16, 1992 (Fig. 4). These data are related to a geographical region from 45° west to 45° east and from 55° north to 80° north. The column amounts for HNO_3 vary between 1 and 3 x 10^{16} mol. /cm^2 and for O_3 between 180 and 420 DU. However, the strong linear correlation between these different stratospheric trace gases indicates that the large variations are mainly due to dynamical processes in the arctic atmosphere during the observing period (see Fig. 4).

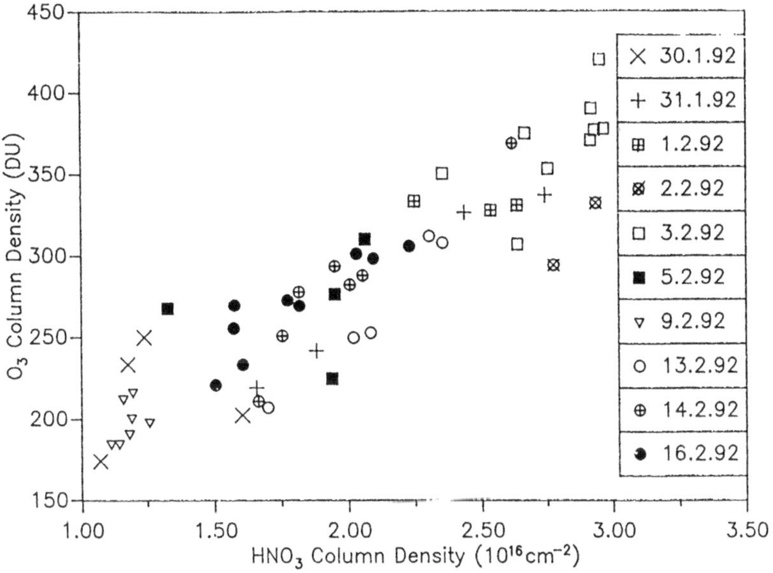

Fig. 4: Column amounts of O_3 in Dobson units above the 350 hPa level measured between January 30 and February 16, 1992 plotted versus the simultaneously observed HNO_3 column amounts; the measurements of each day are marked by a special symbol.

The cooled MIPAS space experiment is in development for the core payload of the 1. European Polar Platform (ENVISAT) which will be launched in 1998.

Conclusions

FTIR instruments meet important requirements for the measurements of atmospheric trace constituents, e.g. simultaneous measurement of various trace species during day and night. Such measurements can be used for the investigation of chemical and dynamical processes in the atmosphere and their separation. Corresponding experiments during the last ten years have proven the extraordinary capabilities of FTIR instrumentation. Up to now no cooled FTIR instrumentation has been flown aboard satellites. Therefore, the MIPAS space experiment will provide new information on the spatial and temporal distribution of atmospheric trace species at the end of this decade.

References

Abbas M.M., V.G. Kunde, J.C. Brasunas, J.R. Herman, and S.T. Massie (1991), Nighttime reactive nitrogen measurements from stratospheric infrared thermal emission observations, J. Geophys. Res. 96:10885

Adrian G.P. , T. Blumenstock, H. Fischer et al. (1991), Column amounts of trace gases derived from ground-based measurements with MIPAS during CHEOPS III, Geophys. Res. Lett. 18:783

Adrian G.P., T.v. Clarmann, H. Fischer, H. Oelhaf (1993), Trace gas measurements with the ground-based MIPAS experiment during the arctic winters 1990 to 1992, Proceedings International Radiation Symposium, Tallinn, August 1992, in press

Blom C., H. Fischer et al. (1993), Airborne measurements during the arctic stratospheric experiment 4. Column amounts of HNO_3 and O_3 derived from FTIR emission sounding, submitted to GRL in Nov. 1992.

Burkert P., F. Fergg and H. Fischer (1983), A compact high resolution Michelson Interferometer for passive atmospheric sounding (MIPAS), IEEE Trans. Geosci. Rem. Sens. GE-21:345

Clarmann T.v., H. Oelhaf, H. Fischer (1993), Retrieval of atmospheric O_3, HNO_3, CFC-11 and CFC-12 profiles from MIPAS-B-89 limb emission spectra, submitted to Applied Optics

Farmer C.B. et al. (1987), Final Report on the first flight of the ATMOS instrument during the Spacelab 3 mission, April 29 through May 6, 1985, JPL-Publication 87-32

Fischer H., F. Fergg et al. (1983), Simultaneous detection of trace constituents in the middle atmosphere with a small He-cooled high resolution Michelson Interferometer, Beitr. Phys. Atmos. 58:260

Fischer H. (1992), Remote sensing of atmospheric trace constituents using Fourier Transform spectrometry, Ber. Bunsenges. Phys. Chem., 96:306-314.

Oelhaf H. and H. Fischer (1983), Measurements of minor constituents in the middle atmosphere from IR limb emission spectra - a feasibility study, Appl. Opt. 24:2515

Oelhaf H., H. Fischer et al. (1991), Remote sensing of trace gases with a balloon-borne version of the Michelson interferometer for passive atmospheric sounding (MIPAS), Proceedings ESA Symp. on European Rocket and Balloon Programmes, Mandelieu, France, May 1991, ESA-Report SP-317

Oelhaf H., T.v. Clarmann, H. Fischer et al. (1993), Stratospheric $ClONO_2$, HNO_3 and O_3 profiles inside the arctic vortex from MIPAS-B limb emission spectra obtained during EASOE 1992, submitted to GRL, Nov. 1992

Revercomb H.F. et al. (1988), Radiometric calibration of IR Fourier Transform spectrometers: Solution to a problem with the High-resolution Interferometer Sounder (HIS), Appl. Optics, 27:3210-3218

THE MEASUREMENT OF GLOBAL CARBON MONOXIDE USING THE ATMOSPHERIC INFRARED SOUNDER (AIRS)

L. Larrabee Strow
Department of Physics
University of Maryland Baltimore County
5401 Wilkens Avenue
Baltimore, Maryland 21228
USA

1. Introduction

One goal of the Earth Observing Systems (EOS) is to measure concentrations of key atmospheric gases on a global scale in order to understand biogeochemical cycles in both the natural, unpolluted atmosphere, and in regions where anthropogenic activities have perturbed the concentrations of gas–phase species. A particular concern in tropospheric chemistry is that increasing levels of carbon monoxide (CO) may lead to a decrease in atmospheric hydroxyl (OH), which would reduce the atmosphere's ability to scavenge other trace gases.

CO has an approximate residence time of 2 months in the atmosphere and consequently is not uniformly mixed. Seasonal variations in the unpolluted atmosphere are on the order of 100 ppb. A large series of in–situ measurements of CO using aircraft flights found a number of regions where CO concentrations in the first two kilometers of the atmosphere ranged from a low of approximately 50 ppb to highs of 200–300 ppb (Marenco et al. 1989). (Much higher levels measured near urban areas were omitted to improve the reliability of the data.) The 200–300 ppb concentrations are widespread, but can be attributed to weather patterns bringing in polluted air from urban areas. Levels of CO in the boundary layer can reach 20 *ppm* (The Conservation Foundation 1987) in urban areas, which are far above natural variations.

CO is an ideal candidate for satellite observations, given its temporal and geographic variations. Global measurements may allow a better understanding of: (1) the sources and sinks of CO, (2) long–term trends in the globally averaged CO concentrations, and (3) trends in regional CO pollution.

This paper examines the capability of the Atmospheric Infrared Sounder (AIRS) (AIRS 1991) to perform scientifically useful global measurements of the CO column content. AIRS is primarily a temperature and humidity sounder for EOS as well as a prototype for the next generation operational NOAA sounder. However, the present configuration of AIRS channels includes spectral coverage of the 1–0 vibration–rotation band of CO at 2140 cm^{-1} with sufficient spectral resolution to resolve individual CO lines. In particular, we will determine the ability of a standard signal processing technique to retrieve the relatively small CO signal from the AIRS radiances.

Two other planned EOS instruments, MOPITT (Drummond 1991) and TES (Margolis et al. 1991), will measure CO at a higher effective spectral resolution than AIRS and consequently will have more sensitivity to CO variations. MOPITT, a gas cell correlation

NATO ASI Series, Vol. I 9
High Spectral Resolution Infrared Remote Sensing
for Earth's Weather and Climate Studies
Edited by A. Chedin, M.T. Chahine and N.A. Scott

spectrometer and filter radiometer, will not be able to characterize the atmospheric state (temperature, humidity, etc.) as well as AIRS, which may limit its ability to retrieve CO concentrations to an accuracy commensurate with its sensitivity to CO spectral features. TES, a high–resolution interferometer, should be able to determine the atmospheric state as well as AIRS. The higher spectral resolution of both of these instruments will allow the retrieval of CO profiles as well as column content values. The main disadvantage of TES will be its low coverage in both space and time due to its extremely high data rate and lack of flight opportunities. Since AIRS may become the next generation NOAA operational temperature sounder, a measurement of CO using AIRS promises a much longer time series of CO concentrations than either TES or MOPITT, and more complete spatial coverage than TES. However, AIRS will not be able to measure vertical profiles of CO concentrations.

2. Radiance Simulation

The simulated radiances used to test our CO retrieval algorithm were generated using GENLN2 (Edwards 1987), a line–by–line radiative transfer program. The radiances were then convolved with the AIRS instrument function, which has a trapezoidal shape near the center frequency with wings extending far beyond the trapezoid cutoff.

2.1. AIRS Instrument

AIRS is a grating spectrometer that nominally operates between 650 and 3000 cm^{-1} with a resolving power ($\nu/\Delta\nu$) of 1200. It has a 1.1° instantaneous field–of–view (IFOV) which corresponds to a 13.5 km diameter footprint at nadir. Scan lines sweep between ±49°. The AIRS design goal is for maximum noise equivalent temperature variations ($NE\Delta T$) of 0.20K RMS per spectral resolution element, or 0.28K RMS per detector element (two detectors per spectral resolution element), at a 250K scene temperature. The detector outputs will have a nominal digitization level of 0.1K.

We performed simulations at both the 0.20K noise level and for a noise level of 0.07K, which is the projected *detector* noise in the CO spectral region. Both of these figures assume a 250K scene temperature. In the CO region, higher scene temperatures will be the norm and consequently the 0.20K noise level is conservative by more than a factor of two. However, other system noise may preclude reaching the 0.07K noise figure.

2.2. Pre–processing Required for the CO Algorithm

The CO algorithm assumes that all of the standard atmospheric and surface parameters have been retrieved over a nominal 50 by 50 km area that includes 9 AIRS IFOV's and one AMSU IFOV. We also assume that a cloud clearing algorithm has been applied to the CO channels, which effectively averages the radiances in the 9 IFOV's and reduces the noise by a factor of 3. This noise reduction is somewhat optimistic. We model this averaging procedure as follows. First, uniformly distributed noise, with an RMS level of 0.28K per detector element, is added to the simulated brightness temperatures in each of the 9 identical IFOV's. We then digitize these 9 values to 0.1K, and average the results to obtain a single brightness temperature for a 50 by 50 km area.

2.3. CO Signal Characteristics

Two characteristics of the CO spectrum make satellite atmospheric measurements of CO possible. First, the CO spectrum lies in a nominal atmospheric window between 2080 and 2200 cm^{-1}, with increasing H_2O at lower frequencies and increasing CO_2 at higher frequencies. Second, CO is a linear molecule with regularly spaced spectral lines that have a sinusoidal appearance in the observed radiances at the AIRS spectral resolution. This sinusoidal signal allows the use of powerful signal processing techniques for noise reduction.

A number of different simulated radiances were generated to study the characteristics of the CO signal and for testing the CO retrieval algorithm. All of these profiles were based on CO concentrations in the U.S. standard atmosphere (Anderson et al. 1987), which is denoted as the "Reference profile" in Fig. 1. This reference CO profile was modified two different ways for algorithm testing, as shown in Fig. 1. The CO profile denoted "Reference profile * 1.1" is the reference CO profile modified to have 10% greater CO concentrations at all altitudes. The second test profile is a modification of the reference profile in the first 4 km to reflect the maximum CO concentrations reported by Marenco et al. (Marenco et al. 1989), where CO amounts in the 300 ppb range in the first two km were commonly observed. As mentioned earlier, this level of CO is far below the range of CO concentrations seen in polluted urban areas. Other atmospheric parameters (temperature, humidity, ozone, etc.) were set to the U.S. standard atmosphere values unless stated otherwise. We always assume that the radiances have been cloud–cleared. If the cloud fraction is too large, a CO retrieval will not be attempted, which should not adversely affect the science requirements for CO measurements given the relatively slow temporal variations of CO.

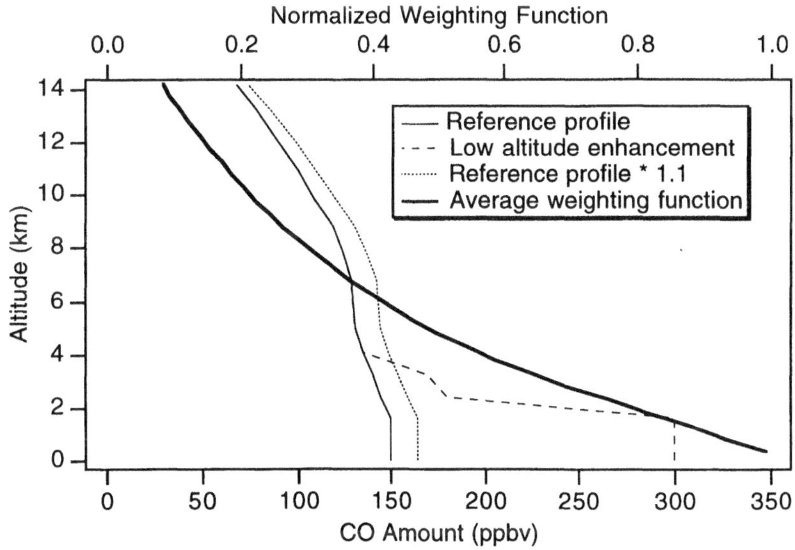

Fig. 1. CO profiles and average weighting function.

Figure 1 also shows the average weighting function at the centers of the CO lines between 2150 and 2190 cm^{-1}. This average weighting function is an approximation, in that it includes all contributions to the radiance, and not that due solely to CO. Weighting functions at CO line centers below 2150 cm^{-1} are more contaminated with contributions from H_2O and O_3 and were left out of this average. This average weighting function peaks at the surface, indicating that most of the CO radiance comes from the first 5-10 km.

Figure 2 shows a portion of the simulated AIRS brightness temperature spectrum in the region of CO emission. The largest features in the spectrum are due to H_2O, but the high frequency oscillations due to CO are evident throughout the spectrum. The CO lines produce a contrast of roughly 3K.

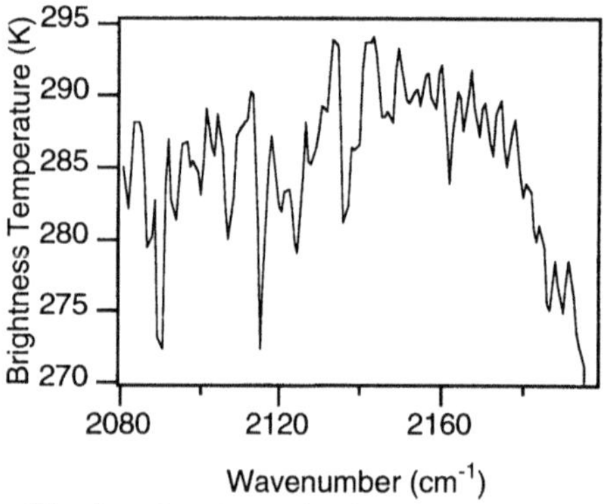

Fig. 2. Simulated AIRS spectrum in the region of CO emission.

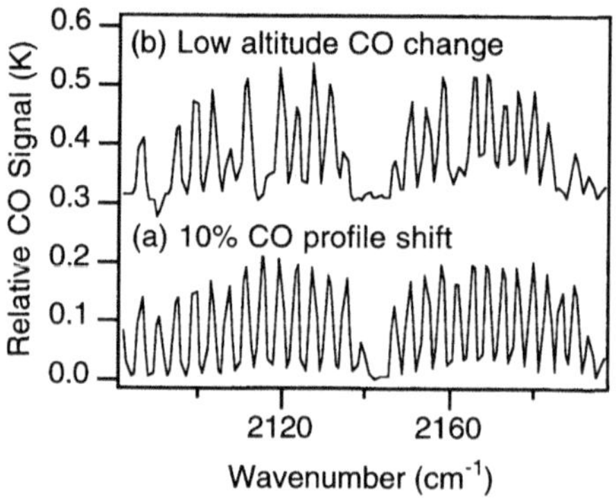

Fig. 3. Change in AIRS brightness temperature spectrum for changes to the reference CO profile as shown in Fig. 1.

Figure 3 shows the changes in the AIRS spectrum when the reference CO profile is modified as shown in Fig. 1. Figure 3a, where the whole CO profile is increased by 10%, has a much smoother envelope compared to the curve in Fig. 3b, where changes in low–altitude CO are masked by H_2O emission in several spectral regions. The effects of uncertainties in the temperature and humidity profiles on the spectrum in Fig. 2 are shown in Fig. 4. The sensitivity to errors in the temperature profile was examined by simply adding 1K to the correct temperature profile at all levels and recomputing the brightness temperature spectrum. This 1K offset error is probably larger than what will be seen in practice. Expected errors in the AIRS retrieved H_2O profile of 10% were modeled by changing the H_2O amount by 10% at all altitudes. The signs of the temperature and humidity errors were chosen so the curves in Fig. 4 would have the opposite sign for clarity. The CO signal curve plotted in Fig. 3a is also included in Fig. 4 for reference. H_2O lines dominate both the temperature and H_2O sensitivity curves. However, even with a 1K temperature offset error, the CO signal contrast is relatively unchanged in regions away from H_2O lines. It is harder to visually sort out the changes to the CO signal for the 10% H_2O error curve.

The sensitivity of the CO signal to the temperature contrast between the surface and the first air layer was also examined. Fig. 5 shows the changes in the AIRS spectra when the reference CO profile is changed to the low altitude enhanced profile, for a series of offsets between the surface temperature and the first air layer temperature. These offsets range from -12K to +8K in 5K steps. Note, for example, that there is effectively no CO signal when the surface temperature equals the first air layer temperature minus 7K. At this surface temperature, the surface and CO emission, averaged over the first several km, both take place at roughly the same average temperature. The temperature contrast between the surface and the first several air layers will generate a significant spread in the accuracy of the individual CO retrievals.

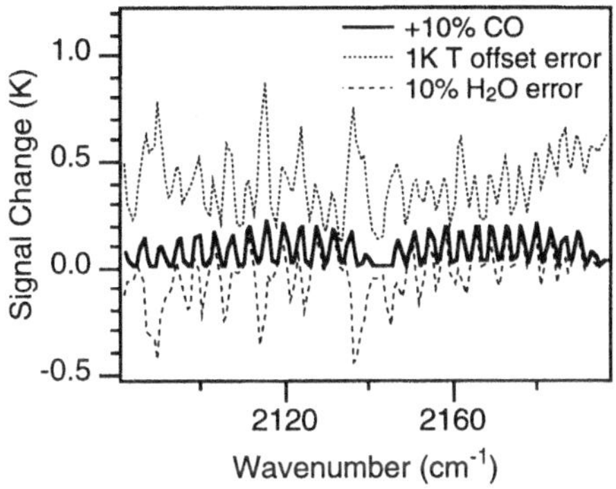

Fig. 4. Effect of a 1K offset in the temperature profile and a 10% error in H_2O amount on the CO signal.

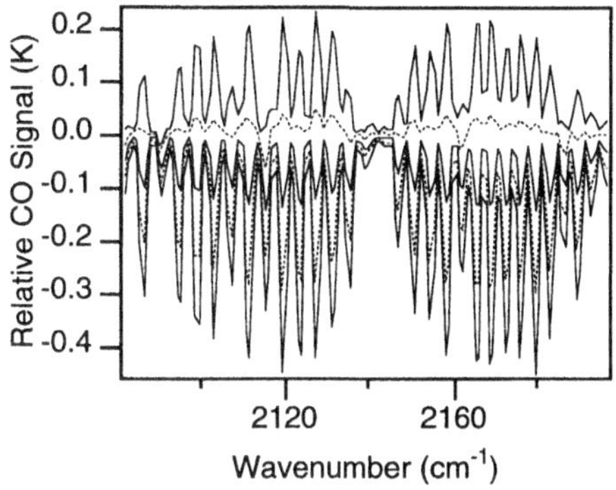

Fig. 5. Effect of varying the surface temperature (in 5K steps) from -12K to +8K of the first air layer temperature. The -12K contrast is at the top of the figure.

Clearly, these variations in CO amount are small compared to the AIRS noise level of 0.20K per resolution element, making it difficult to retrieve these small changes in CO using individual channels. The effects of the two noise levels used in our simulations are shown in Fig. 6, where the change in the CO signal for a low–altitude increase in CO is plotted without noise and with 0.20K or 0.07K RMS noise per resolution element. The 0.20K noise completely overwhelms the CO signal. The CO signal is difficult to discern in the 0.07K noise plot as well.

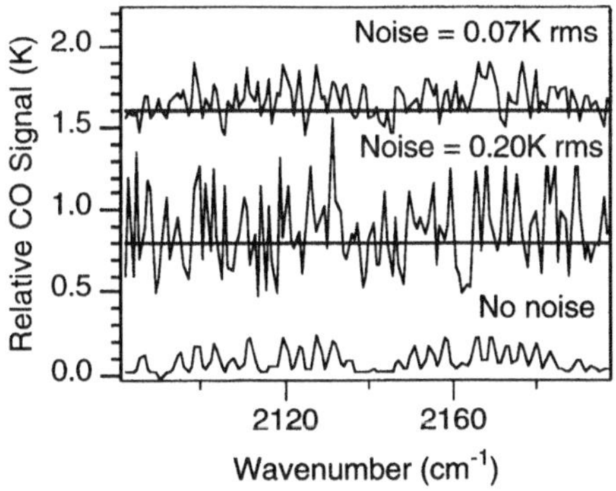

Fig. 6. Effect of noise on the CO signal for the -12K surface contrast.

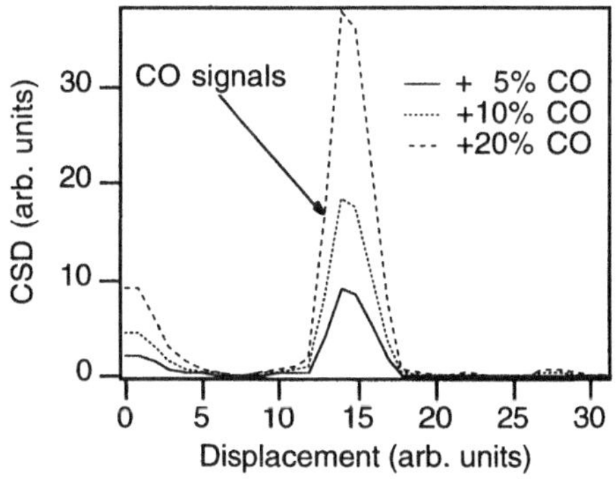

Fig. 7. Noise–free cross–spectral density for CO increases of 5, 10, and 20% over the whole profile.

3. Cross Spectral Density

The striking feature of the noise–free CO signals in Figs. 4–6 is their sinusoidal appearance, although a few periods of the sinusoid are missing due to H_2O contamination and the "missing" line at the center of the CO band near 2140 cm^{-1}. This characteristic of the CO signal, coupled with the high noise of AIRS relative to the expected variations in the CO signal, suggests the use of signal processing techniques for the extraction of a single frequency out of a noisy time series. Our time series becomes the (infrared) frequency dependence of the difference between a brightness temperature spectrum calculated using a reference CO profile and the observed brightness temperature spectrum.

The sensitivity of the DC level of our CO signal to undetected cloud fraction and other factors require the use of a relative measurement of the CO signal contrast, which is just what a signal processing algorithm will accomplish. Of course, the undetected cloud fraction will also affect the CO signal contrast, but not as severely, as will be shown later.

We investigate here a CO retrieval based on a standard technique in signal processing called the Welch method (Candy 1988, Fante 1988), which estimates the power or cross spectral density of a signal in random noise. Our CO signal is defined as the difference between the observed AIRS brightness temperature spectrum, and a spectrum calculated using the AIRS–retrieved atmospheric parameters and a reference CO profile. This CO signal is divided into k (overlapping) sections of m points, where m is some power of two. Each section is Hanning windowed and transformed with an m–point FFT (Fast Fourier Transform). These k FFT's are summed for noise reduction. We then multiply this result with the complex conjugate of the FFT of a noise–free reference CO signal. This noise–free reference CO signal is the difference between a brightness temperature spectrum calculated using the reference CO profile and one calculated using a slightly perturbed profile for the CO concentrations. The AIRS–retrieved atmospheric parameters are used in both of these calculations. This last operation in Fourier space removes that part of the signal that is insensitive to CO, i.e. some of the D.C. signal level, interfering lines, etc. The resulting quantity is sometimes called the cross–spectral density (CSD) in the parlance of signal processing, and is used to determine the CO amount.

In our context, the term spectral in cross–spectral density does not correspond to an optical frequency, but to the second member of the Fourier pair (wavenumber $\leftrightarrow$ cm). The CSD is related to the interferogram of the observed infrared spectrum, hence we use the term displacement for describing the abscissa of the CSD in later figures. We choose to retain the term CSD so the reader can easily refer to the signal processing literature.

We can summarize this algorithm as follows. Let

$$\mathbf{X}_\nu = B\left(\nu, T(z), \rho_{CO}^{ref}(z)\right)^{calc} - B\left(\nu, T(z), [1+\gamma(z)]\, \rho_{CO}^{ref}(z)\right)^{calc} \tag{1}$$

and

$$\mathbf{Y}_\nu = B\left(\nu, T(z), \rho_{CO}^{ref}(z)\right)^{calc} - B(\nu)^{meas} \tag{2}$$

where $B(\nu, \ldots)$ is the brightness temperature, $\rho_{CO}^{ref}(z)$ is the reference CO profile, ν the frequency, T the temperature, and z the altitude. $\gamma(z)$ is the perturbation to the CO profile used to generate a noise–free brightness temperature difference spectrum. This perturbation can have a relatively arbitrary magnitude, its shape should roughly follow the expected variations in the CO profile. Now partition $\mathbf{X}_\nu$ and $\mathbf{Y}_\nu$ into k (overlapping) sections, $\mathbf{x}_k$ and

$\mathbf{y}_k$, of length m. Hanning window $\mathbf{x}_k$, $\mathbf{y}_k$ to produce $\mathbf{x}_k^h$ and $\mathbf{y}_k^h$. The CSD is then given by

$$CSD = \sum_k FFT(\mathbf{y}_k^h) FFT(\mathbf{x}_k^h)^*. \qquad (3)$$

In the above operation, $FFT(\mathbf{x}_k^h)^*$ windows each individual FFT of our signal so only those parts sensitive to CO remain. The sum over k reduces noise, but sacrifices the resolution of the CO signal in Fourier (displacement) space by sectioning the CO spectrum into k smaller sub–units.

The spectral region studied here contains 130 points (or equivalently 65 spectral resolution elements). Through trial and error testing, we chose m=64, which gives k=17 sections if each section overlaps 60 points. This set of parameters is a compromise between signal–to–noise and resolution of the main CO "bump" in the CSD (see Fig. 7 which will be discussed below). If we instead chose an overlap of 32 points there would be only k=3 sections, with results that are only slightly worse, but with far less computation. No attempt has been made to determine if these parameters are truly optimal, or how the CSD measurement technique compares to other signal–processing techniques.

The CSD can also be interpreted as the Fourier transform of the cross–correlation between $\mathbf{x}_k^h$ and $\mathbf{y}_k^h$. Hence we are looking for a correlation between our observed CO signal (suitably windowed for signal–to–noise enhancements) and a CO signal "template" calculated assuming perfect knowledge of the atmospheric state and some arbitrary change in CO concentration over the reference profile. This template is what is shown in Figs. 3 and 5. Because of the sinusoidal nature of the CO lines, the correlation function is sinusoidal and its Fourier transform has a peak at the "frequency" of this sinusoid.

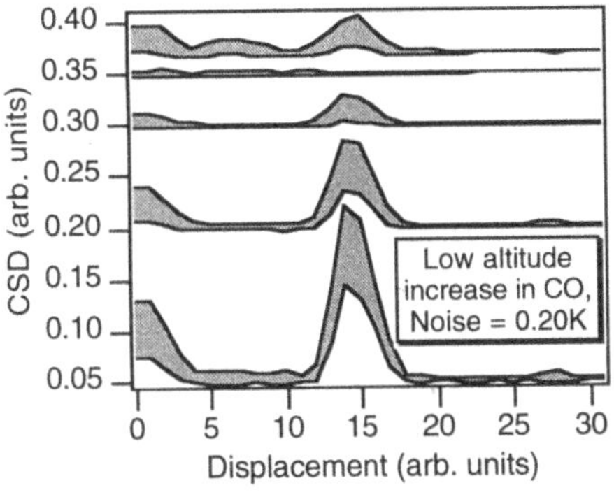

Fig. 8. The CSD and its standard deviation for the various surface temperature contrasts shown in Fig. 5, a 0.20K RMS noise, using the low altitude CO profile change.

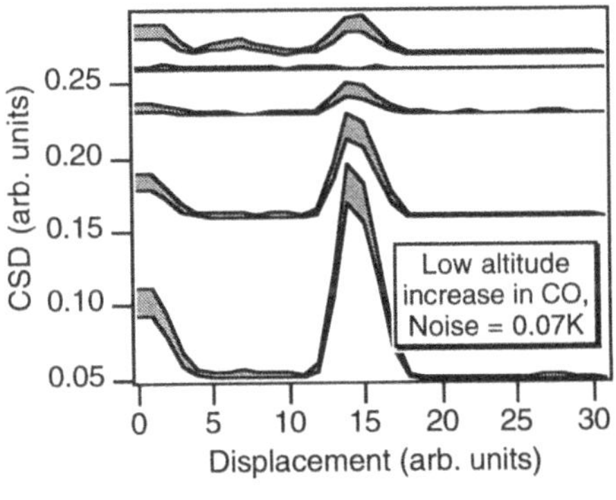

Fig. 9. Same as Fig. 8 but for 0.07K RMS noise.

Figure 7 shows the noise–free CSD calculated for increases in CO of 5%, 10%, and 20% over the standard CO profile. These percentage increases were made over the complete profile. $\gamma(z) = 0.05$, for all z, was used in the CSD calculations. The peak in the CSD near 14 displacement units is due to the "sinusoidal" feature of the CO spectrum. The "DC" portion of the CO spectrum, due to the average value of the sinusoidal contribution and other effects not related to the CO amount, shows up in the CSD near 0 displacement units. We would clearly want to avoid the 0 displacement portion of the CSD when determining the CO amount since it will contain radiometric bias errors, errors resulting from inaccurate cloud clearing, etc. An integral of the CSD somewhere between 10 and 20 displacement units should give the best determination of the atmospheric CO amount. Note that the peak value of the CSD at 14 displacement units is linearly related to the increase in CO amount. Similar results are obtained when CO is only increased in the lower altitudes.

The actual CO amount is determined by first calculating the CSD for the case $\mathbf{Y}_\nu = \mathbf{X}_\nu$. The resulting CSD peak near 14 displacement units (or a suitably integrated value for the CSD in this region) is denoted CSD^{ref}, and is the CSD for a simulated, noise–free measurement of $B(\nu)$ produced using a CO profile given by $[1 + \gamma(z)]\, \rho_{CO}^{ref}(z)$. This simulated CSD assumes perfect knowledge of atmospheric conditions. The true CSD is then computed, using the same $\mathbf{X}_\nu$, but now using the actual measured values for $B(\nu)$ in the expression for $\mathbf{Y}_\nu$. Its peak (or integrated value) is denoted CSD^{meas}. Our estimate for the CO profile is then given by

$$\rho_{CO}(z) = \rho_{CO}^{ref}(z) \left[1 + \frac{CSD^{meas}}{CSD^{ref}} \gamma(z)\right]. \tag{4}$$

Of course the resultant CO profile is referenced to the shape of some best–guess CO profile by one constant, (CSD^{meas}/CSD^{ref}). Consequently the retrieved profile is only changed from the reference by a scaling constant and contains no new information about the relative shape of the observed CO profile. Short–term time and space variations in CO will most likely be due to low–altitude concentration changes near urban areas. Consequently, the most accurate CO retrievals in urban and surrounding areas will result if the CSD is calculated with a CO profile that is perturbed in the lower atmosphere. A different perturbed reference profile may be better in other areas.

4. Simulation Results

A number of simulations were performed to explore the effects of both random noise and systematic errors on the CSD. As discussed earlier, noise levels of 0.20K and 0.07K RMS per spectral resolution element were studied. Actual noise levels should be somewhere between these two values. Figures 8 and 9 show the CSD for the low–altitude increase in CO shown in Fig. 1 for noise levels of 0.20K and 0.07K RMS respectively. The shaded area in each curve is the $\pm$ 1 standard deviation range. The $\gamma(z)$ used for these calculations was just the difference between the reference CO profile in Fig. 1 and the low–altitude enhanced CO profile.

Each graph has 5 CSD's representing different thermal contrasts between the surface and first air–layer temperature. Values for $T_{surface}$-T_{air} equal -12K, -7K, -2K, +3K, and +8K from the top to bottom curves, respectively, in Figs. 8 and 9. The strong dependence of the CSD on the thermal contrast is evident. Clearly a thermal contrast of -7K will not allow a measurement of CO variations of this magnitude. The signal–to–noise ratio of the CO measurements will be approximately the ratio of the CSD peak value to the standard deviation at the peak. These figures show that variations of 150 ppb in low–altitude CO can be measured by AIRS in a 50 by 50 km area with a signal–to–noise ratio that varies from almost zero to as high as 10 or more, depending upon the instrument noise and thermal contrast.

Figures 10 and 11 show the same simulations as Figs. 8 and 9, except for a constant 10% increase in CO over the whole profile. Here $\gamma(z) = 0.05$, for all z. As expected, the CSD is less dependent on the thermal contrast in this case and the signal–to–noise ratio varies from somewhat less than unity to approximately 5. Variations in the CO profile of this type should be rare.

These results show that this technique can, in many cases, reduce the effects of random noise in the AIRS radiance spectra to the point where natural background variations in CO can be observed. Further improvements may be possible by time and space averaging of the CO retrievals. Of course, systematic errors could dominate the CO measurements and their space and time averages.

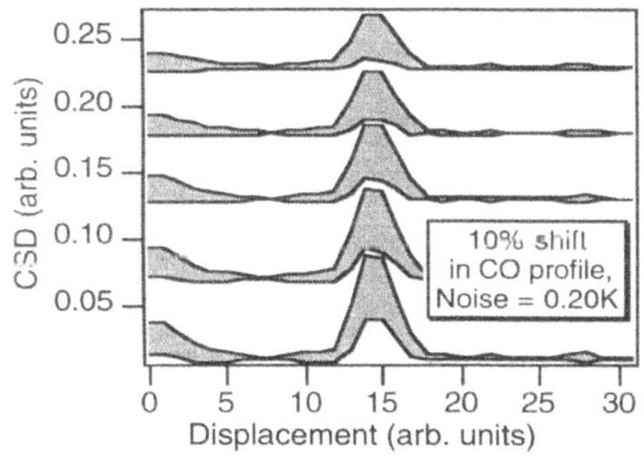

Fig. 10. The CSD and its standard deviation for various surface temperature contrasts, 0.20K RMS noise, for a 10% concentration shift over the whole CO profile.

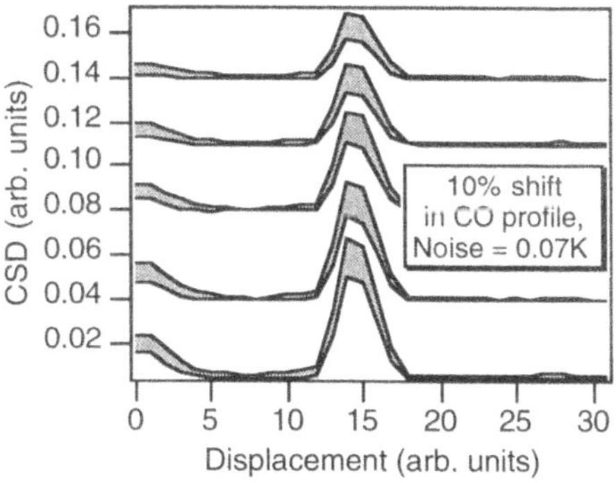

Fig. 11. Same as Fig. 10 but for 0.07K RMS noise.

Up to this point we have assumed that we know the atmospheric state perfectly except for the CO amount. The fact that our CO measurements utilize the same spectrometer and IFOV as the AIRS temperature and humidity measurements means that the CO algorithm will have the best available measurement of the atmospheric state. In addition, any systematic errors in the AIRS temperature measurements may cancel similar systematic errors in the CO measurement since they use the same instrument and IFOV. Some preliminary tests were performed to determine the sensitivity of the CO algorithm to errors in the temperature and humidity profile and to an undetected cloud fraction. These tests were performed for an atmosphere with a 10% increase in CO over the whole profile and a 0.20K RMS random noise, which is represented by the middle curve in Fig. 10.

AIRS is expected to retrieve atmospheric temperatures to an accuracy of 1K in 1 km intervals and humidity to approximately 10% in more than 5 layers. We have chosen an extreme case of this temperature error to test, namely a constant 1K *offset* from the true profile. Similarly we have also determined the effect of a constant 10% error in the humidity profile. It is estimated that AIRS cloud–cleared radiances may contain up to a 3% undetected cloud fraction. We introduce this error by performing a CO retrieval on a radiance computed from

$$R_\nu = 0.97 * R_\nu^{cloud\ free} + 0.03 * R_\nu^{100\%\ cloudiness\ at\ 100\,mbar}. \qquad (5)$$

The CSDs determined from radiances containing these systematic errors in the atmospheric state are presented in Fig. 12. For this particular choice of thermal contrast between the surface and the first air layer (we are using the conditions of Fig. 10, middle curve) the systematic errors all lie within one standard deviation of the noise error. These relatively low errors reflect the fact that the CSD is essentially a relative measurement of the CO signature in the AIRS radiance spectra and can reject large variations in the H_2O spectral features.

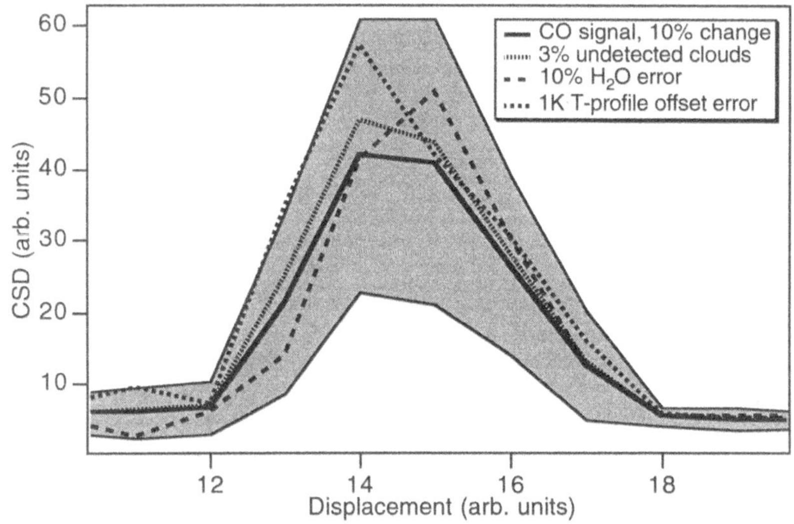

Fig. 12. Effect of various systematic errors on the CSD. CO profile conditions are those of Fig. 10, middle curve.

We also examined the sensitivity of the CO measurement to errors in the temperature contrast between the surface and the first air layer. These results are shown in Fig. 13, where the CSD is computed with perfect knowledge of the contrast and with a temperature contrast error of 0.5K. Since the CO retrieval does not depend on knowing the actual surface temperature, but only the temperature–emissivity product, our choice of a 0.5K error in the average temperature contrast is reasonably conservative. The curves shown in Fig. 13 were for the conditions of Fig. 8, 0.20K RMS noise and the low altitude increase in CO. Note that each curve has been normalized to a common peak value for clarity, unlike those shown in Fig. 8. Under these conditions, these contrast errors of 0.5K contribute smaller errors to the retrieval than random noise.

5. Conclusions and Future Work

We have shown that AIRS can detect variations in atmospheric CO on the order of the natural background variations using the cross–spectral density technique to lower noise and sensitivity to systematic errors. Commonly occurring levels of CO pollution in urban areas should be detected with ease since they are far above the variations examined here. This technique lowers the effects of noise on the measurement to a magnitude similar to the level of systematic errors caused by uncertainties in the state of the atmosphere.

The disadvantage of this technique is the lack of any profile information. However, most observed variations in CO over short time and space scales are in the first several kilometers, and thus the lack of profile data is probably not a severe hindrance to the scientific utility of these data. The advantages of using AIRS for CO measurements are several. First, AIRS will provide a better knowledge of the state of the atmosphere than any other satellite instrument making routine global measurements. This means that AIRS measurements of CO may have smaller systematic errors than those obtained by MOPITT, which cannot directly use the AIRS retrievals for the atmospheric state. Secondly, AIRS is a prototype for a NOAA operational sounder and may fly over a much longer time period than other instruments measuring CO. This may give scientists longer, and more consistent, measurements of CO than those provided by several specialized, and different, instruments. Thirdly, AIRS will provide partial redundancy for MOPITT, should it fail, at a very low incremental cost over what is required for temperature and humidity sounding using AIRS.

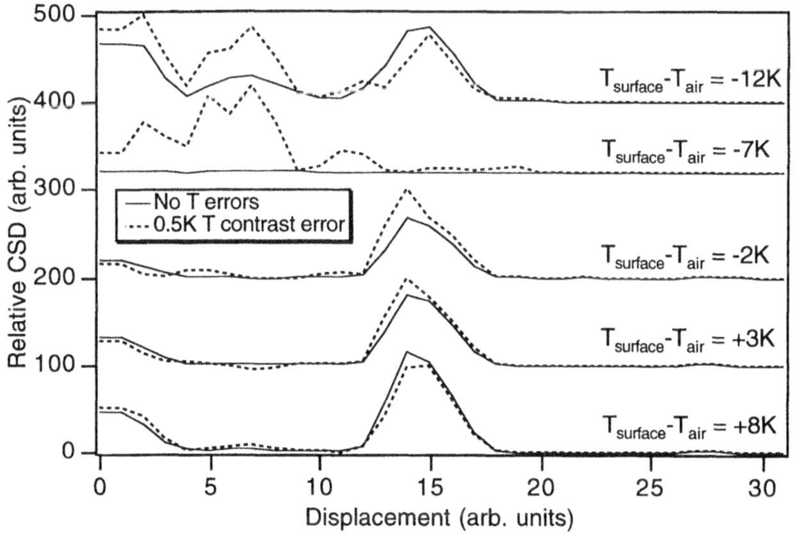

Fig. 13. Effects of temperature contrast errors of 0.5K on the CSD for the conditions of Fig. 8. The curves have been offset for clarity.

A number of additional simulations are required to fully understand the capabilities of the proposed technique. In particular the systematic errors should be studied over a much wider range of atmospheric conditions. Possible ozone contamination in the low frequency side of the CO band requires study. Future simulations should also include solar radiation, aerosols, better noise statistics, cloud contamination models, and a better statistical sample of CO profiles.

Optimization of the CSD algorithm may also improve the results shown here. Other noise reduction techniques used in signal processing applications may also prove superior to the CSD and should be examined.

Acknowledgments

I would like to thank Scott Hannon and David Tobin for their assistance with the line–by–line calculations of the CO radiances.

References

AIRS (1991) Atmospheric Infrared Sounder: Science and Measurement Requirements. Technical Report D6665 Rev. 1, Jet Propulsion Laboratory, Pasadena, CA

Anderson GP, Clough SA, Kneizys FX, Chetwynd JH, Shettle EP (1987) AFGL Atmospheric Concentration Profiles (0–120 km), AFGL–TR–86–0110, AFGL (OPI), Hanscom AFB, MA 01736

Candy JV (1988) *Signal Processing, The Modern Approach.* McGraw–Hill Book Company

Drummond JR (1991) Remote sounding of lower atmospheric composition from satellites. In *Optical Remote Sensing of the Atmosphere.* Optical Society of America, 1991 Technical Digest Series, Volume 18, November 18–21

Edwards DP (1987) Genln2: The new Oxford line–by–line atmospheric transmission/radiance model. Memorandum 87.2, Dept. Atmospheric, Oceanic and Planetary Physics, University of Oxford

Fante RL (1988) *Signal Analysis and Estimation: An Introduction.* John Wiley and Sons

Marenco A, Macaigne M, Prieur S (1989) Meridonal and vertical CO and CH_4 distributions in the background troposphere (70°n–60°s; 0–12 km altitude) from scientific aircraft measurements during STRATOZ III experiment (June 1984). *ATMOS Envir.*, (23):185–200

Margolis JS, Martonchik JV, Beer R (1991) Detection sensitivities and measurement accuracies for tropospheric species using the TES instrument on EOS/B. In *Optical Remote Sensing of the Atmosphere.* Optical Society of America, 1991 Technical Digest Series, Volume 18, November 18–21

The Conservation Foundation (1987) State of the Environment: A View Toward the Nineties. Technical Report, Washington, D.C.

<u>Chapter 6</u>

Next generation IR observations : impact on weather and climate analysis

"Direct use of satellite sounding radiances in numerical weather prediction"

J. EYRE, E. ANDERSSON, A.P. McNALLY - ECMWF, GB

"Retrieval and assimilation : system considerations"
A. C. LORENC - The Met. Office, GB

"On measuring the greenhouse effect of Earth"

G.L. STEPHENS - Colorado State Univ., USA
A. SLINGO, M. WELL

"Multi-year statistics of total cloud amounts from the Nimbus-7, ISCCP and coads data sets".

B.C. WEARE - Univ. of California,

DIRECT USE OF SATELLITE SOUNDING RADIANCES IN NUMERICAL WEATHER PREDICTION

J.R. Eyre, E. Andersson and A.P. McNally
European Centre for Medium-Range Weather Forecasts
Shinfield Park
Reading, RG2 9AX
UK

1. INTRODUCTION

Since 1978, operational satellite soundings of the atmosphere's temperature and humidity structure have been provided by the TIROS Operational Vertical Sounder (TOVS) on the TIROS-N/NOAA series of polar orbiting satellites (*Smith et al.*, 1979). TOVS consists of the High-resolution Infra-red Radiation Sounder (HIRS/2), the Microwave Sounding Unit (MSU) and the Stratospheric Sounding Unit (SSU). Starting with the satellite NOAA-K (currently scheduled for launch in 1995), this system will be upgraded to become the Advanced TOVS (ATOVS), in which MSU and SSU will be replaced by the Advanced Microwave Sounding Unit (AMSU). Through an increased number of channels and higher horizontal resolution, AMSU will lead to improved sounding performance in cloudy areas. However, in clear areas ATOVS will not improve significantly on vertical resolution and hence information content on temperature in the troposphere (see *Eyre*, 1990).

Numerical weather prediction (NWP) requires observations of the global temperature, wind and humidity fields in order to analyse the initial state of the atmosphere from which forecasts are made. Deficiencies in the accuracy of the initial state are major causes of forecast error in the short and medium-range. At present the combined meteorological observing network, including its satellite components, is far from meeting the data requirements for NWP. Satellites offer the only practical means of approaching these needs, particularly over the oceans and other data-sparse areas, but at present the TOVS system does not meet requirements of NWP for temperature and humidity

NATO ASI Series, Vol. I 9
High Spectral Resolution Infrared Remote Sensing
for Earth's Weather and Climate Studies
Edited by A. Chedin, M.T. Chahine and N.A. Scott

information; its vertical resolution is poor (~ 3-4 km) and its coverage deficient in cloudy areas. ATOVS will help to address the latter problem, but the problem of vertical resolution will remain, and it will become more acute as NWP models themselves move to higher vertical resolution (currently ~ 1 km). It is therefore a major requirement for NWP to obtain satellite sounding data of improved vertical resolution. In the next 10-15 years, infra-red instruments of high spectral resolution, with a vertical resolution ~ 1-2 km, offer the best prospect for approaching the goals of observational data for NWP (see *Eumetsat*, 1991).

Over the last few years, as NWP systems have improved, major NWP centres have experienced increasing difficulty in demonstrating consistent positive impact on northern hemisphere forecasts from temperature profiles retrieved from satellite sounding data by conventional techniques (for example, see *Andersson et al.*, 1991; *Kelly et al.*, 1991; *Thoss*, 1991). It is now recognized that the information content of TOVS data (relative to existing "knowledge" of the NWP system) is rather low, mainly as a consequence of their poor vertical resolution. If these data are to have a positive impact in NWP, they must be assimilated with proper regard to their error characteristics and true information content (see *Eyre and Lorenc*, 1989). In order to address these problems, there are currently moves at a number of NWP centres towards more direct use of the radiance information itself within the assimilation system.

In section 2 of this paper, recent research at ECMWF on more direct methods for using TOVS radiance information is discussed. Firstly, studies of some statistical properties of the differences between measured radiances and those calculated from a forecast "first-guess" are presented. Secondly, the concepts of a variational approach to the assimilation of radiance information are introduced. This approach has been developed for TOVS data in a one-dimensional (i.e. vertical) scheme, which is equivalent to a retrieval scheme using the NWP model's forecast profile as a "first-guess" (background) and the expected covariance of forecast error as a constraint. A summary of some forecast impact experiments using this scheme is presented. The same approach is also being developed with a three-dimensional scheme in which global TOVS radiances are used (along with other observations) to analyse simultaneously the state of the global atmosphere. Progress on the development of this scheme is discussed.

The developments of schemes for direct use of radiances in NWP, first for TOVS and then for ATOVS, pose important questions related to the methods to be used for processing and assimilating data from advanced infra-red sounders of high-spectral resolution. Will they be assimilated directly as radiances, will we revert to using independently-retrieved temperature/humidity profiles, or is some intermediate option available? These issues and other problems associated with the use of these new data are discussed in section 3.

2. DIRECT USE OF TOVS RADIANCES AT ECMWF

2.1 Studies of radiance statistics

Since April 1991, ECMWF has been processing routinely (but not assimilating) global TOVS radiances received in near real-time from NOAA/NESDIS (see *Kelly*, 1992). These data have been pre-processed (i.e. adjusted to nadir, etc.) and cloud-cleared by the operational procedures at NESDIS, broadly following the techniques described by *Smith et al.* (1979) and *McMillin and Dean* (1982). Short-range forecasts (at 3, 6 and 9 hours) provided by the ECMWF operational model are interpolated in time and space to the location of each TOVS sounding. For each channel a clear-column "forecast radiance" is calculated from the temperature/humidity profile using the scheme described in *Eyre* (1991). The difference between the measured and calculated radiance can then be used for a number of purposes.

Of concern in any scheme to assimilate TOVS radiance information are biases introduced by errors in the radiative transfer model. These are found to be significant in most TOVS channels, and in some they vary with air mass (i.e. from the tropics to the polar regions) by around 2 K in brightness temperature. It is desirable to correct absolutely for such errors, if the absolute errors can be evaluated, and further work is required on this problem. In the meantime, we have chosen to seek to remove any bias in the measured radiances relative to those calculated from the forecast. It is not our intention here to discuss this technical problem in detail (see *Eyre*, 1992). Suffice it to say that such a scheme has been developed using the data base of global measured-minus-forecast brightness temperatures. Following strict quality control to remove outliers and to check for consistency between window channel radiances and an independent sea surface temperature field, air mass dependent bias corrections for each TOVS channel are obtained using a simple linear regression on the measured brightness temperatures in MSU channels 2, 3 and 4. This appears to remove the major part of the air mass dependent bias in the key tropospheric temperature sounding channels.

Using the bias-corrected data for periods of about 2-4 weeks, it is interesting to look at the residual standard deviation. Figure 1 shows an analysis of the field of local standard deviation of the measured-minus-forecast brightness temperature in MSU channel 2 (weighting function peak around 700 hPa) for NOAA-11 in March 1992. The data are restricted to sea and sea-ice areas, and only to those which are cloud-free soundings are used. The contour interval is 0.1 K and areas of standard deviation greater than 0.4 K are shaded. The most striking features of this figure are the very low values of standard deviation, i.e. 0.3-0.5 K over most of the world. Such plots have been generated each month for a year, and their values are found to vary little from month to month. Note that the variance of the difference between any two data sets with independent errors (a good assumption in this case) is equal to the sum of the variances of error in both data sets. This includes errors in the measurements, errors in the radiative transfer model and errors in the forecast profiles. Such plots give confirmation that the residual errors in the radiative transfer calculations are small, but they also demonstrate that the errors in the forecast profiles, when mapped into brightness temperature space, are also small most of the time. This confirms our impression, based on theoretical investigation and other empirical studies, that the information content of the radiance data relative to the forecast is (statistically) quite small and that it must be treated with great care in the assimilation process. It also helps explain why independently retrieved profiles, in which much larger errors have been introduced as artifacts of the inversion process, are so difficult to use effectively in modern NWP systems.

For all the tropospheric temperature sounding channels, we find that the biases (after correction, as described above) are small, and so are the residual standard deviations. However, for the humidity channels sensing the mid to upper troposphere (HIRS channels 11 and 12), the picture is completely different. Figure 2 shows the bias plot for HIRS channel 11, which also peaks around 700 hPa. Here the contour interval is 0.5 K and shading is for areas > +1 K (light) and < -1 K (heavy). Temperature-dependent biases which are correlated with measurements in MSU channels 2, 3 and 4 have been removed (in practice, chiefly from polar regions). These biases may be due to radiative transfer model bias or to forecast model bias; further work is required to explore this. However, large biases still remain in the sub-tropics (positive, corresponding to the NWP model being more humid than the measurement) and in the region of the ITCZ (negative: NWP model drier than measurement). The measurements used are systematically biased towards clear-sky conditions, and this may be important in certain areas. However, it cannot explain the bias near the ITCZ (it is the wrong sign) and is unlikely to be significant in the areas of sub-tropical anticyclone. The magnitude of the biases, more than 2 K in places, is equivalent to monthly mean biases of more than 20% in the specific humidity. The patterns appear to be similar to

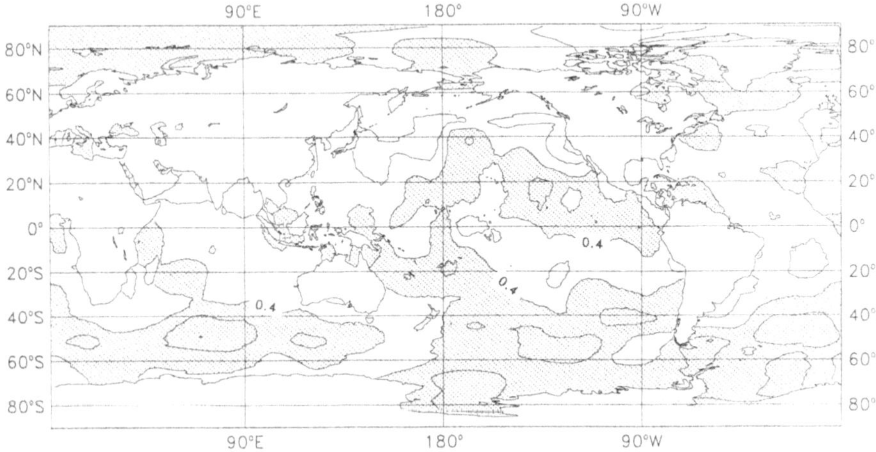

Fig. 1 Field of local standard deviation of the difference between measured brightness temperatures and those calculated from a forecast first-guess for MSU channel 2 for NOAA-11 in March 1992. Contour interval = 0.1 K; shading for values > 0.4 K

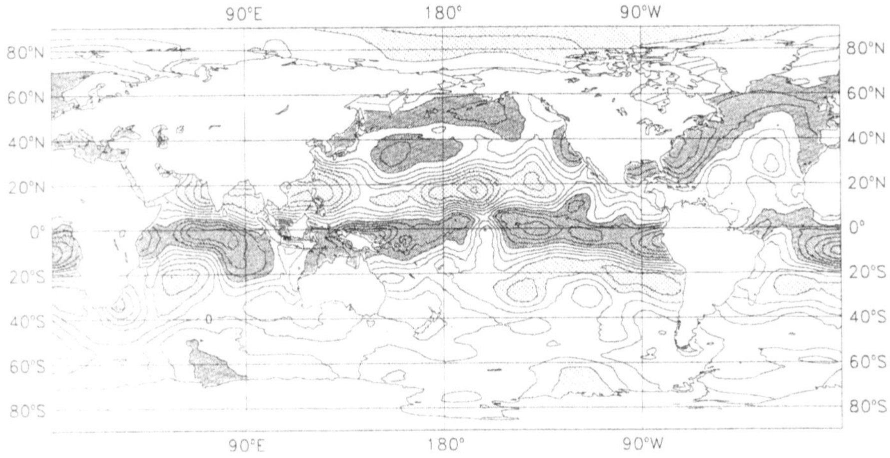

Fig. 2 Field of mean difference between measured brightness temperatures and those calculated from a forecast first-guess for HIRS channel 11 for NOAA-11 in March 1992. Contour interval = 0.5 K; light shading > 1 K; dark shading < -1 K

those found when comparing total precipitable water derived from SSM/I data with ECMWF model equivalents (*Liu et al.*, 1992; *Phalippou*, 1992). These preliminary results suggest that HIRS channels 11 and 12 have an important potential role in diagnosing problems with the humidity fields in the mid and upper troposphere in NWP and climate models.

2.2 The variational approach

The variational approach to the assimilation of data into an NWP system has been described by a number of authors (e.g. *Lorenc*, 1986). When applied specifically to the assimilation of satellite radiance information it takes the following form: we try to minimise a penalty function $J(x)$ with respect to the atmospheric state x, where $J(x)$ measures the degree of fit to the radiances, to background information and possibly also to other observations and to other physical or dynamical constraints. If the errors involved have Gaussian distributions, then the optimal penalty function can be shown to be a sum of quadratic terms:

$$J(x) = (y^m - y\{x\})^T \cdot (O+F)^{-1} \cdot (y^m - y\{x\}) + (x - x^b)^T \cdot B^{-1} \cdot (x - x^b) + J_o + J_c$$

where x is a vector containing the atmospheric state, y^m is a vector of measured radiances, $y\{x\}$ is the radiance vector corresponding to the atmospheric state (i.e. involving radiative transfer calculations), O is the expected error covariance of the measured radiances, F is expected error covariance of the "forward operator" $y\{...\}$, x^b is a background profile (e.g. short-range forecast "first-guess"), B is its expected error covariance, J_o (optionally) represents the penalty for other observations, and J_c (optionally) represents other constraints.

Note that the expression for $J(x)$ is general with respect to the dimension of x: it can be applied to a one-dimensional (vertical) problem analogous to a conventional retrieval, or to a 3-dimensional atmospheric state (e.g. the whole global atmosphere). Note also that there are a number of methods for finding the minimum of such a penalty function but that all require or imply the calculation of the gradient of $\mathbf{y}\{\mathbf{x}\}$ with respect to $\mathbf{x}$. This involves the calculation of a Jacobian or else the evaluation of the adjoint of the radiative transfer model.

2.3 One-dimensional variational analysis (1DVAR)

At ECMWF, a scheme has been developed to apply this approach to the vertical analysis problem at each TOVS sounding location (see *Eyre et al.*, 1992). The method is the same as described by *Eyre* (1989), except that it has been applied to global clear-column radiances received operationally from NESDIS (rather than to raw, cloud-affected radiances, as in the original implementation). The "retrieved" temperature and humidity profiles generated by the 1DVAR are then assimilated into the ECMWF optimal interpolation (OI) analysis scheme. The problem of the correlation of retrieval error with forecast background error is treated approximately using the method proposed by *Lorenc et al.* (1986).

A series of forecast impact studies has been conducted to test and tune the 1DVAR scheme. A full report on these experiments is given by *Eyre et al.* (1992); only a summary of some of the results is reported here. One set of experiments was conducted using TOVS data in the period 1-15 May 1991. Global data (excluding data below 100 mb over land) was assimilated every 6 hours for 14 days. After each day of assimilation a 10-day forecast was run and compared with the operational analysis at the appropriate time as verification. The results were compared with equivalent verification data from ECMWF operational forecasts which, during this period, used NESDIS retrievals only above 100 mb in the northern hemisphere extratropics and at all levels in the southern hemisphere extratropics. Figure 3 shows the mean forecast verification scores for 14 cases, expressed as

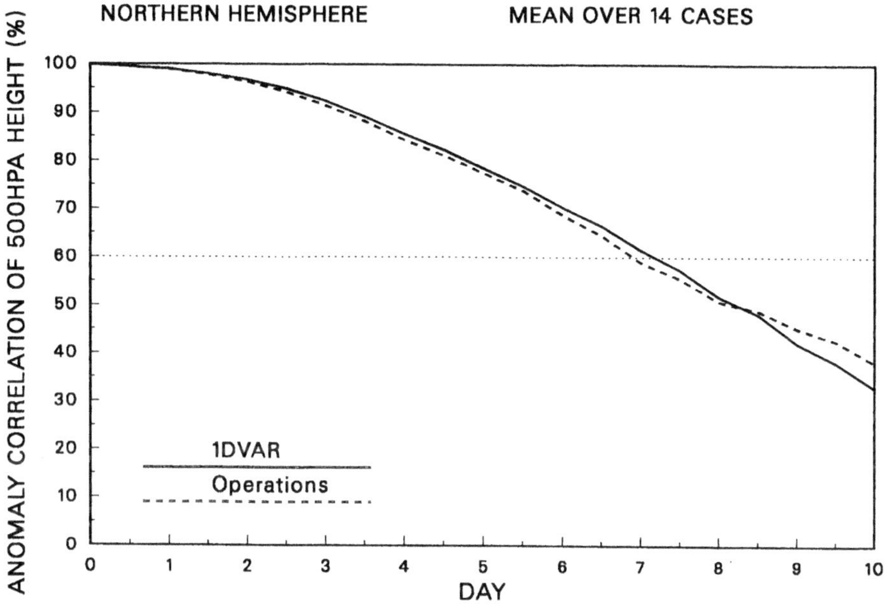

Fig. 3 Anomaly correlation of differences between forecasts and verifying analyses for northern hemisphere extratropics, 500 hPa height. Mean of 14 cases in May 1991.

anomaly correlations at 500 hPa, for the whole of the northern hemisphere. The mean scores are clearly in favour of the 1DVAR assimilation compared with the operational control from day 3 onwards. Figure 4 shows the scores for each of 14 individual forecasts at day 3. All but one are favourable to the 1DVAR. Figure 5 shows the mean anomaly correlations over North America where the positive impact of the 1DVAR is strongest, and reflects the effect of significant analysis changes over the Pacific.

2.4 Three-dimensional variational assimilation (3DVAR)

A major project is under way at ECMWF to develop a system for 3DVAR of all meteorological observations (*Pailleux*, 1990; *Pailleux et al.*, 1991). This system is still under development, but some experiments have already been performed to test the sub-

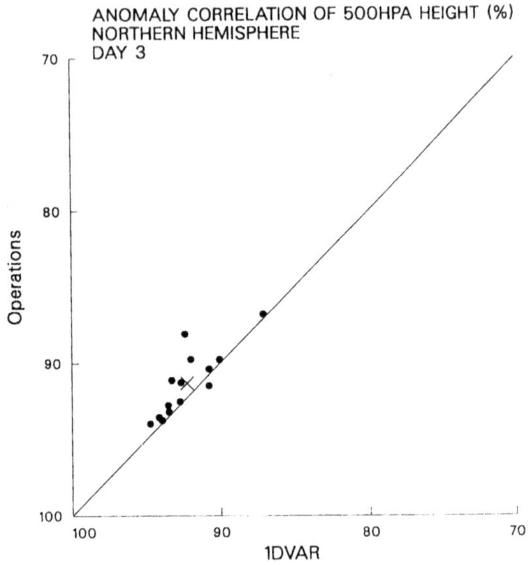

Fig. 4 Scatter plot of anomaly correlation of differences between individual 3-day forecasts and verifying analyses for 500 hPa height for northern hemisphere extratropics. 14 cases in May 1991.

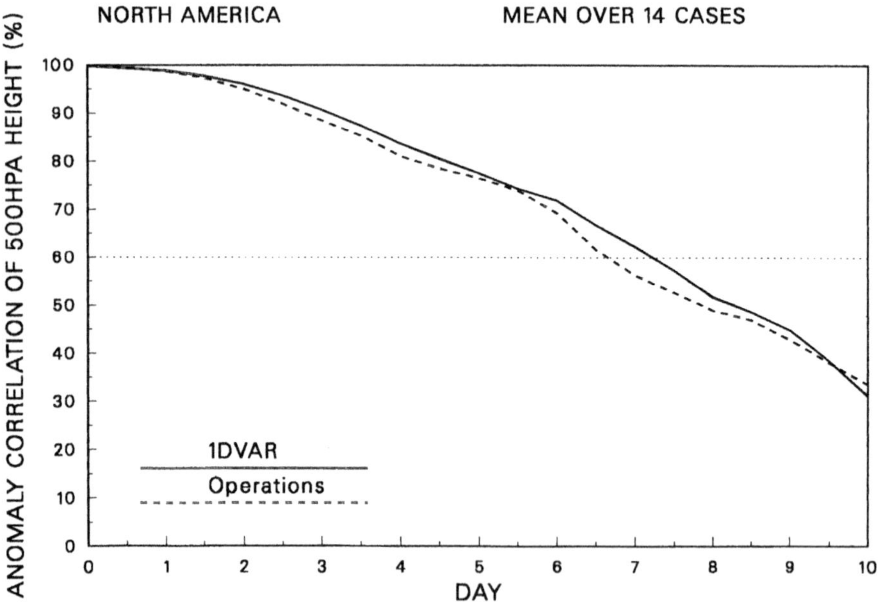

Fig. 5 As Figure 3, except for North America only.

systems for assimilating TOVS clear-column radiances. An "observation operator" is required for TOVS radiances, together with its associated adjoint operator, and the fast TOVS radiative transfer model (*Eyre* 1991) fulfils this role. Experiments have so far been conducted using the NWP model at fairly low resolution (spectral truncation T42 with 19 vertical levels). About 30 000 global TOVS radiances (on average 10 channels at about 3 000 locations) have been assimilated simultaneously. At the present stage of development, progress is satisfactory: the iterative solution of the problem converges to produce analysis increments (i.e. changes to the first-guess) which are of reasonable magnitude and lead to an analysis which fits the measured radiances better than does the first-guess in all channels.

3. USE OF HIGH SPECTRAL RESOLUTION SOUNDER DATA IN NWP

3.1 Options

The main purpose on reporting here on developments on the 1DVAR and 3DVAR projects at ECMWF is to demonstrate that significant progress is being made on methods to use TOVS radiance data more directly in NWP. Similar work is in progress at a number of NWP centres. We can therefore anticipate that such methods will be in common use, first for TOVS and then for ATOVS, by the time data become available from advanced infra-red sounding instruments. Moreover, experience of problems in trying to assimilate conventional retrieved profiles, together with evidence from studies in radiance space (as in section 2.1), suggest that direct radiance methods will be necessary if we are to exploit sounding data effectively.

Given this scenario, what will be the options for using data from the new instruments? One will be to revert to the use of independently-retrieved profiles of temperature and humidity. This is theoretically unattractive given the experience with TOVS and the direction of current developments in data assimilation.

It may be argued that methods for direct assimilation will not be necessary for the advanced instruments because they will have higher vertical resolution and will yield more accurate independent retrievals. Whilst this is true, it is important to bear in mind that NWP systems will not stand still; by the time the advanced instruments with double the present vertical resolution become available, it is probable that NWP models will also have doubled their vertical resolution. Thus the ratio between the vertical scales sensed by the satellite instruments and those resolved by the models will remain roughly the same, and the problems of correctly using the radiance information will remain the same.

The second option will be to adapt the radiance assimilation methods to handle the data from high-resolution instruments (either radiance spectra or interferograms). Although this is the obvious theoretical extension, the data volumes will be more than 100 times current levels and could cause logistical problems. Even if advances in computing and telecommunications keep pace with the increase in data volumes, it will not be economical to process ~2000 channels of data when they contain only ~20 independent pieces of temperature information. The third option, then, is to seek methods for reducing the data volumes <u>in the radiance space</u> whilst retaining most of the available information. Perhaps a reduction to ~200 channels or pseudo-channels will be justified.

3.2 Scientific problems

Clearly there are many problems associated with the assimilation of satellite radiances in NWP. However some are general problems for variational assimilation of all types of data, and some are common to other satellite sounding data, such as TOVS and ATOVS. We highlight here only problems specific to the use of data from advanced infra-red sounders.

They will require new radiative transfer models (and their adjoints), fast enough to allow the required radiative transfer calculations in real-time, yet sufficiently accurate to allow useful information to be extracted from the data. Significant work is required in this area. Also, as important as the data is a proper specification of their error characteristics and those of the corresponding radiative transfer calculations. The error covariance matrices for instruments of ~2000 channels present considerable problems. Research is needed to characterise them, to approximate them and to assess sensitivity to their accuracy.

Because the new sounders are infra-red instruments, the treatment of cloud will continue to be a major problem. The increased spectral resolution and coverage will offer new opportunities to retrieve cloud information, but they will also lead to increased sensitivity to the cloud models used and to errors in their underlying assumptions. Developments will be required either on methods for cloud detection and "cloud-clearing" prior to inversion, or on direct assimilation of cloud-affected radiances.

The problem of data volumes has already been mentioned, and this will be most acute in operational meteorology where near real-time processing is required. To ease these problems, research is needed on how to reduce the data volume in the radiance space with minimal loss of information.

3.3 Fast radiative transfer calculations

While the data volume problem deserves attention, it is already clear that the radiative transfer calculations involved within the assimilation of radiance data from advanced sounders should not be an insuperable problem. Using the latest fast models (e.g. *Eyre*, 1991) the "forward" calculation for ~20 TOVS channels for one profile requires about 1 ms on 1 CPU of a CRAY-YMP. The associated adjoint calculation takes about 3 ms. If we postulate a 3DVAR system requiring 10 iterations (involving forward plus adjoint) for adequate convergence of the minimisation, and using global TOVS data at about 100 km resolution, we will require ~800 CPU seconds per 6 hours per satellite. Increases in computing power by 1997 are likely to reduce this to ~30 CPU seconds, which would be handled easily. There is therefore plenty of scope for increasing the computing requirement to handle advanced sounder data. The number of channels or pseudo-channels will increase by ~10-100. Horizontal resolution could also be increased, but it should be remembered that the vertical/horizontal aspect ratio of atmospheric temperature perturbations is typically in the range 1:50-100. Therefore, even for a sounder with a vertical resolution of 1 km, a horizontal resolution of 50-100 km may be consistent.

4. DISCUSSION

Advanced infra-red sounding instruments of high spectral resolution offer a means of alleviating one of the major weaknesses of present operational systems - their poor vertical resolution. The development and implementation of such sounders should represent a significant contribution to improving the observational data base for NWP.

Recent advances in the use of satellite sounding information in NWP have focused on more direct methods for assimilating the radiances. Such methods are now demonstrating positive results for TOVS and will probably be well-established at several NWP centres by the time advanced sounder data are available. Planning of systems to process and exploit advanced sounder data should take account of these developments in NWP data assimilation.

REFERENCES

Andersson E, Hollingsworth A, Kelly G, Lönnberg P, Pailleux J Zhang Z (1991) Global observing system experiments on operational statistical retrievals of satellite sounding data. Mon Wea Rev 119:1851-1864

Eumetsat (1991) Requirements for an operational interferometer thermal sounder. Eumetsat report, EUM/ESSWG/REP/91/1
Eyre JR (1989) Inversion of cloudy satellite sounding radiances by nonlinear optimal estimation: theory and simulation for TOVS. Q J R Meteorol Soc 115:1001-1026
Eyre JR (1990) The information content of data from operational satellite sounding systems: a simulation study. Q J R Meteorol Soc 116:401-434
Eyre JR (1991) A fast radiative transfer model for satellite sounding systems. ECMWF Technical Memorandum 176
Eyre JR (1992) A bias correction scheme for simulated TOVS brightness temperatures. ECMWF Technical Memorandum 186
Eyre JR, Kelly GA, McNally AP, Andersson E (1992) Assimilation of TOVS radiance information through one-dimensional variational analysis. ECMWF Technical Memorandum 187
Eyre JR, Lorenc AC (1989) Direct use of satellite sounding radiances in numerical weather prediction. Meteorol Mag 118:3-16
Kelly G (1992) Satellite observations for global monitoring. Adv Space Res 12:(7)263-275
Kelly G, Andersson E, Hollingsworth A, Lönnberg P, Pailleux J, Zhang, Z (1991) Quality control of operational physical retrievals of satellite sounding data. Mon Wea Rev 119:1866-1880
Liu TW, Tang Wenqing (1992) Precipitable water and surface humidity over global oceans from Special Sensor Microwave Imager and European Centre for Medium-Range Weather Forecasts. J Geophys Res 97C2:2251-2264
Lorenc AC, Adams W, Eyre JR (1986) The analysis of high resolution satellite data by the Meteorological Office. Proc ECMWF Workshop on "High Resolution Analysis"; Reading; 24-26 June 1985; ECMWF Report 69-87
McMillin LM, Dean C (1982) Evaluation of a new operational technique for producing clear radiances. J Appl Meteorol 21: 1005-1014

Pailleux J (1990) A global variational assimilation scheme and its application to TOVS radiances. Preprints WMO Intl Symp on "Assimilation of Observations in Meteorology and Oceanography"; Clermont-Ferrand, France; 9-13 July 1990; WMO Report 325-328
Pailleux J, Heckley W, Vasiljevic D, Thépaut J-N, Rabier F, Cardinali C, Andersson E (1991) Development of a variational assimilation system. ECMWF Technical Memorandum 179
Phalippou L (1992) Comparison between SSM/I and ECMWF total precipitable water. Proc Conf on "Microwave radiometry and remote sensing"; Boulder, Colorado; 14-16 January 1992.
Smith WL, Woolf HM, Hayden CM, Wark DQ, McMillin LM (1979) The TIROS-N Operational Vertical Sounder. Bull Am Meteorol Soc 60:1177-1187
Thoss A (1991) Cloud motion winds, validation and impact on numerical weather forecasts. Proc Workshop on "Wind extraction from operational meteorological satellite data"; Washington DC; 17-19 Sept 1991; Eumetsat Report

RETRIEVAL AND ASSIMILATION: SYSTEM CONSIDERATIONS

Andrew Lorenc
Meteorological Office (S)
London Road
Bracknell, RG12 2SZ
England

1 INTRODUCTION

The information content of a piece of information can only be defined with regard to what is already known. An expression, following Shannon, for this is given in section 3. But for practical applications, this is not a sufficient measure. We need also to consider the usefulness of the information.

Data assimilation, like the retrieval of information from remote sensing, can be considered as an inverse problem. The equations derived in section 3 are applicable to either problem. It is shown that the error of the forward process, estimating the observed parameters from the model parameters, must be considered in the inverse process, deducing model parameters from observations. In the infrared remote sensing which is the subject of this workshop, the forward process is the radiative transfer equation. If we cannot calculate that accurately, radiances are less useful. In data assimilation, the forward process is a Numerical Weather Prediction (NWP) model (or a very similar General Circulation Model (GCM) if the assimilation is for climate studies). Since assimilation is an inverse problem, if the model errors in prediction of a parameter are large, observations of that parameter are less useful. Thus, by definition, the model can make reasonably accurate predictions (if it has information from earlier observations) for observations which are useful in the assimilation.

This reasoning has lead some NWP centres to use the forecast fields as background information, when retrieving temperature soundings. Satellite radiances from the current TOVS instrument do not contain information about detailed vertical structure, compared to the model prediction. By using the forecast as prior to fill this lack of information (null space), instead of information from another source, information on larger vertical scales can

NATO ASI Series, Vol. I 9
High Spectral Resolution Infrared Remote Sensing
for Earth's Weather and Climate Studies
Edited by A. Chedin, M.T. Chahine and N.A. Scott

be extracted from the radiances without introducing less accurate information on smaller scales.

The Met Office was perhaps the first to use this approach operationally; some results are shown in section 2. Experiments at NMC (Baker 1991) and ECMWF (Pailleux *et al.* 1988, and Eyre this workshop), have confirmed that results are better than with separate retrieval and assimilation. Both centres plan to use this method in operational systems (called *Interactive retrieval* and *1DVAR* respectively) in summer 1992. The French *PERIDOT* system used radiances directly in an analysis with a forecast background in 1985 (Durand 1985). Although the next generation of high-spectral resolution sounders will give information on smaller vertical scale, the scales resolved by the model will also have increased, so this qualitative conclusion will probably still hold.

In section 2 the Met Office's research into the use of satellite soundings is reviewed. Development is concentrated on the use of a forecast background, and on accounting for cloud in the radiative transfer (rather than using "cloud-cleared" radiances). Possible strategies for developments to cope with high spectral resolution sounders are discussed. Clouds make the radiative transfer equation very nonlinear. Equations for the inverse problem, are derived in section 3, and nonlinearities are discussed. Final, in section 5, some conclusions and options are discussed.

2 METEOROLOGICAL OFFICE SATELLITE SOUNDING RETRIEVALS

The Met Office has since 1983 run a Local Area Satellite Sounding System (LASS). Initially the system was based on the International TOVS Processing Package from the University of Madison, run on a dedicated mini-computer (Turner *et al.* 1985), using direct read-out data from the TIROS satellite. A new cloud clearing scheme was developed (Eyre and Watts, 1987), but it proved difficult to demonstrate a positive impact of the data on the quality of operational forecasts. In 1987, the system was changed to use forecast backgrounds in the retrieval (Lorenc *et al.* 1986, Eyre and Lorenc, 1989). This is the scheme operational in March 1992.

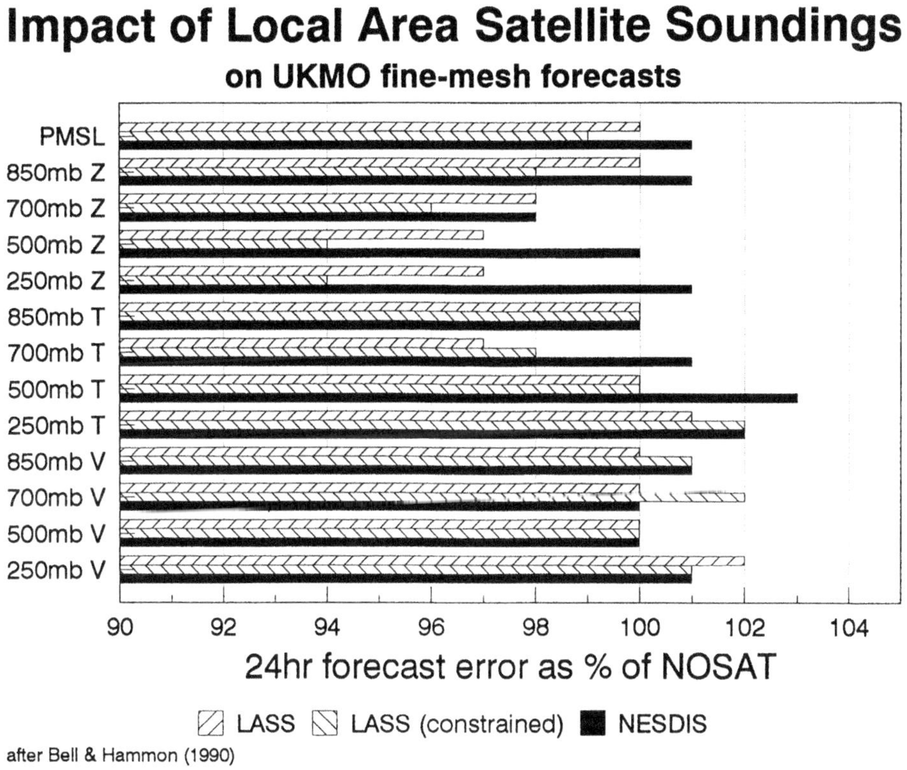

Fig 1. Ratios of rms forecast errors with and without satellite soundings.

Monitoring of the background profiles, and the retrieved profiles, against independent radiosondes does not show a clear impact from the retrieval process; the backgrounds fit the radiosondes about as well as the retrievals do. However impact tests on the quality of forecasts (Bell and Hammon, 1990) do show a slight positive impact. This depends somewhat on how the forecasts are verified. Figure 1 shows the error measured against radiosondes over the whole regional model area, of forecasts made using various forms of the TOVS satellite data, relative to the error in forecasts without the TOVS data. Each set of three bars shows the score for one verification parameter. The top bar in each set is for the operational system, with direct assimilation of the forecast background retrievals. As the retrievals have model information used as background, it is not statistically optimal to use them in this way. Lorenc *et al* (1986) proposed a constraint method to remove the background information (this is discussed further in section 3). The second bar shows results using this method. Results for height verification in particular seem to indicate it is a little better. The third bar shows results from forecasts using the NESDIS retrievals sent out on the Global Telecommunication System (GTS). If anything they have a slight negative impact on forecast quality. In our local area, the forecast backgrounds are usually very good, because even over the Atlantic there is a variety of data. The NESDIS retrievals, made not using these backgrounds, actually damage the model state when we assimilate them. (It might be possible to devise an assimilation method which extracts the useful information while avoiding the damaging bad information from NESDIS's backgrounds, but we have failed to find one).

Since there is little useful information in our local area, we are setting up a scheme (GLOSS) to apply the forecast background retrieval technique to the global cloud-cleared radiances, which are now sent on the GTS in binary (BUFR) code. It is hoped by this means to get useful information in areas where our model backgrounds are less reliable, for instance the Pacific and southern hemisphere. In order to have the computer power for this, and to facilitate closer coupling with the forecast model, GLOSS is being set up on the mainframe Cray YMP computer.

In parallel with the GLOSS development, a scheme for direct use of cloudy radiances is being developed (Eyre, 1989a&b). This technique inverts the radiative transfer equation including clouds. As well as temperature and humidity profiles, cloud parameters can be retrieved (Watts, this workshop). Over the next decade, the accuracy of representation of clouds in forecast models is expected to improve dramatically, so although at present there is no coupling with the model's clouds, it is expected that eventually the retrieved clouds will be used in the model. Because of poor resolution (compared to the instrument's footprint), only limited use will be made of background cloud parameters provided to the retrieval; cloud information from other instruments is needed also. The technique also extends naturally to the microwave sounders planned for ATOVS in 1995. The output from the radiative transfer equation is nonlinearly related to the cloud; the effect of low level temperatures depends strongly on the presence or absence of cloud. So an iterative retrieval method is necessary. Experience with the cloudy radiances in LASS is that about four iterations are necessary.

The retrieval methods described above are one-dimensional. Although background information from a model is horizontally consistent, and most cloud clearing algorithms use information from adjacent fields of view, the actual inversion from radiances to a profile is done for each sounding separately. The nonlinear scheme does not even use any horizontal consistence prior "information" about clouds. If the retrieval were made part of the model assimilation, the derived profiles would be constrained to be consistent with the model's physical parametrisation of cloud, and with other observations, both from the same instrument, and others. Since the assimilation method used at the Met Office (Lorenc Bell and Macpherson 1991) is iterative, it would be feasible to combine the iterative retrieval with the assimilation, by making the iteration module of GLOSS a subroutine of the assimilation. We hope to perform some simple tests along these lines. ECMWF are developing a combined retrieval (of clear-column radiances) and analysis (3DVAR), using the variational method set out in section 3. However it will be technically difficult to extend this to use raw radiances, since derived cloud parameters at the observation positions will have to be made part of the variational control variable, which at present consists only of the model's state parameters (Eyre, personal communication).

It is necessary when planning future operational systems to consider the computer requirements. In setting up GLOSS we estimate that for: ~75000 soundings per satellite per day of ~20 cloud-cleared channels, at ~120km spacing, GLOSS will take ~3500 CPUsec/day, on the CRAY YMP.

To extrapolate this we need the following factors:

2 satellites	2	
40km spacing	9	
cloudy radiances	4	(extra iterations)
ATOVS (1995?)	2	(~40 channels instead of 20)
IASI or AIRS (1998?)	100	(~2000 channels instead of 20)

Thus for ATOVS we need 100 Cray YMP single CPU hours per day. Although the computer is a multi processor, and is to be upgraded, this is still a very significant part of the expected capacity. We will have to justify this with a significant benefit, and probably cut back on the ideal system (e.g. by not processing every sounding; 40km resolution may not be needed globally).

For IASI or AIRS we might need 50 times the computer power needed for ATOVS. By the end of the decade, we are likely to have available massively parallel computers that can provide the required raw CPU power. It will however be difficult to apply these computers to the very large and complex forecasting systems used today in NWP. Reverting to a system dedicated to the retrieval problem may be the most economical way of providing the power. This however rules out our ideas about a closer coupling of the retrieval and assimilation.

3 BAYESIAN DERIVATION OF INVERSION, AND NONLINEARITIES

3.1 **Bayes Theorem**

Notation	
$\mathbf{x}$	atmosphere as represented in model
$\mathbf{x}_t$	model representation of the true state of the atmosphere
$\mathbf{x}_b$	prior estimate of $\mathbf{x}_t$ (e.g. from forecast)
$\mathbf{y}$	observations
$\mathbf{y}_t$	observations that would be given by error-free instruments
$K(\mathbf{x})$	forward operator for calculating $\mathbf{y}$ from $\mathbf{x}$
$\mathbf{K}$	tangent linear operator of K, such that $K(\mathbf{x}+\delta\mathbf{x})=K(\mathbf{x})+\mathbf{K}\delta\mathbf{x}+O(\delta\mathbf{x}^2)$.
P	probability
p	probability distribution function
$P(\mathbf{x})$	= probability that $\mathbf{x}\le\mathbf{x}_t<\mathbf{x}+\mathbf{dx}$ $= p(\mathbf{x})\mathbf{dx}$
$P(A\|B)$	the conditional probability of A, given B.

For satellite retrievals, $\mathbf{x}$ is an atmospheric profile, $\mathbf{y}$ is the set of radiances, and K is the radiative transfer equation. For assimilation, $\mathbf{x}$ is the model state, $\mathbf{y}$ is the set of all observations, and k is an interpolation or extrapolation (including an extrapolation in time).

This derivation follows Lorenc (1986). Probabilities are used in a Bayesian way to describe the state of information. We have some prior information[1] about **x**. We add to this information from observations **y**. We need to know the posterior knowledge about **x**. Operator K does not have a normal inverse. Since we start with our prior knowledge, all probabilities are conditional on knowing $\mathbf{x}_b$. To simplify notation we write $P(\cdot)$ instead of $P(\cdot|\mathbf{x}_b)$. Basic relationships for conditional probabilities give:

$$P(\mathbf{x}\cap\mathbf{y}) = P(\mathbf{x}|\mathbf{y}) \quad P(\mathbf{y}) = P(\mathbf{y}|\mathbf{x}) \quad P(\mathbf{x})$$
$$= p_a(\mathbf{x}|\mathbf{y})\mathbf{dx}\ p_{ofb}(\mathbf{y})\mathbf{dy} = p_{of}(\mathbf{y}|\mathbf{x})\mathbf{dy}\ p_b(\mathbf{x})\mathbf{dx}$$

where

$P(\mathbf{x}|\mathbf{y})$ $=p_a(\mathbf{x}|\mathbf{y})\mathbf{dx}$, is the analysis probability, i.e. the probability that $\mathbf{x}\le\mathbf{x}_t<\mathbf{x}+\mathbf{dx}$, given the background $\mathbf{x}_b$ and the observations.

$P(\mathbf{y})$ $=p_{ofb}(\mathbf{y})\mathbf{dy}$, is the probability of getting observations **y**.

$P(\mathbf{y}|\mathbf{x})$ $=p_{of}(\mathbf{y}|\mathbf{x})\mathbf{dy}$ is the probability of getting observations **y** given that $\mathbf{x}=\mathbf{x}_t$.

$P(\mathbf{x})$ $=p_b(\mathbf{x})\mathbf{dx}$, is the probability that $\mathbf{x}\le\mathbf{x}_t<\mathbf{x}+\mathbf{dx}$, given only the prior knowledge of $\mathbf{x}_b$.

$$p_{of}(\mathbf{y}|\mathbf{x}) = \int p_o(\mathbf{y}|\mathbf{y}_t\cap\mathbf{x})\ p_f(\mathbf{y}_t|\mathbf{x})\ \mathbf{dy}_t$$

$p_o(\mathbf{y}|\mathbf{y}_t\cap\mathbf{x})$ is the instrumental error distribution.

$p_f(\mathbf{y}_t|\mathbf{x})$ is the forward operator error distribution.

$$p_{ofb}(\mathbf{y}) = \int p_{of}(\mathbf{y}|\mathbf{x})\ p_b(\mathbf{x})\ \mathbf{dx}$$
$$= \int\int p_o(\mathbf{y}|\mathbf{y}_t\cap\mathbf{x})\ p_f(\mathbf{y}_t|\mathbf{x})\ \mathbf{dy}_t\ p_b(\mathbf{x})\ \mathbf{dx}$$

---

[1]Note that the whole retrieval philosophy is based on the idea that there is some useful prior information about **x**. Hence the name "prior", or "background", is to be preferred to "first-guess", which is often used. An ideal iterative method should give a result which is independent of the first-guess; it should not be independent of the background.

Bayes' Theorem can be derived from the above:

$$p_a(\mathbf{x}|\mathbf{y}) = p_{of}(\mathbf{y}|\mathbf{x})\ p_b(\mathbf{x})\ /\ p_{ofb}(\mathbf{y})$$

$$= \frac{\int p_o(\mathbf{y}|\mathbf{y}_t \cap \mathbf{x})\ p_f(\mathbf{y}_t|\mathbf{x})\ d\mathbf{y}_t\ p_b(\mathbf{x})}{\int\int p_o(\mathbf{y}|\mathbf{y}_t \cap \mathbf{x})\ p_f(\mathbf{y}_t|\mathbf{x})\ d\mathbf{y}_t\ p_b(\mathbf{x})\ d\mathbf{x}}$$

This p.d.f. describes our total posterior information about $\mathbf{x}$, given $\mathbf{x}_b$ and $\mathbf{y}$. It is exact; no assumptions have been made. The information content, following Shannon's ideas of information entropy (e.g. see Tarantola 1987 p28) is given by

$$\mathcal{I} = \int p_a(\mathbf{x}|\mathbf{y})\ \ln\left(\frac{p_a(\mathbf{x}|\mathbf{y})}{p_b(\mathbf{x})}\right)\ d\mathbf{x}.$$

Note that this definition makes no allowance for the relative importance or usefulness of the different components of $\mathbf{x}$.

3.2 Linear retrieval for Gaussian p.d.f.s

If we assume all the p.d.f.s are Gaussian, and K can be linearized in the region of $\mathbf{x}_b$ and $\mathbf{x}_a$ such that

$$K(\mathbf{x}_a) = K(\mathbf{x}_b) + \mathbf{K}\ (\mathbf{x}_a - \mathbf{x}_b),$$

then, using the properties of Gaussians, we have:
the background error distribution:

$$p_b(\mathbf{x}) = \mathrm{N}(\mathbf{x}_b - \mathbf{x}, \mathbf{B}),$$

the instrumental error distribution:

$$p_o(\mathbf{y}|\mathbf{y}_t \cap \mathbf{x}) = \mathrm{N}(\mathbf{y} - \mathbf{y}_t, \mathbf{O}),$$

the forward operator error distribution:

$$p_f(\mathbf{y}_t|\mathbf{x}) = \mathrm{N}(\mathbf{y}_t - K(\mathbf{x}), \mathbf{F}),$$

where $\mathbf{B}$ $\mathbf{O}$ and $\mathbf{F}$ are covariances.

The observational error distribution, (knowing $\mathbf{x}_t$) is given by:

$$p_{of}(\mathbf{y}|\mathbf{x}) = \int p_o(\mathbf{y}|\mathbf{y}_t \cap \mathbf{x})\ p_f(\mathbf{y}_t|\mathbf{x})\ d\mathbf{y}_t$$
$$= N(\mathbf{y}-K(\mathbf{x}_t), \mathbf{O}+\mathbf{F})$$

where $\mathbf{O}+\mathbf{F}$ (=E) is the observational error covariance.

The observation distribution, (only knowing $\mathbf{x}_b$) is given by:

$$p_{ofb}(\mathbf{y}) = N(\mathbf{y}-K(\mathbf{x}_b), \mathbf{O}+\mathbf{F}+\mathbf{KBK}^T).$$

Substituting in Bayes' Theorem gives:

$$p_a(\mathbf{x}|\mathbf{y}) = p_{of}(\mathbf{y}|\mathbf{x})\ p_b(\mathbf{x}) \ /\ p_{ofb}(\mathbf{y})$$
$$= N(\mathbf{y}-K(\mathbf{x}_t), \mathbf{O}+\mathbf{F})\ N(\mathbf{x}_b-\mathbf{x}, \mathbf{B}) \ /\ N(\mathbf{y}-K(\mathbf{x}_b), \mathbf{O}+\mathbf{F}+\mathbf{KBK}^T)$$
$$= N(\mathbf{x}_a-\mathbf{x}, \mathbf{A}).$$

where $\mathbf{x}_a$ and $\mathbf{A}$ are defined by

$$\mathbf{A} = \mathbf{B} - \mathbf{BK}^T(\mathbf{KBK}^T+\mathbf{E})^{-1}\mathbf{KB}$$

$$\mathbf{x}_a = \mathbf{x}_b + \mathbf{BK}^T(\mathbf{KBK}^T+\mathbf{E})^{-1}(\mathbf{y}-K(\mathbf{x}_b)).$$

3.3 **Variational retrievals**

If K is more nonlinear, or the p.d.f.s are non-Gaussian, then the direct solution derived above cannot be used, although the Bayes' Theorem for the analysis p.d.f. is still valid:

$$p_a(\mathbf{x}|\mathbf{y}) = p_{of}(\mathbf{y}|\mathbf{x})\ p_b(\mathbf{x}) \ /\ p_{ofb}(\mathbf{y}).$$

The expression for p_a which results is usually too complicated to be very useful in describing our knowledge about $\mathbf{x}$; we want an estimate of the "best", most probable, $\mathbf{x}$. In order to avoid a definition of "best" which depends of the choice of model basis for $\mathbf{x}$, we normalize the probabilities with respect to a null prior, $p_{null}(\mathbf{x})$, a probability distribution which describes our knowledge before we knew even $\mathbf{x}_b$. We further assume, for simplicity, that the basis for $\mathbf{x}$ has been chosen so that $p_{null}(\mathbf{x})$ is a constant over the range of values of interest[2]. Then the most probable $\mathbf{x}$ is that which maximizes $p_a(\mathbf{x}|\mathbf{y})/p_{null}(\mathbf{x})$. Since $p_{null}(\mathbf{x})$ and $p_{ofb}(\mathbf{y})$ are independent of $\mathbf{x}$, this is the same as the $\mathbf{x}$ which minimizes a penalty function $\mathcal{J}$ given by

$$\mathcal{J} = -\ln(p_{of}(\mathbf{y}|\mathbf{x})) - \ln(p_b(\mathbf{x})).$$

If we substitute the Gaussian p.d.f.s of the last section into this, we get:

$$\mathcal{J} = \frac{1}{2}(\mathbf{y}-K(\mathbf{x}))^T(\mathbf{O}+\mathbf{F})^{-1}(\mathbf{y}-K(\mathbf{x})) + \frac{1}{2}(\mathbf{x}_b-\mathbf{x})^T\mathbf{B}^{-1}(\mathbf{x}_b-\mathbf{x}) + \text{constant}.$$

If furthermore we make K linearizable, we see why the linear problem with Gaussians is easier to solve: $\mathcal{J}$ becomes a quadratic in $\mathbf{x}$. Using the same algebraic manipulations as are needed to establish the properties of Gaussians used in the last section, gives:

$$\mathcal{J} = \frac{1}{2}(\mathbf{x}_a-\mathbf{x})^T\mathbf{A}^{-1}(\mathbf{x}_a-\mathbf{x}) + \text{constant}.$$

where $\mathbf{x}_a$ and $\mathbf{A}$ are defined by

$$\mathbf{A} = \mathbf{B} - \mathbf{B}\mathbf{K}^T(\mathbf{K}\mathbf{B}\mathbf{K}^T+\mathbf{O}+\mathbf{F})^{-1}\mathbf{K}\mathbf{B}$$

$$\mathbf{x}_a = \mathbf{x}_b + \mathbf{B}\mathbf{K}^T(\mathbf{K}\mathbf{B}\mathbf{K}^T+\mathbf{O}+\mathbf{F})^{-1}(\mathbf{y}-K(\mathbf{x}_b)).$$

If K cannot be linearized over the whole range containing $\mathbf{x}_b$ and possible $\mathbf{x}_a$s, then an explicit solution is not possible. If K is still differentiable, so that

$$K(\mathbf{x}+\delta\mathbf{x}) = K(\mathbf{x})+\mathbf{K}_{\mathbf{x}}\delta\mathbf{x}, \quad \text{as } \delta\mathbf{x}\to 0$$

then we can look for the minimum of $\mathcal{J}$ using a descent algorithm. At the minimum, the gradient of $\mathcal{J}$ with respect to the components of $\mathbf{x}$ is zero:

$$\mathcal{J}' = -\mathbf{K}_{\mathbf{x}}^T(\mathbf{O}+\mathbf{F})^{-1}(\mathbf{y}-K(\mathbf{x})) - \mathbf{B}^{-1}(\mathbf{x}_b-\mathbf{x}) = 0.$$

This formula is exact; we can find the most probable $\mathbf{x}$. The next stage of generalization is to allow the p.d.f.s to be weakly non-Gaussian. That is, we use the Gaussian formulae with $\mathbf{O}_{\mathbf{x}}$ $\mathbf{F}_{\mathbf{x}}$ and $\mathbf{B}_{\mathbf{x}}$ being slowly varying

---

[2] Note that if we change variables from $\mathbf{x}$ to $\mathbf{u}$ by a nonlinear mapping, then $p_{null}(\mathbf{u})$ is not a constant.

functions of $\mathbf{x}$, whose derivatives we can neglect. We also neglect derivatives of $\mathbf{K}_{\mathbf{x}}$. Then if we define $\mathbf{x}_a$ as the $\mathbf{x}$ which minimizes $\mathcal{J}$, i.e.

$$\mathcal{J}' = -\mathbf{K}_{\mathbf{x}a}{}^{T}(\mathbf{O}_{\mathbf{x}a}+\mathbf{F}_{\mathbf{x}a})^{-1}(\mathbf{y}-K(\mathbf{x}_a)) - \mathbf{B}_{\mathbf{x}a}{}^{-1}(\mathbf{x}_b-\mathbf{x}_a) = 0.$$

Then

$$\mathcal{J}'' \cong \mathbf{K}_{\mathbf{x}a}{}^{T}(\mathbf{O}_{\mathbf{x}a}+\mathbf{F}_{\mathbf{x}a})^{-1}\mathbf{K}_{\mathbf{x}a} + \mathbf{B}_{\mathbf{x}a}{}^{-1} = \mathbf{A}^{-1}.$$

Then, in the neighbourhood of $\mathbf{x}_a$,

$$p_a(\mathbf{x}|\mathbf{y}) \propto \mathrm{N}(\mathbf{x}_a-\mathbf{x},\mathcal{J}''^{-1}).$$

If K is sufficiently nonlinear, or the p.d.f.s are sufficiently non-Gaussian, $p_a(\mathbf{x}|\mathbf{y})$ may have multiple maxima. We have then to consider how to decide which is best. It may well be that the "best" $\mathbf{x}$ is that which is most likely to be in a finite, rather than an infinitesimal, region around $\mathbf{x}_t$. This likelihood may then be more related to the integral under the above approximate Gaussian, rather than its peak value. To find the $\mathbf{x}$ which maximizes this integral, we should not search for the maximum of $p_a(\mathbf{x}|\mathbf{y})$, but rather $|\mathbf{A}|^{1/2}p_a(\mathbf{x}|\mathbf{y})$. Note that between local maxima, $|\mathcal{J}''|$ may become zero. In these regions our assumptions about slowly varying $\mathbf{K}$ $\mathbf{B}$ $\mathbf{O}$ and $\mathbf{F}$ are not good. However our approximation $|\mathbf{A}^{-1}|$ remains positive.

3.4 Weight given to Background

If we perturb $\mathbf{x}_a$ to $\mathbf{x}_a+\delta\mathbf{x}_a$, and $\mathbf{x}_b$ to $\mathbf{x}_b+\delta\mathbf{x}_b$, then

$$\mathcal{J}' = \mathcal{J}''\delta\mathbf{x}_a - \mathbf{B}^{-1}\delta\mathbf{x}_b.$$

So to remain a solution of $\mathcal{J}'=0$

$$\delta\mathbf{x}_a = \mathcal{J}''^{-1}\,\mathbf{B}^{-1}\delta\mathbf{x}_b = \mathbf{A}\,\mathbf{B}^{-1}\delta\mathbf{x}_b.$$

This can be used to find the $\tilde{\mathbf{x}}_a$ which would have been found using a slightly different background $\tilde{\mathbf{x}}_b$, without repeating the full nonlinear solution.

If we wish to repeat the analysis, using a slightly different background and perhaps additional observations, then we can use

$$\frac{1}{2}(\mathbf{y}-K(\mathbf{x}))^T(\mathbf{O}+\mathbf{F})^{-1}(\mathbf{y}-K(\mathbf{x})) = \frac{1}{2}(\mathbf{x}_a-\mathbf{x})^T\mathbf{A}^{-1}(\mathbf{x}_a-\mathbf{x}) - \frac{1}{2}(\mathbf{x}_b-\mathbf{x})^T\mathbf{B}^{-1}(\mathbf{x}_b-\mathbf{x}) + \text{const.}$$

Then we have the new penalty given by

$$\tilde{\mathcal{J}} = \frac{1}{2}(\mathbf{x}_a-\mathbf{x})^T\mathbf{A}^{-1}(\mathbf{x}_a-\mathbf{x}) - \frac{1}{2}(\mathbf{x}_b-\mathbf{x})^T\mathbf{B}^{-1}(\mathbf{x}_b-\mathbf{x}) + \frac{1}{2}(\tilde{\mathbf{x}}_b-\mathbf{x})^T\mathbf{B}^{-1}(\tilde{\mathbf{x}}_b-\mathbf{x})$$

+ penalty term for additional observations + constant.

If we define $\mathbf{C}$ and $\mathbf{x}_c$ by

$$\mathbf{C}^{-1} = \mathbf{A}^{-1} - \mathbf{B}^{-1}$$

$$\mathbf{C}^{-1}\mathbf{x}_c = \mathbf{A}^{-1}\mathbf{x}_a - \mathbf{B}^{-1}\mathbf{x}_b.$$

Or equivalently

$$\mathbf{C} = (\mathbf{I}-\mathbf{AB}^{-1})^{-1}\mathbf{A}$$

$$\mathbf{x}_c = \mathbf{x}_b + (\mathbf{I}-\mathbf{AB}^{-1})^{-1}(\mathbf{x}_a-\mathbf{x}_b).$$

Then

$$\tilde{\mathcal{J}} = \frac{1}{2}(\mathbf{x}_c-\mathbf{x})^T\mathbf{C}^{-1}(\mathbf{x}_c-\mathbf{x}) + \frac{1}{2}(\tilde{\mathbf{x}}_b-\mathbf{x})^T\mathbf{B}^{-1}(\tilde{\mathbf{x}}_b-\mathbf{x})$$

+ penalty term for additional observations + constant.

$\mathbf{x}_c$ can be thought of as a linearized representation of the information in the original observations, without any information from $\mathbf{x}_b$. It can be reused in a subsequent analysis (with the same or a different background) as a pseudo-observation.

In practice, things are not this simple. Because of the null-space not properly defined by the observations, $(\mathbf{I}-\mathbf{AB}^{-1})$ is not invertible. A pseudo-inverse can however be defined by excluding the null-space (Purser 1990).

The method used in assimilating the LASS retrievals for the second experiment shown in figure 1, was a simplified version of this (Lorenc *et al.* 1986), appropriate for the horizontal and vertical splitting of the assimilation method (Lorenc *et al.* 1991).

4 DISCUSSION OF THE FUTURE

Experience with the current generation of sounders, in particular the failure despite years of effort to get consistent positive impact in the northern hemisphere, until the forecast background methods were developed, indicates that for data assimilation uses we should stay with the forecast background retrieval method. In this, information from the model is available during the retrieval, and only the new information from the sounding is used in the model. (Note that this conclusion may not apply to other uses, such as atmospheric studies, and climate change detection.)

Improved cloud and moisture information in the soundings from future instruments, and in the NWP models and GCMs currently being developed, means that this two-way interaction will be extended to cloud and moisture parameters.

There are theoretical advantages in combining the retrieval and assimilation even more closely, so that the interaction takes place within a single (iterative) solution of the combined inverse problem.

However this will require a large computer system, capable of simultaneously handling both the retrieval and assimilation steps. It may be more cost effective to perform a forecast background retrieval on one system, and use it on another. A linearized correction to the retrieval, to take account of the background information used, is possible during the assimilation process. This requires that background profile used for the retrieval, and some covariance statistics, to be passed to the assimilation along with the retrieval. (Unfortunately the effect of cloud are nonlinear, so an accurate correction for changes in prior cloud information is not possible in this method).

Until such time as decisions have to be made about computer and system architecture, it would be wise to develop retrieval and assimilation as compatible systems, to keep both options open. Future decisions must take account of the operational benefits achieved, relative to the effort expended.

REFERENCES

Baker, Wayman. 1991. "Joint NESDIS/NASA/NMC effort to develop an advanced satellite retrieval system". A.M.S. Preprints of 9th Conf. on NWP. 113-114.

Bell,R.S., and Olive M Hammon, 1990. "The impact of satellite sounding data in the Meteorological Office fine-mesh model". Preprints, WMO International Symposium on Assimilation of Observations in Meteorology and Oceanography, Cleremont-Ferand, France. July 1990.

Durand, Y. 1985 "The use of satellite data in the French high resolution analysis." Proceedings of the 1985 ECMWF workshop on high resolution analysis, 24-26 June 1985, 89-128.

Eyre,J.R. 1989. "Inversion of cloudy satellite sounding radiances by nonlinear optimal estimation: theory and simulation for TOVS." Quart. J. Roy. Met. Soc., 115, 1001-1026.

Eyre,J.R. 1989. "Inversion of cloudy satellite sounding radiances by nonlinear optimal estimation: application to TOVS data." Quart. J. Roy. Met. Soc., 115, 1027-1038.

Eyre,J.R. and Lorenc,A.C. 1989. "Direct use of satellite sounding radiances in numerical weather prediction". Met. Mag. 118, 13-16

Eyre,J.R., and Watts,P.D. 1987. "A sequential estimation approach to cloud clearing for satellite temperature sounding" Quart.J.Roy.Met.Soc., 113, 1349-1376.

Lorenc, A.C. 1986. "Analysis methods for numerical weather prediction." Quart. J. Roy. Met. Soc. 112, 1177-1194.

Lorenc,A.C., Bell,R.S., and Macpherson,B. 1991. "The Meteorological Office "Analysis Correction" data assimilation scheme". Quart. J. Roy. Met. Soc. 117, 59-89

Lorenc,A.C., Adams,W., and Eyre,J.R. 1985. "The analysis of high-resolution satellite data in the Met Office." Met O 11 Tech. Note 212. Proceedings of ECMWF workshop on high-resolution analysis, June 1985. 69-88.

Pailleux, J., G. Kelly, J-F. Flobert, E. Andersson 1988 "Current work at ECMWF on the use of satellite data." Proceedings of the 1988 ECMWF seminar on data assimilation and the use of satellite data, 5-9 September 1988, vol.II, 215-233.

Purser,R.J. 1990. "Vertical aspects of the assimilation of sounding data". Preprints, WMO International Symposium on Assimilation of Observations in Meteorology and Oceanography, Cleremont-Ferand, France. July 1990.

Tarantola,A. 1987. "Inverse Problem Theory - Methods for data fitting and model parameter estimation". publ. Elsevier. 613pp. ISBN 0-444-42765-1

Turner,J., Eyre,J.R., Jerrett,D., and McCallum,E. 1985. "The HERMES System". Meteorol.Mag., 114, 161-173.

ON MEASURING THE GREENHOUSE EFFECT OF EARTH

Graeme L. Stephens, Anthony Slingo+, and Mark Webb+
Department of Atmospheric Science
Colorado State University
Fort Collins, CO 80523

1. INTRODUCTION

A balance between the amount of sunlight absorbed by the planet and the thermal radiation emitted to space is the primary control of the Earth's climate. Almost 70% of the solar energy reaching the earth is absorbed by the planet (e.g., Vonder Haar and Suomi, 1971), mainly at the surface. This warms the surface which then emits radiation towards space at longer infrared wavelengths. Radiatively active gases, such as water vapour and carbon dioxide, together with cloud particles, absorb radiation emitted from the surface. These constituents also re–emit infrared radiation but at temperatures which are lower than the surface temperature. This absorption and re–emission provides an effective mechanism for 'blocking' the surface emission from escaping to space and helps to maintain higher temperatures than would otherwise occur in the absence of the atmosphere. We call this the greenhouse effect (GE) of the atmosphere based on an analogy with the operation of an actual greenhouse, although this analogy is itself contentious (Bohren, 1987).

Knowledge of GE is fundamental to our understanding of how atmospheres evolved in the past (e.g., Ingersoll,1969) and, in particular, how the Earth's atmosphere is likely to evolve in the future. It is thus crucial that we develop a clear understanding of the different roles of gases and clouds in producing the greenhouse effect, especially given the possibility of feedbacks between the emission of radiation and the concentrations of these constituents. It is also crucial that we develop ways to exploit current satellite data, both to quantify the GE and to provide important tests of current global models.

The purpose of this paper is to outline an observational strategy to provide a quantitative assessment of the role of water vapour in establishing the GE. The strategy is intended to be simple enough that it can be used to test present–day general circulation

+ Hadley Centre for Climate Prediction and Research, Met. Office, Bracknell, UK

NATO ASI Series, Vol. I 9
High Spectral Resolution Infrared Remote Sensing
for Earth's Weather and Climate Studies
Edited by A. Chedin, M.T. Chahine and N.A. Scott

models (GCMs) of climate. The paper hints at the potential of spectral infrared emission measurements in unravelling various complex issues such as the effects of water vapour and its distribution on GE. The focus of this paper is directed toward understanding water vapour and its role on the clear sky greenhouse effect. Discussion of the important role of clouds on this effect (e.g., Stephens and Greenwald, 1991a) is omitted.

2. OBSERVATIONS OF VERTICALLY INTEGRATED WATER VAPOUR

We show below that the principal component of the Earth's greenhouse is atmospheric water vapour. Knowledge about how water vapour is distributed in the atmosphere, how it is transported by atmospheric motions and, perhaps most importantly of all, how it is transformed into clouds is therefore fundamental to understanding the Earth's greenhouse effect. Unfortunately, water vapour is notoriously difficult to measure. Routine observations from weather balloons are limited in both their spatial and temporal coverage, generally accurate to no better than about 10 % and biased to the larger landmasses of the planet. Satellite radiometric observations are near the point where they can provide the spatial coverage necessary to obtain global information about water vapour. Unfortunately, present satellite observing methods suffer their own limitations. Recent microwave observations, such as from the Special Sensor Microwave/Imager (SSM/I, Hollinger, 1987), provide perhaps the most unambiguous measurements of water vapour, but these are restricted to ocean regions and provide information only about the vertically integrated water path (this is known as the precipitable water and we refer to this quantity by the symbol w). Current retrieval methods based on spectral infrared measurements provide, in principle, a more complete global coverage, as well as offering some limited vertical information. However, these observations are biased to clear skies and the accuracy of infrared retrievals is still in question and a matter for current research.

The research described below combines 12 months of microwave water vapour observations (from March 1988 to February 1989, inclusive) with sea surface temperature data and Earth radiation budget data available from the Earth Radiation Budget Experiment (ERBE, Barkstrom, 1984). The microwave water vapour data are a crucial part of the analyses and Fig. 1 provides a broad perspective on the methods used to derive

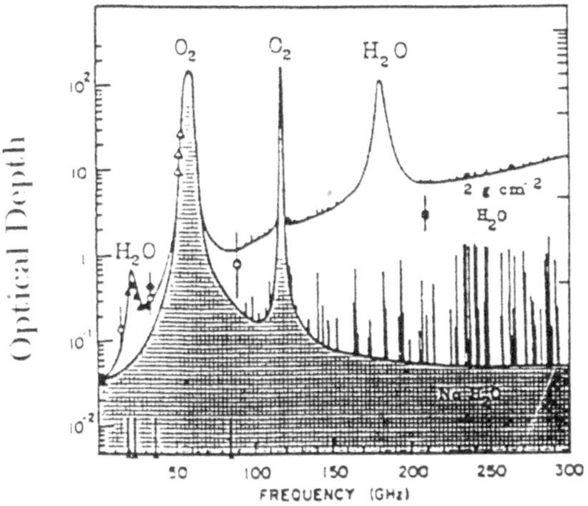

Fig. 1: The absorption spectrum of the atmosphere at microwave frequencies highlighting the location of the four SSM/SSM/II channels at 19.35, 22.235, 37 and 85.5 GHz.

water vapour from microwave emission. The diagram presents the microwave emission spectrum (the function plotted is proportional to the microwave optical depth). The key feature of this spectrum for the data used in this study is the line centred at 22.235 GHz. The SSM/I provides measurements in and around this line and a number of retrieval methods have been developed to estimate w from these radiance measurements. The results of Jackson (1992) who compares four of these methods with selected radiosonde data are summarized in Table 1. The comparisons reveal that the microwave derived values of w agree with radiosonde values typically within 4 kgm^{-2}.

Figure 2 presents the mean of the 12 months analysed (labelled annual in the upper panel) and the 1989 January and 1988 July monthly mean values of w (middle and lower panels) obtained from the Greenwald et al. (1992) retrieval approach plotted respectively as a function of the corresponding mean sea surface temperatures (SSTs). The relationship between w and SST resembles one expected from simple thermodynamic principles (Stephens, 1990) and the general increase of w with increasing SST is an important and necessary condition for the existence of the water vapour feedback

(Stephens and Tjemkes, 1992). However, this relationship varies from season to season, as comparison of the January and July results in Fig. 2 reveal especially for SSTs less than about 290 K. Different relationships between w and SST for the summer and the winter hemispheres exist and are represented by the two branches in the scatter of points for both months considered. These results suggest that transport of water vapour in the middle latitudes during summer exceeds that during winter and this transport significantly enhances w in these regions thus altering the overall relationship between w and SST there. Independent observations of moisture transport that may be studied to confirm this interpretation are presently lacking although the climatologies described by Piexoto and Oort (1992) do not show any marked inter-hemispheric differences in seasonal moisture transports.

Table. 1: The root mean square (RMS), correlation (COR) and bias (BIS) of PWC (in kgm^{-2}) for the retrieval methods of Greenwald et al. (1992) (G), Alishouse et al. (1990) (A), Schluessel and Emery (1990) (SE) and Petty and Katsaros (1990) (PK).

NMC Radiosonde Statistics				
January 1989				
	G	AL	SE	PK
RMS	.25	4.41	4.52	4.96
COR	.925	.956	.947	.944
BIS	.21	.18	-.31	-.94
JULY 1989				
	G	AL	SE	PK
RMS	4.70	4.24	4.96	4.61
COR	.93	.940	.909	.927
BIS	.63	.16	-1.01	-.53

3. A SIMPLE MODEL OF THE EARTH'S GREENHOUSE EFFECT

Before studying the Earth's GE and its relationship to water vapour in any detail, it is first necessary to introduce a quantity to define it. There are several ways to do this. Perhaps the simplest way is to use the difference between the mean radiating temperature of the planet and the surface temperature (Kondratyev and Moskalenko,

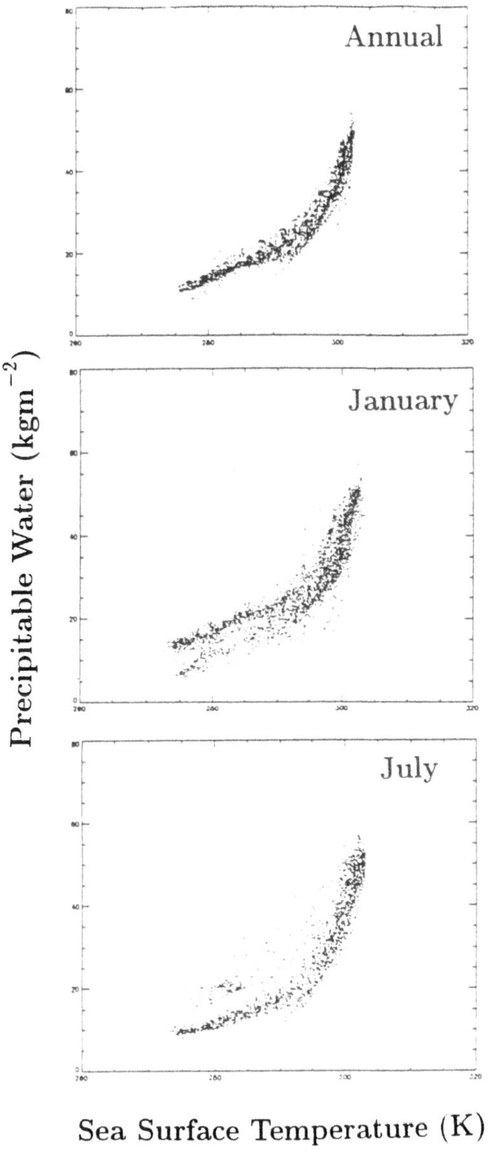

Fig. 2: The correlation of w with SST based on annual mean (upper panel), 1989 January mean (middle panel) and the 1988 July mean data (lower panel).

1984; Stephens and Tjemkes, 1992). However, the relationship between this temperature difference and the concentration of emitting species cannot be simply and conveniently defined. Raval and Ramanathan (1989) introduce the atmospheric absorption,

$$G_a = \sigma T_s^4 - F_\infty \tag{1}$$

as a convenient measure of GE where T_s is the sea surface temperature and F_∞ is the outgoing longwave radiation (OLR) at the top of the atmosphere. This definition, like that introduced below, applies only over the oceans where the surface emission may be taken to approximate closely a black body of temperature T_s. While we expect a general relationship between the absorption path of gas and G_a based on our physical understanding of molecular absorption, it is not clear *a priori* precisely what form this relationship should take. We now introduce a simple model of the greenhouse both as a means of defining the strength of the greenhouse and as a way of establishing such a relationship.

While the model adopted is described briefly in Stephens and Greenwald (1991b), a more detailed account is given here. The model of the greenhouse effect is based on simple considerations of radiative equilibrium. The radiative transfer equations for a non scattering atmosphere are

$$\mu \frac{dI(\tau,\mu)}{d\tau} = I(\tau,\mu) - B(T) \tag{2a}$$

$$-\mu \frac{dI(\tau,-\mu)}{d\tau} = I(\tau,-\mu) - B(T) \tag{2b}$$

where I is the intensity, $\mu = \cos\theta$ where θ is the angle of the beam from the zenith. The optical depth is defined as

$$\tau = \int_z^\infty k\rho_a dz \tag{3}$$

where k is the mass absorption coefficient, ρ_a is the density of the absorbing gas and z is the lowest end point of the path. Casting (2a) and (2b) into equations for upward ($F^\uparrow$) and downward ($F^\downarrow$) hemispheric fluxes leads to

$$\frac{dF^\uparrow}{d\tilde{\tau}} = F^\uparrow - \pi B, \tag{4a}$$

$$-\frac{dF^\downarrow}{d\tilde{\tau}} = F^\downarrow - \pi B, \tag{4b}$$

where

$$\tilde{\tau} = \frac{3}{2}\tau. \tag{5}$$

If we apply these equations to the broadband, then we can replace πB with σT^4 and treat $\tilde{\tau}$ as the greybody optical depth. It follows that in radiative equilibrium

$$F^{\uparrow}(\tilde{\tau} = 0) = Q_o(1 - \alpha) = F_{\infty}$$

where Q_o is the global–annual mean incoming solar radiation at the top of the atmosphere and α is the planetary albedo. It also follows that

$$F_{net} = F^{\uparrow}(\tilde{\tau}) - F^{\downarrow}(\tilde{\tau}) = const$$

which leads to the solution of (4a) and (4b) as

$$\sigma T_s^4 = \frac{F_{\infty}}{2}[2 + \tilde{\tau}_s] \tag{6a}$$

$$F_g^{\downarrow} = F^{\downarrow}(\tilde{\tau}_s) = \frac{F_{\infty}}{2}\tilde{\tau}_s \tag{6b}$$

at the surface where $\tau = \tau_s$ and $\tilde{\tau} = \tilde{\tau}_s$. This leads to a measure of the planetary greenhouse effect (e.g., Stephens and Greenwald, 1991b) as

$$\mathcal{G} = \frac{\sigma T_s^4}{F_{\infty}} = a + b\tau_s \tag{7}$$

where from (6a) and (5), a=1 and b=3/4. This definition identifies the grey body optical depth as the key parameter in defining the strength of the greenhouse. The spectral optical depth is a complex function of wavelength and several ways of spectrally averaging τ exist. The approach used here is one appropriate to the radiative equilibrium arguments introduced to arrive at (7) and follows from the following flux–mean mass absorption coefficient (Mihalas, 1978)

$$k_f = F_{\infty}^{-1} \int k_f F_{\infty,\lambda} d\lambda$$

The contribution to τ by water vapour, carbon dioxide and other minor greenhouse gases present in the Earth's atmosphere obtained using k_f in (3) is shown in Fig. 3. The spectral absorption data used to derive k_f are those tabulated by Rothman (1981). The total greybody optical depth derived from these data is $\tau = 3.9$ corresponding to $w = 28$ kgm^{-2} which is close to the global mean values of w. The value $\tau = 3.9$ is larger than the value of the optical depth derived from (7). This highlights the ambiguity of the quantitative meaning of τ and to understand the value it is necessary to understand how it is derived. The advantage of the flux weighted value is that it allows us to estimate τ from spectral integration and thus the contributions by individual gases. These contributions are shown in Fig. 3 which clearly emphasizes the dominance of water vapour to the total greybody optical depth for the global mean conditions considered.

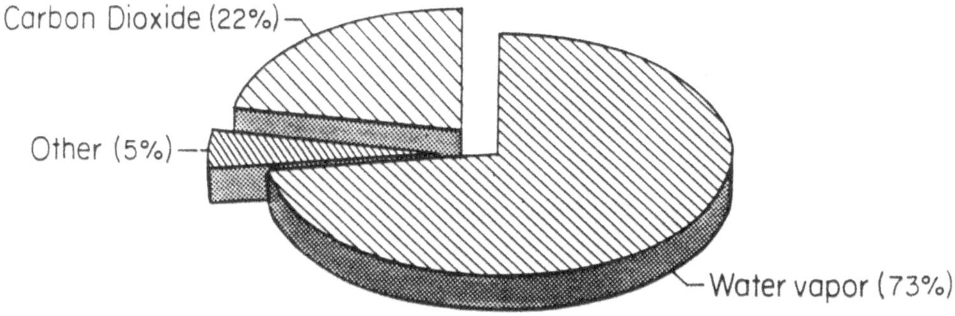

Fig. 3: A pie diagram showing the percentage of the grey body optical depth due to water vapour and other greenhouse gases in the Earth's atmosphere based on typical mean global concentrations of these gases.

4. OBSERVATIONS OF THE GREENHOUSE EFFECT

Let us simply assume that in the case of water vapour, the flux–weighted mean optical depth is

$$\tau_f = \int k_f \rho_a dz \approx k_f w.$$

This assumption, together with (7) then leads to the following (e.g. Stephens and

Greenwald, 1991b)

$$\mathcal{G} = a + cw. \quad (8)$$

The advantage of this relationship over others is that all factors in (8) are independently observed over the global oceans, primarily from satellites. For example, the SST can be obtained from independent analyses of blended ship, buoy and satellite data (e.g. Reynolds, 1984), the OLR available from ERBE data and w follow from microwave measurements described above. The annual, January and July monthly mean values of $\mathcal{G}$ derived in this way are plotted against corresponding mean values of w in Figure 4. The solid line through the scatter of annual mean points depicts the average of these points and the slope parameter $c = 0.00634$ $(\text{kgm}^{-2})^{-1}$ which is estimated via a least squares fit of the data (shown as the solid line in Fig 4) is a measure of how the greenhouse effect changes for given changes in w and is potentially important in analyses of water vapour feedback.

The analyses presented above provides a simple view of the influences of water vapour on the GE. However, a number of factors can potentially influence F_∞ independent of variations in w and thus introduce the scatter exhibited in Fig. 4. For example, increasing water vapour in the middle and upper troposphere only marginally affects w but can significantly alter F_∞. Variations in the vertical profile of temperature may also change F_∞ and thus $\mathcal{G}$. These two factors, we argue, are largely responsible for the observed scatter in the relationships shown.

If we suppose that w is a simple function of SST and express this function as (Stephens, 1990)

$$w = d\exp[r(T_s - 288)], \quad (9)$$

then a simple combination of this expression with (8) yields

$$\mathcal{G} = a + cd\exp[r(T_s - 288)], \quad (10)$$

and the sensitivity of $\mathcal{G}$ to SST follows as

$$g = \frac{d\mathcal{G}}{dT_s} = cdr\exp[r(T_s - 288)]. \quad (11)$$

Figure 5 presents $\mathcal{G}$ as a function of SST and demonstrates a general increasing trend in $\mathcal{G}$ with increasing SST with a non-linear increase as the SST exceeds approximately

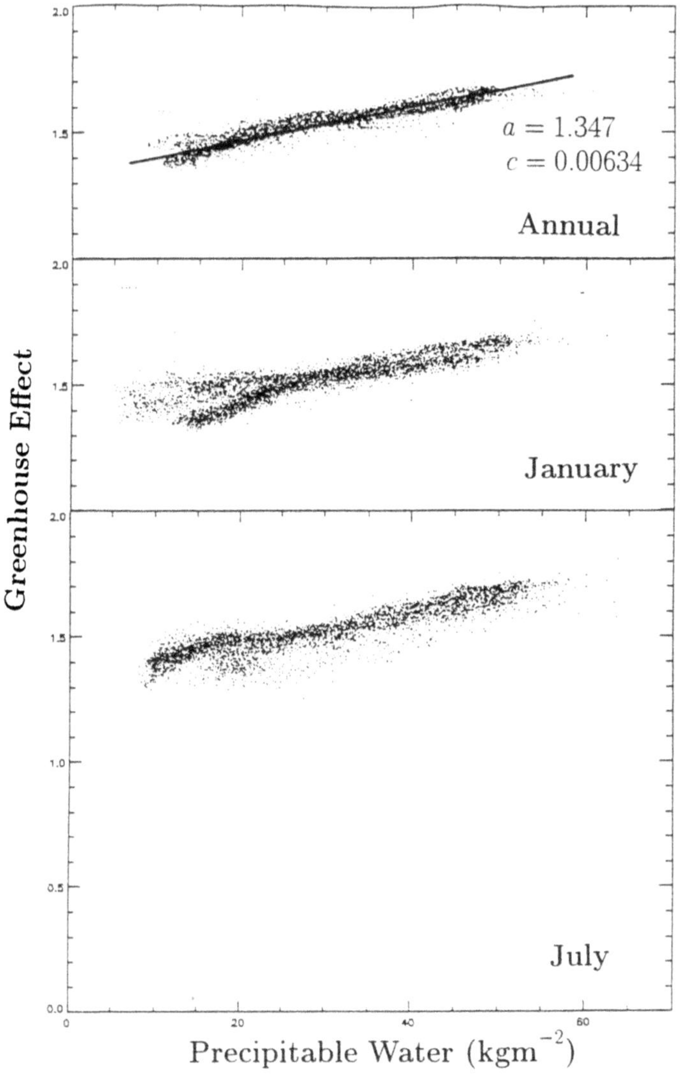

Fig. 4: The correlation of $\mathcal{G}$ with w based on annual mean (upper panel), 1989 January mean (middle panel) and the 1988 July mean data (lower panel). The unit of c is $(kgm^{-2})^{-1}$.

295 K. The relationship (11) and the results of Fig. 5 establish an important property of the Earth's greenhouse; that its sensitivity to changing SST is not uniform over the globe and is significantly greater over regions of warmer SST's. Ramanathan and Collins (1991) refer to this as a 'super-greenhouse' although Fig. 4 suggests that there is nothing special or 'super' in the relationship between $\mathcal{G}$ and w when $w > 30$ kgm^{-2}, which approximately corresponds to SST>295 K. The non-linear relationship observed in Fig. 5 arises primarily by thermodynamic controls on w and not on changes in the radiative transfer at these temperatures. The sensitivity parameter g highlights the relevance of the slope factor c and the r factor in the exponent of (11) in dictating the magnitude of this sensitivity.

5. ANALYSIS OF $\mathcal{G}$ FROM A SIMULATION SYSTEM–SAMSON

(a) Broadband simulations of the greenhouse effect

Slingo and Webb (1992) describe an analysis system (Simulation and Analysis of Measurements from Satellites using Operational aNalyses – SAMSON) which incorporates data from ECMWF operational analyses into a radiative transfer model to simulate the clear sky OLR and compare these simulations to matching ERBE clear sky data. Their analyses show how SAMSON simulations agree well with observations and point to two known systematic errors, a small overestimate in the clear sky ERBE fluxes (producing a bias in the ERBE clear sky OLR of approximately 5 Wm^{-2}, Harrison et al., 1990) together with a tendency for the ECMWF analyses to be too dry in the vicinity of the ITCZ and too moist in the subtropics (Liu *et al.*, 1992)). Over most of the oceans, SAMSON simulations are within 5–10 Wm^{-2}of the ERBE data thus suggesting that it is possible to calculate clear sky OLR with an accuracy comparable to ERBE.

SAMSON is an ideal tool for understanding the effects of variations in the vertical distribution of water vapour and temperature on the kinds of relations shown in Fig. 4. We argue that these effects are largely responsible for the observed scatter in Fig. 4 and further analysis of SAMSON to verify this is a topic of ongoing research. Only preliminary and a limited number of results showing the application of SAMSON to the study of the clear sky greenhouse effect are presented here. These preliminary results are portrayed in Figs 6 and 7. Figure 6a contrasts the relationship between $\mathcal{G}$ and w

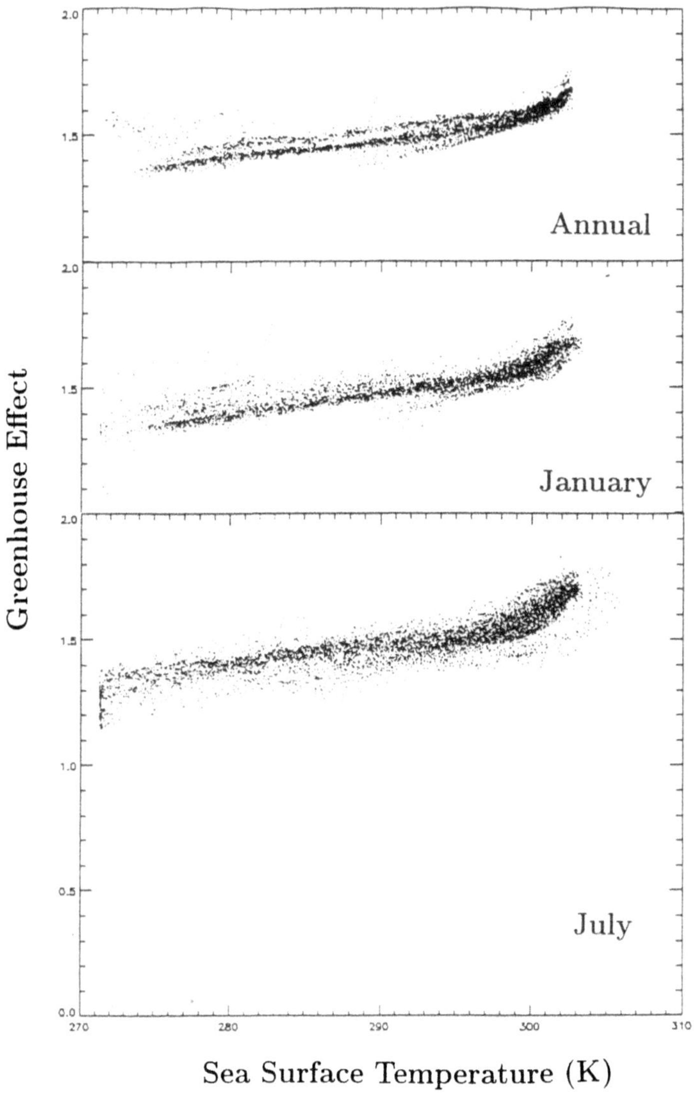

Fig. 5: The correlation of $\mathcal{G}$ with SST based on annual mean (upper panel), 1989 January mean (middle panel) and the 1988 July mean data (lower panel).

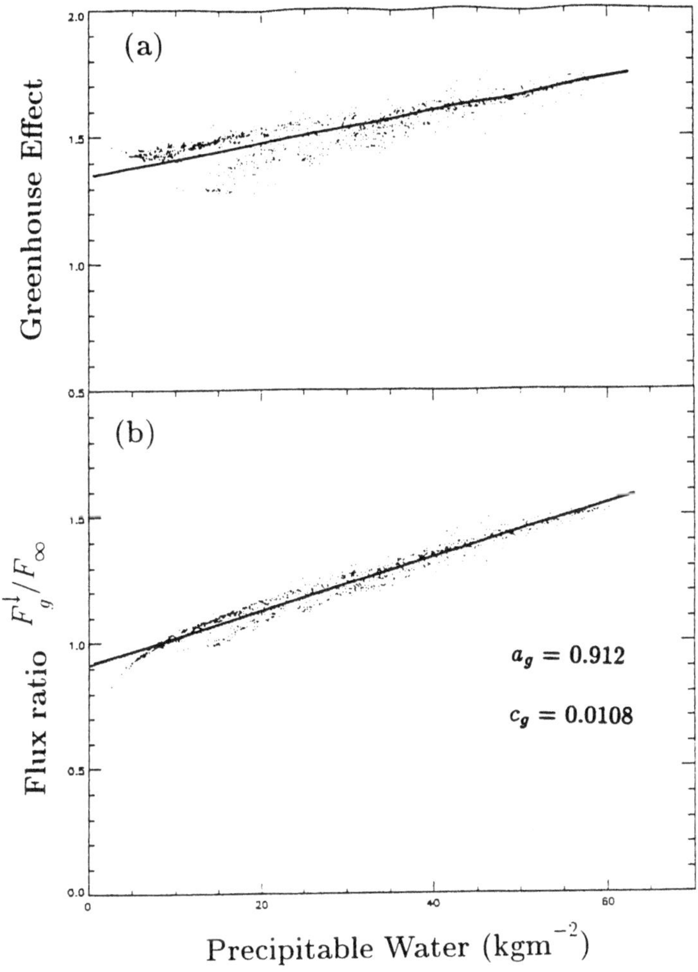

Fig. 6: (a) The correlation of $\mathcal{G}$ with w based on simulations of OLR using ECMWF July 1988 operational analysis for input into a radiative transfer model. The solid line is the fit to the annual mean relation given in Fig. 4. (b) As in (a) but for the ratio of fluxes $F^{\downarrow}/F_{\infty}$. The unit of c_g is $(\text{kgm}^{-2})^{-1}$.

derived from SAMSON simulations of July 1988 clear sky OLR and operational water vapour data (scattered points) and compares this result to the annual mean relationship represented by the solid line in Fig. 4.

If we adopt arguments equivalent to those used to obtain (10) then it also follows from (6b) that the ratio $F_g^{\downarrow}/F_{\infty}$ is a simple function of the precipitable water. This notion is confirmed by the results shown in Fig. 6b where the flux ratio is plotted as a function of w. Further insight into this relationship may also be obtained by studying the distributions of the July mean fluxes $F_g^{\downarrow}$, F_{∞} and the flux ratio $F_g^{\downarrow}/F_{\infty}$ respectively shown in Fig. 7a,b and c. The distribution of $F_g^{\downarrow}$, especially in tropical latitudes, is more zonally symmetric than is the distribution of F_{∞} despite the variations in w that exist in these latitudes (these variations occur as we move from the moist western Pacific, for instance, to the drier regions of the eastern Pacific– Jackson, 1992). These results are consistent with the impression that $F_g^{\downarrow}$ is influenced largely by near surface humidity and temperatures which are more zonally symmetric than are the variations in middle and upper tropospheric moisture which influence F_{∞} . The rms error for simulations of $F_g^{\downarrow}$ based on a least squares linear fit of the ratio in Fig 6b (solid curve) and SAMSON values of both OLR and w is ± 5 Wm^{-2}when compared to the direct SAMSON simulations of $F_g^{\downarrow}$. The shaded regions in Fig. 8 correspond to regions where the predicted values of $F_g^{\downarrow}$ differ by less than ± 5 Wm^{-2}from the simulated fluxes and show how the predicted longwave flux to the ocean is mostly within this rms difference.

The results of Fig. 6b, and any predictive capability that follows from them, if confirmed with independent data, are quite remarkable and suggest that satellite observations of both w and clear sky F_{∞} may provide a way of predicting the clear sky downward flux at the surface with an accuracy comparable to current measurements of these fluxes.

(b) Spectral simulations of the greenhouse effect

Another advantage of SAMSON is that it can also be used to study spectral variations in GE. This information not only reinforces understanding derived from the broadband results, but also identifies spectral regions where GE is most sensitive to changes in either w or SST and perhaps relevant to monitoring long term changes in GE. SAMSON was used with the July 1988 ECMWF analysis data as input to simulate spectral inten-

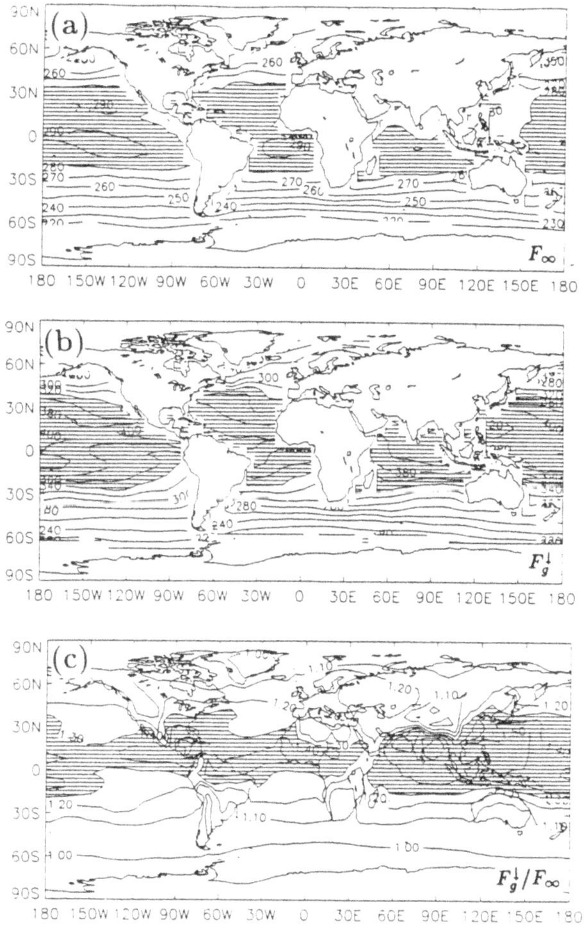

Fig. 7: (a) The distribution of simulated OLR F_∞ (in Wm^{-2}) using ECMWF July 1988 operational analysis for input into a radiative transfer model. (b) As in (a) but for the downward longwave flux $F_g^\downarrow$ at the surface and (c) the ratio of fluxes $F_g^\downarrow / F_\infty$.

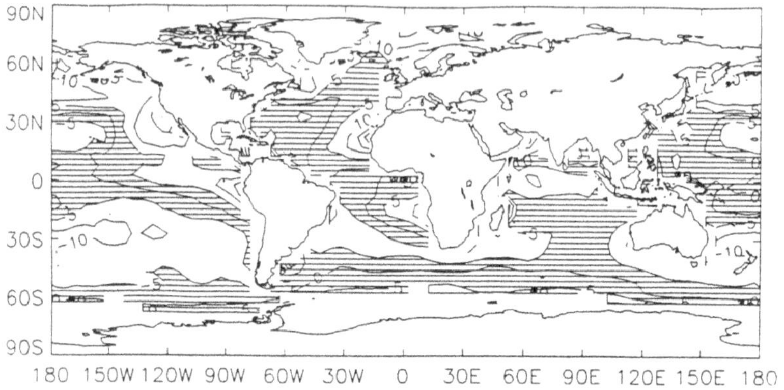

Fig. 8: The difference between the predicted values of $F_g^{\downarrow}$ at the surface (derived from the least squares fit of the ratio $F_g^{\downarrow}/F_{\infty}$, the OLR and w) and simulated values of $F^{\downarrow}$ (contours in units of Wm^{-2}). Shaded areas denote differences of ± 5 Wm^{-2}.

sities corresponding to channels 10 (8.2 μm), 11 (7.3 μm) and channel 12 (6.3 μm) of HIRS. The July mean simulations of channel 12 brightness temperature (derived from calculated spectral nadir radiances) are contrasted against the July averages of HIRS channels 11 and 12 brightness temperatures (Wu *et al.*, 1992) in Fig. 9a,b,c and d. The distribution in the simulated temperatures resembles the measured brightness temperatures and showing regions (shaded in Fig. 9b and d) that are associated with large scale subsidence (Wu et al., 1992). The differences between the simulated and observed brightness temperatures exhibit a well known bias and no attempt has been made to tune the simulations to the observations by accounting for the filter response functions and other factors. The detailed differences between the observed temperatures of this channel and simulated temperatures using a TOVS simulation program which are appropriately tuned against observations are discussed more fully by Eyre et al. (1992, this volume).

Spectral flux simulations of channels 10, 11 and 12 were analysed to examine the behaviour of the spectral greenhouse parameter and its variations with w. We refer to these spectral quantities in terms of the matching HIRS channel number. The GE

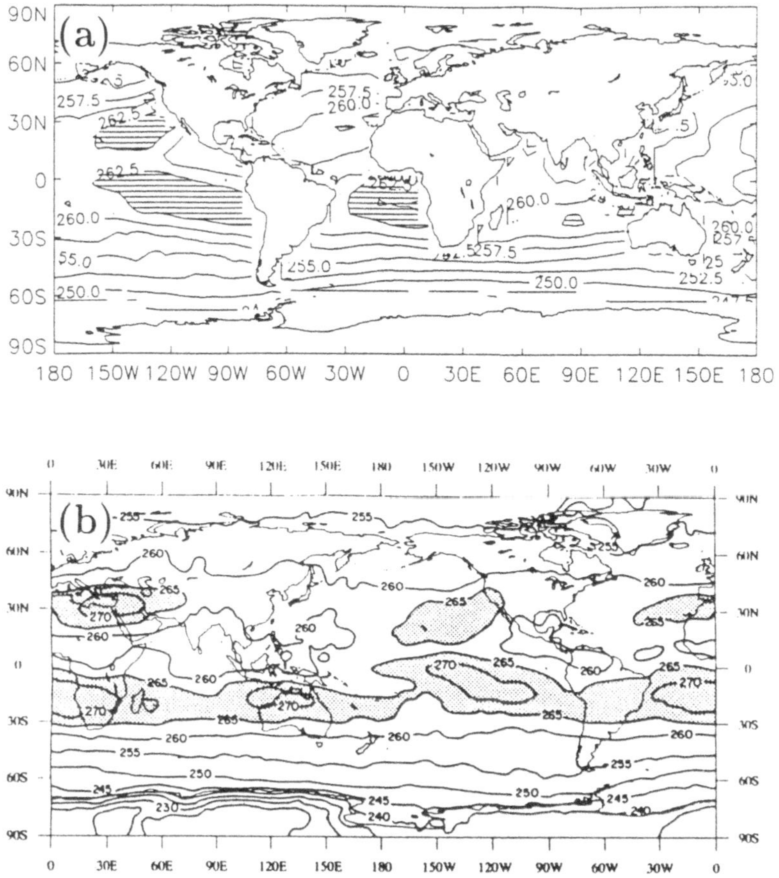

Fig. 9: (a) SAMSON simulations of the HIRS channel 11 brightness temperatures based on ECMWF analyses of July 1988. (b) Observed channel 11 brightness temperatures obtained from the climatological analyses of Wu et al., (1992).

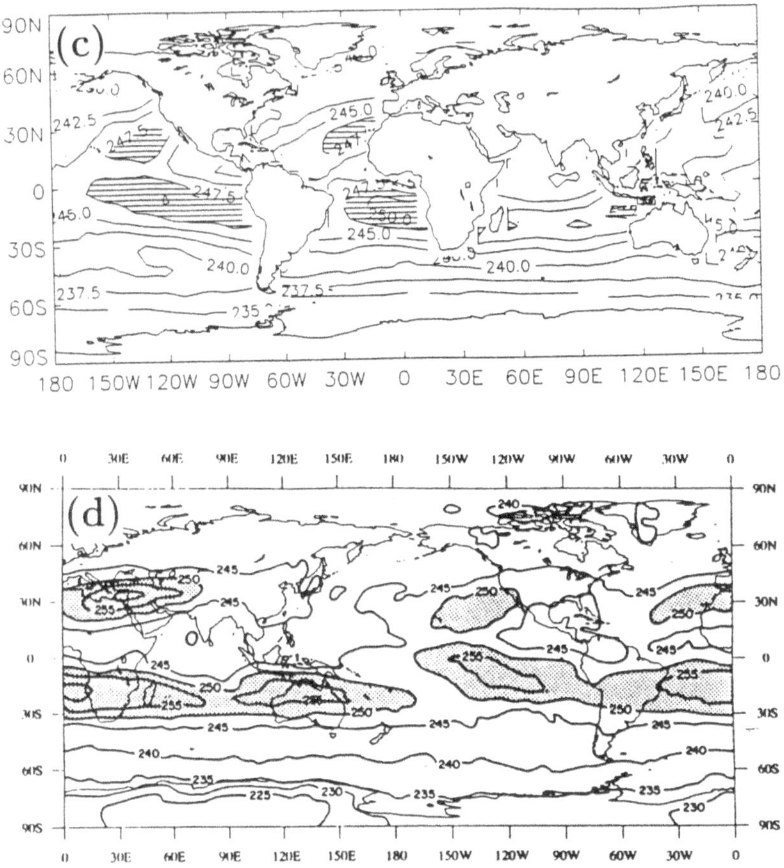

Fig. 9: (c) SAMSON simulations of the HIRS channel 12 brightness temperatures based on ECMWF analyses of July 1988. (d) Observed channel 12 brightness temperatures obtained from the climatological analyses of Wu et al., (1992).

parameters, $\mathcal{G}_{10,11,12}$ are presented as functions of w in Fig. 10 together with an estimate of the slope parameter. A number of aspects of these simulated relationships warrant comment. The relationship between $\mathcal{G}$ and w becomes less distinct as the wavelength varies from the more transparent channel 10 to the more opaque wavelengths of channel 12. The spectral OLR of the latter channel is influenced significantly by middle and upper tropospheric moisture rather than by w as well as by changes in temperature profile at these levels in the atmosphere. The relationship between $\mathcal{G}_{10}$ and w resembles the broadband relationship shown in Figs 4 as previously predicted by Stephens and Greenwald (1991b). Another point to note is that the slope factor c derived for each of the simulated channels varies by an order of magnitude from channel 10 to channel 12. The parameter c_{10} closely matches the broadband sensitivity expressed in Fig. 4. The more opaque channels 11 and 12 are more sensitive to changes in water vapour although there is considerably more scatter to these relationships. These results are at least consistent with the idea that the variations of vertical temperature and moisture tend to account for the observed scatter in the broadband results. A more detailed analyses of these simulations and others is in progress.

6. SUMMARY

This paper outlines an observational strategy that provides both a quantitative assessment of the role of water vapour in establishing the clear sky GE on Earth yet is simple enough to be used to test simulations of the GE by present day general circulation models (GCM) of climate. The virtue of this strategy is that it uses present day satellite observations to make this assessment.

We can conclude from the research presented in this paper that:

(i) under global mean conditions, approximately 75 % of the strength of the greenhouse, which is introduced in this paper in terms of the greybody optical depth of the atmosphere, is due to water vapour. We show how this contribution can be measured in terms of the ratio of the emission from the sea surface and the outgoing longwave radiation and explore the relation between this ratio and the integrated column water vapour amount w using satellite observations of w and outgoing longwave radiation.

(ii) We extend the analysis of the global greenhouse to consider the longwave flux

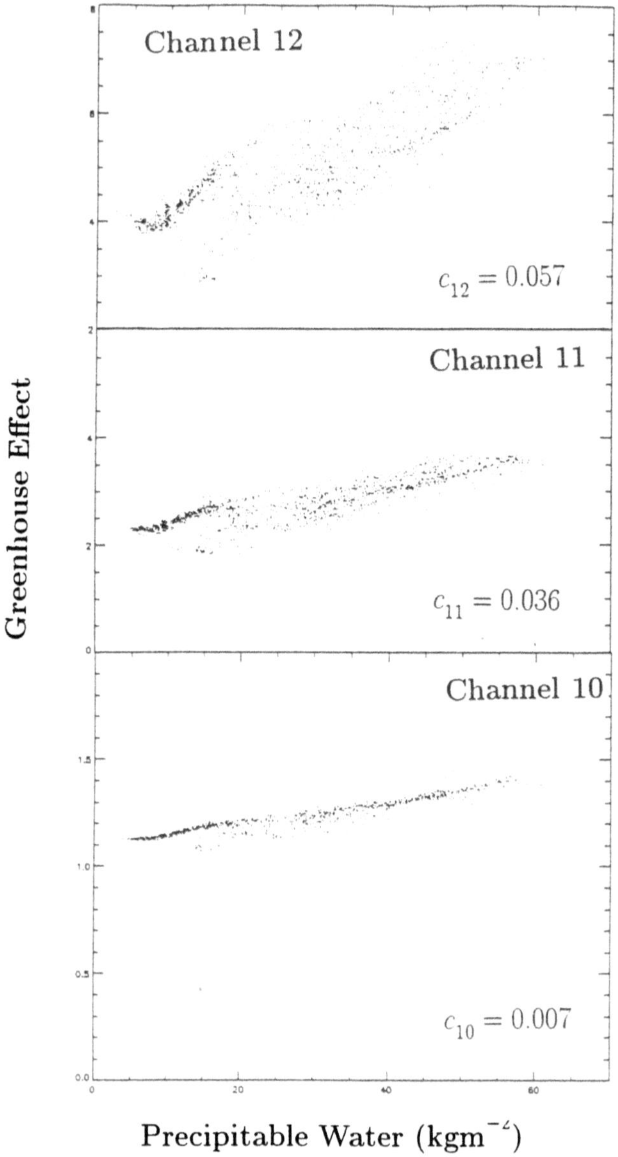

Fig. 10: Simulated $\mathcal{G}_{10,11,12}$ as a function of w together with slope parameters expressed in units of $(\mathrm{kgm}^{-2})^{-1}$.

emitted by the atmosphere to the surface. Using equivalent arguments, we suggest a simple linear relationship between the ratio of broadband fluxes $F_g^{\downarrow}/F_{\infty}$ and w and demonstrate this relationship with simulations. Based on the single month of analyses, this relation, together with measured OLR and w offers a way of predicting monthly mean clear sky fluxes $F_g^{\downarrow}$ within ± 5 Wm^{-2}of the simulated fluxes.

(iii) We provide a preliminary glimpse at the spectral contributions to the GE as a first stage in unravelling various complex issues such the effects of vertical variations in water vapour and temperature on the observed relationships mentioned above. Based on simulations of fluxes in spectral regions characteristic of selected HIRS channels, we are able to show the close similarity between the observed broadband relationships and those of channel 10 (centred on 8.2 μm). We also show how the relationship between $\mathcal{G}$ and w begins to break down in the more opaque regions, characterized by channels 11 and 12.

(iv) Because of the non-linear relation between w and sea surface temperature, the clear–sky greenhouse effect is also non-linearly related to SST. The change in the GE with changing SST, which we refer to as the greenhouse–SST sensitivity, is shown to increase with SST also in a non–linear way with the greatest sensitivity occurring over the warmest oceans. This is in direct contrast to the simple analysis of water vapour feedback which is often described in terms of the Budyko sensitivity parameter B (e.g. North et al., 1981) which is taken to be a global quantity.

The feedback between surface temperature and water vapour is thought to contribute substantially to the projected global warming induced by rising levels of carbon dioxide. The sensitivity parameter g introduced in this paper and which can readily be estimated from available satellite data should not be misconstrued as a measure of feedback between $\mathcal{G}$ and the SST. We note that many factors other than SST influence $\mathcal{G}$ and thus g. Among these factors are the vertical profiles of water vapour and temperature. The parameters used to estimate g in this study correspond to a set of conditions which do not remain fixed but vary in association with changing SST. A fundamental issue, and a topic of a further study, concerns the robustness of relationships shown here and whether the same sensitivities derive from climate states which are different from the present day state observed in this study.

7. ACKNOWLEDGMENTS

Aspects of this research are supported by DOD-NAVY-Office of Naval Research Contract #N00014-92-1385, NOAA Climate Change Program Contract #NA90AA-D-AC822 and by the UK Department of the Environment under contract PECD 7/12/37.

8. REFERENCES

Alishouse JC, Synder SA, Vongsathorn J, Ferraro RR, (1990) Determination of oceanic total precipitable water from the SSM/I. IEE Trans Geosci Remote Sensing 28:811-816

Barkstrom BR (1984) The Earth Radiation Budget Experiment (ERBE). Bull Amer Meteorol Soc 65:1170-1185

Bohren C (1987) Clouds in a glass of beer. In: Wiley & Sons, New York, p 187 (discussion on page 83)

Eyre JR, Andersson E, McNally AP (to be published) Direct Use of Satellite Sounding Radiances in Numerical Weather Prediction. In: Chedin A, Chahine, MT (eds) High Spectral Resolution Infrared Remote Sensing for Earth's Weather and Climate Studies. NATO ASI Series, D. Reidel Publishing Company Dordrecht Boston Lancaster Tokyo

Greenwald TJ, Stephens GL, Vonder Haar TH, Jackson DL (to be published) A physical retrieval of cloud liquid water over the global oceans using SSM/I observations. J Geophys Res

Harrison EF, Minnis P, Barkstrom BR, Ramanathan V, Cess RD, Gibson GG (1990) Seasonal variation of cloud radiative forcing derived from the Earth Radiation Budget Experiment. J Geophys Res 95:18687-18703

Hollinger J, Lo R, Poe C, Savage R, Pierce J (1987) Special Sensor microwave/imager user's guide. Report Nav Res Lab Washington DC

Ingersoll AP (1969) The runaway greenhouse effect: A history of water on Venus. J Atmos Sci 26:1191-1198

Jackson DL (1992) On the role of SSM/I derived precipitable water over the globe and tropical Pacific. M Sc thesis, Colo State Uni, Ft. Collins, CO

Kondratyev KY, Moskalenko NI (1984) The role of carbon dioxide and other minor gaseous components and aerosols in the radiation budget. In: Houghton JT (ed) The Global Climate, p 225-232

Liu WT, Wang T, Wentz F (1992) Precipitable water and surface humidity over the global oceans from SSM/I and ECMWF. J Geophys Res 97:2251-2264

Mihalas D (1978) Stellar Atmospheres. In: Freeman WF (ed) Freeman and Co 2nd edn. Freeman and Co, San Franscisco p 569

North GR, Cahalan RF, Coakley JA (1981) Energy balance climate models. Rev Geophys Space Phys 19:91-121

Petty GW, Katsaros K (1990) New geophysical algorithm for the special sensor microwave imager. 5th Conf on Sat Met and Oceanog Sept. 3-7 London Eng

Piexoto JP, Oort A (1992) Physics of Climate. American Inst of Phys, New York, p 520

Ramanathan V, Collins W (1991)Thermodynamic regulation of ocean warming by cirrus clouds deduced from observations of the 1987 El Nino. Nature 351:27-32

Raval A, RamanathanV (1989) Observational determination of the greenhouse effect. Nature 342:758-761

Reynolds RW (1988) A real-time global sea surface temperature analysis. J Climate 1: 75-86

Rothman LS (1981) AFGL atmospheric absorption line parameters compilation: 1980 version. Appl Opt 20:791-795

Schluessel P, Emery WJ (1990) Atmospheric water vapour over oceans from SSM/I measurements. Int J Remote Sensing 11:753-766

Slingo A, Webb MJ (to be published) Simulation of clear-sky outgoing longwave radiation over the oceans using operational analyses. Q J Roy Meteorol Soc

Stephens GL (1990) On the relationship between water vapour over the oceans and sea surface temperature. J Climate 3:634-645

Stephens GL, Greenwald TJ (1991a) Observations of the Earth's Radiation Budget in Relation to Atmospheric Hydrology. Part II: Cloud Effects and Cloud Feedback. J Geophys Res 96:15,325-15,340

Stephens G L, Greenwald TJ (1991b) Observations of the Earth's Radiation Budget in Relation to Atmospheric Hydrology Part I: Clear Sky Greenhouse Effect and Water vapour Feedback. J Geophys Res 96:15,311-15,324

Stephens GL, Tjemkes S (to be published) Water vapour and its role in the Earth's greenhouse. Australian J Phys

Vonder Haar TH, Suomi V (1971) Measurements of the earth's radiation budget from satellite during a five-year period. Part I: Extended time and space means. J Atmos Sci 28:305-314

Wu X, Bates JJ, Khalsa SJS (in preparation) Upper tropospheric water vapour climatology derived from TOVS data

MULTI-YEAR STATISTICS OF TOTAL CLOUD AMOUNTS FROM THE NIMBUS-7, ISCCP AND COADS DATA SETS

Bryan C. Weare
Atmospheric Science Unit
Department of Land, Air and Water Resources
University of California
Davis, CA 95616
U.S.A.

A primary aim of this work is to establish how compatible are the satellite cloudiness estimates made by the Nimbus-7 Cloud Team (Nimbus-7) and the International Satellite Cloud Climatology Project (ISCCP) with a surface-derived summary. Because there is no reason to believe that the absolute magnitude of cloud fraction estimated by Nimbus-7 and ISCCP and the surface observers should be identical, emphasis should be given to the evaluation of statistics, which are relatively insensitive to specific mean values. Since only six years of Nimbus-7 and five years of ISCCP data are available, these statistics will concentrate on intra-annual variations from regional annual means.

The monthly mean Nimbus-7 cloud data for April 1979 through March 1985 are available on an equal area grid with a resolution of approximate $(500)^2$ km^2 (Stowe et al., 1988, 1989). Total cloud fractions were available for both local noon and local midnight times. The daytime (local noon) cloud estimates are based upon bispectral tests using a window infrared channel and an ultraviolet solar channel. The nighttime (local midnight) estimates necessarily rely only on the infrared information. Threshold tests to determine whether or not subregions were clouded compared satellite brightness temperatures and reflectances with those for the surface obtained from the Snow/Ice Analysis and the Surface Temperature Analysis of the U. S. Air Force 3-D Cloud Nephanalysis. The latter analyses rely upon a combination of conventional surface reports, satellite radiances and climatology (Luces et al., 1975).

The ISCCP analysis methods and preliminary results are thoroughly described in several published papers (Rossow et al., 1989a,b) and an extensive

NATO ASI Series, Vol. I 9
High Spectral Resolution Infrared Remote Sensing
for Earth's Weather and Climate Studies
Edited by A. Chedin, M.T. Chahine and N.A. Scott

technical report (Rossow et al.,1988). These satellite estimates of clouds and their properties are based upon a combination of geostationary and polar orbiting satellite data. At latitudes equatorward of approximately 50° this data has allowed the evaluation of many parameters every three hours; other parameters using solar radiances are available only during the daylight hours. The calculated statistics at each available time are based upon 40-60 **samples** of 8km x 8km resolution narrow-band infrared and visible radiances within equal area boxes which are approximately 2.5°x2.5° at the equator. In order to determine whether a pixel is clouded the thermal or solar radiance is compared to the value judged due to radiation emitted or reflected from the surface. These surface values are estimates from a detailed space-time statistical analysis of the ISCCP radiances themselves, relying upon the fact that surfaces are usually warmer and darker than clouds. The ISCCP C1 data set provides up to 133 variables for each three hour period during each day beginning in July 1983 and running currently through December 1990. The ISCCP C2 data set, which is herein utilized, provides monthly means of up to 72 variables which are averages for data for the eight reporting hours each day.

This work will first illustrate comparisons of Nimbus-7 and ISCCP total cloud statistics with those of the surface-derived Consolidated Ocean Atmosphere Data Set (COADS) in order to gain a better understanding of the reality of the ISCCP estimates and the magnitude and nature of uncertainties. The COADS data and their processing have been described in Weare (1992). These are traditional observations made by personnel of commercial and military shipping, which are archived by international agreement. The only available cloud parameter in the COADS interim release for the 1980s (Woodruff et al., 1987) is the monthly mean fraction of sky covered with cloud for a 2°x2° region. In addition use will be made of the COADS monthly mean sea surface temperatures (SST) and surface relative humidity (SRH). Because the COADS often has relatively few observations over large areas of ocean, all comparisons will be done on a course 10°x10° grid. In addition the variances and correlations utilizing COADS will be adjusted when the number of observations is small using the technique outlined in Weare (1992).

The calculated statistics include the five or six year annual means, intra-annual variances and intra-annual correlations. The annual means of a variable x at location j over Y years are defined by the usual formula

$$\mathbf{x}_{aj} = \sum_{y=1}^{Y} \sum_{m=1}^{12} x_{ymj}/N \quad ;N=Y*12 \qquad 1)$$

Differences in the means of data sets 1 and 2 are judged to be significantly different from zero using a Z-statistic (Brunk, 1965), such that

$$Z = (\mathbf{x}_1 - \mathbf{x}_2)/(\sigma_1{}^2{}_{intra}/N + \sigma_2{}^2{}_{intra}/N)^{.5} \qquad 2)$$

where each intra-annual variance σ^2_{intra} is given by given by

$$\sigma^2_{intra}(x_j) = \sum_{y=1}^{Y} \sum_{m=1}^{12} (x_{ymj} - \mathbf{x}_{aj})^2/(N-1) \qquad 3)$$

If Z is greater than 1.96, then the means are statistically different at the 95% significance level. The variances themselves for pairs of cloud data sets will also be compared. The variance of one data set will be judged to be different than another if the ratio of the two variances is statistically different than one at the 95% level using an F-test (Brunk, 1965), in which F is a number-weighted ratio of the two variances. Correlations r of intra-annual departures also were calculated for variables x and z using the formula

$$r_{intra}(x_j) = \sum_{y=1}^{Y} \sum_{m=1}^{12} (x_{ymj} - \mathbf{x}_{aj})(z_{ymj} - \mathbf{z}_{aj})/(N\sigma_{intra}(x_j)\sigma_{intra}(z_j)) \qquad 4)$$

The 95% significance levels of these correlations were derived by first estimating the approximate number of independent samples N (Leith, 1973) and using a Z-statistic (Brunk, 1965). The number of independent samples was estimated by calculating lag correlations out to three months using a formula similar to Eq. 4. The absolute magnitude of a correlation coefficient $|r|$ is statistically significant at approximately the 95% confidence level if

$$|r| > \frac{e^{(3.92/\sqrt{(N-3)})} - 1}{e^{(3.92/\sqrt{(N-3)})} + 1} \qquad 5)$$

Fig. 1 shows the regional differences between the 5-year means of the Nimbus-7 and ISCCP C2 and COADS total cloud amounts. The Nimbus-7 estimates are usually more than 10% smaller except in the convective areas of

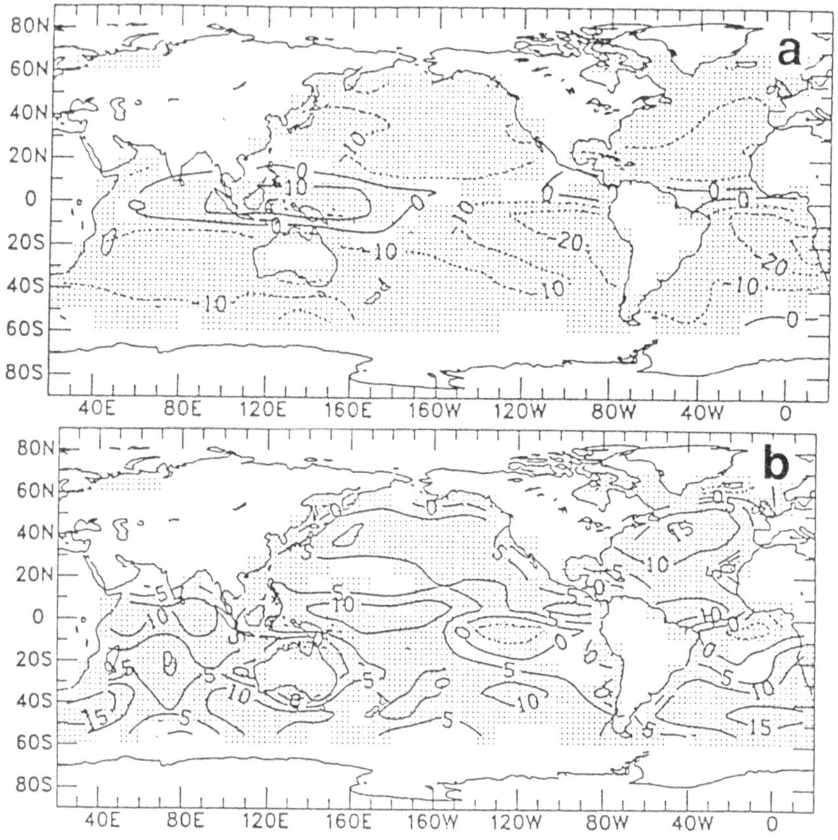

Fig. 1. Differences in annual mean total cloud cover: a) Nimbus-7 minus COADS annual means for January 1980 - December 1984; b) ISCCP minus COADS annual means for July 1983 - June 1988. Stippled areas are significantly different at the 95% confidence level.

equatorial western Pacific. The differences in the persistent stratus regions of the eastern oceans are especially large. Fig. 1b shows that the ISCCP means tend to be 5-10% larger than those of COADS nearly everywhere. Overall, the ISCCP means agree better with those of COADS than do those of Nimbus-7. However, one must keep in mind the imprecise nature the definition of 'cloud amount'.

Fig. 2 illustrates that the ratio of intra-annual variances of departures from

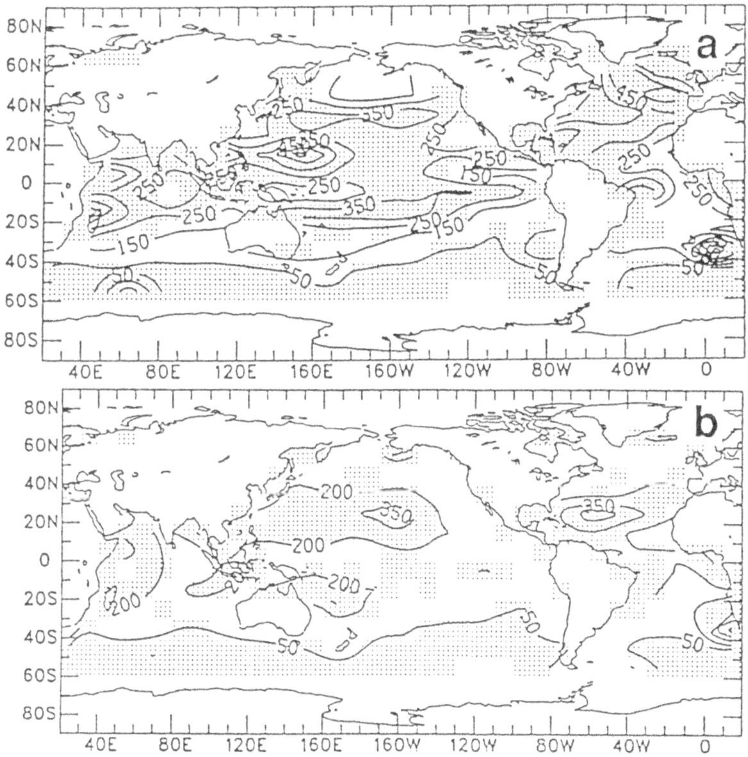

Fig. 2. Ratios of intra-annual variances of total cloud cover: a) Nimbus-7 over COADS times 100 for January 1980 - December 1984; b) ISCCP over COADS times 100 for July 1983 - June 1988. Stippled areas are significantly different from one at the 95% confidence level.

their respective annual means of the Nimbus-7 and ISCCP total cloud amounts relative to those of COADS. In general over much of the ocean the Nimbus-7 seasonal variability is significantly larger at the 95% confidence level. The ratio of ISCCP to COADS variances (Fig. 2b) are often insignificantly different from one. The most prevalent areas in which ISCCP and Nimbus-7 variances are greater than the COADS seem to exist are near 20° N and S in both the Atlantic and Pacific Oceans. Since it is generally agreed that surface observers tend

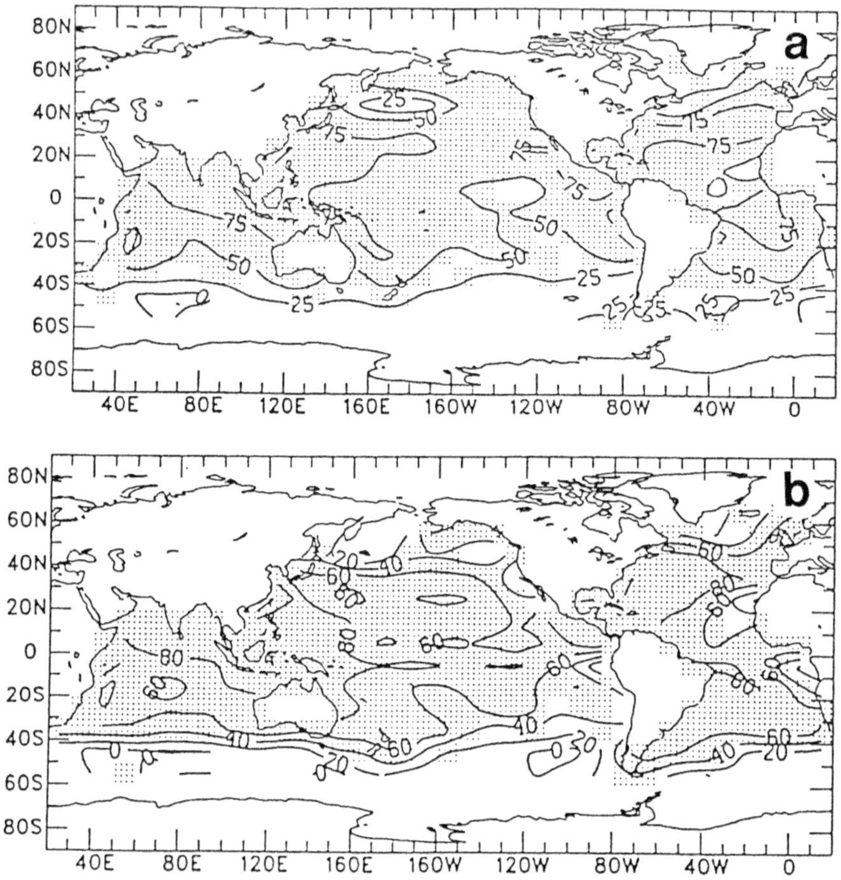

Fig. 3. Intra-annual correlations of total cloud cover :a) Nimbus-7 versus COADS for January 1980 - December 1984; b) ISCCP versus COADS for July 1983 - June 1988. Stippled areas have significant correlations at the 95% confidence level.

overestimate cloud cover in regions of cumulus clouds as would be typical in these regions, the reason for the much larger satellite variances is unclear. However, one explanation might be that in these regions of small cloud diameters the large satellite footprints coupled with the simple yes or no cloud decision might tend to artificially increase the variance. In the other area of significant differences, south of about 40°S, COADS variances are undoubtedly inflated by a lack of ship observations.

Fig. 3, which shows the intra-annual correlations between Nimbus-7 and

ISCCP total cloud amounts and the COADS values. It indicates good agreement for both satellite sets as to the phase of the annual cycle for most regions north of 40°S. Of note is the fact that the ISCCP correlations are greater than 50% in regions of persistent stratus, such as in the eastern oceans or the northern Pacific, where satellite algorithms often have considerable difficulties (Stowe, et al., 1989). On the other hand the largest correlations exist in the convective zones of the northern Indian, western equatorial Pacific and western equatorial Atlantic Oceans, where satellite thermal tests most readily separate warm sea surfaces from the generally high, cold convective cloud tops.

The intra-annual correlations for of COADS SST with the three total cloud data sets (Fig. 4) show that for all three cloud data sets local increases in SST may be associated with either increases or decreases in cloud cover. Overall the pattern of the ISCCP correlations agrees slightly better with that of the COADS than with the Nimbus-7. However, the areas of positive correlations for the Nimbus-7 data in the tropics are more extensive and have larger maxima. This is partially due to the fact that the time period of the Nimbus-7 data includes the very strong 1982/83 El Niño event. Other differences in Fig. 4 are in the North Pacific where the COADS-only data suggest moderate positive correlations, whereas the ISCCP data only weakly suggest this and the Nimbus-7 cloud data indicate moderate negative correlations.

All three cloud data sets show areas of significant positive intra-annual correlations with the COADS surface relative humidity (SRH) as shown in Fig. 5. For the COADS cloudiness the magnitudes of these are often larger than those for SST despite the fact that SRH is relatively difficult to measure at sea, and thus would be expected to contain a relatively large random error. The large COADS values may be due to the fact that usually the COADS SRH and cloudiness measurements are colocated samples, whereas the Nimbus-7 and ISCCP cloudiness may be more representative of large scale monthly averages. On the other hand it is possible that the Nimbus-7 and ISCCP specification algorithms, which largely rely upon separating cloud and surface brightness temperatures, may sometimes misrepresent changes in cloud cover in situations in which column moisture changes are relatively large as might be expected during periods of large SRH

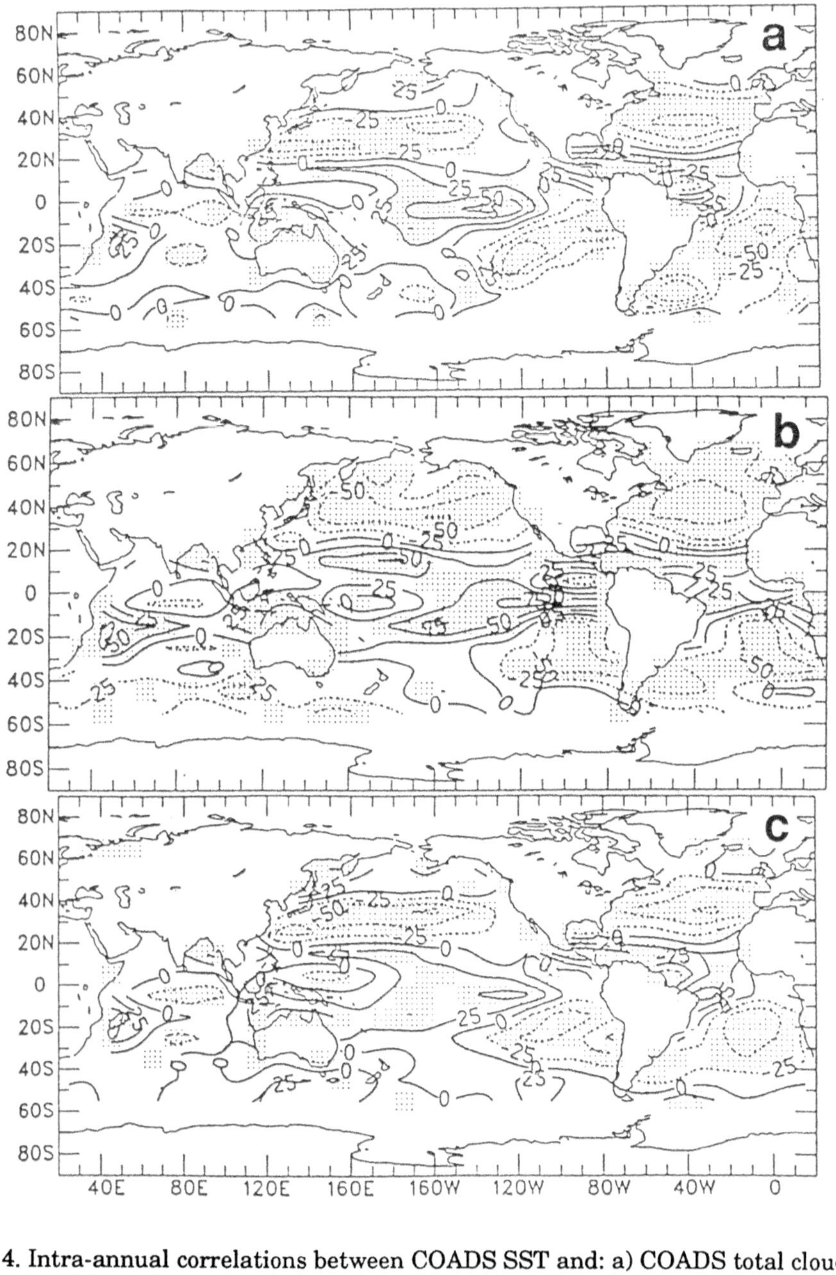

Fig. 4. Intra-annual correlations between COADS SST and: a) COADS total cloud cover for 7/1983-6/1988; b) Nimbus-7 total cloud cover for 1/1980-12/1984 ; and c) ISCCP total cloud cover for 7/1983-6/1988 as in Fig. 3); negative values are dashed.

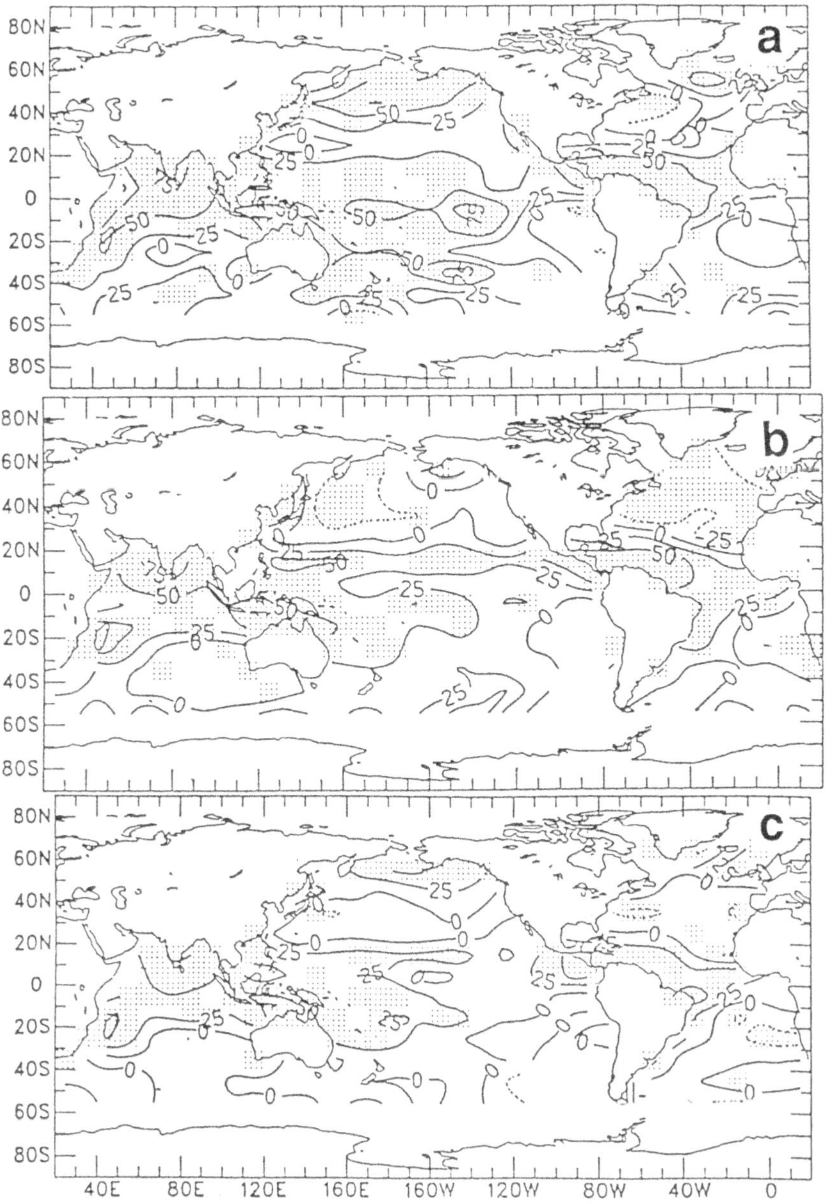

Fig. 5. Intra-annual correlations between COADS SRH and: a) COADS total cloud cover for 7/1983-6/1988; b) Nimbus-7 total cloud cover for 1/1980-12/1984 ; and c) ISCCP total cloud cover for 7/1983-6/1988 as in Fig. 3).

deviations. As with SST the patterns of correlations are quite similar for the ISCCP and COADS data for the area north of 40°S, but differ from those of the Nimbus-7 differs in a number of respects. For instance the other two data sets imply a positive correlation in the far northern Pacific and Atlantic Oceans, whereas the Nimbus-7 data suggest negative correlations for most of these regions.

Overall, Fig. 1-5 show that there is good agreement between the total cloud amount intra-annual statistics derived from the traditional COADS and the Nimbus-7 analyses and very good agreement between the COADS and the ISCCP analyses. Although no cloud data set can be conclusively judged to be 'correct,' this agreement lends confidence to the propriety of the substitution of the better sampled satellite data in uses such as the verification and parameterization of cloudiness parameterization in general circulation models.

References

Luces, S.A., S.J. Hall and J.D. Mortens, 1975: The AFGWC snow cover analysis model. AFGWC Tech. Memo. 75-1, HQ Air Force Global Weather Central, Offutt AFB, 19pp.

Rossow, W.L., L.C. Garder, P.J. Lu, A.W. Walker, 1988: International Satellite Cloud Climatology Project (ISCCP) documentation of cloud data. WMO/TD-No. 266, World Meteorological Organization, Geneva, 78pp. plus two appendices.

Rossow, W.L., L.C. Garder, A.A. Lacis, 1989a: Global seasonal cloud variations from satellite radiance measurements. Part I.:Sensitivity of analysis. *J. Climate, 2*:419-458.

Rossow, W.L., C.L. Brest, L.C. Garder, 1989b: Global seasonal surface variations from satellite radiance measurements. *J. Climate, 2*, 214-247

Rossow, W.L. and R.A. Schiffer, 1991: ISCCP cloud data products. *Bull. Amer. Meteor. Soc., 72*:2-21.

Stowe, L.L., G.G. Wellemeyer, T.F. Eck and H.Y.M. Yeh, 1988: Nimbus 7 global cloud climatology. Part I. Algorithms and validation. *Journal of Climate, 1*: 445-470.

Stowe, L.L, H.Y.M. Yeh, T.F. Eck, C.G. Wellemeyer, H.L. Kyle and the Nimbus 7 cloud data processing team, 1989: Nimbus 7 global cloud climatology. Part II. First year results. *Journal of Climate, 2*:671-709.

Weare, B.C., 1992: Comparisons of multi-year statistics of selected variables from the COADS, Nimbus-7 and ECMWF data sets. *Q. J. Roy. Meteor. Soc.*, 118:965-985.

Woodruff, S.D., R.J. Slutz, R.L. Jenne and P.M. Seurer, 1987: A comprehensive ocean-atmosphere data set. *Bull. Am. Met. Soc.*, 68:1239-1250.

Chapter 7

Forward modelling

"Transmittance coefficient generation for fast radiative transfer models. Application to new satellite sounding systems"

P. BRUNEL, L. LAVANANT, G. ROCHARD - Météo-France, France

"Review of existing spectral line data catalogs"

N. HUSSON, A. CHEDIN, B. BONNET - ARA/LMD, France

"CO_2 Q-branch spectral line shapes for atmospheric remote sensing"

L. LARRABEE STROW - Univ. Maryland, USA

"Optimal spectral estimation algorithms for Michelson interferometer spectrometers".

C. SERIO - Univ. Napoli, Italy

U. AMATO, V. TRAMUTOLI, R. RIZZI, V. CUOMO

TRANSMITTANCE COEFFICIENT GENERATION FOR FAST RADIATIVE TRANSFER MODELS: APPLICATION TO NEW SATELLITE SOUNDING SYSTEMS

P. Brunel, L. Lavanant, G. Rochard
Météo France / Centre de Météorologie Spatiale,
BP.147, 22302 Lannion
FRANCE

ABSTRACT

We have generated polynomial transmittance coefficients internal to the RTTOV fast radiative transfer model, with atmospheric transmittances calculated by using the FASCOD2 line-by-line model. The FASCOD2 computations are made for a small set of earth representative profiles, uniformly mixed-gases and water vapor molecular species separately and different scanning angles and pressure levels. The accuracy of the process is checked for the NOAA11 TOVS channels. A transmittances storage at a medium frequency resolution is tested. This storage allows afterwards any filter convolution of the transmittances and a quick adaptation of the RTTOV model for any new channels.

Key words: radiative transfer models, RTTOV, FASCOD2.

1. INTRODUCTION

The Centre de Météorologie Spatiale (CMS) of Lannion is an operational acquisition station for Meteosat, GOES and NOAA satellites. One of its vocations is to implement and validate radiative transfer models for TOVS and AVHRR channels, by comparing synthetic and measured satellite radiances.

We have implemented the RTTOV fast radiative transfer model developped at ECMWF. Its approach follows the general method used at NOAA and fully described in Eyre,1991. Regressed transmittance

NATO ASI Series, Vol. I 9
High Spectral Resolution Infrared Remote Sensing for Earth's Weather and Climate Studies
Edited by A. Chedin, M.T. Chahine and N.A. Scott

of NOAA7 and NOAA9 satellite AVHRR/TOVS measurements matched with radiosonde profiles. The input geophysical and geometrical conditions are settled as close as possible for the two models: same atmospheric layer boundaries, scanning angles, surface emissivities, same humidity and ozone profiles. The 4A results are internally adjusted by introducing two empirical corrections, for each channel (γ on the individual level transmission and δ on the calculated radiant energy already convolved with the filter). The FASCOD2 routine is not dedicated to specific channels or satellites and no empirical corrections are internally applied. For that reason, it is expected that the 4A radiances will be closer to the measured radiances than the FASCOD2 outputs.

FASCOD2 - MEASUREMENT:															
NOAA7	1	2	3	4	5	6	7	8	13	14	15	18	19	A4	A5
bias	-1.20	-0.69	-0.24	0.43	-0.21	0.35	-0.57	-0.07	2.27	-0.30	0.92	-0.01	0.1	-0.27	-0.05
std. dev.	5.67	3.27	2.62	1.41	0.97	0.76	0.68	0.54	0.64	0.56	0.99	0.40	0.49	0.45	0.63
NOAA9	1	2	3	4	5	6	7	8	13	14	15	18	19	A4	A5
bias	7.08	3.14	3.19	2.45	0.36	1.10	0.74	0.11	-0.92	-1.49	-1.57			-0.14	0.21
std. dev.	3.97	1.88	2.19	1.08	0.64	0.54	0.43	0.61	0.53	0.44	0.43			0.64	0.57

4A - MEASUREMENT:															
NOAA7	1	2	3	4	5	6	7	8	13	14	15	18	19	A4	A5
bias	-3.16	-1.69	-1.63	-0.46	-0.03	0.33	0.71	-0.01	-0.19	-0.33	-0.88	-0.29	-0.14	0.20	0.10
std. dev.	5.42	3.16	2.62	1.32	0.89	0.69	0.65	0.55	0.76	0.60	1.32	0.41	0.48	0.45	0.64
NOAA9	1	2	3	4	5	6	7	8	13	14	15	18	19	A4	A5
bias	4.27	1.89	1.65	1.10	0.48	0.56	0.24	-0.38	0.67	-0.19	-0.14			0.34	0.31
std. dev.	3.88	1.95	2.35	1.13	0.52	0.47	0.54	0.65	0.46	0.41	0.56			0.69	0.61

FASCOD2 - 4A:															
NOAA7	1	2	3	4	5	6	7	8	13	14	15	18	19	A4	A5
bias	1.96	1.00	1.38	0.89	-0.17	0.01	-1.28	-0.08	2.47	0.03	1.81	0.28	0.25	-0.47	-0.15
std. dev.	0.48	0.45	0.20	0.27	0.26	0.21	0.24	0.03	0.28	0.38	0.51	0.02	0.03	0.09	0.04
NOAA9	1	2	3	4	5	6	7	8	13	14	15	18	19	A4	A5
bias	2.80	1.25	1.54	1.34	-0.11	0.54	0.51	0.49	-1.59	-1.29	-1.43			-0.49	-0.10
std. dev.	0.26	0.09	0.24	0.15	0.19	0.18	0.17	0.05	0.37	0.31	0.61			0.12	0.06

Table1: Comparison of the two line-by-line FASCOD2 and 4A models with measurements

Table 1 gives the comparison in mean and standard deviation for respectively 19 different NOAA7

coefficients are generated by using off-line radiative transfer calculations. In the present version, the regressed coefficients come sofar from the ITPP/TOVSRAD fast model (Eyre,1984).
The purpose of this study is to quickly generate new set of regressed transmittances coefficients, for the RTTOV model, adapted to any new sounding systems (presently for IR channels). The method must be adapted to systems with large spectral filters , like Meteosat, TOVS and AVHRR radiometers, but is also expected to work with future interferometers systems with narrow spectral resolution around 0.5 cm-1.
We have developed at CMS a modified version of the RTTOV model where the internal coefficients are generated from transmittances computed with the line-by-line model FASCOD2 for a small number of representative atmospheric profiles (32). The off-line FASCOD2 calculations are time consuming and the results are stored at a medium frequency resolution (about 0.25 cm-1). This permits a quick evaluation of the transmittances and a new internal fast model coefficients, for any new channel.
The choice of the line-by-line model is not a key-point as it results from a study of the differences between the 4A and FASCOD2 line-by-line models in terms of departure from observations. It appears that the differences between the two models are much smaller than the errors on atmospheric profiles or than the approximations due to fast model generations.
We first present the line-by-line models comparison with in-situ measurements. In a second step, we show, for NOAA11 HIRS2 channels, our results with RTTOV when the transmittance regressed coefficients are generated using the line-by-line FASCOD2 model. Then in the final section, we indicate how the method is quickly applicable to any other sounding systems.

2. DEPARTURE OF THE LINE-BY-LINE MODELS FROM OBSERVATIONS

Internal regressed coefficients of fast radiative transfer models like RTTOV are obtained from RT calculations with line-by-line models. The regressed set is adapted to specific sounding systems and by this way these models are much faster than line-by-line models, necessary point for systematic or operational purposes. The fast model is fitted with its based line-by-line model to minimise their differences, so their results are strongly dependent.
For that reason, we have evaluated in terms of departure from observations the results of the 4A and FASCOD2 line-by-line models, available at CMS.
The 4A and FASCOD2 models are associated with two different spectral line parameter catalogs, respectively the GEISA and AFGL HITRAN atlas. More details can be found in Scott,1981 and Clough,1986.
They have been compared on the CATHIA data set (Lavanant,1988) which is an ancillary selection

and 7 NOAA9 situations. The two first parts of the table are a comparison of the models with measurements. As expected, for a specific channel, the biases are very different, the FASCOD2 one being not corrected from instrumental calibrations. However, the respective standard deviations are very close, somewhat higher for FASCOD2 (no γ correction on the transmittances). This is indicated in the third part of the table, when the standard deviations in the comparison of the two models themselves are very small compared to those in the comparison with measurements.

Even if differences exist between the two models, radiative transmittance calculations and spectroscopic characteristics are very close in regards to the other incertitudes: those coming from the atmospheric profile itself (profile extrapolation, surface temperature errors, humidity measurements errors..) or radiative transfer misunderstanding or approximations due to our general lack of physical knowledge.

The FASCOD2 routine was also tested on clear-sky data from the available EVA data set at CMS (Brunel,1991) including radiosonde profiles and matched 3R computed radiances for NOAA10 and NOAA11 satellites. The 3R routine is the fast routine based on the 4A model and applied to TOVS channels. The difference between the 3R and 4A computed radiances is very small (Chedin,1988). This allows the comparison of FASCOD2 and 3R (instead of 4A) without an important loss of accuracy.

We applied a rough bias on the FASCOD2 radiances function of the satellite and the viewing angle to take into account a satellite calibration correction. This bias is the result of a short study between FASCOD2 and 4A on 5 TIGR situations (Chedin,1983).

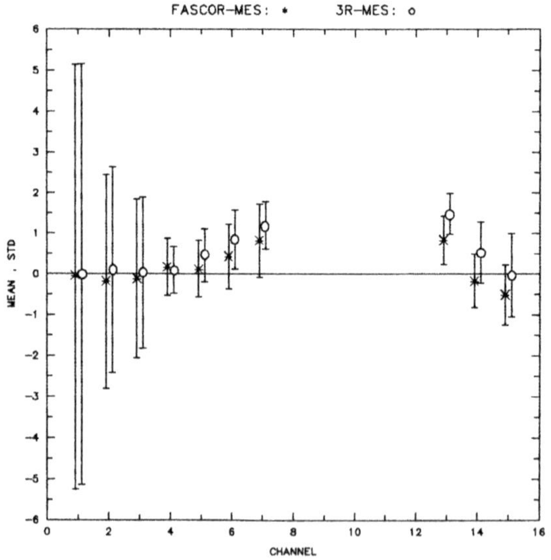

Figure 1: Comparison of FASCOD2(*), corrected from a bias,and 3R (o) with measurements

Figure 1 shows the statistics in mean and standard deviation for 26 combined situations (13 NOAA10, 13 NOAA11). Statistics with stars correspond to the FASCOD validation and the o identification is for the 3R results. The statistics are very similar for the two routines, 3R being somewhat better in standard deviation for channel 7 and the contrary for channel 15.
This indicates that no objective choice between the two line-by-line models to generate the regressed coefficients can be made. We have choosen in the following steps to work with the FASCOD2 routine for convenience.

3. INTERNAL TRANSMITTANCE COEFFICIENTS GENERATION FOR NOAA11 TOVS CHANNELS

We have implemented at CMS the RTTOV fast radiative transfer model developped at ECMWF. All the details of the routine are given in Eyre,1991. Concerning our study, the atmospheric transmittances are computed separately for uniformly-mixed gases, water vapor and ozone, by using different but similar polynomial expansions which terms are functions of temperature, specific humidity and scanning angles. The coefficients of the polynomials are generated off-line by linear regressions on transmittances calculated for 32 atmospheric profiles representative of the earth. The transmittances are calculated for 40 working pressure levels and 5 scan angles, from nadir to 60°, and for specific spectral channels.
The RTTOV polynomial coefficients have been, up to this point, obtained from transmittances computed with the ITPP/TOVSRAD fast model. These two fast models are very similar: their differences are given in Eyre,1991. The main scientific difference concerns the polynomial modelisation for the water vapor computation.
We have generated, at CMS, the polynomial coefficients for uniformly-mixed gases (MG) and water vapor (WV), by computing the transmittances of the representative profiles with FASCOD2. Although no modifications have been applied to the RTTOV code itself, the adapted version based on FASCOD2 is referred for comprehension simplicity (ex: comparison of the 2 versions) as RTCMS in the following steps.
Two independant sets of transmittances, for NOAA11 TOVS channels, were calculated with FASCOD2, one for uniformly-mixed gases only, the second for water vapor only. The ozone coefficients remain unchanged in the two versions.
The accuracy of the RTCMS version was checked by comparing the brightness temperatures (Tb) computed directly by FASCOD2 with those from RTCMS, on the 32 representative profiles. Two ways are considered:

- first, the FASCOD2 brightness temperatures are calculated from the transmittance product of the separate MG and WV FASCOD2 transmittances (τ_{MG} ,τ_{WV}). Table 2 shows the results in mean and standard deviations of the difference:

$Tb(\tau_{WV} * \tau_{MG})_{RTCMS} - Tb(\tau_{WV} * \tau_{MG})_{FASCOD2}$

These results show the accuracy of both the polynomial modelisation and the linearity assumption for the regression coefficients. The errors are negligible for channels principaly affected by MG spectroscopic lines but are larger for channels including WV lines (around 0.5K in standard deviation for the channels, 5-7, 11-12).

channel	1	2	3	4	5	6	7	8	10	11	12	13	14	15	18	19
M	-0.01	-0.02	-0.02	0.01	0.04	0.01	0.01	0.00	-0.01	-0.11	-0.28	0.00	0.00	0.01	-0.01	0.00
S.D.	0.04	0.04	0.05	0.25	0.53	0.55	0.45	0.11	0.19	0.45	0.50	0.31	0.43	0.37	0.08	0.13

Table 2: Polynomial coefficients validation

- second, in an independant run, FASCOD brightness temperatures were directly calculated with all molecular species (MG + WV) together. Table3 shows the results of the differences:

$Tb(\tau_{WV} * \tau_{MG})_{RTCMS} - Tb(\tau_{all\ mol.})_{FASCOD2}$

These statistics correspond to the combined quality of the RTCMS routine (Table 2 results) and of the split of the spectroscopic lines into molecular species. The results are also good (same as table 2) for channels with absorption by only one molecular species. We observe a degradation for the channels dominated by the two species. This effect is due to the different integration of the transmittances over the filter and do not exist for monochromatic computations. So, we should obtain better results by cutting the filters in several parts but the computing time will increase in connection.

channel	1	2	3	4	5	6	7	8	10	11	12	13	14	15	18	19
M	0.66	1.19	1.04	0.24	-0.63	-0.83	-1.23	0.01	0.01	0.30	-0.94	0.03	-0.04	0.03	-0.03	0.00
S.D	0.18	0.50	0.51	0.49	0.87	0.78	0.64	0.10	0.19	0.52	1.19	0.32	0.42	0.37	0.08	0.14

Table 3: Effect of the split into molecular species

We used the EVA data set of collocated radiosondes and satellite measurements to validate, with observations, the RTTOV and RTCMS routines in clear-sky conditions. In this file, the surface temperatures are the result of an AVHRR imager pixels processing in the TOVS F.O.V.

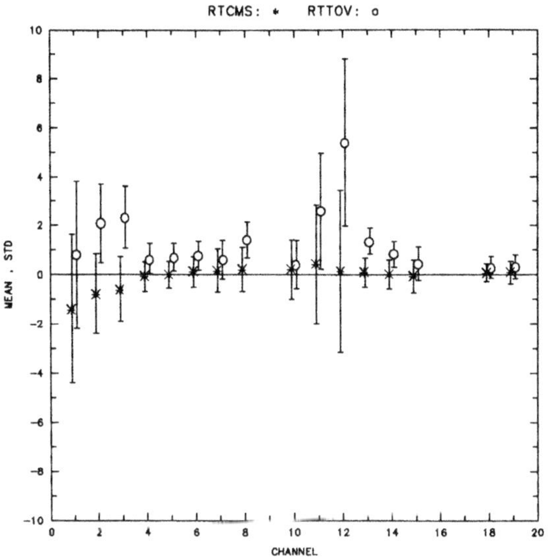

Figure 2: RTTOV (o) and RTCMS (*) validation on clear-sky situations of the EVA data set (1546 cases).

The results (figure 2) are presented for NOAA11 during the year 1990/1991 (1546 cases) and over sea (surface emissivity of 1). They include two adjustment corrections, given in table 4, (γ on transmissions and δ on brightness temperatures). For the RTCMS routine, these corrections were calculated with the data of the year 1990. Although the bias are different (no δ correction for the RTTOV model), the standard deviations are very similar. Moreover, the application of the same δ coefficients on the two models will efficiently reduce the biais difference. This demonstrates the ability of the FASCOD2 model to generate correct coefficients for the fast routine.

RTTOV:

channel	1	2	3	4	5	6	7	8	10	11	12	13	14	15	18	19
γ	.845	1.0	1.0	.982	.967	.981	.953	1.0	1.36	1.08	1.08	1.15	1.05	1.16	1.0	1.0
δ	0.0	0.0	0.0	0.0	0.0	0.0	0.0	0.0	0.0	0.0	0.0	0.0	0.0	0.0	0.0	0.0

RTCMS:

channel	1	2	3	4	5	6	7	8	10	11	12	13	14	15	18	19
γ	0.8	0.9	0.94	0.94	0.87	0.90	0.82	1.0	0.85	0.89	1.16	1.02	0.92	1.05	0.93	1.02
δ	3.1	3.4	3.8	0.5	0.4	-0.15	0.15	0.7	0.8	0.0	0.0	0.15	0.10	0.20	0.0	0.0

Table 4: γ and δ corrections for the RTTOV and RTCMS versions

4. APPLICATION TO FUTURE INSTRUMENTS

In the previous section, we have directly made the convolution of the FASCOD2 monochromatic transmissions with the NOAA11 TOVS filters, with no possibilities after this step to generate polynomial coefficients for other channels. All the off-line FASCOD computations are time-consuming (140h CPU on the IBM 3090 of the 'Centre National d'Etudes des Télécommunications') and we are looking to the possibility to store, at a medium frequency resolution, the FASCOD transmissions for the representative profiles.

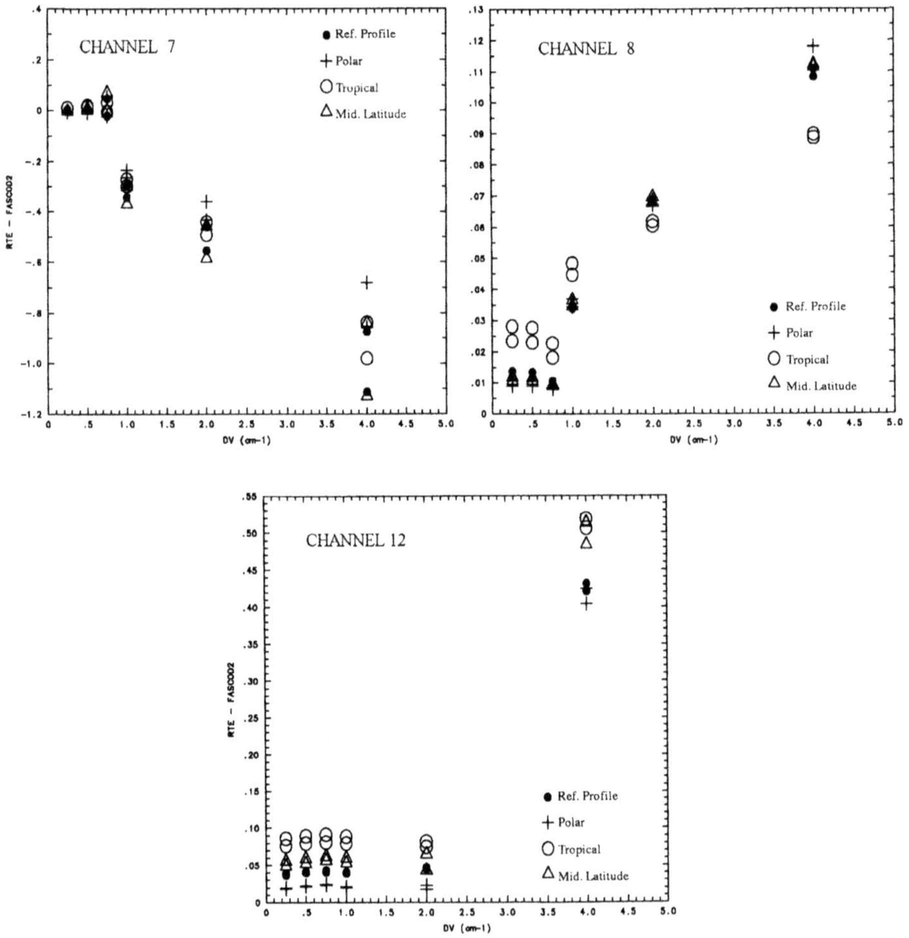

Figures 3: Variation of the brightness temperature difference with the storage resolution for: channel 7, channel 8, channel 12.

For each new channels, the medium frequency resolution transmittances are afterwards convolved over the filter and new polynomial coefficients generated. These steps are very fast and the design permits a quick adaptation of the fast model to new satellites. Moreover, the design allows possible extension of the fast routine to future interferometer systems, if the transmittance storage is made at the ad-hoc frequency resolution.

Brightness temperatures Tb_{RTE} were computed with an RT equation similar to that in FASCOD, but using medium frequency resolution transmittances in input instead of monochromatic transmittances. The accuracy of the transmittance storage was checked for different storage resolutions, on the NOAA11 TOVS channels, by comparing Tb_{RTE} to the brightness temperatures Tb_{FAS} directly computed by FASCOD2.

Figures 3 show the variation of Tb_{RTE} - Tb_{FAS} with the storage resolution, from 0.25 cm-1 to 4 cm-1, for channels 7, 8 and 12 which results are representative of the other channels. All the results are given for 4 different profiles (polar, tropical, mid-latitude, mean 'reference profile' of an atmospheric profile data set) and for 2 scanning angles (0°, 60°). For the 0.25 cm 1 and 0.5cm-1 , the errors are always less than 0.1K. For window channels the errors still remain in the order if 0.1K(with a possible storage up to 1 or 2 cm-1). For channel 7, the errors increase dramatically for larger resolution and reaches values of 1K at 4cm-1. For the water vapor channel 12, the results seem very dependant on the profile class, even at 0.25 cm-1, with larger errors for the tropical profile. For this channel, smaller storage resolution should be investigated.

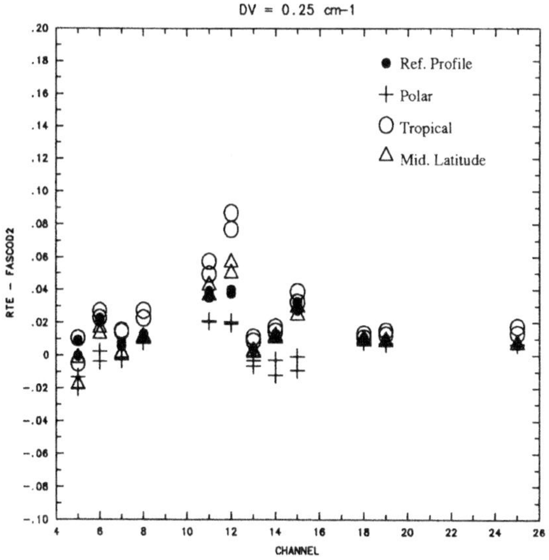

Figure 4: Brightness temperature errors for the NOAA11 HIRS2 channels , at a 0.25 cm-1 storage resolution.

Figure 4 summarizes the errors, at 0.25 cm-1, for the NOAA11 HIRS and AVHRR channels. If a constant storage resolution value has to be taken for all spectrum areas, the 0.25 cm-1 resolution seems to be a good compromise and shall be efficient for interferometer purposes. At this resolution, the transmittance file dimension, for 40 pressure levels, 5 incident angles, 3 molecular species, 32 atmospheric profiles, will be of 800Mbytes.

5. CONCLUSION

We have demonstrated the ability of using FASCOD2 radiative transfer computations to generate polynomial coefficients internal to the fast RTTOV routine. To avoid the problem of large computing time inherent in line-by-line calculations, the monochromatic transmittances are averaged at a medium frequency around 0.25 cm-1 and stored one time for all. This resolution does not degrade the results by more than 0.1K, for TOVS channels. The adaptation of the fast routine to new channels will afterwards be possible in a minimum computing time.

6.ACKNOWLEDGMENTS

We would like to thank Dr J.R. Eyre from ECMWF for the RTTOV routine providing and all his advice.

Thanks also to M. Journe and M. Leclerc from the CNET administration for providing us the CNET IBM facilities to run FASCOD2.

7.REFERENCES

Brunel P., L. Lavanant.1991, E.V.A. Report 1. CMS Internal Report.

Chedin A., N. Scott.1983, Improved initialisation inversion procedure ('3I'). First Int. TOVS Study Conf.

Chedin A., N. Scott, N. Husson....1988,Satellite meteorologie and atmospheric spectroscopy: recent progress in earth remote sensing from the satellites of the TIROS-N series. J. Quant. Spectrosc. Radiart. Transfer. 40.

Clough S.A., F. Fneizys, E. Shettle....1986, Atmospheric radiance and transmittance: FASCOD2. Proc. Sixth Conf on Atm. Radiance.

Eyre J.R. 1984, User guide to TOVSRAD: A program for calculating synthetic HIRS2 and MSU equivalent black-body temperatures. Met. O. branch Memorandum n°75

Eyre J.R. 1991, A fast radiative transfer model for satellite sounding systems. ECMWF technical Memorandum n° 176.

Lavanant L., P. Brunel, M. Derrien....1988, CATHIA: a data set for cloud classification and HIRS cloud clearing. Proc. Fourth Int. TOVS Study Conf.

Scott N., A. Chedin. 1981, A fast line-by-line method for atmospheric absorption computations: the Automatized Atmospheric Absorption Atlas. J. Appl. Meteor.,20.

REVIEW OF EXISTING SPECTRAL LINE DATA CATALOGS

N. Husson, A. Chedin, B. Bonnet
ARA / Laboratoire de Météorologie Dynamique du CNRS
Ecole Polytechnique
91128 Palaiseau Cedex, France

ABSTRACT

Four spectroscopic line parameter compilations exist for general use by the remote sensing and climate research community which are periodically updated and enhanced :

- the HITRAN (High Resolution Transmission) molecular absorption data base, from Phillips Laboratory, Department of the Air Force (USA) ;
- the GEISA (Gestion et Etude des Informations Spectroscopiques Atmosphériques : Management and Study of Atmospheric Spectroscopic Information) data bank, from ARA Group at Laboratoire de Météorologie Dynamique, CNRS (France) ;
- the ATMOS (Atmospheric Trace MOlecule Spectroscopy) molecular line list compiled at Jet Propulsion Laboratory (USA) for the ATMOS experiment ;
- the JPL (Jet Propulsion Laboratory, USA) catalog and atlas of microwave and submillimeter transmission.

The present status (more than 720,000 entries, corresponding to 40 molecules and 86 isotopic species, between 0 and 22,656 cm^{-1}, as a whole) of these data bases is reviewed.

I. INTRODUCTION

Recent progress in the analysis of planetary spectra combined with a new generation of vertical atmospheric sounders will result in an improved knowledge of the vertical atmospheric structure and of surface properties. For example, high spectral resolution infrared sounders like AIRS (Atmospheric Infrared Sounder ; JPL publication, 1991) and IASI (Improved Atmospheric Sounding in the Infrared ; Chedin & Scott, 1990 ; Cayla, 1992) are being

NATO ASI Series, Vol. I 9
High Spectral Resolution Infrared Remote Sensing
for Earth's Weather and Climate Studies
Edited by A. Chedin, M.T. Chahine and N.A. Scott

developed to meet the needs of both the operational meteorology and atmospheric science communities. Achieving the anticipated improvements in retrieval accuracy from high spectral resolution infrared and microwave data will require adequate tools for the calculation of atmospheric radiative transfer. Among them, is a good knowledge of spectroscopic line parameters, which are particularly important given the increasing impact of physical retrieval techniques for space-based remote sensing of the planetary atmosphere structure. Consequently, several computer accessible compilations of spectroscopic line parameters, extending from infrared to microwave, are available for molecules of interest for atmospheric remote sensing.

In this paper, published data bases for high resolution transmission and emission simulation will be reviewed. At present, the two public data banks HITRAN and GEISA represent, as a whole, over 720,000 entries cataloging parameters of 40 molecules and 86 isotopic species, in the spectral region 0-22,656 cm^{-1}.

II. SPECTRAL LINE DATA CATALOGS DESCRIPTION

A previous assessment of existing database of spectroscopic parameters has been reported in Appendix B (Husson et al., 1986a) of the 1985 Atmospheric Ozone Assessment Report (Watson, 1986). At the present time, four compilations are periodically updated for general use by the remote sensing community :

- the AFGL/HITRAN (Air Force Geophysics Laboratory ; Phillips Laboratory / High-Resolution Transmission) Data Bank (McClatchey et al., 1973 ; Rothman et al., 1987, 1992) ;
- the GEISA (Gestion et Etude des Informations Spectroscopiques Atmosphériques : Management and Study of Atmospheric Spectroscopic Information) Data Bank (Chedin et al., 1980 ; Husson et al., 1986b, 1991, 1992a, 1992b) ;
- the ATMOS (Atmospheric Trace MOlecule Spectroscopy) Molecular Line-List, compiled by Brown et al. (1987) for the ATMOS experiment (Farmer et al., 1987) ;
- the JPL (Jet Propulsion Laboratory) catalog and atlas of microwave and submillimeter transmission by Poynter and Pickett (1985).

The initial emphasis of each of the four compilations varied. The AFGL/HITRAN compilation was intended for terrestrial atmospheric problems, the GEISA compilation for terrestrial and Giant Planet atmospheres and the JPL catalog for astrophysical studies. The ATMOS line list was tailored to the specific needs of ATMOS. Presently, the different databases include many of the same molecular transitions relevant to remote sensing.

II.1 Spectroscopic data catalogs parameters

The spectroscopic parameters are provided for all significant transitions of molecular and atomic species of interest. Table I (from Table B-10 of Husson et al., 1986a) summarizes the parameters that are currently incorporated into the compilations.

Table I Spectroscopic data catalogs parameters (from Husson et al., 1986a)

PARAMETER †			DATABASE		
	Type	Unit	ATMOS compilation	JPL catalog	HITRAN GEISA
MOL	Molecular specie index		x	x	x
ISO	Isotopic variant index		x	x	x
ν	Resonant frequency	cm^{-1}	x	x	x
S	Intensity of transition	$cm^{-1}/(mol\ cm^{-2})$	x	x	x
R	Transition moment	Debye			x
g	Upper state degeneracy			x	
γ	Air-broadened halfwidth	cm^{-1}/atm	x		x
γ_s	Self-broadened halfwidth	cm^{-1}/atm			x
E"	Lower state energy	cm^{-1}	x	x	x
n	Temperature-dependence of halfwidth				x
δ	Pressure shift of transition	cm^{-1}/atm			x
v'	Upper vibrational quanta		x	x	x
v"	Lower vibrational quanta		x	x	x
Q'	Upper rotational quanta		x	x	x
Q"	Lower rotational quanta		x	x	x
e	Error estimates for major parameters			x	x
Ref	citation for major parameters				x

† Reference temperature 296 K

Until 1984, the HITRAN and GEISA compilations included for each absorption line of a given molecule :
- the wavenumber (in cm^{-1}),
- the intensity [in $cm^{-1}/(molecule\text{-}cm^{-2})$],
- the Lorentzian collision halfwidth
  (halfwidth at half-maximum, in cm^{-1}/atm),

- the lower state energy of the associated transition (in cm^{-1}),
- the quantum identification (vibrational, rotational, electronic level, hyperfine, and splitting designation if necessary),
- the entry date, isotope and molecule codes.

A common 80 character "card-image" format was adopted for describing these parameters.

In 1984, because the importance of the temperature variation of the halfwidth, $\gamma(T)$, of a line on the accuracy of radiative transfer computation, the line halfwidth temperature-dependence has been introduced as a new parameter in GEISA (Chedin et al., 1985 ; Husson et al., 1986b).

In addition, in 1986, five new parameters have been included in HITRAN (Rothman et al., 1987) :
- the transition moment (in debyes),
- the self-broadened halfwidth (in cm^{-1}/atm),
- the pressure shift of the line (in cm^{-1}/atm),
- approximate error estimates for frequency, intensity and halfwidth,
- references for frequency, intensity and halfwidth.

These new parameters have been included in GEISA (Husson et al., 1992a) from HITRAN (Rothman et al., 1992).

The reference temperature for the intensity, the collision (mainly with air) and self halfwidths, and the pressure shift, is 296K.

The major part of all the above listed line parameters is also included in the ATMOS line list and in the JPL catalog.

II.2 Present time status of HITRAN and GEISA data bases

The HITRAN data base (the so-called "AFGL tape") finds its origin in a first compilation, oriented towards the Earth's atmosphere and initiated in 1973, at the US Air Force Geophysics Laboratory by McClatchey et al. (1973) and Garing et al. (1973). In 1976, a similar work was started at Laboratoire de Météorologie Dynamique (LMD) of CNRS, in France, with the development of the GEISA data bank described by Chedin et al. (1980). From its origin, the major orientation of GEISA has been, first, to introduce molecules of interest not only for the Earth's atmosphere, but also for planetary atmospheres, in particular for the giant planets, and, second, to develop software that makes possible various kinds of extractions covering applications in the field of atmospheric physics as well as in the field of molecular spectroscopy.

TABLE II Summary of the molecular species cataloged in the line parameter portions of HITRAN 91-92 and GEISA 91

Molecules	# of isotopes		# of bands		# lines		FMIN	FMAX
	H	G	H	G	HITRAN91-92	GEISA-91	(cm^{-1})	(cm^{-1})
H2O	4	5	134	136	48523	49296	0.007	22597
CO2	8	9	592	598	60790	60948	442	9649
O3	3	3	76	76	168881	168881	0.026	3226
N2O	5	5	140	140	24125	24125	0.838	5131
CO	5	5	31	102	3600	13205	3	8465
CH4	3	2	48	39	47415	40514	0.010	6185
O2	3	3	18	18	2254	2254	0.000	15928
NO	3	3	13	50	7385	7385	0.000	3966
SO2	2	2	7	7	26225	23659	0.017	2527
NO2	1	1	9	9	55468	55468	0.498	2938
NH3	2	2	9	23	5817	6784	0.231	2154
PH3	1	1	2	5	2886	4635	18	2446
HNO3	1	1	13	13	143021	143021	0.035	1388
OH	3	3	27	103	8676	8676	0.003	9997
HF	1	1	6	6	107	107	41	11536
HCL	2	2	17	17	371	371	20	13458
HBR	2	2	16	16	398	398	16	9759
HI	1	1	9	9	237	237	12	8487
CLO	2	2	8	8	6020	6020	0.015	890
OCS	4	4	6	23	737	4171	0.381	4118
H2CO	3	3	10	10	2702	2702	0.000	2999
C2H6	1	1	2	2	4749	8944	765	3000
CH3D	x	1	x	9	x	6457	7	3146
C2H2	2	2	9	13	1258	1258	638	3397
C2H4	x	1	x	1	x	203	903	1093
GEH4	x	1	x	1	x	824	1937	2225
HCN	3	3	8	41	772	2575	2	18408
C3H8	x	1	x	1	x	9019	700	800
C2N2	x	1	x	7	x	2577	204	2182
C4H2	x	1	x	5	x	1405	190	654
HC3N	x	1	x	20	x	2027	474	691
HOCL	2	2	6	6	15565	15565	0.024	3800
N2	1	1	1	1	120	117	1992	2626
CH3CL	2	2	6	6	6687	6687	2907	3173
H2O2	1	1	2	2	5444	5444	0.155	1500
H2S	1	3	1	6	661	4058	1	1574
HCOOH	x	1	x	1	3388	3388	1061	1161
COF2	1	1	7	3	46894	18242	725	1982
SF6	1	1	1	1	11520	11520	940	952
C3H4	x	1	x	1	x	3390	290	360
Total lines					709308	726557		

The recent updates of HITRAN and GEISA have been finalized at nearly same time, and include very similar data sources for many molecules. Because of its origin, certain molecules, related with planetary atmospheres, are specific to GEISA, such as : C_2H_4, GeH_4, C_3H_8, C_2H_2, C_4H_2, HC_3N, HCOOH, C_3H_4.

Independent of their line parameter files, HITRAN and GEISA include files of cross-sectional data on heavy molecular species such as the chlorofluorocarbons (CFCs) which are not amenable to line-by-line representation, at the present time.

II.2.1 Line parameter files

The molecular species cataloged in the line parameter portions of HITRAN91-92 (Rothman et al., 1992) and GEISA91 (Husson et al., 1991, 1992a) are summarized in Table II, which gives the molecular formula, total number of isotopes, bands and lines in HITRAN91-92 (referred to as H) and GEISA91 (referred to as G) respectively, in columns 1 to 4. The discrepancy between the numbers of bands in HITRAN and GEISA can find its origin in different conventions for counting between the authors.

Molecules specific to GEISA are identified with the sign (X) in the related HITRAN column. Among them is CH_3D which appears as an isotope of CH_4 in HITRAN but as an independent molecular species in GEISA. The two last columns list the minimum and maximum range of the line positions for the overall data set HITRAN91-92 - GEISA91.

II.2.2 Cross section files

In both GEISA and HITRAN , a cross section file has been created, independent of the line parameter file, the content of which can be summarized as follows :

★ New cross sections from NCAR (Massie et al., 1985, 1991 ; McDaniel et al., 1991) :

In 20 spectral regions , from 765 to 1335 cm^{-1}, and at 6 temperatures from 203 to 293K, cross sections of the following CFC's species have been cataloged in both data bases : CF_3Cl, CHF_2Cl, $C_2Cl_3F_3$, $C_2Cl_2F_4$, $CFCl_3$, CF_2Cl_2, C_2ClF_5, CF_4. This represents, as a whole, 80,699 points per temperature for each database.

In 4 spectral regions, from 555 to 1765 cm^{-1}, cross sections of the following species are cataloged in HITRAN, specifically :

N_2O_5 4 spectral regions from 555 to 1765 cm^{-1},
4 temperatures, 497 points per temperature.

HNO_4	1 spectral region from 770 to 830 cm^{-1}, one temperature, 5476 points.
$CHCl_2$	1 spectral region from 785 to 840 cm^{-1}, one temperature, 5020 points.
CCl_4	1 spectral region from 786 to 806 cm^{-1}, one temperature, 1826 points.

For more details, see Husson et al. (1992a) and Rothman et al. (1992).

★ $ClONO_2$ cross sections (Ballard et al., 1988).

In 3 spectral regions, from 740 to 1790 cm^{-1} and at 2 temperatures, cross sections of $ClONO_2$ have been cataloged in both databases (13,311 points per temperature, for every database).

II.2.3 *CFC-12 data from a line-by-line calculation in the 923 cm^{-1} region (Deroche et al., 1987)*

Spectroscopic line parameters (line frequency, relative intensity, assignment and lower state energy) for nine bands of the isotopic species : $CF_2{}^{35}Cl_2$, $CF_2{}^{35}Cl^{37}Cl$, $CF2^{37}Cl_2$ are cataloged in GEISA specifically as an independent file containing 375,962 entries. For more details, see Husson et al. (1992a).

II.3 The ATMOS compilation

During its first mission in 1985 onboard Spacelab 3, the ATMOS instrument, a high speed Fourier transform spectrometer operating in the middle infrared (2.2-16 μm), recorded more than 1500 solar spectra at ≈ 0.0105 cm^{-1} resolution (Farmer et al., 1987). The analysis and interpretation of these data in terms of composition, chemistry, and dynamics of the Earth's upper atmosphere requires good knowledge of the molecular line parameters for those species giving rise to the absorptions in the atmospheric spectra.

The ATMOS compilation (Brown et al., 1987) was tailored to the specific needs of the ATMOS experiment. This in-house compilation of line parameters for the 1 - 10,000 cm^{-1} region was initially based on parameters in the AFGL/HITRAN and GEISA compilations, but eventually included parameters of specific species targeted for analysis by ATMOS.

The ATMOS compilation is divided into two lists :

- the main list contains data based on published analyses of laboratory spectra,
- the supplemental linelist contains data based on very preliminary results, with no better existing representation, but necessary for ATMOS data processing.

At present time, the main ATMOS list consists of the HITRAN 1992 linelist with some changes for the following molecules (Brown, 1992) : H_2O, O_3, N_2O, NO_2, HNO_3, HF, HDO, HOCl, CH_3Cl, C_2H_6, because of new or complementary data available.

Molecular species which currently have parameters in the supplemental line list are shown in Table III (from Brown, 1987), which gives the molecular formula, total number of lines, minimum and maximum range of the line positions in units of cm^{-1}, the total number of isotopes and bands. The supplemental line list has 86,105 lines from 697 to 4135 cm^{-1}.

Table III Summary of the ATMOS supplemental linelist in september 1986 (from Brown, 1987).

Molecule	# Lines	F min (cm^{-1})	F max (cm^{-1})	# Isotopes	# Bands
O3	15117	926.1130	1544.9600	3	5
CH4	3279	1185.0918	4135.8174	1	3
HNO3	10522	762.8553	1342.9993	1	4
OCS	245	2885.0200	4116.6875	1	2
H2CO	587	1707.0925	1781.3870	1	1
HO2	4450	1055.1858	3719.3393	1	3
HONO	2935	779.1260	3603.6958	1	5
HO2NO2	341	802.4781	802.6542	1	1
N2O5	3106	725.0000	1807.9499	2	3
ClONO2	4035	779.7114	1309.9910	2	2
CH3F	739	987.9178	1089.1345	1	1
CH3Cl	1738	697.4966	1377.7244	2	3
CF4	1736	1281.3152	1286.4141	1	2
CCl2F2	5772	1070.0378	1161.2793	1	2
CCl3F	4480	834.0952	1099.9978	1	2
CH3CCl2	251	1382.5000	1385.0000	1	1
CCl4	1826	786.0010	805.9979	1	1
COF2	1908	1920.0474	1977.1528	1	1
CFClO	413	764.1111	764.7198	1	1
C2H6	75	2976.7524	2996.9800	1	1
CHF2Cl	69	828.9552	829.1376	1	1
COCl	484	848.6393	850.9967	1	1
CH3Br	6498	871.3259	3147.9634	1	2
CH3I	2528	793.4666	971.6544	1	1
HCOOH	3388	1060.9622	1161.2512	1	1
CHFCl2	9583	785.0000	1099.9919	1	1

II.4 The submillimeter, millimeter and microwave spectral line catalog (so-called JPL catalog)

This computer accessible catalog of submillimeter, millimeter and microwave spectral lines in the frequency range between 0 and 10,000 GHz (i.e. wavelengths longer than 30 μm) has been developed at JPL by Poynter and Pickett (1985).

The selection of the species included in the catalog has been made to fill the needs of both astronomers and atmospheric scientists. The catalog may also be of use for laboratory studies and for laser or far-IR calibration.

Table IV, from Poynter and Pickett (1985), lists the 151 atomic and molecular species included in the JPL catalog with the number of lines corresponding to each of them. This represents a total of 237,120 spectral lines.

III. SPECTRAL LINE DATA CATALOGS AVAILABILITY

The public HITRAN and GEISA data bases are regularly updated and are currently available for dissemination. Copies of both data bases can be obtained on classical large magnetic data tapes. The current version of HITRAN is available on both floppy diskettes, in compressed form, and on CD-ROM disks. In the case of GEISA, special extractions for some particular molecules or spectral ranges, and copies of the associated software can be made on floppy diskettes ; the same facility will be offered for the complete data catalog in the very near future. Updated versions of both data bases are issued every two to four years.

As already mentioned, the ATMOS in house molecular line parameters compilation has been initiated to provide rapid response to the needs of the ATMOS experiment. Updates are made two or three times a year, at the ATMOS Data Analysis Facility at JPL, essentially in the 580-4500 cm^{-1} region where the ATMOS experiment requires parameters. The other regions are included as a service, upon request, to non-ATMOS investigators.

The JPL catalog is available both on magnetic data tape and as a set of microfiche films.

IV. CONCLUSIONS AND RECOMMENDATIONS

The status of the spectroscopic database as a whole and current laboratory research is related to the accuracies with which atmospheric structure and surface parameters can be determined from spectral measurements.

Table IV JPL catalog species list (from Poynter and Pickett, 1985)

Species name	Lines	Species name	Lines	Species name	Lines
H-ATOM	1	O2-V1	153	O3-SYM-O-18	3184
D-ATOM	1	CH3OH	709	O3-ASYM-O-18	7304
HD	1	H2CO-18	449	O3-S-O18-V2	2387
C-ATOM	2	HO2	6174	O3-A-O18-V2	4213
C-13-ATOM	7	O-18-O	383	HCCCN	154
CH	324	H2S	311	CLO	2610
N-ATOM	2	PH3	142	HCCC-13-N	152
N-ATOM-D-STATE	6	H2O2	883	HCC-13-CN	146
O-ATOM	2	HDS	1138	HC-13-CCN	144
OH	299	HCl	17	HCCCN-15	99
NH3	446	HCl-37	17	DCCCN	156
CH3D	80	CH3CCH	813	HOCL	3919
NH3-V2	442	CH3CN	1441	C2H3CN	3697
OD	912	CH3CC-13-H	822	CL-37-O	2649
N-15-H3	235	CH3C-13-CH	813	CH2CHC-13-N	118
H2O	424	C-13-H3CCH	821	CH2C-13-HCN	117
NH2D	5036	CH3CCD	822	C-13-H2CHCN	116
H2O-V2	292	CH2DCCH	223	CH2CDCN	322
HO-18	295	CH3CN-15	2755	HOCL-37	3923
HDO	461	CH2CO	835	C2H5CN	3684
D2O	427	CHDCO	886	CH3CH2C-13-N	1183
CCH	114	CS	51	CH3C-13-H2CN	1252
CN	385	SIO	31	C-13-H3CH2CN	1183
CN-V1	437	CH3CHO-A	1347	C2H5CN-15	1621
HCN	40	N2O	61	CH2DCH2CN-S	1166
HNC	33	CH3CHO-E	691	CH2DCH2CN-A	1286
CO	26	C-13-S	29	OCS	99
HC-13-N	45	SI-29-O	66	SIS	97
HCN-15	35	NH2CHO	3476	OC-13-S	99
DCN	54	CS-34	34	SI-29-S	98
HNC-13	34	SI-30-O	67	OC-34-S	99
HN-15-C	33	H2CS	517	O-18-CS	99
DNC	39	C2H5OH	1236	SI-30-S	99
C-13-O	27	HCOOH	1888	SIS-34	99
HCO+	33	NO2	6360	HNO3	24767
CH2NH	2957	H2C-13-S	110	S2	34
HCO	2454	HC-13-OOH	1194	SO2	9622
NNH+	34	DCOOH	628	HCCCCCN	99
CO-18	27	HCOOD	612	HCCCCC-13-N	99
HC-13-O+	34	SO	330	HCCCC-13-CN	99
DCO+	41	SO-V1	261	HCCC-13-CCN	99
H2CO	611	H2CS-34	111	HCC-13-CCCN	99
C-13-H2NH	439	O3	6271	HC-13-CCCCN	99
CH2N-15-H	440	O3-V2	4337	HCCCCCN-15	99
CH2ND	1834	O3-V1,3	9459	DCCCCCN	99
NO	1909	O3-2V2	1779	HBR-79	19
NND+	41	O3-SYM-O-17	13277	HBR-81	19
HCO-18+	34	O3-ASYM-O-17	24138	BR-79-O	850
H2C-13-O	601	S-34-O	280	BR-81-O	850
O2	174	SO-18	179	CL-35-NO3	4198
				CL-37-NO3	4182

Today's accuracies for the 3 most significant line parameters : wavenumber, intensity, halfwidths, are summarized in columns 2 (microwave) and 3 (infrared) of Table V. An evaluation of these parameters in terms of remote sensing requirements is presented in Table VI.

Table V Present status of the today's accuracies for 3 most significant spectroscopic parameters

LINE PARAMETERS	ACCURACY	
Wavenumber Intensity Halfwidth	30 kHz (microwave) 2 to 5 % (relative) 2 to 5 % (absolute)	0,0001 cm^{-1} (infrared) 10 to 15 % (absolute *)

* For absolute intensities, 5 % can be achieved with particularly careful measurements.

Table VI Evaluation of the 3 most significant line parameters in terms of remote sensing calculation requirements

QUALITY INDEX →	A	B	C
LINE PARAMETER ↓			
Wavenumber	85 %	17 %	8 %
Intensity	60 %	30 %	10 %
Halfwidths	5 %	55 %	40 %

A The data are good for many applications.
B The data are good for some applications, but need improvement.
C The data require major improvement.
(720,000 entries, 40 molecules, 86 isotopic species, in the spectral range 0-22,656 cm^{-1}).

Quality indexes are given in the columns 2 to 4, in terms of percent of the whole database entries, as the following :

A the data are good for many applications ;
B the data are good for some applications, but need improvement s;
C the data require major improvements.

This evaluation has been made on the basis of all the entries included in the present HITRAN and GEISA releases. One conclusion is that many of the spectroscopic parameters needed for the future space-based programs require greater accuracy that can be achieved at the present time.

The major requirements for further work in laboratory spectroscopy for atmospheric remote sensing have already been discussed. Specific details for many of these requirements can be found in a report by Smith (1985), in the UARS Spectroscopic Requirements document (Roche, 1985), and in the 1985 Global Ozone Research and Monitoring Project (Watson, 1986). A summary list can be given as follows. Importance of the needed quantities depends on the particular application.

1. *Line positions and intensities*

   Spectral parameters for missing infrared bands of major and trace constituents.

2. *Linewidths*

   - Improved knowledge of air-broadened halfwidths and their temperature dependence for nearly all the optically active atmospheric species.
   - Self-broadened halfwidths.

3. *Line shapes*

   Laboratory and theoretical studies of deviations from the Lorentz line shape (see Chedin and Scott, 1984).

4. *Unresolved bands*

   Accurate intensities and band models representations for unresolved bands.

5. *Pressure induced bands*

   More accurate absorption coefficients for the important pressure induced bands of oxygen and nitrogen.

6. *Water vapor continuum*

   Improved models for the temperature and pressure dependence of the water vapor continuum.

7. *Non-LTE radiative transfer*

   Improved parameters for transitions involved in non-LTE (non local thermodynamic equilibrium) radiative transfer in the upper atmosphere.

8. *Other effects*

   Quantification of effects such as pressure induced line shifts.

Workshops of the Atmospheric Spectroscopy Applications Working Group (ASA-WG) of the International Radiation Commission (IRC) have offered, in particular, the opportunity to discuss the future development and management of spectroscopic databases for radiative transfer, in 1987 (at Rutherford Appleton Laboratory) and in 1990 (in Moscow).

The final and most important recommendations can be summarized as follows (see ASA-WG Workshop Proceedings, 1987) :
A mechanism should be established to address the issues of :

- the number of databases, including the use of supplemental lists for preliminary data ;
- a peer review of data being put into the databases. Intensive collaboration and collaborative efforts should exist between several laboratories to reduce systematic errors. Sensitivity studies should be made to better show the effects of the errors on particular simulations ;
- an interaction between the database managers and the user community.

Increasingly more sophisticated satellite atmospheric remote sensing experiments are being proposed by NASA (USA), ESA (Europe), CNES (France), Japan, ..., for possible implementation beyond the later part of this century. The increasing number of accurate observations from space will require better knowledge of spectroscopic parameters at conditions prevalent in the upper stratosphere, mesosphere and possibly at even higher altitudes. Remote sensing of the troposphere from space will also require improved knowledge of halfwidths, line shapes, continua, and aerosol extinction, as well as line parameters for additional molecules which are not important in the stratosphere. Improved knowledge of these same spectroscopic parameters, along with accurate representations of absorption and emission in broad spectral regions covering many molecular bands, will also be required for climate studies.

REFERENCES

Atmospheric Spectroscopy Applications Workshop Proceedings, December 1987. Edited by J. Ballard, Rutherford Appleton Laboratory, RAL-88-009.

Ballard J, Johnston WB, Gunson MR, Wassell PT (1988) : Absolute absorption coefficients of ClONO2 infrared bands at stratospheric temperatures. G. Geophys. Res., 93, 1659-1665

Brown LR, Farmer CB, Rinsland CP, Toth RA (1987). Molecular line parameters for the atmospheric trace molecule spectroscopy experiment. Appl. Opt. 23, 5154-5182

Brown LR (1992). Atmos line list in 1992. Private communication

Cayla FR (1992) : IASI an infrared interferameter for operations and research. Proceedings of NATO advances Research Workshop on "High Spectral Resolution Infrared Remote Sensing for Earth's Weather and Climate Studies", Paris, France, 23-26 March 1992, Springer, Berlin Heidelberg New York, to be published.

Chedin A, Husson N, Scott NA, Jobard I,. Cohen-Hallaleh I, Berroir A. (Oct. 1980) : La banque de données GEISA. Description et logiciel d'utilisation. Laboratoire de Météorologie Dynamique du CNRS, Internal Note 108, Ecole Polytechnique, F- 91128 Palaiseau Cedex

Chedin A, Scott NA (1984) : The impact of spectroscopic parameters on the comparison of the Jovian atmosphere discussed in connection with the recent laboratory, earth and planetary observations programs. JQSRT 32, 453-461

Chedin A, Husson N, Scott NA, Cohen-Hallaleh I, Berroir A (Feb. 1985) : The GEISA data bank : 1984 version. Internal Note 127, Laboratoire de Météorologie Dynamique, Ecole Polytechnique, Palaiseau, France.

Chedin A., Scott NA (1990) : Improved atmospheric sounding in the infrared. Internal Note 157, Laboratoire de Météorologie Dynamique, Ecole Polytechnique, Palaiseau, France.

Deroche JC, Goldman A (1987) : Line-by-line calculation of CF_2Cl_2 (F-12) absorption profile in the 923 cm-1 region. Tenth Colloquium on HIgh Resolution Molecular Spectroscopy, Dijon, France, 14-18 sept. 1987

Farmer CB, Raper OF, O'Callaghan FG (Oct. 1987) : Final Report on the First Flight of the Atmos Instrument during the Spacelab-3 mission, April 29 through May 6 1985. JPL Publication 87-32

Garing JS, McClatchey RA (1973) : Atmospheric absorption line compilation. Appl Opt, 12, 2545-2545

Husson N, Barbe A., Brown LR, Carli B, Goldman A, Pickett HM, Roche, Rothman LS, M.A.H. Smith MAH (1986a). Atmospheric Ozone 1985. Assessment of our understanding of the processes controlling its present distribution and change. Vol. III, Appendix B, W.M.O. Global Ozone Research & Monitoring Project. Report n° 16, 911-946

Husson N, Chedin A, Scott NA, Bailly D, Graner G, Lacome N, Levy A, Rossetti C, Tarrago G, Camy-Perret C, Flaud JM, Bauer A, Colmont JM, Monnanteuil N, Hilico JC, Pïerre G, Loete M, Champion JP, Rothman LS, Brown LR, Orton G, Varanasi P, Rinsland CP, Smith MAH, Goldman A (1986b) : The GEISA spectroscopic line parameters data bank in 1984. Ann. Geophys., Fasc. 2, Series A, 185-190

Husson N, Bonnet B, Scott NA, Chédin A (1991) : The GEISA Data Bank, 1991 version. Internal Note LMD n° 163. Laboratoire de Météorologie Dynamique du CNRS, Ecole Polytechnique, Palaiseau, France

Husson N, Bonnet B, Scott NA, Chédin A (1992a) : Management and Study of Spectroscopic Information. JQRST, 48, 509-518.

Husson N, Bonnet B, Chédin A, Scott NA, Arié E, Ballard J, Bjoraker G, Brown LR, Camy-Peyret C, Deroche C, Fayt A, Flaud JM, Goldman A, Graner G, Hillman JF, Hilico JC, Lellouch E, Perrin A, Remedios JJ, Rinsland CP, Rothman LS, Tipping (1992b) : The GEISA spectroscopic line parameters data bank in 1992. Ann. Geophys., in preparation

JPL (1991) Airs Atmospheric Infrared Sounder Science and measurement requirements. JPL Publication, D6665 REV.1 9/91. Jet Propulsion Laboratory, Pasadena

Massie ST, Goldman A, Murcray DG, Gille JC (1985) : Approximate absorption cross sections of F12, F11, ClONO2, N2O5, HNO3, CC14, CF4, F21, F113, F114 and HNO4. Appl. Opt., 24, 3426-3427

Massie ST, Goldman A, McDaniel AH, Cantrell CA, Davidson JA, Shetter RE, Calvert JG (1991) : Temperature dependent infrared cross sections for CFC11, CFC12, CFC13, CFC14, CFC22, CFC113, CFC114 and CFC115. NCAR Technical Note TN-358+STR

McClatchey RA, Benedict WS, Clough SA, Burch DE, Calfee RF, Fox K, Rothman LS, Garing JS (1973) : AFCRL atmospheric absorption line parameters compilation. Environmental Research Paper 434, AFCRL-TR-73-0096 : Air Force Cambridge Research Laboratories

McDaniel AH, Cantrell CA, Davidson JA, Shetter RE, Calvert JG (1991) : The temperature dependent infrared absorption cross sections for the chlorofluorocarbons : CFC11, CFC12, CFC13, CFC14, CFC22, CFC113, CFC114 and CFC115. J. Atmos. Chem., 12, 211-227

Poynter RL, Pickett HM (1985) : Submillimeter, millimeter, and microwave spectral line catalog. Appl. Opt. 24 : 2235-2240

Roche AE (1985) : UARS Spectroscopy Requirements : Report of the UARS Spectroscopy Working Group

Rothman LS, Gamache RR, Goldman A, Brown LR, Toth RA, Pickett HM, Poynter RL, Flaud JM, Camy-Perret C, Barbe A, Husson N, Rinsland CP, Smith MAH (1987) : The Hitran database : 1986 edition. Appl. Opt. 26 : 4058-4097

Rothman LS, Gamache RR, Tipping RH, Rinsland CP, Smith MAH, Benner DC, Malathy Devi V, Flaud JM, Camy-Peyret C, Perrin A, Godman A, Massie ST, Brown LR, Toth RA (1992) : The HITRAN Molecular Database : Editions of 1991 and 1992. JQRST, 48, 469-507.

Smith MAH (Ed) (1985). Review of spectroscopic parameters for upper atmospheric measurements, NASA Conference Pubication 2396, 28 pp., NASA, Langley Research Center, Hampton, VA

Watson RT (1986) : Atmospheric Ozone 1985. Assessment of our understanding of the processes controlling its present distribution and change. Vol. I, Introduction & Science Summary. Global Ozone Research and Monitoring Project. Report n° 16, 1-25.

CO_2 Q–BRANCH LINESHAPES FOR ATMOSPHERIC REMOTE SENSING

L. Larrabee Strow
Department of Physics
University of Maryland Baltimore County
5401 Wilkens Avenue
Baltimore, Maryland 21228
USA

1. Introduction

The high spectral resolution of today's infrared remote sensing instruments puts great stress on our understanding of atmospheric spectral lineshapes. This is especially true for nadir sounders that use channels between spectral lines and in the far line wing, where the weighting functions are narrowest. The atmospheric emissions in these wing regions are sensitive to the molecular lineshape, which is difficult to measure in the laboratory over the wide range of atmospheric conditions. Since satellite temperature sounders utilize CO_2 emission features we will concentrate on CO_2 spectral lineshapes and their effect on atmospheric spectra in this paper. Two regions of the CO_2 spectrum are often chosen for sounding channels; (1) Q–branch wings in the 15 μm region, and (2) the ν_3 R–branch bandhead at 4.3 μm. Both of these regions are extremely sensitive to aspects of CO_2 lineshapes that have only been studied in some detail in the last several years. For example, absorption coefficients in the CO_2 Q–branch wing deviate from a Lorentz lineshape by up to 50% because of line mixing. Even larger deviations from a Lorentz lineshape are found in the 4.3 μm R–branch wing. This paper will concentrate on Q–branch spectral lineshapes which we are now able to model with high accuracy. Physically–based lineshape models that provide a basis for accurate spectral calculations under wide ranging conditions are still under development for the 4.3 μm region.

The theory of Q–branch line mixing and relevant laboratory measurements will be briefly reviewed here, followed by examples of line mixing in atmospheric spectra. Approximation techniques important for the incorporation of line mixing in atmospheric radiative transfer calculations are then addressed. Some of the pitfalls of these approximation techniques are discussed in detail.

2. Theory

When spectral structure is dense enough to overlap due to pressure broadening, interference effects may cause deviations from directly additive individual line profiles, provided that the particular rotational levels involved in the transitions are coupled by the collisions. At high pressures this effect, called line mixing, causes the band profile to visibly narrow. The most dramatic effect of line mixing at atmospheric pressures is a large decrease in the band wing absorption. This collision–induced interference, also known as line coupling, line mixing, cross relaxation, rotational collisional narrowing, and Q–

NATO ASI Series, Vol. I 9
High Spectral Resolution Infrared Remote Sensing
for Earth's Weather and Climate Studies
Edited by A. Chedin, M.T. Chahine and N.A. Scott

branch collapse, has been the subject of a number of recent investigations utilizing both Raman (Rosasco et al. 1983, Lempert et al. 1984, Lavorel et al. 1987) and infrared spectra (Strow & Gentry 1986, Gentry & Strow 1987, Hoke et. al. 1988, Cousin et al. 1986, Bulanin et al. 1984). The effects of line mixing on the atmospheric spectrum of O_2 in the microwave were initially examined some years ago by Rosenkranz (Rosenkranz 1975) and others (Smith 1981, Lam 1977) and are well understood.

Line mixing in Q-branch spectra is most evident within $\sim 10\ cm^{-1}$ of the Q-branch lines. This allows us to neglect duration of collision effects and use the impact approximation to describe the collisions. In this limit, the absorption coefficient at frequency ν for overlapping lines may be written as (Ben-Reuven 1966, Smith 1981)

$$k(\nu) = \frac{N}{\pi} \mathbf{IM} \left\{ \sum_{j,k} d_j \langle\langle j| \left[(\boldsymbol{\nu} - \boldsymbol{\nu}_o) - iP\mathbf{W} \right]^{-1} |k\rangle\rangle d_k \rho_k \right\}, \tag{1}$$

where d_j and d_k are the dipole moment matrix elements for the radiative transitions j and k, ρ_k are density matrix elements that give population differences between levels involved in the doubled state $|k\rangle\rangle$, $\boldsymbol{\nu}$ is a diagonal matrix with $\langle\langle j|\boldsymbol{\nu}|k\rangle\rangle = \nu\delta_{jk}$, $\boldsymbol{\nu}_o$ is the diagonal matrix $\langle\langle j|\boldsymbol{\nu}_o|k\rangle\rangle = \nu_j\delta_{jk}$ and ν_j is the transition frequency associated with the doubled state $|j\rangle\rangle$, P is the total pressure, N is the absorber number density, and $\mathbf{W}$ is the relaxation matrix in doubled state space or what is sometimes called line space. If mixing is only allowed between the Q-branch lines the diagonal elements of $\mathbf{W}$ are the Q-branch pressure broadening coefficients and $|j\rangle\rangle$, $|k\rangle\rangle$ represent Q-branch transitions.

The off-diagonal matrix elements of $\mathbf{W}$ are the negative of the rate at which collisions transfer intensity from one line to another. If these matrix elements are sufficiently small the spectrum becomes the sum of non-interacting Lorentzian lines. The rotational energy levels of CO_2 are separated by much less than thermal translational energies at atmospheric temperatures, so collisions are energetically able to allow the transfer of rotational and translational energy. The combination of closely spaced transitions (spacings on the order of the inverse of the time between collisions) and the high rate of rotationally inelastic collisions leads to significant line mixing. This mixing narrows the profile of the overlapping Q-branch lines with increasing pressure. The narrowed spectrum can be calculated if the off-diagonal elements of the $\mathbf{W}$ matrix are known in addition to the standard isolated Lorentz line parameters.

The calculation of the $\mathbf{W}$-matrix elements from first principles for atmospheric applications would be exceedingly complex, computationally expensive, and require good CO_2-N_2,O_2 intermolecular potentials. Instead, we estimate the off-diagonal $\mathbf{W}$-matrix elements by modelling the state-to-state rotationally inelastic collision rates in each vibrational level using a simple rotational energy gap law (Brunner & Pritchard 1982, Rosasco et al. 1983) The off-diagonal $\mathbf{W}$-matrix elements are then equated to the relevant inelastic state-to-state rates. The procedures used to calculate $\mathbf{W}$ follow those used by Rosasco and others (Rosasco et al. 1983, Lempert et al. 1984) for line mixing in Raman Q branches and adapted to infrared Q branches by Strow and Gentry (Strow & Gentry 1986, Gentry & Strow 1987). It is assumed that elastic reorientation collisions and vibrational relaxation, dephasing do not contribute significantly to the widths. Consequently, the pressure broadened halfwidth for a single line can then be calculated from the sum of all collision induced transitions that depopulate the rotational energy levels involved in the transition.

The rotational relaxation rates for collisions within a single vibrational level state, $K_{j'j}$, are modeled using the following empirical energy–gap scaling law,

$$K_{j'j} = a_1 \left(\frac{|\Delta E_{j'j}|}{B_o} \right)^{-a_2} \exp\left(\frac{-a_3|\Delta E_{j'j}|}{kT} \right) \tag{2}$$

for $j' > j$. Detailed balance gives $K_{j'j}$ for $j' < j$. $\Delta E_{j'j} = E_{j'} - E_j$ is the rotational energy gap. B_o, the rotational band constant, is included to give a_1 the units of a rate. The free parameters in this expression (a_1, a_2, and a_3) are determined from a least–squares fit of the observed pressure–broadened linewidths over a wide range of J to the following expression (good for Π–Σ bands), which relates the rotational relaxation rates to the widths of isolated spectral lines, or

$$W_{jj} = -\frac{1}{2}\left(\sum_{j' \neq j}^{even} K_{j'j}^{\Sigma(e \leftarrow e)} \right) - \frac{1}{2}\left(\sum_{j' \neq j}^{even} \beta K_{j'j}^{\Pi(f \leftarrow f)} + \sum_{j' \neq j}^{odd} (1-\beta) K_{j'j}^{\Pi(e \leftarrow f)} \right), \tag{3}$$

where W_{jj} is the linewidth. The superscript on $K_{j'j}$ indicates the symmetry of the vibrational level involved (Π or Σ) and the parity of the rotational levels (f or e). The first sum in this expression is over all rotational levels in the Σ vibrational state, which contains only even j values due to boson spin statistics. The second and third sums are over the Π vibrational state rotational levels, which include $f \leftarrow f$ transitions for even j' and $e \leftarrow f$ collisions for odd j'. This expression is valid for a $\Pi - \Sigma$ vibrational transitions which include all of the stronger 15 μm CO_2 bands, except for the $\Delta - \Pi$ band centered at 741 cm^{-1}. Note that the Q-branch transitions all connect to the Π^f level. The above functional form for $K_{j'j}$ is used for both the upper and lower vibrational state manifold of rotational levels.

The above expression for W_{jj} is modified slightly from previous studies (Strow & Pine 1988b) by the introduction of the parameter β, which gives the relative rate of $f \leftarrow f$ versus $e \leftarrow f$ collisions in the Π vibrational state. For self–broadened spectra $\beta = 0.5$ gives excellent agreement with experiment. However, our recent laboratory results (to be discussed later in this paper) show that $\beta = 0.63$ gives much better agreement with experiment for N_2–broadened CO_2 spectra, which means that Π state $f \leftarrow f$ collisions in the CO_2–N_2 system are more likely than $e \leftarrow f$ collisions. Finally, $\mathbf{W}$ is obtained from the Π state rates by setting (Green 1989)

$$W_{j'j} = -\beta K_{j'j}^{\Pi(f \leftarrow f)}. \tag{4}$$

Line mixing can only occur if ($f \leftarrow f$) collisions in the Π state and ($e \leftarrow e$) collisions in the Σ state, between the same J levels, occur simultaneously. Since the Π state collision rates are smaller than the Σ state rates by a factor of β the Π state rates will determine the overall amount of line mixing, consistent with the above expression for $W_{j'j}$.

Once the $\mathbf{W}$–matrix is determined the final step in calculating the collisionally narrowed profile is the evaluation of Eq. 1. At first glance it appears that the inversion of a rather large matrix will be required at each frequency in the spectrum. Fortunately, a matrix inversion for each frequency can be avoided by following the procedures discussed by Koszykowski (Koszykowski et al. 1985, Gordon & McGinnis 1968) and applied by Gentry and Strow (Gentry & Strow 1987) to an infrared CO_2 Q branch. First, we rewrite Eq. 1 in the following simple form,

$$k(\nu) = \frac{N}{\pi}\mathbf{IM}\left\{\mathbf{d}\cdot\mathbf{G}(\nu)^{-1}\cdot\boldsymbol{\rho}\cdot\mathbf{d}\right\}, \tag{5}$$

where $\boldsymbol{\rho}$ is the diagonal matrix of population differences and $\mathbf{d}$ is a vector of transition amplitudes. $\mathbf{G}$ is defined by

$$\mathbf{G} = \boldsymbol{\nu} - \mathbf{H} \tag{6}$$

where $\mathbf{H}$, which is independent of frequency, must then equal

$$\mathbf{H} = \boldsymbol{\nu}_o + iP\mathbf{W}. \tag{7}$$

$\mathbf{H}$ is diagonalized by the complex matrix $\mathbf{A}$ to give

$$\mathbf{A}^{-1}\cdot\mathbf{H}\cdot\mathbf{A} = \mathbf{L} \tag{8}$$

where $\mathbf{L}$ is the diagonal eigenvalue matrix. $\mathbf{G}$ only differs from $\mathbf{H}$ by a constant added to each diagonal element so that $\mathbf{G}$ and hence $\mathbf{G}^{-1}$ can also be diagonalized with $\mathbf{A}$. The resulting expression for the absorption coefficient is then

$$k(\nu) = \frac{N}{\pi}\mathbf{IM}\left\{\sum_i \frac{(\mathbf{d}\cdot\mathbf{A})_i(\mathbf{A}^{-1}\cdot\boldsymbol{\rho}\cdot\mathbf{d})_i}{\nu - l_i}\right\}, \tag{9}$$

where the l_i are the diagonal elements of $\mathbf{L}$. After the initial diagonalization of $\mathbf{H}$ and inversion of $\mathbf{A}$ the absorption coefficient can be calculated at any frequency from the above algebraic expression. It should be noted that Eq. 9 is an exact expression for $k(\nu)$ and involves no approximations. This method for calculating $k(\nu)$ considerably reduces the computational time because only a single matrix diagonalization and inversion is needed to calculate the entire spectrum.

3. Laboratory Measurements of Q–branch Line Mixing in CO_2

Line mixing has been observed in laboratory spectra of more than five different CO_2 Q branches. We will examine here the 791 cm^{-1} Q branch, which we have studied in detail in support of the Cryogenic Limb Array Etalon Spectrometer (CLAES) on the Upper Atmosphere Research Satellite (UARS). CLAES uses this particular Q branch for limb temperature sounding. The behavior of the 791 cm^{-1} Q branch is typically of any Π–Σ band, including all the important atmospheric CO_2 Q branches. The strongest atmospheric CO_2 Q branch which is not a Π–Σ transition is the Π–Δ Q branch at 741 cm^{-1}.

Figure 1 shows the pressure dependence of the absorption in the 791 cm^{-1} Q branch for 1% CO_2 in N_2. These experimental spectra were recorded with a tunable diode laser spectrometer in our laboratory. Also shown is a Lorentz calculation of the Q–branch absorption for the 720 torr spectrum. The typical effects of line mixing at atmospheric pressures are clearly seen; line mixing causes the observed spectrum to be super–Lorentz inside the Q branch and sub–Lorentz in the Q–branch wing. The effects of line mixing are much greater, in a relative sense, in the Q–branch wing than inside the band. As will be shown later, the ratio of the mixing to Lorentz absorption coefficients in the far

Q–branch wing is a constant approximately equal to 0.5. As the pressure is decreased line mixing will become less important near the Q–branch line centers, but will continue to have the same relative effect on the absorption coefficients in the Q–branch wing.

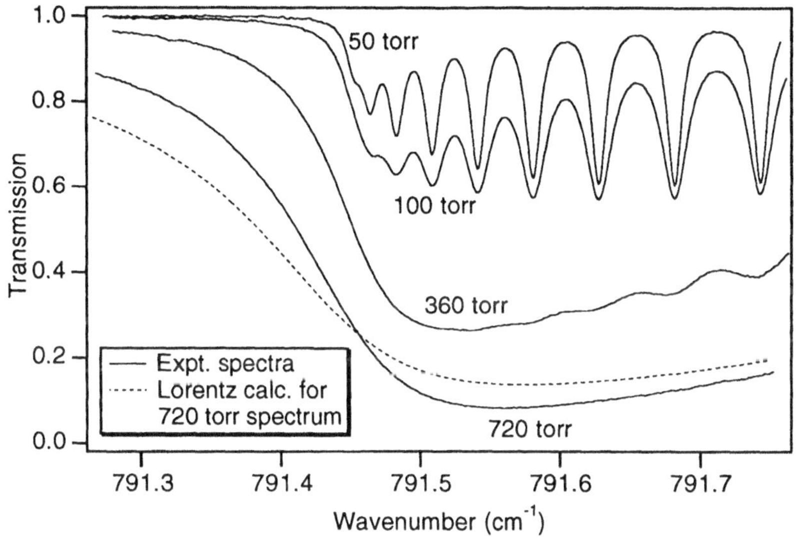

Fig. 1. Tunable diode laser spectra of the 791 cm^{-1} Q branch of CO_2, N_2–broadened at room temperature. A calculation of the 720 torr spectrum using a Lorentz lineshape is also shown.

Figures 2 and 3 show experimental tunable diode laser spectra of the 791 cm^{-1} CO_2 Q branch for both self– and N_2–broadening. Excellent agreement between theory and experiment is evident for self–broadening with $\beta = 0.5$. This means that $f \longrightarrow f$ collisions and $f \longrightarrow e$ collisions are equally likely in the Π vibrational state. This is not the case for N_2–broadening, where $\beta = 0.50$ results in too little line mixing. Instead, $\beta = 0.63$ gives good, although not perfect, agreement with experiment. Residual errors are on the order of 1%, which is close to the accuracy of the data.

The higher value of β for N_2–broadening means that $f \longrightarrow f$ collisions are more likely than $f \longrightarrow e$ collisions. A more extreme case of this behavior is seen in He–broadened CO_2 Q branches where $\beta \sim 0.75$. A theoretical basis for $\beta \neq 0.5$ for He–broadened spectra can be found in the work of Clary (Clary 1983), who calculated rotational state–to–state rates in both the Σ and Π vibrational levels for CO_2-He collisions. Their results indicated that β is J dependent, with a J–averaged value on the order of what we observe. It is difficult to directly compare our results with theirs since we observe a thermal average of relative collision velocities as well as a complicated weighted average of the J dependent β's. They calculated individual state–to–state rates for one relative collision velocity. The J dependent nature of β may be responsible for the residual disagreement between theory and experiment seen in Fig. 3.

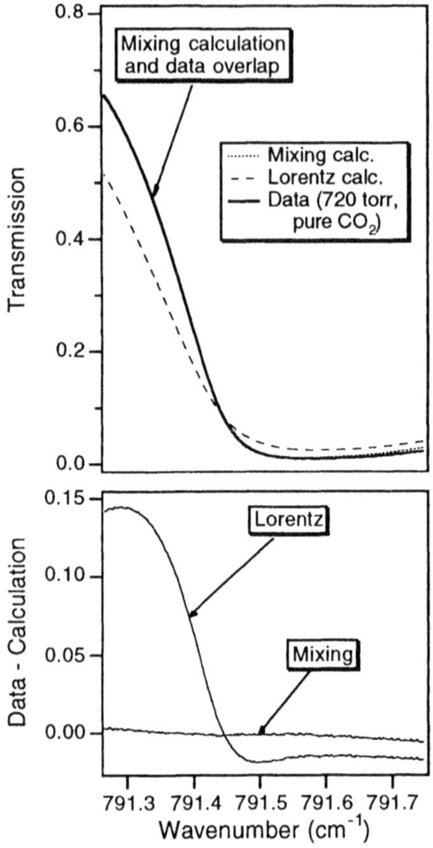

Fig. 2. Experimental and calculated spectra of the self-broadened 791 cm^{-1} CO_2 Q branch

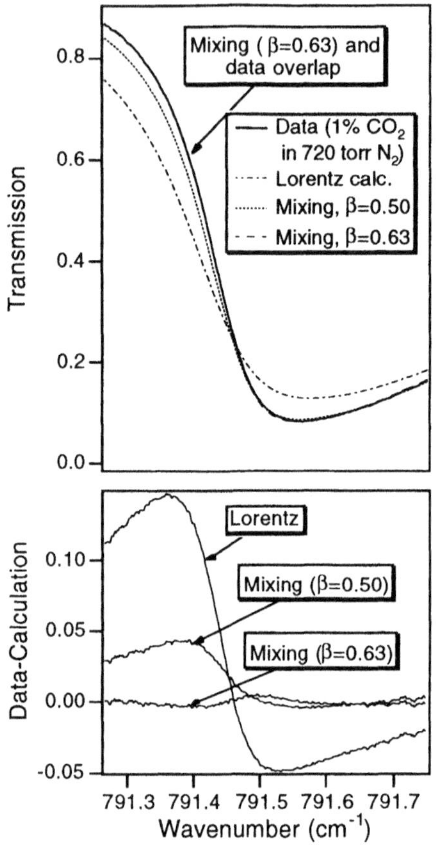

Fig. 3. Same as Fig. 4 except for N_2 broadening.

Good agreement with experiment is obtained using $\beta = 0.63$ for all other Π–Σ Q branches as well. The theoretical approach outlined here has also been shown to accurately model experimental measurements of Q–branch line mixing in pure CO_2 over a wide temperature range from 220K to 350K (Zheng 1989). Line mixing is more complicated for the Π–Δ Q branch at 741 cm^{-1} since both vibrational states involved in the transitions are l–doubled. We are presently developing a model of line mixing in the 741 cm^{-1} Q branch that is also consistent with the Π–Σ Q branches.

4. Atmospheric Observations, Simulations

4.1. ATMOS Limb–Viewing Spectra

Line mixing in atmospheric spectra has been observed both in limb viewing and nadir experiments. Figure 4 shows a comparison of our calculation of the 617 cm^{-1} CO_2 Q–branch transmission using the Genln2 radiative transfer code (Edwards 1992) with the limb transmission observed by the ATMOS experiment that flew on Spacelab 3. The ATMOS spectra were taken during solar occultations using a fast scanning modified Michelson interferometer. Measurements were made over the 2–16 μm spectral range at an unapodized resolution of about 0.01 cm^{-1} for limb tangent heights of 5–130 km. The atmospheric profiles used in the Genln2 calculations for physical quantities and gas mixing ratios were those supplied by the ATMOS team. Their retrievals of pressure and volume mixing ratio were made using narrow spectral intervals containing temperature insensitive lines. Pressure and temperature profiles were retrieved using isolated CO_2 lines and N_2 quadrupole lines. (These particular lines and spectral regions are not influenced by the lineshape considerations that are studied here.) A constant mixing ratio for CO_2 of 330 *ppmv* was assumed.

Each transmittance calculation was performed using three different line shapes for modeling the CO_2 absorption. The first used the simple Voigt lineshape assuming Lorentzian line wings and not including line mixing. In the second the lineshape included line mixing for the Q–branch being studied and all other CO_2 lines were modelled using the simple Voigt lineshape. For the third calculation, the lineshape included line mixing for the Q–branch being studied and the line wings of CO_2 lines from the ν_2 band were modified using the χ–factor of Cousin et al. (Cousin et al. 1985). It is important to note that the ATMOS calculations presented here are simulations of the ATMOS spectra and *not* fits to the observed data.

The spectral region around 617 cm^{-1} becomes opaque for limb view tangent heights much below 20 km. The ATMOS spectra were simulated at a tangent height of 22.4 km, estimated to be accurate to ± 0.25 km. It is clear that there is a large difference in the transmittance in the CO_2 Q–branch wing between the calculation that assumes a Voigt lineshape and that which includes line mixing, and that the latter is closer to the observation. Local line mixing in this band accounts for essentially all the non–Lorentzian CO_2 lineshape behavior. If the ν_2 line wings are cut off 25 cm^{-1} from the line centers, instead of using the Cousin χ–function, the calculated transmittances are too high by about 15%. Most of the "bumps" in the ATMOS-CALCULATION part of Fig. 4 are due to inaccuracies in O_3 lines. The ATMOS-CALCULATION curves were smoothed to approximately 2X the resolution of the spectra in order to reduce noise and accentuate the continuum errors rather than errors in local lines. The signal–to–noise ratio of the ATMOS measurement in this region is relatively poor as it lies close to the edge of the ATMOS wideband interference filter.

Our ability to accurately model line mixing in the ATMOS spectrum gives us confidence in our treatment of the temperature dependence of line mixing for CO_2–N_2 collisions, which have not been studied in the laboratory. A more detailed explanation of this calculation has been presented elsewhere (Edwards & Strow 1991).

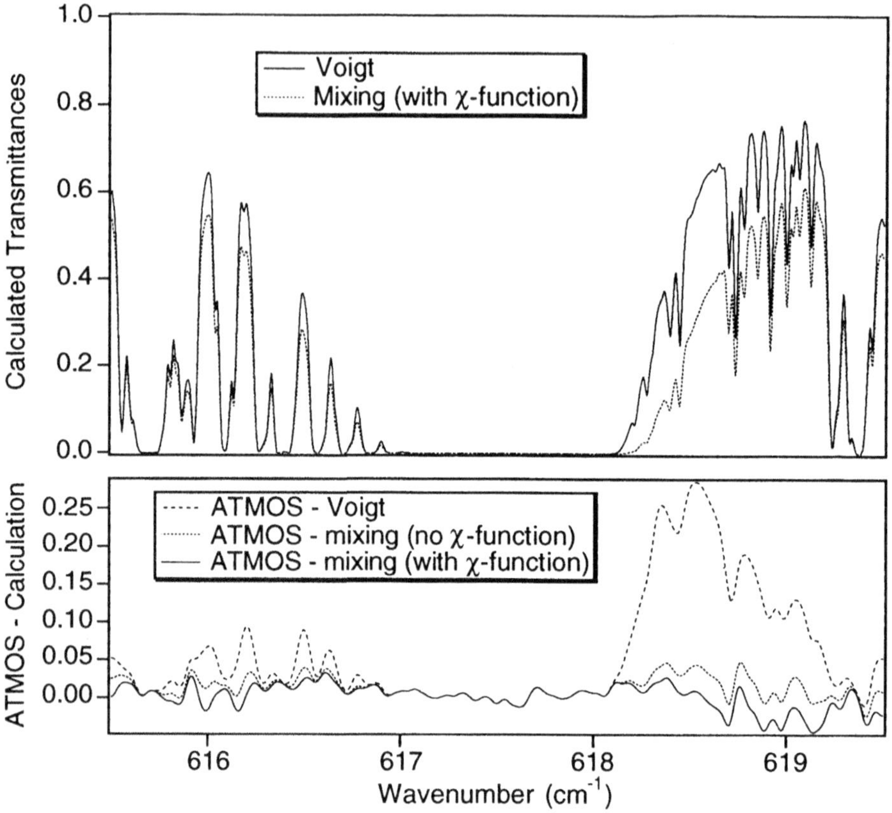

Fig. 4. Line mixing in the 617 cm^{-1} CO_2 Q–branch observed in the limb by ATMOS on Spacelab 3.

4.2. HIS Nadir–Viewing Spectra

Q–branch line mixing also strongly affects nadir–viewing spectra and must be taken into account for any temperature sounder that has channels near the major CO_2 Q branches. The University of Wisconsin HIS interferometer (Revercomb et al. 1989), which flies on a high–altitude aircraft and has an unapodized resolution of 0.36 cm^{-1}, provides excellent spectra for studies of atmospheric line mixing. The HIS spectrum shown in Fig. 5 was taken over the ocean near the California coast. An average atmospheric profile taken from local NWS balloon sondes were used for the radiative transfer calculations. As with the ATMOS spectra, Genln2 was used for this work.

The large deviations between the observed HIS spectrum and our calculation without line mixing that are starred in the bottom plot of Fig. 5 are caused by line mixing. Errors in the observed brightness temperatures of up to 10K in the 617 cm^{-1} Q branch are evident. Smaller, but still significant errors can be seen in the 720 cm^{-1} and 741 cm^{-1} Q branches. The 791 cm^{-1} Q branch is too weak to cause significant errors for a nadir view. Other errors in this plot are due primarily to H_2O and may be due to the rather large errors in the H_2O abundance amounts used in the radiative transfer calculation. Also shown in Fig. 5 is an observed minus calculated curve for the case where line mixing is included in the radiative transfer code. Clearly, line mixing reduces brightness temperatures errors down to the level of other errors caused by an uncertain knowledge of the atmospheric state, and incorrect spectral line parameters.

Line mixing in the intense 667 cm^{-1} Q branch is not evident because all radiation detected in this spectral region is being radiated at the same temperature, namely the temperature of the local environment of the interferometer. The very flat baseline of the error curve in Fig. 5 was only obtained after switching from a constant emissivity to an emissivity model for seawater calculated by Masuda et al. (Masuda et al. 1988).

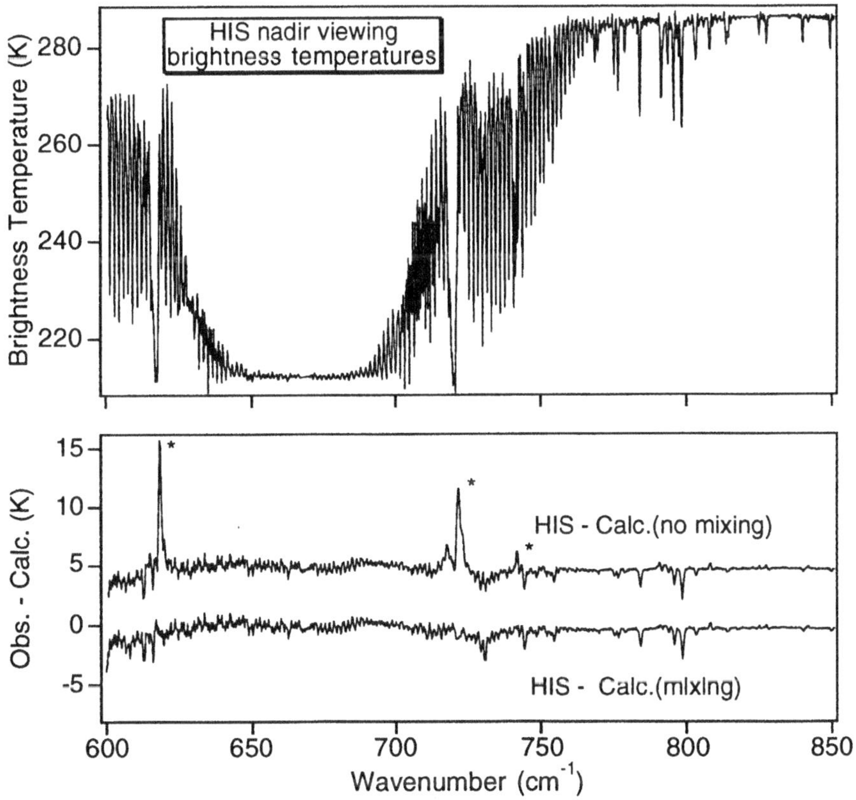

Fig. 5. Line mixing in a nadir viewing atmospheric spectrum recorded by the HIS instrument. The HIS - CALC. (no mixing) curve has been offset by +5K.

4.3. Effect of Line Mixing on AIRS

The primary *observable* effect of line mixing under atmospheric conditions is a reduction of the absorption coefficient in the Q–branch wing over a 1–10 cm^{-1} region. This means that sounders with an instrument function much broader than individual spectral lines can still be strongly influenced by line mixing. For example, the ITRA (Intercomparison of Transmittance and Radiance Algorithms) study (Chedin et al. 1988) found that even for the High Resolution Infrared Radiation Sounder (HIRS) channel 1 (ν_o =668 cm^{-1}), with a half–power band width of 3 cm^{-1}, line mixing accounts for deviations of about 0.02 to 0.03 in the channel transmittance. Line mixing also strongly impacts the Atmospheric Infrared Sounder (AIRS), (AIRS 1991), whose spectral response function is approximately 0.5 cm^{-1} wide in the thermal infrared. Figure 6 shows the effect of line mixing on a simulated AIRS spectrum for the U.S. Standard Atmosphere. The 617 cm^{-1} Q branch was included in this figure although AIRS has no channels in this spectral region. The largest sensitivity to line mixing, as shown in Fig. 6b, is approximately 7K in the 720

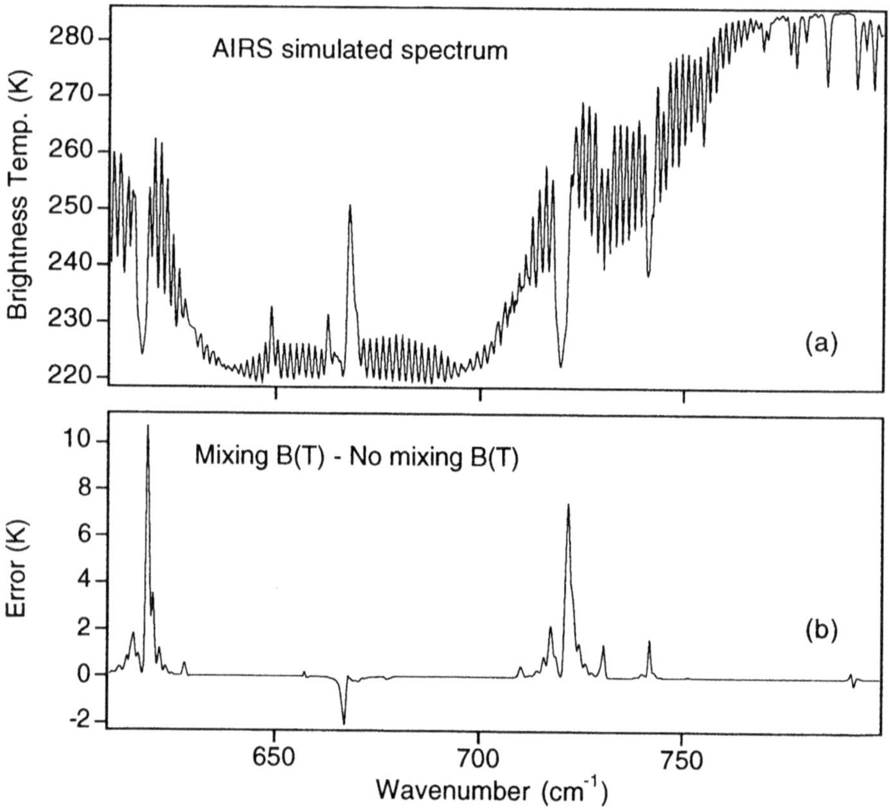

Fig. 6. Effects of line mixing on a simulated AIRS brightness temperature spectrum.

cm^{-1} Q–branch wing. This is more than an order–of–magnitude larger than the desired accuracy for the forward model that AIRS will use for temperature retrievals. Errors in the 667 cm^{-1} Q branch are smaller, and of different sign, since the emission in this region is somewhat above the tropopause where the lapse rate has changed sign and is small when averaged over the channel weighting function.

5. Approximation Methods

Rosenkranz (Rosenkranz 1975) introduced a first–order approximation for line mixing that is exact in the low pressure limit, and is very simple to include in radiative transfer codes. Limb atmospheric spectra rarely probe atmospheric levels above 100 torr in pressure. Similarly, nadir–viewing emissions of CO_2 Q branches in the thermal infrared either come primarily from pressures levels below several hundred torr, or are very weak and unimportant. Consequently, absorption coefficients calculated using the first–order approximation for line mixing are often sufficiently accurate for atmospheric applications. We will describe the first–order approximation for line mixing here and discuss its accuracy and some potential dangers inherent in the use of this approximation.

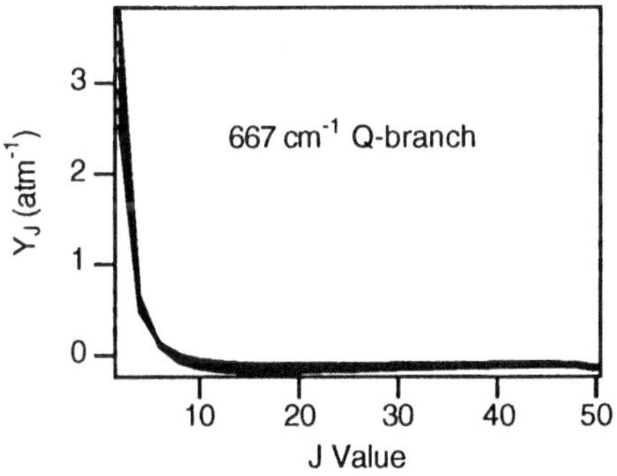

Fig. 7. First–order line mixing coefficients for 6 temperatures between 200K and 300K for the 667 cm^{-1} CO_2 Q branch.

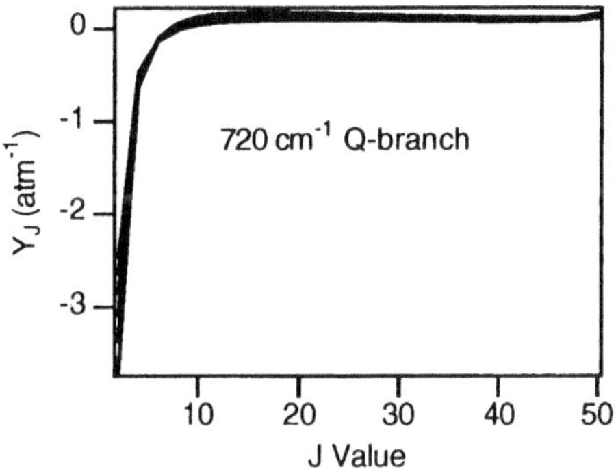

Fig. 8. Same as Fig. 7 but for the 720 cm^{-1} Q branch.

Following Rosenkranz, the low-pressure approximation for the absorption coefficient is given by,

$$k(\nu) = \frac{N}{\pi} \sum_j S_j \left(\frac{P\gamma_j + (\nu - \nu_j)\, PY_j}{(\nu - \nu_j)^2 + (P\gamma_j)^2} \right), \tag{10}$$

which introduces the first-order mixing coefficients,

$$Y_j = 2 \sum_{k \neq j} \frac{d_k}{d_j} \frac{W_{kj}}{\nu_j - \nu_k}, \tag{11}$$

where $\gamma_j = W_{jj}$ is the Lorentz halfwidth/atm for $Q(j)$ and $S_j = \rho_j d_j^2$ is the line strength. This method for calculating the mixing is very convenient once the $Y_j' s$ are known and is easily incorporated into existing radiative transfer codes. If the pressure is low enough this expression for $k(\nu)$ must be convolved with a Doppler lineshape. This convolved lineshape can be expressed in terms of the complex probability function, $w(x, y)$, as

$$k(\nu) = N \sum_j k_{o,j} \left(\mathbf{RE}\left\{w(x_j, y_j)\right\} + P Y_j \,\mathbf{IM}\left\{w(x_j, y_j)\right\}\right), \qquad (12)$$

where

$$k_o = \frac{S}{\alpha_D}\sqrt{\frac{\ln(2)}{\pi}}, x = \frac{\nu - \nu_o}{\alpha_D}\sqrt{\ln(2)}, y = \frac{P\gamma}{\alpha_D}\sqrt{\ln(2)},$$

α_D is the Doppler halfwidth, and ν_o is the line center frequency. The real part of $w(x, y)$ is used to calculate the normal Voigt profile while the imaginary part gives the mixing contribution to $k(\nu)$. The algorithm of Humlicek (Humlicek 1979) can be used for calculating $w(x, y)$. A new **W**–matrix, and Y_j's must be determined for each temperature. This is a rather involved procedure for atmospheric calculations so we have parametrized the Y_j's at temperature $T[K]$ using

$$Y_j(T) = \left(\frac{200}{T}\right)^{0.75} * \left(A_j + B_j(T - 200) + C_j(T - 200)^2 + D_j(T - 200)^3\right). \qquad (13)$$

The first term in this equation provides most of the temperature dependence since this term describes the temperature dependence of the widths use to derive the Y_j's.

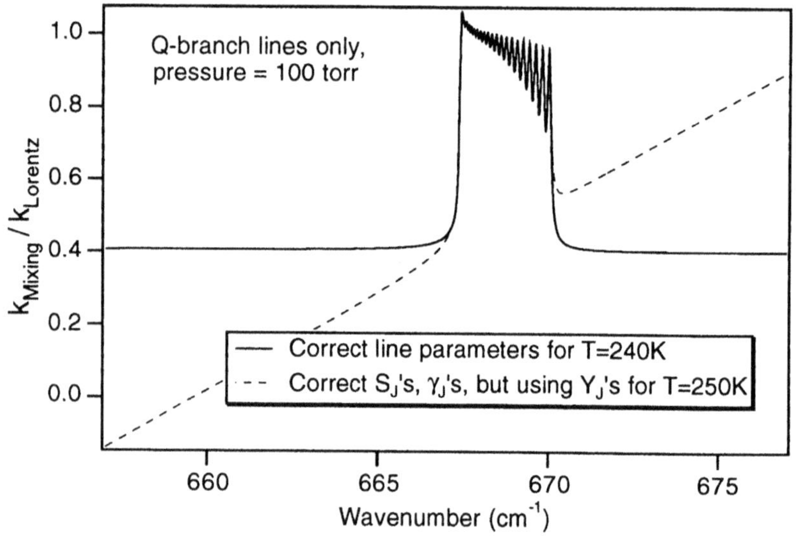

Fig. 9. Sensitivity of line mixing to inconsistent line parameters.

The dependence of the mixing coefficients on J and temperature are shown in Fig. 7 for the 667 cm^{-1} CO_2 Q branch and in Fig. 8 for the 720 cm^{-1} CO_2 Q branch. Each figure has six curves for evenly spaced temperatures between 200 and 300K, which are difficult to individually discern in these plots. Notice that the the Y_j's have different signs for these two Q branches, which is determined by whether the Q–branch lines shade to higher or lower frequencies. Secondly, we see that the Y_j's are largest for $J = 2$, and fall quickly to a much smaller value that varies little with J.

Figure 9 (solid line) illustrates the effects of line mixing on the 667 cm^{-1} Q branch at T=240 K and a 100 torr N_2-broadening pressure. This spectrum includes only Q-branch lines so that local absorption due to the P– and R–branch lines does not mask the effects of line mixing. This plot clearly shows that away from the Q-branch line centers the ratio of the line mixing absorption coefficient, k_M, to the Lorentz coefficient, k_L, has a constant value of about 0.4 at 240K. This behavior is true in general and is not just a feature of the first–order approximation. This ratio does not depend upon pressure and is only weakly dependent on temperature. This simple relationship between the Lorentz and mixing absorption coefficients in the Q-branch wing is easily derived using Rosenkranz's first–order theory (Strow & Reuter 1988a).

The first–order calculation of line mixing is not without problems. These include lowered accuracy, and more importantly extreme sensitivity to consistency among the Y_j's, the line widths, and the line strengths. This sensitivity can be understood by first noting that each Y_j is multiplied by $(\nu - \nu_j)$ in Eq. 10. Consequently any errors in the Y_j's are amplified in the Q–branch wing where $(\nu - \nu_j)$ becomes large. A more complete understanding of this problem can be gained by examining the ratio of k_M to k_L in the Q–branch wing.

Following Strow and Reuter (Strow & Reuter 1988a) the ratio of k_M to k_L in the far wing of the Q branch can be written as,

$$\frac{k_M}{k_L} = 1 - \frac{\sum_j S_j Y_j \delta_j - (\nu - \bar{\nu}) \sum_j S_j Y_j}{\sum_j S_j \gamma_j}. \tag{14}$$

where $\bar{\nu}$ and δ_j are defined such that

$$\nu - \nu_j = \nu - \bar{\nu} + \delta_j \; , \; \delta_j \ll \nu - \bar{\nu}. \tag{15}$$

Here, $\bar{\nu}$ is some constant frequency near the center of the Q branch and the $\bar{\nu} - \delta_j$ terms are the center frequencies of the Q branch lines. The first term in the numerator of Eq. 14 equals zero. We can rewrite this term as

$$\sum_j S_j Y_j = 2 \sum_{j,k \neq j} \frac{\rho_j d_j d_k W_{kj}}{\nu_j - \nu_k}, \tag{16}$$

using $S_j = \rho_j d_j^2$. Detailed balance requires that

$$\rho_j W_{kj} = \rho_k W_{jk} \tag{17}$$

so for every term in the sum

$$\frac{\rho_j d_j d_k W_{kj}}{\nu_j - \nu_k} \tag{18}$$

there will be a corresponding negative term which cancels the first term. Thus

$$\sum_j S_j Y_j = 0. \tag{19}$$

The remaining terms in Eq. 14 are constants independent of frequency, and thus the ratio of k_M to k_L is also independent of frequency in the Q–branch wing as shown in Fig. 9.

Clearly, if Eq. 19 is not satisfied errors in the absorption coefficient will result, which can become rather large in the Q–branch wing. We have found that Eq. 19 must be satisfied to a precision several orders–of–magnitude better than the accuracy of the individual terms in this sum. Calculational errors of this type can arise a number of different ways. First, if the Y_j's are calculated from a **W**–matrix with maximum dimension $J = 50$, then *all* Y_j's up to $J = 50$ *must* be used in the calculation of the absorption coefficient. Leaving some of the Y_j's out of the calculation is the same as setting them equal to zero and thereby invalidating Eq. 19. Secondly, if a certain set of Q–branch strengths and widths are used to generate a **W**–matrix, these *exact* same strengths and widths must be used in the calculation of the absorption coefficient. A subset of this requirement is that the temperature dependence of the Y_j's must be consistent with the temperature dependence of the strengths and widths to a high precision or otherwise Eq. 19 will again not be satisfied and large errors can result.

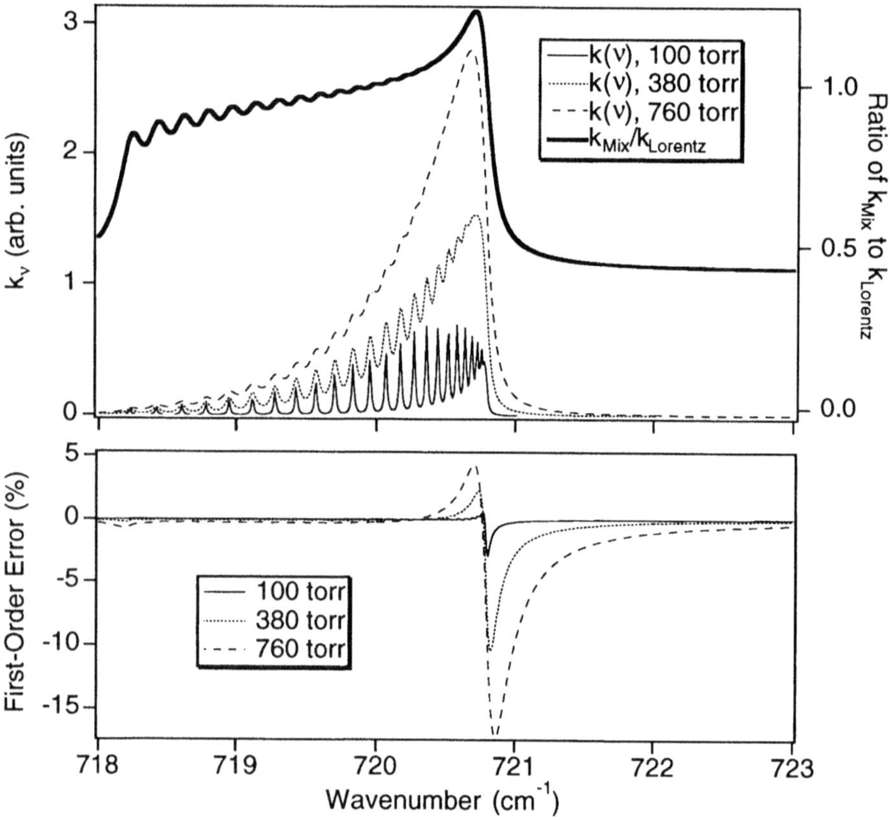

Fig. 10. Accuracy of the first–order approximation for line mixing for the 720 cm^{-1} CO_2 Q branch.

To illustrate the errors possible if the Y_j's are not consistent with the strengths and widths consider the dashed line in Fig. 9. Here we plot the ratio of k_M to k_L using the correct strengths and widths for $T = 240K$, but use Y_j's for $T = 250K$ instead of the Y_j's for $T = 250K$. Clearly this causes extremely large errors in the absorption coefficient in the wing, that grow in size with $(\nu - \nu_i)$. Changes to the absorption coefficient using line widths in error by 10K, for example, would be far lower than those seen in Fig. 9. We suggest that preliminary test calculations, such as those shown in Fig. 9, be performed to ensure that all of the line parameters are consistent, especially with regard to their temperature dependence.

Figure 10 shows the accuracy of the first–order approximation for the 720 cm^{-1} Q branch at room temperature for three atmospheric pressures. The ratio of k_M to k_L is also shown here for reference. The first–order theory is relatively accurate for 100 torr, with a maximum error on the order of 3%. The errors grow to a maximum of approximately 16% at 760 torr in the Q–branch wing. However, as mentioned earlier, few atmospheric applications will observe emissions in the troposphere at pressures where these errors are significant. The first–order approximation can be extended to second order, with a concomitant increase in complexity, if an application warrants it.

6. Conclusions

Q–branch line mixing in atmospheric spectra can now be modelled to an accuracy comparable to our knowledge of the standard line parameters. A simple first–order approximation allows line mixing to be easily added to existing radiative transfer codes, although a small number of applications may require a more accurate second–order treatment of line mixing.

The first–order mixing coefficients, as a function of temperature, for the major CO_2 Q branches are available from the author. They can also be obtained from Genln2 (Edwards 1992), a line–by–line radiative transfer code written by David Edwards, which is distributed by the National Center for Atmospheric Research. FASTCODE 3 from the Air Force Phillips Laboratory also contains line mixing parameters.

The focus of future work will be on Π–Δ bands and far–wing lineshapes in the 4.3 μm ν_3 R branch that are affected by both line mixing and duration of collision effects. Studies of far wing lineshapes in the 15 μm region should also receive some attention, although they are expected to be rather small compared to line mixing effects.

Acknowledgments. I would like to thank Scott Hannon, David Tobin, and David Edwards for their many contributions to this work. Thanks are also due to Bill Smith et al. at the University of Wisconsin for providing the HIS aircraft spectrum. This work was supported by the National Aeronautics and Space Administration.

References

AIRS (1991) Atmospheric Infrared Sounder: Science and Measurement Requirements. Technical Report D6665 Rev. 1, Jet Propulsion Laboratory, Pasadena, CA

Ben-Reuven A (1966) Impact broadening of microwave spectra. *Phys. Rev*, 145:7–22

Brunner T, Pritchard D (1982) Fitting laws for rotationally inelastic collisions. In *Advances in Chemical Physics, Vol. 50*, John Wiley & Sons, p 589

Bulanin M, Dokuchaev A, Tonkov M, Filippov N (1984) Influence of line interference on the vibration-rotation band shapes. *J. Quant. Spectrosc. Radiat. Transfer*, 31:521–543

Chedin A, Fischer H, Künzi K, Spänkuch, D, and Scott N (Editors), 1988 Report on the ITRA (Intercomparison of Transmittance and Radiance Algorithms). International Radiation Commission, May 1988

Clary D (1983) Rotational and vibrational-rotational relaxation in collisions of $CO_2(01^10)$ with He atoms. *J. Chem. Physics*, 78:4915–4923

Cousin C, Le Doucen R, Boulet C, Henry A (1985) Temperature dependence of the absorption in the region beyond the 4.3 μm band head of CO_2. 2: N_2 and O_2 broadening. *Appl. Opt.*, 24:3899–3907

Cousin C, Le Doucen R, Boulet C, Henry A, Robert D (1986) Line coupling and the temperature and frequency dependencies of absorption in the microwindows of the 4.3 μm CO_2 band. *J. Quant. Spectrosc. Radiat. Transfer*, 36:521–538

Edwards DP, Strow LL (1991) Spectral line shape considerations for limb temperature sounders. *J. Geophys. Res.*, 96(D11):20,859–20,868

Edwards DP (1992) Genln2: A general line–by–line atmospheric transmittance and radiance model. NCAR Technical Note 367+STR, National Center for Atmospheric Research

Gentry B, Strow LL (1987) Line mixing in a N_2-broadened CO_2 Q branch observed with a tunable diode laser. *J. Chem. Physics*, 86:5722–5731

Gordon RG, McGinnis RP (1968) Line shapes in molecular spectra. *J. Chem. Physics*, 49:2455–2456

Green S (1989) Pressure broadening and line coupling in bending bands of CO_2. *J. Chem. Physics*, 90:3603–3614

Hoke ML, Clough SA, Lafferty WJ, Olson BJ (1988) Line coupling in oxygen and carbon dioxide. In *Proceedings of the International Radiation Symposium*, Lille, France

Humlicek J (1979) An efficient method for evaluation of the complex probability function: the Voigt function and its derivatives. *J. Quant. Spectrosc. Radiat. Transfer*, 21:309–313

Koszykowski ML, Farrow RL, Palmer RE (1985) Calculation of collisionally narrowed coherent anti-stokes Raman spectroscopy spectra. *Optics Letters*, 1:478–480

Lam KS (1977) Application of pressure broadening theory to the calculation of atmospheric oxygen and water vapor microwave absorption. *J. Quant. Spectrosc. Radiat. Transfer*, 17:351–382

Lavorel B, Millot G, Bonamy J, Robert D (1987) Study of rotational relaxation fitting laws from calculation of SRS N_2 Q branch. *Chem. Phys.*, 115:69–78

Lempert W, Rosasco GJ, Hurst WS (1984) Rotational collisional narrowing in the NO fundamental Q branch, studied with cw stimulated Raman spectroscopy. *J. Chem. Physics*, 81:4241–4245

Masuda K, Takashima T, Takayama Y (1988) Emissivity of pure and sea waters for the model sea surface in the infrared window regions. *Remote Sensing of Environment*, 24:313–329

Revercomb HE, Smith WL, Knuteson RO, Woolf HM, Howell HB (1989) Comparisons of FASTCODE spectra with HIS observations. *Annual Review Conference on Atmospheric Transmission Models*, Air Force Geophysics Laboratory, Hansom Air Force Base, MA

Rosasco GJ, Lempert W, Hurst WS, Fein A (1983) Line interference effects in the vibrational Q-branch of N_2 and CO. *Chem. Phys. Lett.*, 97:435–440

Rosenkranz PW (1975) P. W. Rosenkranz. Shape of the 5 mm oxygen band in the atmosphere. *IEEE Trans Antennas Propag.*, AP-23:498–506

Smith EW (1981) Absorption and dispersion in the O_2 microwave spectrum at atmospheric pressures. *J. Chem. Physics*, 74:6658–6673

Strow LL, Gentry BM (1986) Rotational collisional narrowing in an infrared CO_2 Q branch studied with a tunable-diode laser. *J. Chem. Physics*, 84:1149–1156

Strow LL, Reuter D (1988a) Effect of line mixing on atmospheric brightness temperatures near 15 μm. *Appl. Opt.*, 27:872–878

Strow LL, Pine AS (1988b) Q–branch line mixing in N_2O: effects of ℓ-type doubling. *J. Chem. Physics*, 89:1427–1434

Zheng L (1989) Lui Zheng. An infrared spectral line shape study of CO_2 using diode lasers, M.S. Thesis, University of Maryland Baltimore County, Baltimore, MD

OPTIMAL SPECTRAL ESTIMATION ALGORITHMS FOR MICHELSON INTERFEROMETER SPECTROMETERS.

C. Serio[1], U. Amato[2], V. Tramutoli, R. Rizzi[3] and V. Cuomo
Istituto di Metodologie Avanzate di Analisi Ambientale CNR
Tito Scalo, Potenza
Italy

1. Introduction

Spectra recorded by Michelson interferometer spectrometers are usually distorted because the spectrometers have a finite resolving power. Indeed, although the radiance spectrum is obtained in principle by a Fourier transform of the interferogram, various factors intervene to make the recovered spectrum an imperfect representation of the true spectrum. The most important ones are: a)- aperture effects; b)- tilt and aberrations; c)- phase and compensation; d)- truncation; e)- noise.

Generally the first two sources of error can be corrected by a suitable design of the hardware (Vanasse 1981), while phase and compensation errors can be eliminated, in principle, by both suitable design and measuring two-sided interferograms (Vanasse 1981). Finally, truncation errors and noise are usually handled by software: apodization (truncation error) and filtering (noise). This paper will mostly deal with the latter two types of error.

The methods presently used for reconstructing the spectrum are, in the large majority of cases, purely digital and operate from sampled interferograms. Only a finite number of samples corresponding to a given maximum path difference are measured and, as a consequence, the information beyond the truncation point is lost and the

[1]C. Serio is with Dipartimento Scienze Fisiche, Università di Napoli Federico II, Italy
[2]U. Amato is with Istituto per Applicazioni della Matematica CNR, Napoli, Italy
[3]R. Rizzi is with ECMWF, Reading, UK and EUMETSAT, Darmstadt, Germany

NATO ASI Series, Vol. I 9
High Spectral Resolution Infrared Remote Sensing
for Earth's Weather and Climate Studies
Edited by A. Chedin, M.T. Chahine and N.A. Scott

recovered spectrum (Fourier transform of the interferogram) exhibits an oscillatory behaviour which can seriously affect its quality. This is the truncation error. One method for overcoming this problem is to assume that all data beyond the end point of the finite-length interferogram are zero, then choose an appropriate window function (apodization operation) in order to improve the Fourier transform. Apodization also permits to reduce the effect of measurement errors: one usually uses weighting functions with weights decreasing to zero as the path difference becomes large. However truncation error and noise play a conflicting role. With each given window function it is never possible to minimize their effects simultaneously. Thus one has to consider a compromise between resolution and variance of the final spectral estimates. Such a problem admits an optimal (in the Least Squares sense) solution (Tikhonov and Arsenin 1977, Jenkins and Watts 1968), but its value in spectral analysis is very limited, since it depends on the true interferogram. For this reason, up to now, approaches used to get smoothed or apodized spectral estimates have been typically highly empirical (Norton and Beer 1975).

In this paper the problem of obtaining *optimal* spectral estimates from interferogram samples is investigated. Two methods are discussed and intercompared. The first method relies on autoregressive stochastic processes (Amato et al 1990, 1992), whereas the second one (Amato and Serio 1991a, Amato et al 1991b) is based both on theory of inverse problems and Generalized Cross Validation criterion (Wahba 1977); it retains all the computational aspects which are proper of the Fast Fourier Transform and then can cope with real-time applications involving large amount of data.

This paper is organized as follows: section 2 is devoted to a brief summary of basic equations and definitions in Fourier Spectroscopy; the paragraph also offers a brief outline of the meaning of apodization, stressing the relationship of such a practice with the convergence of the Fourier Transform. Section 3 deals with the autoregressive approach, whereas section 4 describes the second method based on the Generalized Cross Validation criterion. Numerical examples performed on synthetic data are shown in section 5. Conclusions are drawn in Section 6.

2. Mathematical Background

For an ideal Michelson interferometer spectrometer the interferogram signal is linked to the light source radiance spectrum (supposed limited to a given band $(\sigma_1 \div \sigma_2)$) by a cosine Fourier transform (Bell 1972):

$$I(k\Delta x) = \int_{\sigma_1}^{\sigma_2} B(\sigma)\cos(2\pi\sigma k\Delta x)d\sigma; \quad k = 0, \ldots, N-1 \tag{1}$$

where $B(\sigma)$ is the radiance spectrum; $I(x)$ denotes the interferogram; σ denotes wavenumber and x the optical path difference; Δx the sampling interval and N the number of interferogram samples. According to the sampling theorem (e.g., Bracewell 1965) the interferogram is to be sampled at the rate $\Delta x = (2\Omega)^{-1}$, Ω being the bandwidth ($\Omega = \sigma_2 - \sigma_1$), but with the important auxiliary condition $\sigma_1 = \ell\Omega$, $\ell = 1, 2, 3, \ldots$. Then, introducing the dimensionless wavenumber f: $f = \frac{1}{2}\frac{\sigma_2 - \sigma}{\sigma_2 - \sigma_1}$, ℓ odd; $f = \frac{1}{2}\frac{\sigma - \sigma_1}{\sigma_2 - \sigma_1}$, ℓ even, Eq. 5 can be written

$$I(k\Delta x) = \frac{1}{\Delta x}\int_0^{\frac{1}{2}} B(f)\cos(2\pi k\Delta x f)df \tag{2}$$

providing that the conditions expressed by the sampling theorem are met.

This is the basic form of the relation between interferogram and spectrum that we shall consider in the following. Furthermore, for terminology, the name *apodizing function* for the function multiplying the interferogram amplitudes and *instrumental line shape* for the Fourier Transform of the apodizing function will be employed in the following. Moreover, we shall assume to work with calibrated and even interferogram functions.

Our problem consists in recovering the spectrum, $B(f)$, when the interferogram, $I(x)$, is known. This can be accomplished out by Fourier transforming (2). Indeed, after defining the two new even functions, $\tilde{I}(x)$, $\tilde{B}(f)$: $\tilde{I}(-x) = \tilde{I}(x) = I(x); \tilde{B}(-f) = \tilde{B}(f) = \frac{B(f)}{2}$, the solution reads

$$\tilde{B}(f) = \sum_{k=-\infty}^{\infty} \tilde{I}(k)\exp(-2\pi i f k) \tag{3}$$

where, without any loss of generality, we have set $\Delta x = 1$ in (2), so that $\tilde{I}(k\Delta x) \equiv \tilde{I}(k)$. Taking into account that the interferogram signal is of finite-extent, (3) becomes

$$\tilde{B}(f) = \sum_{k=-(N-1)}^{N-1} \tilde{I}(k)\exp(-2\pi i f k) \tag{4}$$

where N always indicates the size of the sample $\{I(k)\}$.

The question now posed is, to what extent is the finite summation in (4) a meaningful approximation of the infinite summation in (3)? It must be observed that the finite sum in (4) is the rectangle-apodized version of the infinite sum in (3), that is

$$\tilde{B}(f) = \sum_{k=-(N-1)}^{N-1} \tilde{I}(k)\exp(-2\pi i f k) = \sum_{k=-\infty}^{\infty} W(k)\tilde{I}(k)\exp(-2\pi i f k) \qquad (5)$$

with $W(k)$ being the rectangular apodizing-function. Furthermore, the finite sum in (4) is nothing but the partial sum of the infinite series (3). From this point of view we can pose the question above in terms of the convergence properties of the partial sum of Fourier series. It is well-known that the partial sum is the least mean-square error approximation to the infinite sum (Jenkins and Watts 1968; Tikhonov and Arsenin 1977; Harris 1978). However, mean-square estimates tend to oscillate about their means, and do not exhibit uniform convergence. Furthermore, the approximation in a neighborhood of a point of continuity may get worse if more terms are added to the partial sum. Finally, if noise is considered the partial sum (4) does not convege in any statistical sense to a limiting value as N tends to infinity (Jenkins and Watts 1968; Tikhonov and Arsenin 1977). It is this oscillatory behaviour we are trying to control by the use either of apodization, i.e. other apodizing functions than the rectangular one or of autoregressive approach.

By convolution theorem, the equivalent of (4) or (5) in the wavenumber-domain is

$$\tilde{B}(f) = \int_{-\infty}^{\infty} B_t(\tilde{f})w(f-\tilde{f})d\tilde{f} \qquad (6)$$

where B_t indicates the *true* spectrum and w the Fourier transform of the apodizing function W, i.e. the instrumental line shape. Eq. (6) is the key point of processing finite-extent interferogram samples. We see that the contribution to the spectral estimate $\tilde{B}(f_o)$ does not come only from a narrow spectral range, $f_o - \Delta f \div f_o + \Delta f$, centered at f_o (Δf is the spectral resolution) but from the overall spectrum, although convolved as indicated in (6). Thus, depending on the instrumental line shape, the spectral estimate $\tilde{B}(f_o)$ may be highly biased in energy. Again, if a suitable apodization is used, the contribution to the spectral estimate, $\tilde{B}(f_o)$, coming from spectral regions outside $f_o \pm \Delta f$ becomes rapidly neglegible.

The spectral estimation (4) (or (5)) is a function of continously varying dimensionless wavenumber f. There is no special significance to specify equally spaced wavenumbers as there is in the *Discrete Fourier Transform* case which is usually computed by Fast Fourier Transform. This is a source of major confusion in literature about spectral estimation. Eq. (4) and discrete Fourier transform are usually confused. Discrete Fourier transform realizes the discrete sampling of (4) at harmonic wavenumbers $f_l = l/(2N)$; $l = 0, \ldots, N-1$:

$$\tilde{B}(f_l) = \sum_{k=-(N-1)}^{N-1} \tilde{I}(k) \exp(-2\pi i f_l k) \qquad (7)$$

which taking into account the eveness of $\tilde{I}$, can be written in the more usual form of discrete Fourier cosine transform

$$B(f_l) = 2\tilde{B}(f_l) = 2\left(\sum_{k=0}^{N-1} C(k)\cos(-2\pi i f_l k)\right) \qquad (8)$$

with $C(0) = I(0)$ and $C(k) = 2I(k)$; $k = 1, \ldots, N-1$. Eq. (8) gives the rectangle-apodized discrete Fourier transform version of (3). The harmonic wavenumbers f_l are the points where the sinc function, i.e., the instrumental line shape corresponding to the rectangle apodizing-function, has its zeros. Thus, each one of the N spectral estimates, $B(f_l)$, obtained from (7) (or (8)), being located at points where the instrumental line shape has its zeros, does not contribute to bias the others in the convolution (6). However, they are still biased from the spectral content at non-harmonic wavenumbers. The property above is often referred to as the *orthogonal properties of the sinc function.* There is nothing magical with such a properties, it only reflects the *orthogonal property* of the trigonometric basis in which the interferogram samples are expanded in order to obtain the spectral estimates. This is a mathematical fact (not a physical property) whose elegance is indisputable. Yet, it usually produces biased results. Unbiased results are only obtained providing that the underlying spectrum consists of δ-functions located exactly at the harmonic wavenumbers $f_l = l/(2N)$. In such a case, as far as noise is concerned, the spectral estimates obtained by (8) are also statistically independent, providing that the noise constitutes an uncorrelated stationary Gaussian process.

However, if a broadband spectrum is to be investigated, e.g., the infrared emission spectrum of the earth, the spectral estimates at harmonic wavenumbers may be greatly affected by the adjacent ones as it is illustrated in Fig. 1 which shows a plot of two sinc functions centered at the two adjacent harmonics $f_2 = 2/(2N)$ and $f_4 = 4/(2N)$ respectively. The sinc function (normalized at the peak value) were computed for $N = 2000$, which is approximately the number of interferogram samples observed by Michelson spectrometers designed for the remote sensing of the earth from satellite. As it possible to see the two main lobes are clearly separeted. However the largest secondary maximum of one fall just inside the main lobe of the other so that spectral leakage will dominate the two estimates.

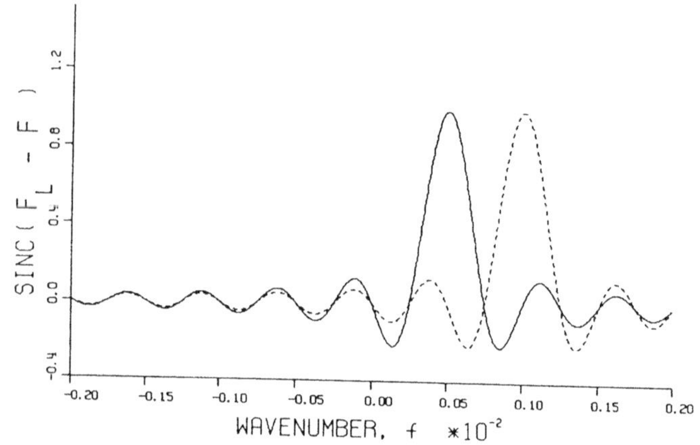

Fig. 1: Sinc functions centered at the harmonic wavenumber $f_l = .0005$ (solid line) and at $f_l = .001$ (dashed line).

To sum up, although the approach of obtaining unapodized spectrum is fairly elegant, in many practical situations this elegance must be compromised in order to obtain meaningful results. One such compromise consists of applying windows to the sampled interferogram function or smoothing the spectral samples.

3. Parametric Autoregressive Modeling Approach

Eq. (1) is only a particular case of a general theorem, the Wiener-Khinchine theorem (Bracewell 1965), which states that the relation between the covariance function and the spectrum of any process is a Fourier transform. The Wiener-Khinchine theorem provides the necessary link between the interferogram $I(x)$ and the spectrum $B(\sigma)$, since an interferogram is the covariance or autocorrelation of the incident wave amplitude. Furthermore, it enables the use of the autoregressive theory, which permits to recover spectra once the covariance functions are known and to exploit the symmetry properties which have to be fulfilled from the covariance matrix of a stationary signal, the signal being the electromagnetic wave amplitude incident on the interferometer.

An autoregressive process of order p is a stochastic process whose covariance function satisfies a difference equation of order p:

$$I(k) = \alpha_1 I(k-1) + \alpha_2 I(k-2) + \ldots + \alpha_p I(k-p); \quad k > 0 \tag{9}$$

where for simplicity we continue to write $I(k\Delta x) = I(k)$, with Δx the sampling interval. The spectral estimate is (Box and Jenkins, 1976):

$$B(f) = \frac{2P_{p+1} \cdot \Delta x}{\| 1 - \sum_{j=1}^{p} \alpha_j \exp(-2\pi i f j \Delta x) \|^2} \tag{10}$$

The unknown P_{p+1} and the coefficients $\{\alpha_1, \ldots, \alpha_p\}$ appearing both in (9) and (10) can be computed by solving the following set of linear equations:

$$\mathbf{A} \cdot \alpha = \mathbf{P} \tag{11}$$

where the matrix $\mathbf{A}$ is the covariance matrix; $\alpha = (1, \alpha_1, \ldots, \alpha_p)^T$ and $\mathbf{P} = (P_{p+1}, 0, \ldots, 0)^T$. The autoregressive approach (hereafter AR-approach) consists of fitting the covariance function of an autoregressive model of order p to the measured interferogram. Whatever the chosen value of p may be, equation (11) determines a certain sort of extrapolation of the interferogram to optical path difference larger than the maximum delay in the measured data set. The extrapolated values beyond the optical path difference $p\Delta x$ are computed by (9) with $k > p$ and taking as initial conditions the first p measured interferogram samples. The autoregression coefficients are obtained from (11). Then the spectral estimator (10) is just the Fourier transform of such infinite-extent interferogram sample (Childers 1978). Thus, the AR-approach solves the problem of dealing with finite-extent data just by extrapolating suitable values up to infinity by means of a model. This is how the AR-approach provides the solution for the truncation error. Of course the more adequate the autoregressive model fitted to the data is, the better the spectral estimates are.

A crucial point of the AR-approach consists in properly selecting the model order p fitted to the data, since it determines the smoothness degree of the final spectral estimates. Towards this objective the following procedure (Amato et al 1992), which requires that all possible orders p, $p = 0, 1, \ldots, N-1$, be fitted to the data, is employed.

From the difference equation of order p fitted to the data and considering the first p interferogram samples as initial conditions, the predicted values, $I^*(k)$, $k = p+1, \ldots, N-1$, are computed and compared against the observations to obtain the square of the standard error

$$S^2(p) = \frac{1}{N-p-1} \sum_{k=p+1}^{N-1} (I^*(k) - I(k))^2 \tag{12}$$

Then the optimal order, p_{opt}, is selected that minimizes the quantity:

$$SIC(p) = N \log(S^2(p)) + 2p \tag{13}$$

The first term on the right-hand side of (13) is a measure of how well a difference equation of order p fits the data. The second term is a penalty function for the p autoregression coefficients $\{\alpha_{11}, \ldots, \alpha_{pp}\}$ that need to be estimated. Relation (13) has its theoretical foundation in the Information Theory. The SIC-statistic belongs to a general class of statistics whose minimization leads to reliable criteria for defining best fitting models. Its derivation is too involved to be presented here. The interested reader should refer to the original work by Akaike (1974). According to Akaike (1974) the introduction of the term $+2p$ into the definition of SIC is crucial in order that the criterion works well.

To summarize, the procedure consists of fitting recursively difference equations from order 0 up to $N-1$ to the data. At each step the extrapolation, or prediction of the interferogram, from the lag $p+1$ up to $N-1$, is computed and compared with the measured values.

4. Regularizing the Fourier Transform

In presence of measurement errors the inverse transform operation of (2) which leads to the result (4) amplifies spurious oscillations in the final spectrum. To limit such oscillations a procedure that permit to stabilize the Fourier transform operation will be discussed in this paragraph.

In this procedure the solution for the unknown, i.e., the spectrum $B(f)$, in the integral equation (2) is seeked by solving the following variational problem

$$\min_{B} \mathcal{L}, \quad \mathcal{L} = \frac{1}{N} \sum_{k=0}^{N-1} \left(I(k) - \int_{0}^{\frac{1}{2}} B(f) \cos 2k\pi f df \right)^2 + \lambda \mathcal{D}(f) \tag{14}$$

where for simplicity we have again written $I(k)$ for $I(k\Delta x)$ and have set $\Delta x = 1$. Always in eq. (14) $\mathcal{D}$ is a smoothing functional and λ is the regularization parameter.

The role of the smoothing functional consists in smoothing the oscillations of the solution when the interferogram is affected by error. In general it is assumed (m-th order differential regularization)

$$D(f) = \frac{1}{4} \int_{0}^{\frac{1}{2}} \left[\frac{d^m B(f)}{df^m} \right]^2 df \tag{15}$$

where m is the order of differentiation ($m = 0$ means no differentiation or usual square of the L^2 norm). If we look for the solution in the form

$$B(f) = 2 \left(c_0 + 2 \sum_{i=1}^{N-1} c_i \cos 2\pi i f \right) \tag{16}$$

then the following theorem holds (Amato and Serio 1991)

THEOREM. *The regularized solution of the problem (14) for $m > 0$ is*

$$\begin{cases} c_0^{(m)} = I(0) \\ c_k^{(m)} = \frac{I(k)}{1+(2k\pi)^{2m}\lambda N} \quad k = 1, \ldots, N-1 \end{cases} \tag{17}$$

Furthermore, the optimal value of the regularization parameter, λ, can be estimated by using the Generalized Cross Validation (GCV) criterion (Wahba 1977). For the problem at the hand, the optimal value of λ is solution of the implicit equation (Amato and Serio 1991)

$$\sum_{k=1}^{N-1} d_k^2 I^2(k) \sum_{k=1}^{N-1} d_k^2 - \sum_{k=1}^{N-1} d_k^3 I(k) \sum_{k=1}^{N-1} d_k = 0 \tag{18}$$

where

$$d_k = \frac{(2k\pi)^{2m}}{(2k\pi)^{2m} N\lambda + 1} \tag{19}$$

Let $I(k)^\delta = I(k) + \delta_k$ be the interferogram affected by error and $\delta \equiv \|\delta_k\|_{l_2}$ be the corresponding norm of the error. Moreover, let $B^\varepsilon(f) = B(f) + \varepsilon(f)$ be the retrieved spectrum when error δ affects the interferogram and $\varepsilon \equiv \|\varepsilon(f)\|_{L_2}$ be the corresponding norm of the error. It is possible to prove that $\|\varepsilon\|_{L_2} = O(\delta)$ for the regularization technique, provided that $\lambda = O(\delta)$ (details of the proof are given in Amato and Serio 1991 for a more general case). It follows that the regularized spectrum converges to the true one when error affecting the interferogram tends to vanish. Moreover, according to Groetsch 1983, the order of convergence is optimal.

The solution (17) can be also interpreted as a particular class of apodizing functions. Unlike the other apodizing functions available in literature, the one yielded by the mathematical method we have discussed above are optimal (in the Least Square sense) and adaptive in the sense that their length depends only on the input data. Furthermore, the only exogenous parameter is the order of differentiation m. Hereafter we shall refer to the whole procedure above to obtain optimally apodized spectrum as ASE algorithm ASE being a shortening for Advanced optimal Spectral Estimation.

5. Numerical Examples

In order to evaluate the performance of the technique above, one test spectrum was generated. Our exercise, then, consists in retrieving such test spectrum from finite-extent interferograms.

The test spectrum $B_t(\sigma)$ was generated using the radiative transfer code FASCOD2. The spectrum extends from $\sigma_1 = 600$ cm^{-1} (16.7 μm) to $\sigma_2 = 1200$ cm^{-1} (8.34 μm); inside this spectral region there is the 15$-\mu$m CO_2 absorption band which is the primary band for temperature sounding. The spectrum was generated in such a way to have 2049 equispaced spectral ordinates ($\sigma/\Delta\sigma \approx 3000$) that corresponds to the following spectral resolution: $\Delta\sigma = 0.29$ cm^{-1}. Furthermore, the spectrum was computed for the 1962 U.S. Standard Atmosphere. Every spectral ordinate, $B_t(\sigma)$, was obtained by applying a rectangular window (directly in the spectral domain) of size $\sigma \pm \Delta\sigma$. In this way the spectral ordinate at σ has contribution only from the narrow interval $\sigma \pm \Delta\sigma$, that is, the spectral ordinates $B(\sigma)$ are not biased in energy. Thus, the product $B_t(\sigma)\cdot\Delta\sigma$ gives exactly the energy in the channel centered at σ and does not overlap adjacent channels. In turn, this means that the 2049 spectral ordinates $B_t(\sigma)$ carry on information from non-overlapping portions of the spectrum.

The interferogram fuction corresponding to the spectral region above was computed by:

$$I(x) = \int_{\sigma_1}^{\sigma_2} B(\sigma)\cos(2\pi x\sigma)d\sigma$$

Such an integral was computed using again FASCOD2 and performing a line-by-line calculation. Then, the function $I(x)$ was sampled using the sampling interval: $\Delta x = (2(\sigma_2-\sigma_1))^{-1}$, to give the partial or truncated interferogram signal $I(k)$, $k = 0,\ldots,2048$. Such a discrete function simulates the output of an ideal Michelson interferometer spectrometer.

To simulate the effects of measurement errors, Gaussian white noise, $u(k)$, with zero mean and standard deviation σ_u was generated and added to the partial interferogram, $I(k)$, to give the noise corrupted interferogram signal. The following value of σ_u was considered: $\sigma_u = I(0)/10000$, that is a signal-to-noise ratio equal to 10000 at zero delay.

Taking the inverse Fourier Transform of $I(k)$ we obtain an estimate of the spectrum inside the proper spectral region and with the proper resolution as well. Let $\tilde{B}(\sigma)$ indicate such a spectrum. In general the latter spectrum is different from the corresponding test spectrum $B_t(\sigma)$ (also in the noiseless case), since $\tilde{B}(\sigma)$ is now affected by the truncation error. As it was pointed out in section 2, the truncation error derives by processing finite-length interferogram, i.e., the interferogram extends from 0 up to a maximum path difference, which, by necessity, must be finite. The consequence of processing finite-extent records will be a spectral leakage. It should be noted that, although the amount of leakage is influenced by the sampling interval, Δx, leakage is not caused by the sampling itself, but only by truncating the interferogram at a given point different from infinity.

Fig. 2 compares $B_t(\sigma)$ to $\tilde{B}(\sigma)$ (only the detail around the 15-μm CO_2 absorption band is shown). Spurious oscillations in the case of $\tilde{B}(\sigma)$ are quite evident. The differences between the test spectrum and the unapodized one reach about 20% at locations where the test spectrum shows either absorptance minima or absorptance maxima.

Fig. 3 permits to evaluate the improvement in the convergence of the Fourier transform which is reached by applying the AR-algorithm. Comparing with Fig. 2 it is

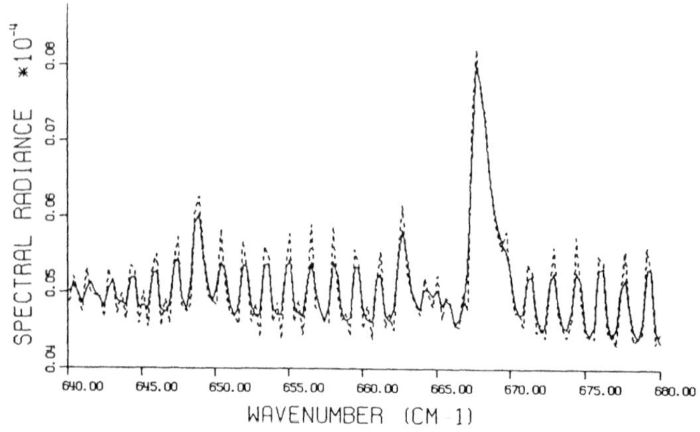

Fig. 2: Comparison between the test spectrum (solid line) and the one computed from the truncated interferogram (dashed line) by the unapodized Fourier transform. Only the spectral range around the 15-μm CO_2 absorption band is shown in detail. Units on y-axis are W cm^{-2} $(cm^{-1})^{-1}$ sr^{-1}.

possible to see that now the recovered spectrum is less affected by spurios oscillations, although differences at relative minima and maxima are still remarkable.

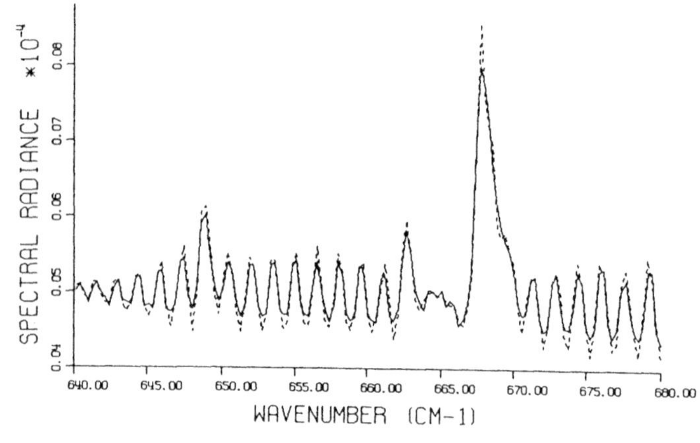

Fig. 3: As Fig. 2 but now the retrieved spectrum was obtained by the AR-method.

The optimal order p in this case turned out to be $p_{opt} = 1700$ as indicated from Fig. 4 which shows the plot of $SIC(p)$ vs. p.

It is evident from Fig. 4 that SIC-curve exhibits more local minima corresponding to solutions whith a different amount of smoothing (the case shown in Fig. 3 corresponds to the deepest minimum). This behaviour is due to the regular spacing of the CO_2 lines ($\approx$ 1.6 cm^{-1}) which causes the absorption line positions to be related near harmonically. That is, if the M-th line is at $M\sigma_o$, then the $(M+1)$-th one is at $(M+1)\sigma_o$. Due to the absorption process each line will subtract a cosine varying increment from the interferogram and the arguments of the cosine terms are related harmonically. In the neighborhood of $x = M(\Delta\sigma)^{-1}$, $\Delta\sigma$ being the line spacing, the cosine arguments will be integer multiples of 2π and a large variation, the so-called signatures, in the interferogram will be seen. Thus, *signatures* in the interferogram will appear located at $(\Delta\sigma)^{-1}$ interval.

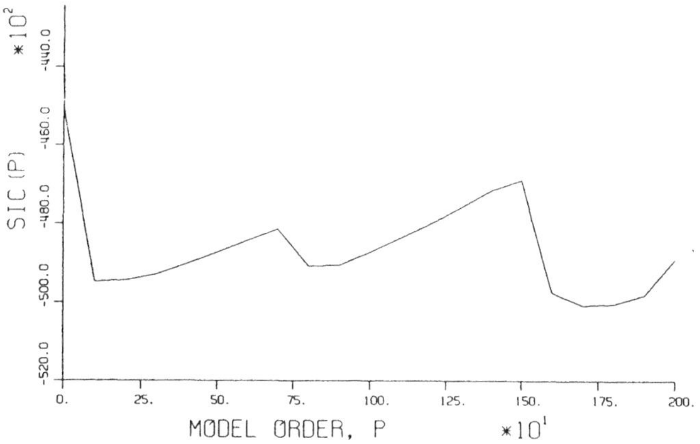

Fig. 4: SIC(p) vs p for the interferogram of Fig. 5.

This effect is illustrated in Fig. 5 which shows the interferogram computed for the spectral region of interest. Three signatures are quite visible and become smaller as a function of the optical path difference, x. It can be shown that the decreasing amplitude of the signatures is due to the finite width of the lines.

Finally, Fig. 6 permits to evaluate the improvement in the convergence of the Fourier transform which is reached by applying the ASE-algorithm (the differentiation order $m = 2$ was used for such an exercise). Now, spurious oscillations nearly disappear and the bias at relative minima and maxima become quite neglegibile, indicating that most of the spectral leakage present in the unapodized spectrum has been removed.

6. Conclusions

The authors have analysed the capability of getting improved spectral estimates with two optimal spectral estimation algorithms. Applications were shown in the context of Fourier interferometry applied to remote sensing of atmosphere.

As far as the ASE-algorithm is concerned our discussion pointed out that such a tool is quite effective to address the problem of optimally estimating the spectrum from interferogram signals.

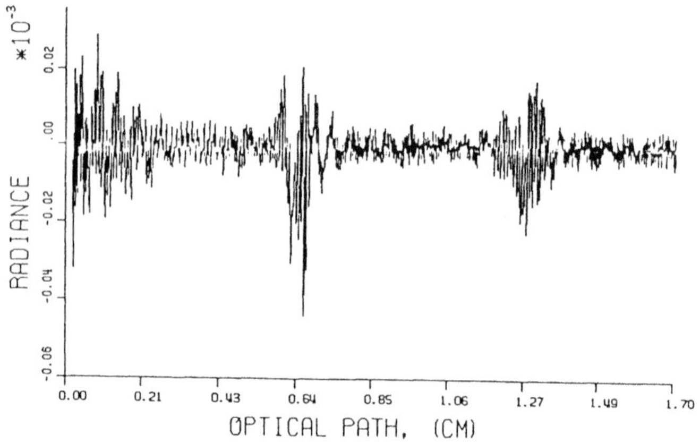

Fig. 5: Truncated interferogram for the spectral region 600–1200 cm^{-1}. The value at zero delay is not shown. Units on y-axis are W cm^{-2} sr^{-1}.

It has been demonstrated that the problem of improving the Fourier transform convergence in presence both of noise and truncation error admits a fully ***analytical solution*** in the frame of the regularization theory with differential functionals. Furthermore, it has been demonstrated that GCV criterion permits to solve the problem of properly selecting the regularization parameter. After that it has been demonstrated numerically that the algorithm here developed can be successfully used to get improved spectral radiance estimates from interferogram signals.

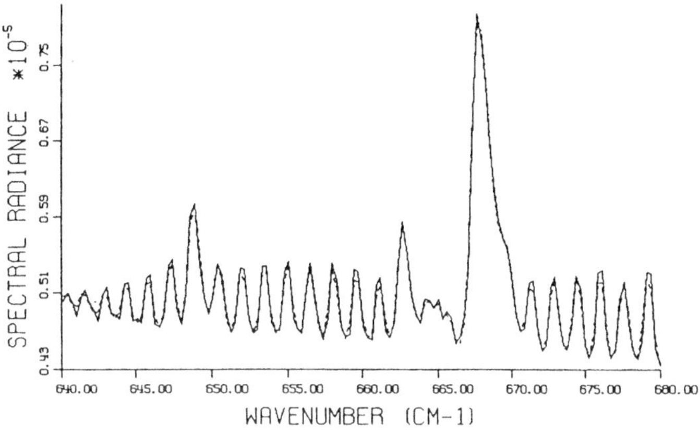

Fig. 6: As Fig. 2 but now the the retrieved spectrum (dashed line) was obtained by ASE-algorithm.

As far as the AR-approach is concerned we have found that such a tool estimator does not provide valuable improvement upon the usual estimators, its variance being equivalent to the values obtained using Fourier estimators applied to the same number of samples. The problem is that the AR-method works best when the process of interest admits an autoregressive representation and this is not the case for the infrared emission spectrum of the earth.

Acknowledgements: Work supported by ASI and CNR.

References

Akaike, H., 1974: A new look at the statistical model identification. *IEEE Transactions, Automatation and Control* **AC-19**, pp. 716–723.

Amato, U., M. Bonzagni, R. Rizzi and C. Serio, 1990: Autoregressive spectral estimates in Fourier spectroscopy applied to remote sensing of the atmosphere. *Proc. V International TOVS Study Conference, Tolouse, France,* CIMSS, Madison, pp. 1-10.

Amato, U., M. Bonzagni, R. Rizzi and C. Serio, 1992: Spectral Estimates in Fourier spectroscopy applied to remote sensing of the atmosphere: a parametric autoregressive modeling approach. *Int. J. Rem. Sens.* **13**, pp. 1807–1827.

Amato, U. and C. Serio, 1991a: Retrieving a periodic function from its Fourier transform corrupted by error. *Colloquia Mathematica Societatis Jànos Bolay* **57**, *Approximation Theory, Kecskemét (Hungary)*, North-Holland, pp. 9–39.

Amato, U., M.F. Carfora, V. Cuomo, F. Esposito and C. Serio, 1991b: Optimal spectral deconvolution of interferogram samples. *Proc. VI Internatioanl TOVS Study Conference, Airlie VA, USA,* CIMSS, Madison, pp. 17-23.

Bell, R.J., 1972: *Introductory Fourier transform Spectroscopy.* Academic Press, New York.

Box, G.E.P. and G.M. Jenkins, 1976, *Time Series Analysis.* Holden-Day, San Francisco.

Bracewell, R, 1965: *The Fourier transform and its applications.* McGraw-Hill, New York.

Childers, D., G. (Editor) 1978, *Modern Spectrum Analysis.* (New York: IEEE Press).

Groetsch, C.W., 1983: *The Theory of Thikhonov regularization for Fredholm equations of the first kind.* Pitman, Boston.

Harris F.J., 1978: On the Use of Windows for Harmonic Analysis with the Discrete Fourier Transform. *Proceedings of IEEE,* Vol. 66 No. 1, pp. 51-83.

Jenkins, G.M. and D.G. Watts, 1968: *Spectral Analysis and its applications.* Holden-Day, S. Francisco.

Norton R.H. and R. Beer, 1976: New Apodizing Functions for Fourier Spectrometry. *J. Opt. Soc. Am.* Vol. 66, No. 3, pp. 259-264.

Tikhonov, A.N. and V.Y. Arsenin, 1977: *Solutions of ill-posed problems.* J. Wiley & Sons, New York.

Vanasse G.A., 1981: *Spectrometric techniques*, Vol. II. Academic Press, New York.

Wahba, G., 1977: Practical approximate solutions to linear operators equations when the data are noisy. *SIAM J. Numer.Anal.,* **14**, 820-831.

Printing: Druckhaus Beltz, Hemsbach
Binding: Buchbinderei Schäffer, Grünstadt

GPSR Compliance

The European Union's (EU) General Product Safety Regulation (GPSR) is a set of rules that requires consumer products to be safe and our obligations to ensure this.

If you have any concerns about our products, you can contact us on ProductSafety@springernature.com

In case Publisher is established outside the EU, the EU authorized representative is:

Springer Nature Customer Service Center GmbH
Europaplatz 3
69115 Heidelberg, Germany

Batch number: 10844669

Printed by Printforce, the Netherlands